Developments in Petroleum Science, 31

microbial enhancement of oil recovery - recent advances

DEVELOPMENTS IN PETROLEUM SCIENCE

Advisory Editor: G.V. Chilingarian

Volumes 1, 3, 4, 7 and 13 are out of print.

2. W.H. FERTL – Abnormal Formation Pressures
5. T.F. YEN and G.V. CHILINGARIAN (Editors) – Oil Shale
6. D.W. PEACEMAN – Fundamentals of Numerical Reservoir Simulation
8. L.P. DAKE – Fundamentals of Reservoir Engineering
9. K. MAGARA – Compaction and Fluid Migration
10. M.T. SILVIA and E.A. ROBINSON – Deconvolution of Geophysical Time Series in the Exploration for Oil and Natural Gas
11. G.V. CHILINGARIAN and P. VORABUTR – Drilling and Drilling Fluids
12. T.D. VAN GOLF-RACHT – Fundamentals of Fractured Reservoir Engineering
14. G. MÓZES (Editor) – Paraffin Products
15A. O. SERRA – Fundamentals of Well-log Interpretation, 1. The acquisition of logging data
15B. O. SERRA – Fundamentals of Well-log Interpretation, 2. The interpretation of logging data
16. R.E. CHAPMAN – Petroleum Geology
17A. E.C. DONALDSON, G.V. CHILINGARIAN and T.F. YEN (Editors) – Enhanced Oil Recovery, I. Fundamentals and analyses
17B. E.C. DONALDSON, G.V. CHILINGARIAN and T.F. YEN (Editors) – Enhanced Oil Recovery, II. Processes and operations
18A. A.P. SZILAS – Production and Transport of Oil and Gas, A. Flow mechanics and production (second completely revised edition)
19A. G.V. CHILINGARIAN, J.O. ROBERTSON Jr. and S. KUMAR – Surface Operations in Petroleum Production, I
19B. G.V. CHILINGARIAN, J.O. ROBERTSON Jr. and S. KUMAR – Surface Operations in Petroleum Production, II
20. A.J. DIKKERS – Geology in Petroleum Production
21. F. RAMIREZ – Application of Optimal Control Theory to Enhanced Oil Recovery
22. E.C. DONALDSON, G.V. CHILINGARIAN and T.F. YEN – Microbial Enhanced Oil Recovery
23. J. HAGOORT – Fundamentals of Gas Reservoir Engineering
24. W. LITTMANN – Polymer Flooding
25. N.K. BAIBAKOV and A.R. GARUSHEV – Thermal Methods of Petroleum Production
26. D. MADER – Hydraulic Proppant Fracturing and Gravel Packing
27. G. DA PRAT – Well Test Analysis for Naturally Fractured Reservoirs
28. E.B. NELSON (Editor) – Well Cementing
29. R.W. ZIMMERMAN – Compressibility of Sandstones

Developments in Petroleum Science, 31

microbial enhancement of oil recovery - recent advances

Proceedings of the 1990 international conference on microbial enhancement of oil recovery

Edited by

ERLE C. DONALDSON
402 S. Taylor Avenue, Wynnewood, OK 73098, U.S.A.

ELSEVIER, Amsterdam — Oxford — New York — Tokyo 1991

ELSEVIER SCIENCE PUBLISHERS B.V.
Sara Burgerhartstraat 25
P.O. Box 211, 1000 AE Amsterdam, The Netherlands

Distributors for the United States and Canada:

ELSEVIER SCIENCE PUBLISHING COMPANY INC.
655, Avenue of the Americas
New York, NY 10010, U.S.A.

ISBN 0-444-88633-8

This book is printed on acid-free paper.

Printed in The Netherlands

DEDICATION

These Proceedings are dedicated to the late Dr. Claude ZoBell whose pioneering work with microbial release of oil from sands has blossomed into world-wide field application of the diverse technologies known as microbial enhancement of oil recovery which are presented in this volume.

TABLE OF CONTENTS

PREFACE

This is the third conference that has been sponsored by the Bartlesville Project Office of the U.S. Department of Energy (as well as two workshops):

(1) The first conference, in 1982, was designed to explore the nature of the international interest in microbial enhanced oil recovery (MEOR). We found that there was a great deal of interest in MEOR, but that it was directed principally to laboratory studies;

(2) Therefore the objective of the Second Conference, in 1984, was to discuss field applications of MEOR. Several small companies were using MEOR and there were rumors that some of the major oil companies also were looking at it, but papers on field activities did not materialize;

(3) This conference, once more, was instituted to examine field activities in MEOR. The U.S. Department of Energy has sponsored several field projects and the details from some of these were presented, as well as a few from industry. The balance of the program was concerned with new developments in research.

MEOR had its beginning with the suggestion by Beckman in 1926 that certain microbes might be useful for enhancement of petroleum recovery. At the time no one was particularly interested in MEOR because there was a surplus of oil. In the 1930's and 1940's, Dr. ZoBell (1947) led the way with several interesting experiments and ideas that formed the actual basis for MEOR. In the 1950's it was Drs. Updegraff (1954) and Hitzman (1956) who published papers on MEOR in the United States.

Later, in 1967, Dr. Davis published a book on petroleum microbiology where he reviewed field work being conducted in Eastern European countries, and patents and papers, that had appeared in the literature. In the 1970's we began to hear about the work that Dr. Lazar (1978) was conducting in Romania.

In 1979, the U.S. Department of Energy entered the field by holding a forum on MEOR in San Diego, California (1979). It was held in San Diego because Dr. ZoBell (then retired) said that he would attend the forum if it could be held there. He did attend, and the results of that 3-day forum developed into the extensive research program of the U.S. Department of Energy which has led to this conference.

Today's oil production technology leaves one third to one half of the original oil in place in the reservoir at abandonment of secondary recovery (waterflooding). This leaves a very large target for microbial enhanced oil recovery which was shown by the research papers of this conference to be capable of producing up to 50% of the residual oil. The field trials show that the normal projected oil production decline curve can be reversed, or leveled off, by MEOR. The art of MEOR is too young to show the ultimate field oil recovery that can achieved by MEOR. The trends thus far, however, are encouraging enough to stimulate more small field trials and even entire oilfield applications.

This conference has shown that a variety of applications are possible to correct oilfield problems as well as to enhance oil recovery. Among these is the suppression of hydrogen sulfide production which alone is a tremendous advance because of the large quantity of sour oil production. If hydrogen sulfide production can be curtailed it would increase the value of the produced oil, decrease it toxicity, and largely decrease it corrosiveness. All of these would be welcome both in the field and at the petroleum refinery where special precautions must be taken to process sour crude oil.

Another very important discovery is the ability of certain bacteria to eliminate paraffin deposition around the producing well and in the tubulars. This is a welcome improvement for many producers who have considerable difficulty in controlling paraffin deposition.

MEOR research has apparently taken a turn to more applied aspects than the very basic processes of MEOR. A considerable amount of discussion, however, was devoted to analyses of the probably mechanisms of MEOR and the physiology of MEOR bacteria.

The papers were organized into: Introduction, Research, Field and an Appendix in which Dr. Lazar present a condensed review of 30 field trials that have been fairly well documented.

This Appendix can serve as a quick review of published field results and a guide for selection of an appropriate field procedure (or protocol) for a specific field. The papers in each section were numbered; for example, Research Paper No. 5 was labeled R-5 and Field Paper No. 3 is F-3. This convention was adopted to afford the reader more rapid reference to the papers.

REFERENCES

Beckman, J. W., 1926. The Action of Bacteria on Mineral Oil. Ind. Eng. Chem. News, 4 (Nov 10): 3.

Davis, J. B., 1967. Petroleum Microbiology. Elsevier Science publishers, Amsterdam, New York, 604 pp.

Hitzman, D. O., 1956. Recovery of Oil from Oil Sand and the Like. U.S. Patent No. 2,907,389.

Lazar, I., 1978. Microbial Methods in Secondary Oil Recovery. European Symp. on EOR. Heriot-Watt University, Edinburgh, (Aug 29-Sep 1), VII: 5-7.

Science Applications Staff Report, 1979. Conference on Microbiological Processes Useful in EOR. Final Report, Sand Diego, California, CONF 790871 (UC-92), NTIS, Springfield, VA 22161, 54 pp.

Updegraff, D.M. and Wren, G.B., 1954. The Release of Oil from Petroleum-Bearing Materials by Sulfate-Reducing Bacterial. Appl. Microbiology, 2: 309-322.

ZoBell, C. E., 1947. Bacteria Release of Oil from Oil-Bearing Materials, Parts I and II. World Oil, 126(13), 36-47 (I); 127(1): 35-41 (II).

Erle C. Donaldson

INTRODUCTION

The introduction to the Proceedings is composed of four introductory papers beginning with the address by the Director of the Department of Energy Bartlesville Project Officer, Mr. Thomas C. Wesson. This is followed by the Keynote Address which was presented by Dr. Donald O. Hitzman, President of INJECTECH INC.

Dr. Vivian Moses presented a provocative speech which, in a way, was a challenge to the Conferees - why is it that MEOR is still in the formative stage when proof of its success has so often been shown from both laboratory and field results?

Finally, Drs. Qin and Yang presented an overview of MEOR developments in China which was more appropriate as an introductory presentation.

Reprinted from: *Microbial Enhancement of Oil Recovery - Recent Advances*, edited by E.C. Donaldson

CH. I–1

ADDRESS BY THE DIRECTOR OF THE DEPARTMENT OF ENERGY, BARTLESVILLE PROJECT OFFICE

Thomas C. Wesson

Thank you, Erle. It's always a pleasure to attend an event at the University of Oklahoma--the facilities and hospitality are first rate. A lot of time and effort goes into preparing a meeting like this, and you and Leon Crowley have done a great job in arranging a successful conference.

The Department of Energy appreciates the University of Oklahoma's commitment and support for MEOR. The faculty and staff have been leaders in MEOR research, both in laboratory and field projects, and the University has been a co-sponsor of several MEOR conferences since 1982.

It is particularly gratifying to see the strong international participation here today--we have MEOR experts from a dozen countries and all six inhabited continents. If it had oil wells, I'm sure someone would have been here from Antarctica, too. Your attendance here denotes the global nature of MEOR, an aspect that has been growing with every gathering over the last decade.

Since our last meeting, a MEOR workshop sponsored by DOE and NIPER at Bartlesville in 1987, we have seen some encouraging signs. The agenda at this conference is one indication--42 technical presentations, compared to 19 in 1987. More research is being done, more field projects are being performed, and we are obtaining more results across a wider range of MEOR topics.

DOE funding for MEOR research has been substantially increased--more than tripled since 1987, in fact. And MEOR research, itself, has made significant progress in furthering our knowledge of the basic processes of MEOR, as I'm sure we'll see during the technical sessions over the next few days.

But the news is not all good:

1. The price of oil is still uncomfortably low, which inhibits the application of enhanced recovery projects;
2. U.S. domestic oil production is still dropping; where it was at one time 2/3 of domestic consumption, the U.S. is now producing less than 1/2 of its consumption;

3. A recently completed DOE study reveals that a stripper well abandonment rate of 17-18,000 per year threatens future access to a significant portion of our oil resources.

This last fact is a particularly sobering one for the future of EOR in general. Successful application of EOR requires economic access to the reservoir, and there is no economic access to a field of abandoned, plugged wells. It's imperative that we curtail well abandonments in order to save the future of EOR. We must use every resource available to us, and appropriately enough, one of those particularly relevant resources is MEOR itself!

MEOR processes are especially useful here because:

1. They are applicable to shallow waterflooded "stripper fields", a group of fields that contain a lot of remaining oil but have been resistant to other EOR technologies;
2. They use the existing field infrastructure (wells and equipment) and personnel;
3. They are not capital intensive.

(These three characteristics make these processes especially suited to independents.)

and,

4. They are environmentally benign--using non-toxic injectants while consuming little energy themselves in the process.

We can see, then, that there are significant problems on the immediate horizon for the future of domestic oil production, and MEOR holds promise for providing some of the solutions. DOE will examine the results of this conference as we have previous such meetings to see how MEOR will measure up to the challenge. We will be looking for:

1. Areas for future research that will refine and build upon results already obtained, or broaden MEOR applicability;
2. Laboratory determination of the feasibility of techniques suitable for field validation and testing;
3. Improved understanding of the mechanisms of microbial systems that have been used in field trials.

I mentioned a moment ago that MEOR was especially suitable for independent operations. That has particular significance today, because the Department has recently mounted an initiative to assist independent operators to increase production and profitability from their marginal wells. For a number of years DOE supported mainly long-term, high-risk research, but several events have led to a shift to research with more immediate results that can be applied directly in the oil field.

An intensive Departmental survey of the nation's oil industry, our well abandonment study, and the input from more than a dozen rounds of hearings on the National Energy Strategy pointed out the critical necessity of maintaining access to those reservoirs containing large amounts of remaining oil. Much of this resource is produced by independent operators who do not have the capital and resources to mount expensive projects. Processes and techniques are needed that can be quickly and inexpensively applied in the field to produce more oil for less money. That is where we see a prominent role for MEOR.

So, we will be looking at the results of this conference in light of our new initiatives. There are critical problems that must be solved, and we are looking to MEOR for some of those solutions.

Thank you for your participation here this week. I look forward to a fine meeting. We have a terrific program lined up. I suggest we get on with it.

Reprinted from: *Microbial Enhancement of Oil Recovery - Recent Advances*, edited by E.C. Donaldson

CH. I–2

MICROBIAL ENHANCED OIL RECOVERY - THE TIME IS NOW

D. O. HITZMAN
INJECTECH, Inc., P. O. Box 360, East Main Street, Ochelata, Oklahoma 74051 USA

ABSTRACT

The United States desperately need a technology today that will effectively release and recover more oil from our known oil reservoirs. Microbial Enhanced Oil Recovery (MEOR) is a low cost technology which now offers a solution for recovering oil before the abandonment of our oil fields. It is now time for the oil industry to seriously consider using MEOR in their field operations.

INTRODUCTION

Microbial Enhanced Oil Recovery (MEOR) is a technology which can recover more oil at low cost. This positive statement should excite the oil industry and the nation since the American Petroleum Institute reports that the U. S. production declined by half a million barrels a day in 1989, twice the decline experienced in the previous two years. This decrease in production caused the nation's dependency on imported oil to rise to 54% of demand in January 1990, which breaks the record set in 1977 at the peak of the energy crisis. With oil prices still too low to encourage increased exploration and drilling, the oil import level is expected to increase even more, which in turn will increase our trade deficit. This means the oil industry desperately needs a technology now that will effectively release and recover more oil from our known oil reserves. MEOR is a field-tested technology that can fill this need and is a "near term" solution for recovering oil before the abandonment of our oil fields.

TECHNOLOGY NEEDS

Over 500 billion barrels of oil-in-place have been discovered in the U. S. of which 145 billion barrels have been produced through 1987. Proved reserves that can be produced by present technology and at current oil prices are about 27 billion barrels leaving nearly 2/3 of the discovered oil–about 340 billion barrels–remaining in known reservoirs at the conclusion of conventional production. This enormous oil potential is the target for improved oil recovery technologies which must include MEOR.

Just as important as the massive amount of oil which remains unrecovered is the fact that such remaining oil faces mass abandonment at current oil price levels. At current prices, the U.S. is plugging and abandoning its marginal producing wells at a rapid and increasing rate. The abandonment of stripper wells has increased 175% since 1980, and it is estimated that at $16/bbl, fully 2/3 of the domestic oil resources could be abandoned by 1995. As the wells are abandoned, the access to the oil and any practical prospect of applying future cost-effective technologies is lost, since all techniques for producing the oil requires existing wells for implementing any recovery technology. It is economically unrealistic to consider redrilling wells to apply a technology. This means that within 15-20 years the U. S. could have access to less than 25% of its remaining known oil resources.

These remaining oil resources which will be lost consist of two broad classes_mobile and immobile oil. The mobil oil resources_almost 100 billion barrels in total - is distributed across all the oil producing regions of the nation with the greatest volume in areas where reservoir heterogeneities and oil characteristics contribute to substantial trapping and by-passing of such oil. Nineteen states have more than 0.5 billion barrels of mobile oil, while ten have more than 1 billion barrels. The immobile oil_resources totaling more than 240 billion barrels_is equally distributed, with 20 states having at least 1 billion barrels and 8 having more than 5 billion barrels. The greatest concentration of immobile oil lies in California and the Southwest, while at least 27 states contain sufficient quantities to warrant serious recovery investigations.

A new recovery technology is required to unlock the enormous potential of the remaining mobile and immobile oil resources. The magnitude of this potential can be cited from a survey of reservoirs in Texas, Oklahoma, and New Mexico. In these three states it is estimated that without technological advances, the incremented oil recovery would range from 2 billion barrels at $16/bbl to 9 billion barrels at $32/bbl. However, with technological advances the recovery would rise to 6 billion barrels and 16 billion barrels at the same respective prices. All these potential estimates assume that production from the existing wells continue, since with abandonment of the wells the reserves are inaccessible. Therefore, the oil industry is very receptive to apply at an early stage what may be considered even marginally economically attractive technologies for oil recovery.

This 1990 meeting will address the scientific advancements made in MEOR technology by presenting information on new discoveries and work which have occurred since our last meeting. However, as important as these new scientific investigative results are to MEOR, it must be recognized that there should also be a corresponding advancement in developing the industrial acceptance of the MEOR technology. It is now time for us to convince the oil industry that MEOR can recover more oil from reservoirs before their abandonment.

MEOR HISTORY

MEOR is a technology that has a history based on over 60 years of research and field studies. The early work by C. E. ZoBell and D. Updegraff (USA), S. I. Kuznetsov, and L. D. Shturm (USSR), M. Dostalek and M. Spurny (Czech.), J. Karaskiewicz (Poland), I.Jarnyi and M. Dienes (Hungary) and I. Lazar (Romania), shows the international scope of the work. This work was expanded in the 1950's mainly by investigators (V. F. Coty, H. Yarborough and D. O. Hitzman) in the major oil companies in the United States. Unfortunately, much of the work done by the oil companies remains proprietary and is available only as patents. Other academic investigators have studied individual aspects of such microbial systems in laboratory environments and provided answers to basic questions concerning the metabolism and physiology of the microorganisms associated with oil reservoirs. Much of the recent MEOR research and field studies have been carried out by academic institutions and by independent laboratories who have received funding from the DOE. It is important to recognize and to acknowledge the role that DOE has played in advancing and sponsoring MEOR field tests and to encourage their continued support of MEOR studies.

In spite of this long history of MEOR investigations, the results which lead to the development of the technology have gone unrecognized by industry. This lack of industrial awareness has occurred for several reasons. In many cases the research and field information has been presented in fragmented pieces in obscure foreign publications or in journals of biological sciences which remained unavailable or unread by workers in the oil industry. In addition, MEOR requires the awareness and interactions of many scientific disciplines, each of which approaches the introduction and operation of this technology differently. The petroleum microbiologist views it from a biological perspective while the reservoir engineer, owner-operator, and economist have different perspectives and objectives. Thus a cohesive and comprehensive picture of MEOR technology has been difficult to present to the oil industry because of the many disciplines, applications, and situations involved its development.

It is now time to bring together these different perspectives and it is our responsibility to inform the oil industry that MEOR could be the answer to their needs for a cost-effective oil recovery technology. The work being presented at this meeting has been built on the base of all the preceding research and field studies. It is important that this point be stressed since MEOR is still considered by the oil industry to be technology that is new, untested, and only of research interest.

MEOR TECHNOLOGY

MEOR technologies have the common basis of introducing or stimulating viable microorganisms in an oil reservoir for the purpose of enhancing oil recovery. However, this broad generic definition of MEOR has lead to some confusion since MEOR is not a single methodology but is a broader technology which can be designed for different and selective applications. A list of the major application areas of MEOR will show the diversity and scope of the technology. Each of these proposed applications may employ different microbial cultures which have different requirements for their introduction and growth in the reservoir. The application, conditions, and cultures can be targeted to meet a particular oil recovery situation. It is convenient to divide the MEOR technology into the following application groups:

<u>Applications of MEOR Technologies</u>

1. Single well stimulations
2. MEOR waterfloods
3. Paraffin removal
4. Viscosity modification
5. Water diversion
6. Heavy oil modification

MEOR discussions should be directed to the specific applications and methodology to be employed to prevent misunderstanding in comparisons and evaluations of the test results.

In addition to this classification of MEOR technologies, each of the mentioned applications can be further divided by identifying the microbial mechanism which is proposed or best suited for each application. The classification of the MEOR technologies by the proposed oil releasing mechanism shows the range of microbial effects which can be identified or expected to occur and to which the MEOR systems can be directed.

MEOR Oil Releasing Mechanisms

1. Gas generation
2. Acid production
3. Surfactant production
4. Physical oil displacement
5. Biopolymer production
6. Hydrocarbon modification
7. Viscosity modification

Thus, MEOR is a multi-application technology based on known oil releasing mechanisms which has matured from the generalized approaches associated with the early investigations. This means that MEOR offers the field operator a wide diversity of oil releasing options, each designed to function in different problem areas.

This versatility of MEOR technology differentiates it from most other EOR technologies. Each of the conventional EOR processes is usually targeted for a specific situation which limit its utility and application. However, it is ironic that while MEOR can and should be considered for a variety of oil recovery problem areas, the options and versatility of the MEOR methodologies leads to a natural resistance for its consideration for any one purpose. The potential of MEOR has been diluted by its proclaimed broad range of applications and, therefore, opportunities for its employment have been lost because its role for specific applications is unclear.

It is important to clarify for the oil industry exactly what a MEOR technology can offer. In this regard some of the concerns and reservations associated with MEOR should be discussed. MEOR relies on applying known oil releasing mechanisms to recover oil. All of the proposed major mechanisms by which microorganisms can cause oil release have been demonstrated by conventional EOR techniques. Thus, gas and acid production, surfactant and polymer additions, viscosity modifications, etc., are known and are recognized oil releasing mechanisms. These mechanisms are acceptable for conventional EOR projects, so there should be no reason to believe these same physical effects would not cause oil release if produced by microbial action. Therefore, we must convince the oil industry that the effects of microbial growth and products are in accordance with acceptable EOR mechanisms and practice and that microorganisms have the capabilities to produce large enough quantities of such products by their growth and action in the reservoir to influence the oil-water and reservoir characteristics.

Some find it difficult to believe that microorganisms can proliferate to such an extent that massive changes will occur in an oil reservoir. However, it is known and accepted that entire reservoirs have turned sour due to the microbial production of H_2S. It is unfortunate that such a detrimental example must be used to demonstrate that reservoirs can be altered by microbial action. However, this example does illustrate that microorganisms will grow in and will influence an oil reservoir. If this situation does occur then it is reasonable to expect that other organisms can also survive in the reservoir and would also change conditions which would favor oil release.

Examples of reservoir souring combined with results from MEOR field tests provide data on the rate and extent of microbial invasion and penetration of reservoirs and microbial tolerance to reservoir conditions. By combining information from such known microbial responses in reservoirs with laboratory data and available MEOR field data it is possible to make positive statements concerning expected results of MEOR field tests (Hitzman, 1982, 1988).

MEOR Field Test Results and Observations

1. Microorganisms can grow under normal reservoir conditions.
2. Microorganisms will migrate through reservoirs.
3. The rate and extent of microbial penetration depend on reservoir characteristics.
4. Microbial numbers and products respond to nutrient availability.
5. Microbial products are produced in the reservoir and influence the reservoir fluids and characteristics.
6. Microbial types which are established or stimulated in the reservoir will depend on inoculum, nutrients, and reservoir conditions.

7. The microbial consortium that will dominate the reservoir is dependent on nutrient and reservoir conditions.
8. While MEOR technologies have broad applications, each MEOR test must be designed for a specific reservoir.
9. MEOR projects require close cooperation and coordination between microbiologists and reservoir engineers.

In addition, it is expected that if the correct oil releasing chemicals or mechanisms are activated or produced by the microbial growth, then oil should be released or recovered in the same manner as proposed by conventional EOR technology. These field results and observations provide encouragement and optimism for believing that successful MEOR projects will result from a correctly designed and engineered field project.

However, there are several areas which are still unresolved by field observations and are often cited as concerns by the oil industry when MEOR projects are discussed. One of the most important involves possible plugging of a well by microbial growths. Field reports vary from severe plugging by microbially contaminated waters to no measurable plugging even when massive concentrations of cultures and nutrients are added. In many cases of reported well plugging the common comment is that the injection waters are of poor quality and contain observable oil or are unstable due to the incompatability of mixed injection waters. Such conditions result from poor operating conditions of the waterflood. Thus, although microorganisms certainly can contribute to well plugging, the chemical composition and physical state of the injection waters must also be considered. While it is extremely difficult to separate the effects of purely biological growths from poor water quality, it can be assumed that as the quality of the water deteriorates the microbial contribution to plugging will increase. The data indicate that reservoir characteristics such as permeability, porosity, and fractures, together with the overall water quality are the governing factors in well plugging.

In general, field observations from MEOR projects present a more favorable picture of plugging response than do laboratory tests. This is important when it is considered that laboratory tests are usually more favorable than field results. The field data suggest that the heterogenitity and fracture patterns in most wells and reservoirs is greater than suspected which results in the wells accepting microorganisms and nutrients without experiencing any plugging. However, it is always prudent to conduct laboratory tests of total water quality in conjunction with planned MEOR tests.

Several reasons can be considered for the reported differences between laboratory results and field observations in MEOR studies. One of the important factors is that once a microbial population is established in the field the system operates within a large unconfined environment. Such a dynamic system is difficult to duplicate or simulate in the laboratory within small cores or reactors where constraints are present which result in limitations of microbial growth and product formation or removal. This situation does not occur to the same extent with conventional EOR chemical treatments which may be actually helped by such constraints under laboratory conditions. Physical and chemical changes also occur within the reservoir as result of interactions of the multiplying microorganisms with the reservoir matrix that cannot be duplicated in the laboratory. Additionally, in conventional EOR systems it could be expected that the large surface areas of the reservoir matrix will increase the adsorption of the limited amount of the injected chemicals and will decrease their effectiveness. The situation is different with MEOR techniques since microorganisms are self duplicating units which will be stimulated and increase in effectiveness within the reservoir. Thus, the effects of microbial actions are magnified by growth in the reservoir, while in other EOR systems the effect of the additives tend to decrease with time and distances.

These important differences must be considered in evaluating the potential for MEOR to contribute and become a cost-effective oil recovery technology. Thus, although some areas of the MEOR technology may remain controversial, a review of the investigations and experience gained from field studies show that there are positive advantages for considering MEOR technologies when evaluating reservoirs for treatment.

	MEOR processes	EOR processes
1.	Offers mutiple application possibilities.	Each application technology specific.
2.	Multiple oil recovery mechanisms available.	Single oil recovery mechanism targeted.

3.	Treatment effectiveness increases with penetration and duplication.	Effectiveness decreases with penetration and adsorption.
4.	Low start up and capital costs.	High start up and capital costs.
5.	Low operating costs.	High operating costs.

Thus, MEOR offers the oil industry a technology that has:

1. A long history of research and field application data.
2. A variety of applications directed toward specific field problems.
3. Proven mechanisms that can increase oil release.
4 Immediate application for reservoirs before their abandonment

It can be expected that new developments in MEOR will add even more advantages to the present technology. As experience is gained in the field the technology will be expanded to new situations and as operations are improved the costs will be reduced. This historically is the case in all biological processes as shown by developments in the fermentation and pharmaceutical industries. Thus, it is reasonable to expect that even more oil can be recovered from reservoirs by MEOR techniques as our understanding of the biological and reservoir systems increases.

MEOR NEEDS

There are more than 800,000 producing oil and gas wells in the U.S. and many of these wells are candidates for MEOR treatment. We must actively promote and encourage the application of MEOR technology. Our audience should include the smaller independent operator who has an immediate need for MEOR, the major oil companies who have applications in large scale projects that will result in significant increased recovery from their larger depleting fields, and to the DOE who can offer the nation a program that will lead to less oil imports. The technology should be included in our universities' curriculum so our petroleum engineering graduates will be knowledgable in its operation. The information and data now available will show that MEOR will play a role in their future work. The oil industry must be persuaded to fully evaluate the potential of MEOR for any property where immediate abandonment is considered. New and innovative developments by smaller organizations experienced in MEOR should be exploited and utilized. We should encourage the major oil companies to participate together in carefully designed field operations that will fully evaluate the technology for field-wide unitized operations. Such large scale projects could be conducted by industry with government cooperation since the technology would have national importance. International projects should be proposed for field tests since the technology also will be employed world-wide. We have played a key role in development of MEOR and now we must continue our efforts to obtain recognition of the technology by those who will use it.

The oil industry is being offered a technology which can be implemented in the field before the wells and reservoirs are abandoned. This alternative to abandoning wells should be of greatest interest to stripper well operators although its greatest potential is for major waterflood operations. Available data indicate that the costs of applying the technology are low while the risks to the wells and reservoirs are minimal. The incentives to apply this technology must be throughly investigated before considering abandoning an oil well and it is our responsibility to make sure that industry is aware of the potential of MEOR.

The nation needs a technology now to recover our oil resources before well abandonment. MEOR is a technology that can be employed now for this purpose. It is now the time for the oil industry to seriously consider using MEOR in their field operations.

REFERENCES

Hitzman, D. O., 1982. Petroleum Microbiology and the History of its Role in Enhanced Oil Recovery, Proceedings of the International Conference on Microbial Enhancement of Oil Recovery, Conf. - 8205140, Shangri-La, Afton, Oklahoma, pp. 162 - 218.

Hitzman, D. O., 1988. Review of Microbial Enhanced Oil Recovery Field Tests. Proceedings of the Symposium on Applications of Microorganisms to Petroleum Technology, Conf. - 870858 (DE 88001232), Bartlesville, Oklahoma, pp. VI-1 - VI-42.

Reprinted from: *Microbial Enhancement of Oil Recovery - Recent Advances*, edited by E.C. Donaldson

CH. I–3

MEOR IN THE FIELD: WHY SO LITTLE?

V. MOSES
School of Biological Sciences, Queen Mary and Westfield College (University of London), London E1 4NS and Archæus Ltd., Queen's Building, Kidderpore Avenue, London NW3 7ST (United Kingdom)

ABSTRACT

In spite of many years of effort, MEOR has made almost no impression on the oil industry as a way of improving the production of oil. Laboratory experiments over and over again have shown the ability of bacteria to generate a wide range of potentially useful chemical products in more-or-less simulated reservoir conditions. Field trials from the early 1950s until the present day purport to demonstrate the utility of MEOR in the real world. And yet there has been virtually no response from an industry which, while not without its conservative aspects, has pioneered one new development after another. Perhaps it is worth trying to understand why MEOR has been so completely ignored because unless its advocates become more successful at promoting its merits than they have in the past, prospects for the future will appear very limited.

1 THE PROBLEM

At the present time the world's daily crude oil production is about 55 million barrels. How much of that is assisted by microbial methods? 50 barrels? 100? 500? Whatever the actual proportion, it is minute. Why is it that after nearly 45 years of continuous experimentation in a dozen countries following ZoBell's original proposals (or more than 60 years counting from Beck in 1926), microbiology still contributes so little to the industry? Do MEOR methods really not work, or not work well enough to be considered effective? Are they too expensive to develop and operate, too unreliable in use or incompatible with current industry practice? Is the problem one of poor communication between the scientists, often in university laboratories, who have undertaken most of the development and the engineers who manage the fields and make operating decisions? Or is it all the result of organisational and funding problems which make it difficult to transfer the technology from the laboratory to the field? Unless and until some convincing answers can be provided MEOR may remain an intellectual curiosity rather than a production method. This paper will explore the reasons why.

2 SOME POSSIBLE ANSWERS

It is not difficult to conceive of answers but it may not be easy in any particular case to decide which are the most applicable.

2.1 MEOR methods are irrelevant

That cannot be true. MEOR seeks to address some of the major production and reservoir problems encountered by the industry. There is, however, a problem of nomenclature. Some people talk of "MEOR" as if it were a single, almost magic, methodology, a unique technique. Of course, it is not at all like that. MEOR represents a collection of methodologies, analogous in many ways to those of non-microbiological procedures. Others confuse *in situ* methods, in which bacteria and feedstock are injected into the reservoir to generate products directly in the matrix, with the so-called *ex situ* use in the oilfield of biological products manufactured in a factory fermenter somewhere and transported to the wellhead like any other chemical. *In situ* MEOR implies placing bacteria downhole, usually with the intention of achieving one of the following: (a) channel blocking/profile improvement, the control of water movement either to improve the sweep of a waterflood, to control coning or for a number of other specialist applications; (b) polymer flooding; (c) surfactant flooding; (d) matrix (and possibly fracture) acidising; (e) single well stimulation by correction of various types of reservoir damage.

2.2 MEOR techniques are too expensive

Nobody really knows for sure because they have not been used sufficiently widely, but such calculations and estimates as have been made suggest the opposite. Bacterial polymer and surfactant generated downhole from inexpensive feedstocks is expected to be significantly less expensive to produce, quantity for quantity, than equivalent chemicals injected from the surface. For one thing, there is no capital cost for fermenters or labour costs to run them; for another, the products are used in a crude and unpurified state. In theory, the broth from a fermentation carried out on the surface at or near the wellhead could itself be injected directly downhole but the logistics of doing so would not be simple nor would they eliminate the investment in labour and equipment. Few MEOR field trials have been run for long enough to see whether unexpected additional costs arise as the operation continues but detailed published reports make no mention of them.

2.3 MEOR is not necessary because there are other and better ways

It depends, of course, on whom you talk to. MEOR is a set of techniques for generating various types of oilfield chemicals at low cost. There are existing industries which already generate oilfield chemicals. Some of them, like xanthan gum, are microbiological in origin; compared with polyacrylamide, xanthan is considered to be an effective product for oilfield use but it is regarded as being too expensive. Views differ among users and suppliers about the efficacy of some existing oilfield chemicals: some people think that they are well understood, effective and offer good value. Others, apparently, are not so satisfied. Views differ also about the value of microbial products which might be generated downhole and how they would compare with the chemical or microbiological products already on the market or under development; the cost implications of downhole generation appear favourable as we have already noted but experience in their use is limited. In the absence of unanimity it is hardly surprising that there is a body of opinion among certain operators and suppliers of services that, all things considered, MEOR may not have much of value to offer.

2.4 MEOR techniques are not well understood

That, alas, is indeed often true. For those who have spent their lives in the practice of microbiology it may not be obvious that a reservoir engineer listening to a presentation of an MEOR method does not immediately perceive all of the implications of this, to him, novel approach. Petroleum microbiologists, like other specialists, have their own style and vocabulary for communicating with one another and employ a host of assumptions and allusions which they correctly assume will be understood by other microbiologists. It may not be easy to convey all the implications and nuances of an MEOR proposal to someone not used to thinking in microbiological terms. It is rare indeed that an oil man has the opportunity or inclination to try to understand fully the significance of MEOR technologies in the way the microbiologist does. He may, indeed, perceive them in some other way, being cognisant of real problems of reservoir souring and the risks of uncontrolled formation plugging. He must justly be cautious about injecting bacteria, a form of life which he will have encountered mainly as one to be avoided or suppressed. Not infrequently, therefore, reservoir engineers feel uncomfortable at the prospect of making a decision to employ a technology of which they have limited understanding and no direct experience.

Making contact with reservoir engineers and operations managers can be another problem. In some parts of North America access to independents may be relatively easy but only in a comparatively few locations worldwide are major fields located conveniently near to MEOR research laboratories; often they are hundreds or thousands of miles away, with their managers out of range and difficult to identify and reach. Access to the right people in the major oil companies is usually more difficult and the larger the company the greater often are the barriers to meeting the people who actually make the field decisions.

2.5 MEOR methods do not work -- or they are too uncertain and unreliable

Well, we don't really know, do we? There has been a wealth of laboratory work to suggest that they *should* work but not enough reliable and informative field data to be sure that they actually do. That is perhaps the main problem and one to which we will need to return in order to chart a way forward.

3 PROVING THE TECHNOLOGY

Now that many of the prospective MEOR methods have been extensively explored in the laboratory, the major problem of acceptance becomes one of proving to the industry that MEOR works and is cost-effective in use. How can that be accomplished?

3.1 Encouraging familiarity with microbiology

Petroleum microbiology is not only about downhole MEOR. There are many other ways in which the science can be useful to the industry and not only in the negative sense of controlling sulphate-reducing bacteria. A valuable means of gaining credibility will therefore be to design microbiological procedures for use in a positive and constructive way to help resolve some of the problems less arcane than those of tertiary oil recovery. Identification of such opportunities is commercially very sensitive and, while each interested party will naturally have to find his own path to success, the growth of proven microbial technology in the industry at large will enhance the willingness of engineers to accept it as a legitimate means of solving their engineering problems.

3.2 Clearly defining the scope of MEOR

We noted earlier that as a technology the mechanisms and potential benefits of MEOR are often not well understood. That becomes a matter of education, one in which petroleum microbiologists must leave no stone unturned in order to disseminate their experience widely through the industry. Some notable efforts have already taken place with one- or two-day courses offered to industry personnel. More is necessary. A partial difficulty is that petroleum microbiologists themselves sometimes seem unclear as to what MEOR means and what are the detailed mechanisms by which it might aid oil recovery. Confusion on the part of microbiologists is unlikely to lead to clarity on the part of industry personnel.

3.3 Field trials

But of course the proof of the pudding is in the eating: to win acceptance by the industry the efficacy of MEOR as a successful technology must be *repeatedly and clearly demonstrated under field conditions.* Most of the field tests reported briefly in the scientific, technical or trade literature fail to present a clear case either because the reservoir and production problems were not properly recognised and the experimental work itself poorly designed and carried out, or because the reporting is inadequate; often it is impossible from the published information to distinguish which of these is the culprit. In those pilots that have been been well reported the data are far more convincing but unfortunately there are very few of them. What is needed for convincing field trials?

(i) Analysis. Unless the downhole problems and conditions are well understood and properly described, a microbial field test is likely not to be designed in the best way to resolve production and reservoir problems. It will not be recognised as offering a credible solution to a defined problem. Small wonder, then, that all too often the results of MEOR tests are at best open to varying interpretation if not to outright dismissal.

(ii) Purpose and mechanism. To be useful, a field trial needs a precise objective and a defined way of reaching it. It is not enough blindly to produce more oil: more oil alone without an understanding of why it happens leads nowhere except to another blind test. Having recognised a field problem as one possibly amenable to microbiological resolution, it is critical to define clearly how and why microbiological intervention will be useful: the purpose of each trial must be explicitly agreed by the collaborating parties. Benchmarks for success need to be established by which the trial will later be judged. Many operators will wish to carry out a prior computer simulation before being willing to allow tests in their fields: appropriate software may need to be developed and microbial parameters provided for inputting.

(iii) Operator cooperation. The willing and continued collaboration of the operator is essential to success; there have already been enough stories of tests aborted because of a breakdown in collaboration for any one of a variety of reasons. How can one promote enhanced cooperation? By making sure that there is a firm contractual agreement from which the operator clearly benefits. We will return to this point later in discussing who pays.

(iv) Management. Willing collaboration by the operator will hopefully also mean good field management during the trial. There also needs to be correspondingly good management of the project as a whole as well as of the laboratory aspects. Project management generally is crucial to success and academic laboratory scientists will frequently possess neither the skills nor the experience to exercise properly this important function.

(v) Conducive reservoir conditions. It sounds obvious enough that reservoir and field properties should be ones likely to lead to success rather than failure but those simple rules are apparently not universally observed. There seems little point in testing under poor field conditions, trying to resolve a difficult problem or one not well characterised and attempting to do so, moreover, with inadequate resources. Yet there may be little choice: this is what is available and you can take it or leave it. That tests have actually been conducted in less than minimal circumstances is beyond question though they are rarely publicised. Comments emerge during discussions at conferences (like this one) of tests conducted, for example, in wells which did not reach the pay or in those for some other reason producing no oil. The only justification can be that someone has been persuaded to fund the activity and it is better to do something, even something patently unsatisfactory, than do nothing. Unfortunately, such activities do little to improve either the image or the understanding of MEOR.

(vi) Funding. Running tests requires resources: for the expense or opportunity costs of identifying collaborators and candidate fields; for such preliminary evaluations of the field as may be necessary; for development work in the laboratory and the fine-tuning of existing techniques for the specific test site(s) under consideration; for feedstock and producing the bacterial inoculum; for modification or improvement of on-site field facilities; for managing the field operations during the trials and undertaking production and microbiological monitoring during the test runs; and for evaluating and interpreting the data. Resources imply money in cash or in kind: where will it come from? Possible sources include both the public sector (local, national and supranational official agencies) and the private sector (industry, investors, others). They might include contributions from the participants, again in cash or in kind. The successful outcome of the trial might itself generate resources in the form of incremental oil which can perhaps be set against the costs. It is unlikely that successful tests can be mounted without prior resolution of these questions.

Putting in place the necessary financial and organisational structure is not simple. It will certainly take account of the rewards the investment is likely to yield. That brings us directly to:

4 COMMERCIALISING MEOR

The negotiations implicit in securing both funding and operator participation are likely inevitably to define downstream commercialisation. Different categories of contributor will have differing reasons for participating.

4.1 Operators

Field operators will most probably be concerned primarily to secure incremental increases in production from their present and future activities. In some cases, however, the operating company may wish to have a more extended and continuing interest in the technology once it has been proven; as a condition of participating it may require an exclusive franchise or agency for an agreed geographical region which, in some cases, might be very large. This is perhaps more likely to be the case with majors than with small independents. The difference is important: an independent will have a limited radius of action in his own local area while a major may control enormous interests in many countries throughout the world. There are obvious implications for the successful marketing of MEOR.

4.2 Service companies

Sooner or later, MEOR is likely to involve a service company in the application of the new technologies in that company's geographical area(s) of operation. If the service company participates at the proving stage it is likely to demand exclusivity. Even if the company is not part of an activity concerned with initially proving the technology it might later be interested in adding MEOR methodologies to its existing menu of available services. Its view will be coloured both by the expected profitability of MEOR and by its perceived impact on existing products and services: would taking MEOR on board enhance or diminish corporate profits in the short-, medium- and long-term?

4.3 Investors

People, such as venture capitalists, who invest money in new enterprises look for a significant return on investment, bearing in mind the attendant circumstances. They may, for instance, be willing to invest in the development and marketing of MEOR technologies and will judge the financial risks and rewards of doing so against those of making alternative investments elsewhere. Others, like banks, who lend against some form of collateral require assurances that their loans will be repaid.

4.4 Government agencies

Various public sector bodies have differing regimes for offering technology development grants (which, of course, are also forms of investment); their underlying motivation is usually to stimulate the development of the new technology in order to benefit the industry and economy in the region or district for which they have jurisdiction. The relevant bodies include organs of the state and provincial governments in North America, agencies and ministries of national governments worldwide and supranational organisations like the European Community. Each has its own defined area of interest together with a set of rules governing the award of grants and the placing of investments in new technologies. Sometimes cash funding is awarded outright on a direct cost, cost plus or matching funds basis. Other organisations in effect lend funds to be repaid out of the downstream profits of a successful development.

4.5 Routes to commercialisation

Whatever the chosen vehicle for commercialisation, it may be very difficult in its absence to put together a structure with its attendant funding to enable the field testing to be carried out. Careful thought therefore needs to be given to possible alternative structures and the implications they may have both for each of the prospective partner's aspirations for the tests as well as for the later marketing of the technology and share of the rewards. How the parties actually decide on the way to bring the technologies to market will clearly depend on a reconciliation of their various interests.

5 CONCLUSIONS

With such a wealth of opportunity the microbiologist with MEOR protocols he wishes to promote through field testing is going to need help. Very many factors need to be brought into favourable conjunction with one another or the enterprise is likely to make little forward progress. A combination of the scientific and technical understanding and experience, together with a business structure to explore and build ways of exploiting it, seem to be a *sine qua non* for success. There may be many routes to the end of the rainbow, to the pot of gold or the barrels of oil one expects to find there and, while many alas will fail, some advocates of MEOR will no doubt be able to build and exploit an organisation leading eventually to a successful market operation.

Reprinted from: *Microbial Enhancement of Oil Recovery – Recent Advances*,
edited by E.C. Donaldson

CH. I–4

A BRIEF INTRODUCTION TO THE PROGRESS OF MEOR IN CHINA

QIN TONG LUO
Research Institute of Petroleum Exploration and Development
Beijing, China

YANG GUANG HUA
University of Petroleum, China
Beijing, China

ABSTRACT

Since many of the main oil fields in China such as Daqing, Shengli, Karamai, etc., have been waterflooded for more than twenty years, the water content of the producing fluids runs as high as 75-80%. Consequently, tertiary recovery is urgently needed to improve the producing conditions and to increase the ultimate yields of these oil fields. Microbiological tertiary recovery, among other tertiary methods has been receiving considerable attention by our oil industry. Hence, we present an overview of the activities in this field in China that was introduced at the First International MEOR Conference in 1982.

INTRODUCTION

Interest has been focusing on the feasibility of MEOR in the Chinese petroleum industry. The first endeavor made was in increasing the viscosity of injection water using biopolymers. In the early 1980's, attempts were made to prepare the reagents directly from petroleum by biological methods. Some success had been made in this direction and the results were reported at the 1st International Conference on MEOR held at Shangri-la in 1982. Since then, more work has been done on conventional biotechnology for the preparation of biopolymers (mainly Xanthan type polymers), instead of petroleum-derived water thickeners due to technical and economical reasons. Later, the direct injection of screened microorganisms into the oil reservoir was initiated together with the application of biopolymers and biosurfactants in EOR as well.

Thus, efforts in MEOR in China that have been carried out by our research institutes, universities and various oil fields are along three main directions:

1. The direct injection of microorganism into the reservoir to increase the ultimate recovery of an oil reservoir by a huff-puff process in a single well;

2. The preparation of suitable biosurfactants and the evaluation of their physio-chemical behaviors particularly related to the enhancement of the ultimate yields of oil reservoirs;
3. The production of biopolymers and their field applications.

An extensive survey of the development of the Petroleum Microbiology in China was made by S.X.W. Wang, 1988.

In addition, the transport characteristics of surfactants and ferment mixtures through artificial and natural core samples were investigated in the laboratory of the Research Institute, and other research organizations, to obtain fundamental insights into the recovery processes.

All these research activities were closely coordinated by the China National Petroleum Corporation (CNPC), formerly the Ministry of Petroleum, China.

INJECTION OF SELECTED SPECIES OF MICROORGANISMS INTO OIL-PRODUCING WELLS

Screening of strains of microorganisms used for injection into oil reservoirs for EOR is a cooperative effort carried out by the Daqing Oilfield and the Beijing Institute of Microbiology of the Academy of Science, China (Zhang, 1990). Promising results were obtained in the laboratory. Physical modelling of the displacement behavior both at atmospheric and elevated pressure were performed in Daqing.

The criteria for the screening for suitability of a bacteria species for MEOR are as follows:

1. Determination of the change in properties of the crude oil before and after fermentation: (a) Engler distillation with the increase of the percentage of IBP-100°C as the degree of degradation of the crude; (b) chromatographic analysis to detect the change of the ratio of iso-C_{19}/C_{17} and iso-C_{20}/C_{18} as the indication of the degree of degradation, the higher this ratio, the higher the degree of degradation (where: iso-C_{19} --- 2,6,10,14-tetramethyl-pentadecane, iso-C_{20} --- 2,6,10,14-tetramethylhexadecane), and (c) group analysis of the crude to find the changes of the content of paraffins aromatics, resins and asphaltenes before and after fermentation, this gave us some clue as to which part of the crude oil was subjected to attack by the bacteria.
2. Determination of the change in properties of the ferment liquor: (a) evaluation of the interfacial tension change during the evolution of the active surfactant. The lowering of the interfacial tension between the rock and oil by the action of the surfactants is a desired property for MEOR; and (b) determination of the organic acidity of the ferment liquor.

3. By physical modelling of the displacement process of oil in laboratory through core samples, both under atmospheric and elevated pressure, to simulate the conditions underground.

After extensive laboratory tests, it was found that certain blends of the bacteria obtained in Daqing Oilfield and in the Beijing Institute of Microbiology gave the best results as shown in Table 1.

Table 1. The effect of bacterial species in LAZAR Model Test.

NO.	Period of pure oil effluent		In the producing period of oil & water		MEOR		Residue Oil Recovery	Total Gas Evolution
	Oil(ml)	Rec.%	Oil(ml)	Rec.%	Oil(ml)	Rec.%	%	(ml)
6	181	34.2	200	37.7	66.4	12.5	38.2	1250
7	290	51.8	108.1	19.3	104.4	18.6	55.9	4750
8	115	24	153	31.9	165	34.3	69.8	1250

Note: 6: 6A+Su3-2=Su1-2-3+Su5-2+wh1
7: 26A+Su2-1+Su1-2-3+Su5-2+wh1
8: 5GA+Su2-1+Su-2-1+Su1-2-3+Su5-2+sh1

Where Su denotes species from Daqing Oilfield and A species from Beijing Institute of Microbiology

As can be seen from Table 1, with species 6A+Su, gas evolution reached 1250 ml in ten days, oil production 44.5 ml in 37 days, increase of percentage recovery 8.3%, percentage recovery of residue oil, 30%. After a four month fermentation period followed by waterflooding, 66.4 ml of oil was produced which was an increase of the percentage recovery by 12.5%; percentage recovery in terms of residual oil reached 38.2%. With 26A+Su, gas evolution 4750 ml in ten days, oil production 48 ml in 37 days, increase of percentage recovery 8.3%, percentage recovery increase in terms of residual oil 28.9%, with waterflooding after four months, oil production 104.4 ml, increase of percentage recovery 18.5%; percentage recovery of residual oil 55.9%. With 5GA+Su, gas evolution 1250 ml in ten days, oil production 80 ml in 37 days, increase of percentage recovery 16.7%; percentage recovery of residual oil attained to 44.2%. After four months with waterflooding, oil production reached 165 ml, percentage recovery 34.3%; percentage recovery of residual oil 69.8%.

A field pilot test with the above bacteria species is scheduled for late this year in the Daqing Oilfield.

THE USE OF BIOSURFACTANTS IN ENHANCED OIL RECOVERY

The production and evaluation of biosurfactants with respect to the adequacy to EOR are performed in the Shanghai Institute of Organic Chemistry of the Academy of Science, China and the Daqing Oilfield. They have been conducting a rather extensive program in this direction.

The program consists of:

1. The screening of biospecies for the production of satisfactory biosurfactants for EOR and the investigation of the fermentation processes in laboratory.
2. Structure characterization of biosurfactants and means of their structure modification.
3. The biochemical engineering study of the fermentation process.
4. The selection of flooding fluid mixture composition with biosurfactants.
5. The physical modelling of the flooding process with biosurfactants in the laboratory.
6. The blending of the various biosurfactants and their core evaluation in the laboratory.

Some types of glycolipids were obtained as extracellular material in the growth of some microorganisms in a solution containing carbohydrates, vegetable oils, water insoluble alkanes and necessary inorganic salts. Two of such examples are the production of sophoroselipids from screened strains of *Torulopsis* sp (Li, 1986) and the production of rhamnolipids from *Pseudomonas* sp (Li, 1988). Both glycolipids were obtained in laboratory fermenters and their physicochemicl properties as well as their chemical structures were determined. A semi-commercial plant for the production of sophoroselipids was erected and the products were obtained. These biosurfactants reduce the interfacial tension between hexadecane and water to a certain extent, but it remains too high for EOR. Addition of some co-surfactants to the rhamnolipids solution can decrease the interfacial tension between hexadecane and water to an order satisfying the requirements of chemical flooding. More detailed studies on its ability to reduce the interfacial tension between the actual reservoir oil and the formation brines are now being conducted in a field laboratory to give more reliable information about its feasibility for field application (Zhou and Tong). Results are encouraging. The preparation of additional types of biosurfactants is continuing. It is expected that a new candidate of reagents used in chemical flooding may be selected from these.

THE APPLICATION OF BIOPOLYSACCHARIDE IN EOR

This program is being conducted separately in Nankai University, the Shandong University and the Beijing Institute of Microbiology of the Academy of Science, China. The polysaccharide is mainly Xanthan. Main research interests are:

1. The production (on a sizable scale) of Xanthan by biological means. Considerable success has been reported in this field.
2. The characterization of the physio-chemical properties of Xanthan by conventional means and also the core tests.
3. The rheology, characteristics of flow and the permeation behavior of Xanthan plug through porous media.
4. The structure analysis of Xanthan, the structure modification and the subsequent examination of the changes in its rheology behavior.
5. Physical modelling of flooding of oil through core samples with Xanthan mixtures.
6. Field pilot test for the improvement of the profile of the injection fluid front with cross-linked Xanthan fluids.

The biopolymers prepared by all three institutions are of the same type. The microorganisms and culture used in its preparation are nearly the same with minor modifications; however, the procedures of purification and concentration are somewhat different. All products are obtained in powdered form. Their physical and chemical properties have been appraised and are very similar to those of KALCO products.

Two well tests in the Nanyang Oilfield have been completed, and a third well test is under way. Results of these tests are promising, showing the possibility of field application. The preparation of more heat resistant biopolymers is now being contemplated.

THE ENVIRONMENTAL CONCERNS OF MEOR IN CHINA

The main efforts to control the activity of the bacteria for MEOR and the evaluation of their toxicity and environmental influences are contemplated by the Beijing Institute of Microbiology. Reports will be submitted to the Bureau of Environmental Control for assessment and approval.

CONCLUSIONS

MEOR has been investigated in the Chinese petroleum industry under a long term unified program with the participation of research institutes, universities and oil fields in two directions: (1) injection of suitable species of microorganism directly into the reservoir; and (2) application of biological products as additives in the injected water for the improvement of waterflooding. Progress has been made along both lines, showing that it may be feasible. Studies of MEOR in the laboratory as well as in field pilot tests will be continued in the coming decades in China. It is expected that

MEOR will prove to be feasible for field application under certain circumstances.

ACKNOWLEDGEMENT

The authors would like to express their thanks to the China National Petroleum Corporation and the Deqing Oilfield for their permission to quote their research results in MEOR.

REFERENCES

Li, Z.Y., 1986. Chinese Journal of Biotechnology 2(1) 47-51; 2(2) 62-66.
Li, Z.Y., 1988. Acta Chimica Sinica 16, 262-268.
Tong, G. Research Report Institute of Photosensitive Materials, Academy of Science.
Wang, X.W., 1088. The Accomplishment of Industrial Microbiology, The Development of Petroleum Microbiology in China, Science Press, Beijing, China, p. 105-162.
Zhang, C.Y., 1990. Research Report of Daqing Oilfield.
Zhou, Y. Research Report Institute of Percolation Fluid Mechanics of the Academy of Science.

petroleum. In these processes financial returns by definition must remain marginal.

MICROBIAL BIOTECHNOLOGY AND OIL RECOVERY

Beckman proposed in 1926 that bacterial metabolites would assist in the release and transport of oil in geological structures. Later, Zobell (1946, 1947, 1953) described and patented processes by which bacterial fermentation products (CO_2, solvents and acids) released oil from sandpacks in laboratory experiments. These studies demonstrated that bacterial surfactants were at least partially responsible for the enhanced oil recoveries observed in the laboratory. However, the concurrent production of H_2S by the sulphate reducing bacteria used in these studies proved intolerable for the continued development of field applications based on these organisms.

The studies of Zobell highlighted the parallel between the compounds used to improve waterflood efficiency, in chemical and miscible EOR processes and the products of microbial fermentation of carbohydrate. Gases, solvents, surface active compounds, polymers, organic acids and biomass are all regular and predictable products of microbial metabolism. The same compounds are used in petroleum recovery.

Subsequent studies can be grouped into three broad areas. These areas are the production by microorganisms of products and effects similar to those described for chemical and miscible EOR, selective degradation of petroleum components to improve flow characteristics and processes for the injection, dispersion and propagation of microorganisms in petroleum reservoirs.

There is no question that microbial growth and the products of microbial growth have an effect on the chemical and physical properties of petroleum. Well plugging and oil souring, thought regrettable, are direct and definitive evidence of the potential for microorganisms to proliferate in the reservoir environment and produce a significant effect. Laboratory and field studies have shown that microorganisms can assist in the release of oil from capillary forces and improve the sweep efficiency of reservoir and introduced fluids. The microbial production of gases can improve flow characteristics, solvents can reduce interfacial tension, and organic acids can result in the dissolution of carbonates in source rock and increase permeability. Polymers of microbial origin can increase the viscosity of water in waterfloods and divert reservoir fluids to previously unswept areas of a reservoir. Surface active compounds such as glycolipids produced by microorganisms faciltate oil release. Emulsification, an important property of microorganisms capable of assimilating hydrocarbons, is a recognized mechanism for improving oil recovery. Hydrocarbons also stimulate both the production and release of extracellular surfactant chemicals which could be used to release trapped oil. Microbial growth on rock pore surfaces dislodges oil directly by physical displacement and can plug high permeability zones thus increasing production and the viscosity of oil can be reduced by selective partial degradation of whole crude oil or flow restored by the removal of paraffins.

The thrust to continue research and development of microbial EOR is compelling. Microbial EOR biotechnology shares the common advantage of all biotechnologies ie the extraordinary growth potential of microorganisms, and the extreme diversity of microbial biometabolites than can be produced by microorganisms using relatively cheap feedstocks. Further, it is proposed that bacteria can be used <u>in situ</u> to produce metabolites thus avoiding product losses due to adsorbtion to rock surfaces and other losses due to the physical and chemical conditions of the reservoir that reduce the amount of active agent available for EOR.

Successful application of microbial EOR requires the selection of microorganisms producing effective quantities of desired metabolites, processes which permit the injection and dispersion of these microorganisms, and their supporting nutrients, predictable proliferation and metabolic activity and an ability for the new ecosystem to persist for periods consistent with economic profitability. Each of these stages presents major intellectual and practical problems. Research which ignores any facet of any stage is doomed to failure in the field. Examples of this fill the literature and patent files. These poorly conceived systems will never produce a drop of oil from the field.

A major reason for the failure of proposed systems is insufficient consideration of the conditions which characterize petroleum reservoirs and the physiology of microorganisms which thrive in these conditions. It must be emphasized that the activity of bacteria in reservoirs depends on the physical and chemical conditions they encounter. Eh, pH, temperature, salinity, ionic strength, pressure, sources of energy and nutrients and the natural microbial ecology all affect microbial attachment, proliferation, metabolism and persistence.

EXTREME ENVIRONMENTS

The popular concept is that microbial growth begins and ends with aerated, nutrient-rich media at approximately neutral pH, one atmosphere of pressure and ambient temperatures. In fact most microorganisms live in environments more closely allied to those used in chemical reactions than those traditionally associated with microbial growth.

It is easy to appreciate the range and complexity of the microorganisms which colonize petroleum reservoirs if the reservoir is considered as a subterranean thermal environment containing petroleum oil (alkanes, aromatics, NSO compounds), and plant or other organic material. The physical environment is composed of solid rock, an aqueous phase, an organic phase and a gas phase.

It must be emphasized that the extremes of temperature, pH, mineralization, carbon loading, heavy metals and geopressure etc found in petroleum reservoirs are extreme to us but optimal ecological conditions for many microorganisms.

If due attention was given to developments in the microbiology of extreme environments, then statements such as "most bacterial growths are inhibited by salt concentrations of about 2 percent" and "the pH range within which bacteria are optimally able to grow generally covers a narrow range around pH7" will disappear from publications on microbial EOR. Both statements appeared in a 1990 publication on microbial EOR and detract from the scientific credibility of those conducting studies in this biotechnology.

An increasing knowledge of the properties of microorganisms from extreme environments and the realization that these organisms are ubiquitous in nature has created an awareness of their biotechnological potential in oil recovery. Experience in the microbiology of extreme environments has shown that it is inappropriate to use molasses as the primary carbon and energy source for bacterial growth in many reservoirs. Molasses results in a marked decrease in the diversity of microorganisms which can be isolated from petroleum reservoirs and severely limits the viability of those isolated.

The abandonment of molasses based processes which mimic chemical and miscible EOR has resulted in the development of novel methods for microbial EOR. In addition processes have been developed which may prolong primary and secondary recovery by limiting premature breakthrough of aqueous fluids and gas.

ISOLATION OF MICROORGANISMS FROM PETROLEUM RESERVOIRS

The strategy for optimizing microbial growth and diversity in the laboratory instituted in the Microbiology Research Unit is to formulate media for each environment. This is achieved by a systematic evaluation of the chemical composition of the reservoir brine and the composition of the petroleum; determination of the chemical solubilities and availability of each species and examination of culture filtrates for components that could be used in microbial food chains. Based on this data the final media composition is formulated showing due regard for the physical conditions which exist in the natural environment. Growth supplements are then added as required by specific microbial groups.

This approach has provided some startling findings. A good example occurred in the preparation of media for the Fitch units in Houston, Texas.

The recorded pH for the brine from this reservoir was 6.85. Based on the concentrations of Ca^{2+} and HCO_3^- it was concluded that the sample was super-saturated with respect to calcium carbonate. This is to be expected because some of the dissolved CO_2 would have effervesced into the atmosphere before the pH reading was completed. At the concentration of calcium ions and bi-carbonate ions detected in the brine, a pH of 5.92 and a CO_2 pressure of 4 atmospheres would have been required to dissolve the calcium carbonate at 25°C. At 90°C, the actual bottom hole temperature, a pH of 4.93 and a CO_2 pressure of 10 atmospheres would have been required to dissolve the calcium carbonate. Preparation of media with the recorded pH would have created a selective environment against the reservoir bacteria.

This example provides a salutary warning that the formulation of media employed to grow microorganisms from extreme environments will determine the species diversity that can be isolated.

In a similar way, the buffering of cultures requires due care and attention. In selecting a buffer it is necessary to consider the desired pH (matched to the pK of the buffer compound), possible inhibitory or toxic effects of the buffer and possible binding of metal cations by the buffer. The most appropriate buffer for media used to isolate microorganisms from petroleum reservoirs is bicarbonate/carbonate overpressured with carbon dioxide to achieve the desired pH. This is the buffering system most commonly associated with the petroleum reservoir environment. Use of phosphate buffers often is inadvisable as they can result in decreased viability of isolates and decreased species diversity. A simple buffering system can be achieved by using sea shells in media overpressured by CO_2. An added advantage of this system is the large surface area provided by the shells for the development of biofilms. This system results in the isolation of a greater species diversity. It should be noted that incorporation into media of bicarbonate and carbonate without sealing the vessel and pressurizing with CO_2 will not provide a buffering system. In an open vessel the pH will drift toward 10 as carbon dioxide from the media equilibrates with atmospheric CO_2.

The Microbiology Research Unit has constructed apparatus capable of maintaining high pressure and temperature to facilitate the isolation of micro-organisms from the reservoir environment and testing of oil release mechanisms.

BIOLOGICAL STIMULATION OF OIL RECOVERY (BOS)

The development of the microbial EOR process applied successfully in field evaluation has been based on two key observations; that microorganisms capable of producing significant amounts of surfactant lose this capacity on routine subculture or when introduced into simulated reservoir environments, and that microorganisms undergo substantial morphological change when exposed to reservoir conditions for prolonged periods of time. Other considerations in the development of BOS were problems of microbial injection and dispersion in the reservoir and the most frequently neglected issue in microbial EOR studies, persistence.

The BOS concept became to adapt the outer cell layers of naturally occurring reservoir bacteria to act as surface active agents. The process is achieved through the generation of ultramicrocells which because of their size also aid injection and dispersion.

Field application is achieved in the sequence; analyse the reservoir and petroleum chemistry, formulate a suitable base media, enrich the base media, evaluate the microbial ecology of the reservoir environment, determine the conditions which generate ultramicrocells with surface active properties, test the microbial consortium of oil release from sandpacks, optimize the conditions that generate surface active microcells, test for oil recovery from the high-pressure oil reservoir simulation equipment, inject well.

FIELD STUDIES

Field studies are continuing as the scope and applicability of the BOS process is evaluated. Graph 1* shows the changes which resulted from the introduction of the BOS system into the Alton field in Queensland, Australia. Point A indicates the day on which the well was shut-in to generate a baseline. Point B indicates the day on which injection of production water was performed to act as a fluid control and Point C the introduction of the BOS system. The increase in oil production after the application of the BOS system is clearly evident from February 1989. An increase in production is still evident 16 months after the system was injected.

ACKNOWLEDGEMENTS

The author wishes to thank the staff of the Microbiology Research Unit and the staff of the Faculty of Applied Science, Uni. of Canberra. Biological stimulation of oil production is the result of 10 years research and development initiated by the Australian Government Commonwealth Industrial Research Organisation (CSIRO) and the Bureau of Mineral Resources Geology and Geophysics. Initial studies were supported by the National Energy Research, Development and Demonstration program and TMOC Resources/AGL Petroleum. Since 1987 the project costs have been met predominantly by BWN Live Oil Pty Ltd, 7 Abbott Road, Hallam, Victoria, 3803, Australia which holds BOS patents and an exclusive licence to market the BOS process.

*The results shown in Graph 1 were presented to the SPE/DOE Seventh Symposium on EOR, Tulsa, OK, 25 April 1990.

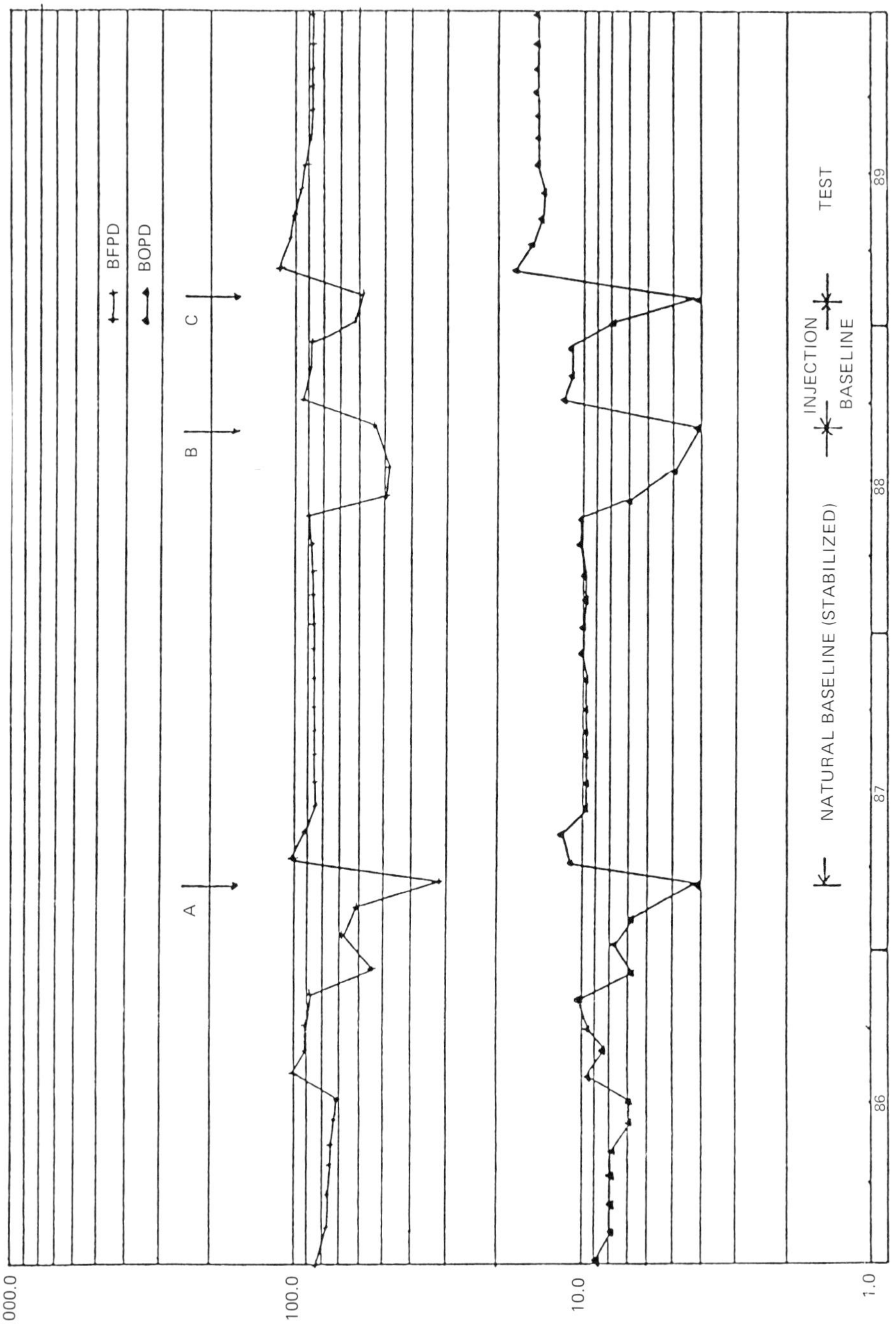

Graph 1. Alton production.

Reprinted from: *Microbial Enhancement of Oil Recovery - Recent Advances*, edited by E.C. Donaldson

CH. B-2

ANALYSIS OF BIOMASS IN ROCK CORES FOLLOWING NUTRIENT STIMULATION

D. GEVERTZ, R.M. SOLIS and W.A. WOOD
Salk Institute Biotechnology/Industrial Associates, Inc. (SIBIA), P.O. Box 85200, San Diego, CA 92186

ABSTRACT

Chemical methods for the determination of biological macromolecules were investigated for use in quantitating the distribution of biomass within a rock core. Cores (3 x 1 inch) were coated with epoxy, brought to residual oil saturation and inoculated with oil reservoir brine. A carbohydrate-based medium was pumped into the core until the permeability reduction reached 95-99%, after which the core was cleaved into six equal transverse sections and crushed for analysis. The following methods were used to measure biomass: 1) bicinchoninic acid (BCA) assay for protein; 2) the anthrone method for carbohydrate; 3) chromogenic _Limulus_ test for lipopolysaccharide; and 4) the orcinol assay for RNA. Measurements of standards were linear in all four assays in the presence of rock, although the sensitivity was lowered by one-half in most cases. With these methods, measurable changes in the biomass profile were apparent when a change in medium composition took place, such as elimination of a specific nutrient from the growth medium. Slight alterations in the concentration or type of nutrient also resulted in altered profiles. These assays are useful for defining where biomass is located within a core and for comparing the amount of biomass generated by different treatments.

INTRODUCTION

The application of tertiary methods for recovery of oil from a reservoir is warranted when the efficiency of waterflooding is reduced to the point where it is no longer economically feasible (Morrow and Heller, 1985). As waterflooding progresses, the ratio of oil to water in produced water decreases, because water preferentially sweeps zones of highest permeability in a reservoir, leaving behind oil in the low permeability zones. Enhanced oil recovery processes can then be used to mobilize oil from zones of low permeability. An underlying principal behind microbial enhanced oil recovery (MEOR) is the ability of bacteria to multiply _in situ_, generating sufficient biomass in the form of cells and polymer to block high permeability zones (Raiders _et al_., 1989; Torbati _et al_., 1986). Selective plugging of high permeability zones facilitates the recovery of residual oil by directing waterflooding to targeted low permeability regions.

Small scale simulations of MEOR processes are carried out in the laboratory using rock cores to model reservoir conditions. Petroleum microbiologists have studied in detail microbial movement, attachment, and growth throughout cores under idealized conditions, using pure cultures of bacteria (Jang _et_ al., 1983; Jenneman, 1985; Raiders et al., 1986). Cells have been shown to attach to surfaces of glass beads or sand grains (packed into artificial cores) and to form biofilms consisting of cells and carbohydrate polymer (Geesey et al., 1987; MacLeod et al., 1988). Biofilm formation resulted in a reduction of permeability of greater than 99% in cores (Shaw et al., 1985). Microbial growth and distribution was measured by dividing the core into discrete sections, disassembling the packed material, and performing a direct count of cell number using epifluorescence microscopy or plating on appropriate media. The amount of biopolymer produced was quantified by carbohydrate assay (Geesey et al., 1987). Cells and polymer have also been visualized by electron microscopy. Metabolic activity was measured radioisotopically and cell distribution determined by using a fluorometric DNA assay (MacLeod et al., 1988).

The process of biomass measurement becomes more difficult when working with consolidated rock. Growth is commonly monitored indirectly, by the resultant drop in permeability associated with production of microbial biomass. Such measurements, however, usually yield little information concerning cell numbers or distribution within a core. Alternative methods are needed to complement permeability measurements, so as to give a more complete and accurate assessment of microbial biomass within a rock matrix. Phospholipid measurements and carbohydrate assays, along with plating on solid media, have been used for this purpose in cores of Berea sandstone (Raiders et al., 1989).

We have applied additional methods routinely used for the estimation of microbial content in soils and sediments to determine microbial distribution in samples of ground core material (Karl, 1986; Litchfield and Seyfried, 1979; Parkes, 1987). These techniques directly estimate the amount of biological macromolecules present in heterogeneous samples, avoiding the need for cultivation of bacterial cells. Target molecules may include lipid, protein, DNA, RNA and ATP. Assays for determining protein, carbohydrate, RNA and lipopolysaccharide, were found to have adequate sensitivity and linearity for use with core material, in spite of the presence of rock, field brine and crude petroleum. These four methods were then applied to locate and quantitate biomass within a core.

MATERIALS AND METHODS

Preparation of cores

Three by one inch sandstone cores were used throughout this study. Berea sandstone cores were obtained from Cleveland Quarries (Amherst, OH). Horizontal cores of Burbank sandstone were drilled from the Burbank oil field, operated by Phillips Petroleum Company (Bartlesville, OK). The cores were fitted with end plates made of Lexan or Ryton. These plates were previously drilled and tapped to accept a nylon Swagelock fitting for 1/8" tubing. End plates were attached to the core with activated epoxy resin (Epoxy-patch, 1 C white, Hysol Aerospace and Industrial Products Division, Pittsburgh, CA) followed by attachment of the Swagelock fittings. The lateral surface and end plates, including the threaded end of the Swagelock fitting, were coated with epoxy. Epoxy was applied manually, while rotating the core with a motor driven mandrill. Three layers of epoxy were applied in all, allowing each coat to dry between applications. The core was weighed and all further manipulations were carried out in an anaerobic chamber (Coy Products, Ann Arbor, MI).

Brine and oil saturation

After equilibrating for several days in the anaerobic chamber, the core was attached to an assembly consisting of tubing, valves, pumps and gauges. The core was evacuated for 30 minutes, after which approximately 100 ml of sterile brine was imbibed into the core. The ends of the core were capped at this point, and the core was temporarily removed from the chamber for the purpose of weighing to determine the pore volume. Upon return to the chamber, the core was placed in a 45°C incubator for the remainder of the experiment. Filtered deoxygenated oil was pumped through the core using a piston displacement pump (Model E-120-S, Eldex Laboratories, Menlo Park, CA) until no brine appeared in the oil effluent (approximately ten pore volumes). Ten pore volumes of brine were pumped through the core with a peristaltic pump, to reach residual oil saturation. Specific permeability was determined according to Darcy's Law.

Inoculation and growth

Inoculation of the core was carried out by pumping through 50 pore volumes of unfiltered oil reservoir brine, which contained approximately 10^4 cells per ml. The brine used in these experiments was collected in sterile anaerobic bottles from the Burbank field in Oklahoma. Cores were incubated at 45°C throughout the experiment to simulate reservoir conditions. After inoculation, nutrient was pumped into the core and the change in pressure recorded daily until the reduction in permeability reached 95-99%. Prior to analysis, cores were flushed with five pore volumes of filtered brine to remove residual nutrient.

Bacteria and media

In one experiment, an *Aeromonas* species isolated from biofilm was used as a source of inoculum. This organism was grown and maintained on M9 salts medium

(Miller, 1972) supplemented with 0.2% (w/v) glucose. M9 medium was also used in core experiments with this organism. Indigenous cells present in brine were used for the majority of the experiments described. Brine-based medium was made by the addition of the desired carbon source and inorganic salts directly to field brine. The medium was sterilized by filtration under anaerobic conditions.

Sectioning and preparation of core samples

Usually, cores were sectioned outside the anaerobic chamber. The core was mounted in a rotating mandrill and circumferences marked at 0.5 inch intervals, to give six equal sections. After placement in a vice, the epoxy coating was sawed along the circumference lines to near the rock surface. The core was mounted in a pipe cutter and pressure carefully applied at one or more points along the circumference line, whereupon the core fractured to produce uniform disc-shaped cross sections. The epoxy was removed from individual sections using plier-type cutters. The rock matrix was then crushed with a mortar and pestle to yield 8-10 grams of powdered core material from each section.

Analysis of core material

Carbohydrate was determined by the anthrone method, with minor modifications, and using glucose as a standard (Hanson and Phillips, 1981). Samples of rock (0.1-1.0 gram) from each section were suspended in 2 ml of water and assayed directly. Particulates were removed by centrifugation before reading the absorbance at 625 nm. Initially, the Lowry assay was used to measure protein (Lowry *et al.*, 1951). One gram of rock was suspended in 1 ml of 1 N NaOH (final concentration) and boiled for 15 minutes to solubilize the protein (Hanson and Phillips, 1981). The assay was then performed directly on the rock slurry. Centrifugation was used to remove rock prior to reading the absorbance. Problems later arose with the Lowry method, *i.e.*, instability of the color formed, and this method was replaced with bicinchoninic acid assay (Smith *et al.*, 1985). In this case, rock samples (0.2 grams per section) were boiled in 0.2 ml of 1 N NaOH (final concentration) for 15 minutes followed by centrifugation to remove rock and clarify samples. The enhanced bicinchoninic acid assay (BCA, Pierce Chemical Co., Rockville, IL) was performed on 0.1 ml of the supernatant fluid. Interference by rock was seen in some cases and additional steps had to be taken to eliminate interfering compounds (Brown *et al.*, 1989). These consisted of precipitating the protein with trichloroacetic acid and resolubilizing the protein with base and sodium dodecyl sulfate prior to assay. Bovine serum albumin (BSA) was used as the standard. Lipopolysaccharide was determined by the quantitative chromogenic Limulus amebocyte lysate (LAL) test (Whittaker Bioproducts, Walkersville, MD) (Watson et al., 1970). A 0.1% (w/v) suspension of rock in 0.05% (v/v) Tween-20 (in pyrogen-free water) was mechanically mixed end on end for 30 minutes at room temperature to help release cells from rock. Samples were allowed to sit 1-5 minutes at room temperature to allow large particulates to settle. Dilutions of the upper portion of the rock slurry were made in pyrogen-free water for analysis. The assay was performed in microtiter plates, according to the manufacturer's directions. Plates were read with a Titertek Multiskan MC microtiter reader (Flow Laboratories, McLean, VA) at 405 nm. Purified Escherichia coli endotoxin from strain 0111/B4, supplied by Whittaker, was used as the standard. RNA was determined colorimetrically by the orcinol method, with minor modifications (Hanson and Phillips, 1981). Samples (0.5-1.0 g) were suspended in 1 ml of distilled water prior to assay. The hydrolysis step with dilute $HClO_4$ was omitted. Ribose was used as the standard. Samples were centrifuged prior to reading, to remove rock. All assays were done in duplicate and average values reported.

RESULTS

Standard assays

The effects of rock, brine, and oil (components of core samples) on standard assays were studied in order to determine changes in sensitivity and linearity. This was done by adding 0.5 ml of reservoir brine and 0.1 ml of crude oil to a 0.5 gram sample of crushed, uncleaned rock from the Burbank field. The desired amount of standard was added and assays performed according to standard protocols. A comparison was made of the performance of the assay in the presence and absence of core components. Figure 1, panels a and b, show the results for two such assays; the anthrone assay for carbohydrate, and the orcinol assay for RNA. The BCA assay for protein, in panel c, Figure 1, was performed in a slightly different manner. Crushed oil-saturated Burbank rock was substituted for rock, brine, and oil due to the small volume of the assay. The assay with and without the addition of rock was performed identically, i.e., each sample was boiled for 15 minutes in the presence of 1 N NaOH. In all three assays, rock, brine and oil decreased the sensitivity of these assays by approximately two-fold; however, the linear relationship between amount of standard and absorbance was maintained. Inhibition or enhancement of values obtained in the Limulus assay was tested by assaying core material alone and by spiking core material with a known amount of endotoxin standard. In all cases, no inhibition or enhancement was observed at the dilutions tested. These assays were subsequently used to estimate biomass present in core material.

The anthrone assay and BCA protein procedure were also tested with rock of different composition (oil saturated) to determine the broader applicability of these methods. Cores of dolomite (from the Shiel formation in Texas County, OK), chalk (from the Eldfisk formation in the North Sea) and sandstone (from the Bean Blossom formation in Garfield County, MT) were used for this purpose. Both methods gave linear responses using standards in the same range as used above. The values obtained in these assays, however, were approximately 25-50% lower than those obtained with Burbank rock (data not shown). The ability to make measurements with these methods on different core materials indicates that these methods are generally applicable. It is assumed this is also true for the Limulus test because the sensitivity of the assay is so great that samples can be diluted to a point where rock is no longer present.

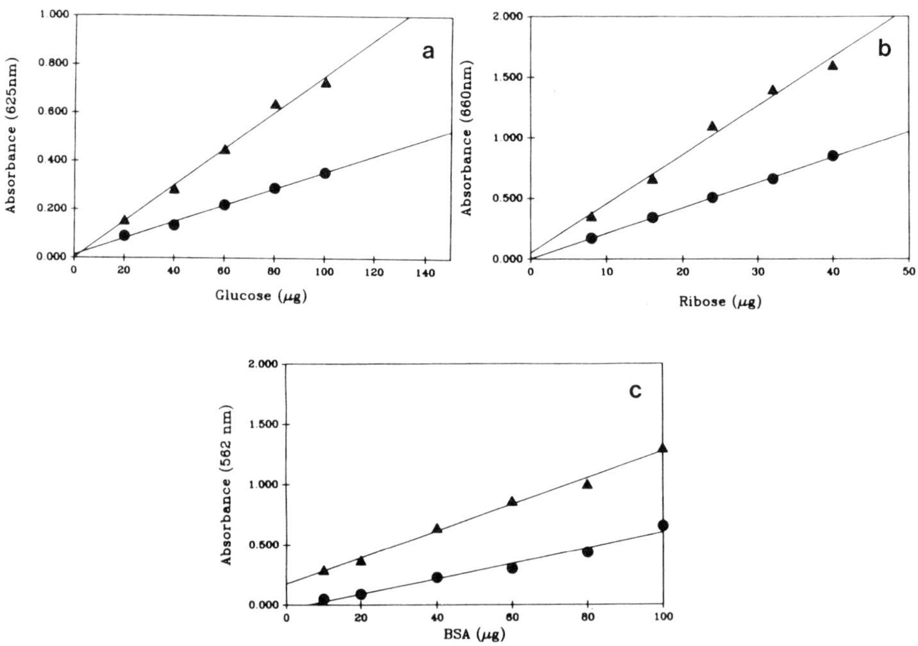

Fig. 1. Standard curves in the presence and absence of rock: (a) the anthrone assay; (b) the orcinol assay; and (c) the BCA assay. Symbols: ▲ standard alone; ●, standard with rock.

Analysis of Berea cores

Initial experiments were done with Berea cores to perfect techniques for running core experiments and gain experience in conducting analyses. In both the experiments described, protein was measured using the Lowry method. The biomass profile shown in Figure 2 was obtained after *Aeromonas* was inoculated into a core followed by the introduction of M9 minimal salts medium supplemented with 0.2% glucose. When the permeability reduction reached greater than 95%, as indicated by pressure measurements, the core was sectioned for analysis. Profiles were constructed by plotting absorbance values obtained for each section with a given method. All three assays indicate that the greatest amount of biomass was located in the first one-half inch of the core. A slow but gradual decrease in biomass measurements for sections 2-6 was apparent, indicative of less growth further into the core.

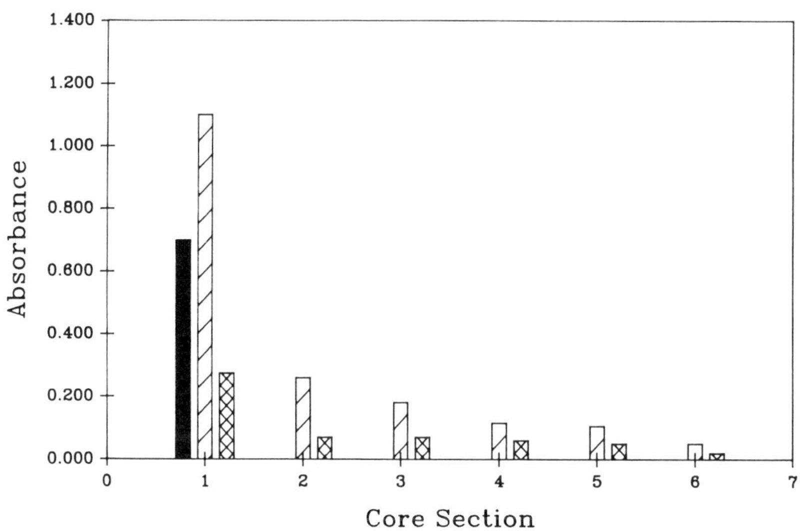

Fig. 2. Biomass pattern of a Berea core inoculated with *Aeromonas*. Symbols: ■, carbohydrate assay; ▨, RNA assay; ▩, protein assay.

A similar experiment was performed in much the same manner, using field brine rather than a laboratory culture as the inoculum. Again, defined medium was pumped through the core until a permeability reduction of greater than 95% was achieved, at which point the core was sectioned and analyzed. The results in Figure 3 indicate that the majority of the biomass was once again present in the first section of the core, and lesser amounts thereafter.

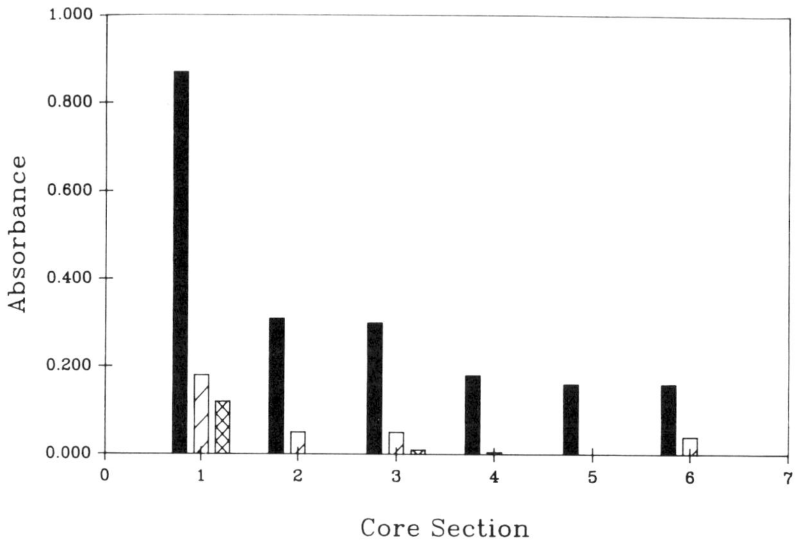

Fig. 3. Biomass pattern of a Berea core inoculated with reservoir brine. Symbols: ■, carbohydrate assay; ▨, RNA assay; ▩, protein assay.

Analysis of burbank cores

The remainder of the experiments were conducted in cores drilled out of the Burbank field in northeast Oklahoma, operated by Phillips Petroleum Company. The permeability of these cores was approximately 100-300 mDarcy. Injection brine used for the inoculum was collected from the tank battery of Tract 5 in the North Burbank unit. The results in Figure 4 compare the biomass profiles obtained when a specific nutrient is present or absent from the growth medium. Two cores were treated identically, except that a specific growth component was excluded from the medium in the core depicted in Figure 4b. Medium was pumped through each core until the same degree of permeability reduction was achieved (ca. 95%), which took approximately one month. Although pressure measurements indicated that both cores had plugged to the same degree, biomass measurements indicated otherwise. Figure 4 compares values obtained in carbohydrate, protein and *Limulus* assays on both cores. Numbers are expressed as μg/gram of rock or, in the case of the chromogenic *Limulus* test, endotoxin units (EU)/gram of rock. All values were calculated based on the response of standards under identical conditions with control rock. Protein was measured using the BCA assay. The core containing the added nutrient component Figure 4a showed more biomass present in the first one-half inch of rock, as indicated by the Limulus test and protein determination. Sections 2-5 of this core also showed more Limulus-positive material as compared to the other core, although this is not readily apparent due to the reduced scale of the graph. The carbohydrate assay, however, showed a different pattern, with higher values for core 2 than core 1 in most sections. This may indicate that more polymer and less cells were produced in core 2, and hence, the same degree of plugging was observed for both cores at the end of the experiment. Alternative explanations for the observed permeability reductions are: (a) that trapped gas was present in core 2, leading to higher pressure measurements, or (b) that the pore geometry of the second core was different from that of the first.

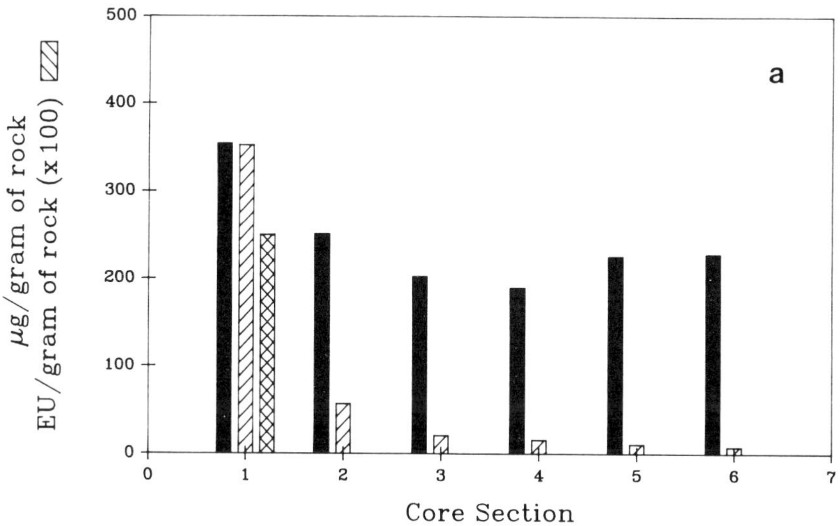

Fig. 4a. Comparison of biomass distributions upon elimination of one component from the medium: core with growth component. Symbols: ▬, carbohydrate assay; ▨, LAL assay; ▩, protein assay.

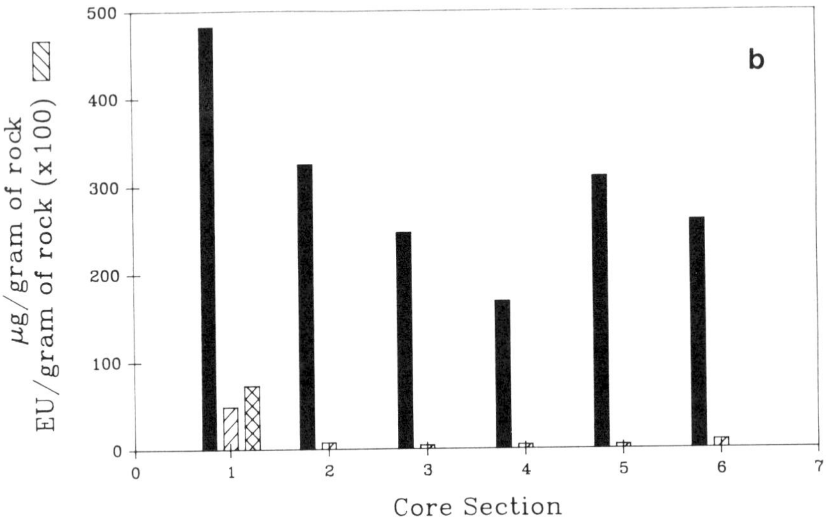

Fig. 4b. Comparison of biomass distributions upon elimination of one component from the medium: core without growth component. Symbols: ■, carbohydrate assay; ▨, LAL assay; ⊠, protein assay.

Other experiments examined the effect of differing levels or types of nutrients on the distribution of biomass within a core. In an experiment performed by Dr. Gary Jenneman at Phillips Petroleum Company, Bartlesville, Oklahoma, medium containing carbohydrate at different concentrations was pumped through two cores until approximately the same level of permeability reduction was reached. Figure 5 illustrates how a four-fold difference in the carbohydrate concentration can have an effect on the biomass distribution within a core. The biomass profile for the core with the increased amount of carbohydrate panel a showed higher levels of protein, carbohydrate, and lipopolysaccharide throughout the core when compared with the profile obtained at a lower carbohydrate concentration panel b.

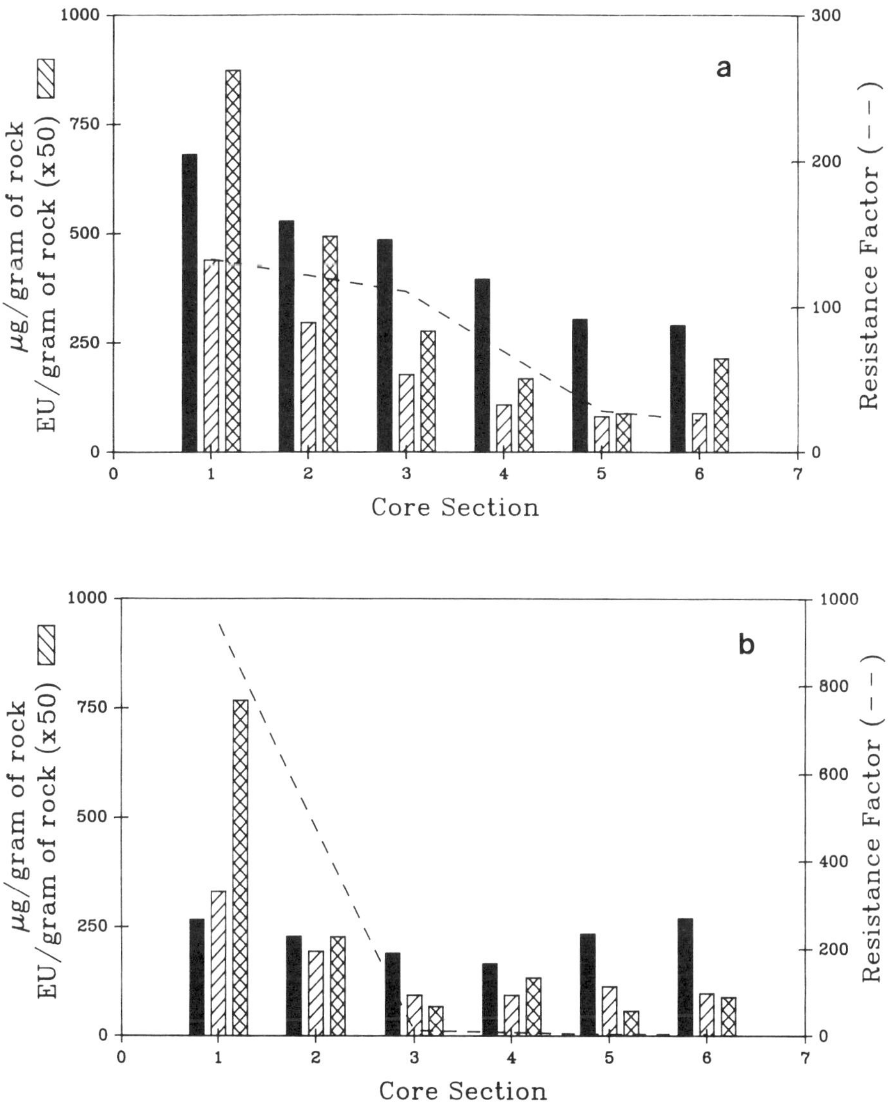

Fig. 5. Biomass profiles of cores with differing levels of carbohydrate: (a) core with increased level of carbohydrate; and (b) core with low level of carbohydrate. Symbols: ■, carbohydrate assay; ▨, LAL assay; ⊠, protein assay.

DISCUSSION

Selective plugging of high permeability zones within a reservoir is necessary to achieve maximal oil recovery. This is best achieved in an MEOR process where cells are stimulated to grow deeply in a formation where production of biomass and products will have the greatest impact. If growth occurs primarily at the well bore, then face plugging will result, and no additional oil will be recovered, rendering the reservoir unproductive. Therefore, laboratory simulations of potential MEOR processes are useful for predicting how and where growth will occur. A successful laboratory process can then be modeled for use downwell.

Chemical methods for the determination of biomass were used to examine microbial distribution within rock cores. Reproducible results with core material were obtained using the *Limulus* test for lipopolysaccharide, the BCA protein assay for protein, the orcinol assay for RNA, and the anthrone method for determination of carbohydrate. These assays were useful for evaluating the effects of different types and concentrations of nutrients on the amount of biomass generated in cores. No single method was a reliable indicator of biomass distribution, for reasons discussed below. However, by combining results from several methods, a composite picture was constructed. This type of information is essential when evaluating the importance of different factors in an MEOR process.

One problem with the carbohydrate and protein assays is the potential for interference by components of the medium used. If either protein or carbohydrate from the medium become adsorbed to the rock surface during the course of the experiment, artificially high values will be obtained. Similarly, the major limitation of the *Limulus* test is the inability to detect gram-positive bacteria. Other investigators have reported that the majority of cells present in oil reservoirs are gram-negative (McKinley *et al.*, 1988), which also seems true in this system. If the addition of nutrients should cause a shift in the population toward gram-positive bacteria, then their presence would go undetected in this particular test. Therefore, comparison of results from a variety of methods can be used to verify patterns obtained with a single parameter.

Several other colorimetric methods were tried with core samples but were abandoned. The acid dichromate procedure for total carbon was found to be inaccurate with rock samples (Johnson, 1949). The diphenylamine assay for DNA (Hanson and Phillips, 1981) was linear with core material; however, the sensitivity was not great enough to detect cells in routine samples. The composition of the rock can have an effect on the assay, as in the case of the orcinol assay for RNA. This assay was useful when working with Berea cores;

however, negative results were obtained for all core sections with Burbank cores. An immunological assay for estimating cell number (Faraji-Shadan and Bowman, 1989) was attempted, but components in the samples interfered with antibody binding. Similarly, when a method using a fluorescent dye to measure DNA was tested (Paul and Myers, 1982), the presence of oil in core samples rendered the assay useless.

Macromolecules other than those described in this study, have also been used to determine biomass. Lipids have been used for the purpose of quantitation and identification of bacteria in environmental samples (White *et al.*, 1979; White, 1983). Lipid analysis is conducted by extraction with organic solvents usually followed by separation using gas chromatography. The chromatographic profiles obtained, in addition to indicating biomass, may also be used to ascertain the presence of specific cell types which contain "signature" lipids (Kaneda, 1977; Nichols *et al.*, 1985; Taylor and Parkes, 1985). Also, hydrolysis of phospholipids can be carried out and determination of free phosphate made as an indicator of microbial growth (Findlay *et al.*, 1989). Teichoic acid can be used to quantitate the gram-positive population (Hanson and Phillips, 1981), complementing results obtained using the *Limulus* test. The microbial community can be more closely examined using specific markers rather than generic assays (Conway de Macario *et al.*, 1982; Smith, 1982).

A conversion of biomass measurements to cell numbers would be highly desirable. Unfortunately, this is difficult to do in a complex mixture of cells since different types of cells may have differing amounts of macromolecules. The physiological state of cells can also have an effect. Therefore, calibrations of methods should be made using the same mixture of cells as those present in core experiments. The best way to accomplish this is to use enrichment cultures grown under the same conditions. Cell numbers can be obtained by epifluorescence microscopy and their chemical response in various assays measured directly. Caution should be used in interpreting results due to limitations in the ability to accurately estimate cell number.

In spite of limitations discussed, biomass measurements are valuable tools for determinating the outcome of a laboratory experiment. Innovative methods for detection of biomass in cores will aid in the understanding of microbial growth in rock cores. This information will ultimately be used to develop a well-controlled MEOR process.

ACKNOWLEDGEMENTS

We wish to thank Dr. Gary Jenneman at Phillips Petroleum Company, Bartlesville, Oklahoma, for running the carbohydrate experiments described in this text. This work was funded by Phillips Petroleum Company, Bartlesville, Oklahoma.

REFERENCES

Brown, R.E., Jarvis, K.L. and Hyland, K.J., 1989. Protein measurement using bicinchoninic acid: Elimination of interfering substances. Anal. Biochem., 180: 136-139.
Conway de Macario, E.C., Wolin, M.J. and Macario, A.J.L., 1982. Antibody analysis of relationships among methanogenic bacteria. J. Bacteriol., 149: 316-319.
Faraji-Shadan, F. and Bowman, P.D., 1989. An enzyme immunoassay for estimating small numbers of cells. Anal. Biochem., 177: 259-262.
Findlay, R.H., King, G.M., and Watling, L., 1989. Efficacy of phospholipid analysis in determining microbial biomass in sediments. Appl. Environ. Microbiol., 55: 2888-2893.
Geesey, G.G., Mittelman, M.W. and Lieu, V.T., 1987. Evaluation of slime-producing bacteria in oil field core flood experiments. Appl. Environ. Microbiol., 53: 278-283.
Hanson, R.S. and Phillips, J.A., 1981. Chemical Composition. In: P. Gerhardt, R.G.E. Murray, R.N. Costilow, E.W. Nester, W.A. Wood, N.R. Kreig and G.B. Phillips (Editors), Manual of Methods for General Bacteriology American Society for Microbiology, Washington, D.C., pp. 328-364.
Jang, L.K., Chang, P.W., Findley, J.E. and Yen, T.F., 1983. Selection of bacteria with favorable transport properties through porous rock for the application of microbial-enhanced oil recovery. Appl. Environ. Microbiol., 46: 1066-1072.
Jenneman, G.E., McInerney, M.J. and Knapp, R.M., 1985. Microbial penetration through nutrient-saturated Berea sandstone. Appl. Environ. Microbiol., 50: 383-391.
Johnson, M.J., 1949. A rapid micromethod for estimation of nonvolatile organic matter. J. Biol. Chem., 181: 707-711.
Kaneda, T., 1977. Fatty acids of the genus _Bacillus_: an example of branched-chain preference. Bact. Revs., 41: 391-418.
Karl, D.M., 1986. Determination of _in situ_ microbial biomass, viability, metabolism and growth. In: J.S. Poindexter and E.R. Leadbetter (Editors). Bacterial in Nature: Volume 2. Methods and Special Applications of Bacterial Ecology, Plenum Press, New York, New York. pp. 85-176.
Litchfield, C.D. and Seyfried, P.L., 1979. Methodology for biomass determinations and microbial activities in sediments. ASTM STP 673, Philadelphia: American Society for Testing and Materials.
Lowry, O.H., Rosebrough, N.J., Farr, A.L. and Randall, R.J., 1951. Protein measurement with the Folin phenol reagent. J. Biol. Chem., 193: 265-275.
MacLeod, F.A., Lappin-Scott, H.M. and Costerton, J.W., 1988. Plugging of a model rock system by using starved bacteria. Appl. Environ. Microbiol., 54: 1365-1372.
McKinley, V.L., Costerton, J.W. and White, D.C., 1988. Microbial biomass, activity, and community structure of water and particulates retrieved by backflow from a waterflood injection well. Appl. Environ. Microbiol., 54: 1383-1393.
Miller, J., 1972. Experiments in Molecular Genetics. Cold Spring Harbor Laboratory, Cold Spring Harbor, New York, p. 431.
Morrow, N.R. and Heller, J.P., 1985. Fundamentals of enhanced oil recovery. In: E.C. Donaldson, G.V. Chilingarian and T.F. Yen (Editors), Enhanced Oil Recovery, 1: Fundamentals and Analysis, Elsevier, Amsterdam, The Netherlands, pp. 47-74.
Nichols, P.D., Smith G.A., Antworth, C.P., Hanson, R.S. and White, D.C., 1985. Phospholipid and lipopolysaccharide normal and hydroxy fatty acids as

potential signatures for methane-oxidising bacteria. FEMS Microbiol. Ecol., 31: 327-335.

Parkes, R.J., 1987. Analysis of microbial communities within sediments using specific biomarkers. Symposium of the Society for General Microbiology, 41. Ecology of Microbial Communities, Cambridge University Press, New York, New York, pp. 147-178.

Paul, J.H. and Myers, B., 1982. Fluorometric determination of DNA in aquatic microorganisms by use of Hoechst 33258. Appl. Environ. Microbiol., 43: 1393-1399.

Raiders, R.A., McInerney, M.J., Revus, D.E., Torbati, H.M., Knapp, R.M. and Jenneman, G.E., 1986. Selectivity and depth of microbial plugging in Berea sandstone cores. J. Ind. Microbiol., 1: 195-203.

Raiders, R.A., Knapp, R.M. and McInerney, M.J., 1989. Microbial selective plugging and enhanced oil recovery. J. Ind. Microbiol. 4: 215-230.

Shaw, J.C., Bramhill, B., Wardlaw, N.C. and Costerton, J.W., 1985. Bacterial fouling in a model core system. Appl. Environ. Microbiol., 49: 693-701.

Smith, D.A., 1982. Immunofluorescence of sulfate-reducing bacteria. Arch. Microbiol. 133: 118-121.

Smith, P.K., Krohn, R.I., Hermanson, G.T., Mallia, A.K., Gartner, F.H., Provenzano, E.K., Fujimoto, E.K., Goeke, N.M., Olson, B.J. and Klenk, D.C., 1985. Measurement of protein using bicinchoninic acid. Anal. Biochem., 150: 76-85.

Taylor, J. and Parkes, R.J., 1985. Identifying different populations of sulphate-reducing bacteria within marine sediments using fatty acid biomarkers. J. Gen. Microbiol., 131: 631-642.

Torbati, H.M., Raiders, R.A., Donaldson, E.C., McInerney, M.J., Jenneman, G.E., and Knapp, R.M., 1986. Effect of microbial growth on pore entrance size distributions of sandstone cores. J. Ind. Microbiol., 1: 227-234.

Watson, S.W., Novitsky, T.J., Quinby, H.L., and Valois, F.W., 1977. Determination of bacterial number and biomass in the marine environment. Appl. Environ. Microbiol., 33: 940-946.

White, D.C., Davis, W.M., Nickels, J.S., King, J.D., and Bobbie, R.J., 1979. Determination of the sedimentary microbial biomass by extractible lipid phosphate. Oecologia (Berl.), 40: 51-62.

White, D.C., 1983. Analysis of microorganisms in terms of quantity and activity in natural environments. In: J.H. Slater, R. Whittenbury, and J.W.T. Wimpenny (Editors). Microbes in Their Natural Environment. Cambridge University Press, New York, New York, pp. 37-66.

Reprinted from: *Microbial Enhancement of Oil Recovery - Recent Advances,*
edited by E.C. Donaldson

CH. R–3

BACTERIA-MINERAL SURFACE INTERACTIONS IN A LABORATORY MODEL RESERVOIR ENVIRONMENT

F.K. HIEBERT, R. L. FOLK, C. H. OPPENHEIMER, L.W. LAKE
Departments of Geological Science, Marine Studies, and Petroleum Engineering
The University of Texas at Austin, P.O. Box 7909, Austin, TX 78712

ABSTRACT

Bacteria were cultured in two types of microcosms that model reservoir conditions. Mineral or rock components of the microcosms were calcite crystals, quartz sand, or sandstone chips. The fluid component was sterile nutrient-enriched seawater. Crude oil was used as the only source of organic carbon. The microcosms were inoculated with a mixed-community culture of hydrocarbon-utilizing bacteria and incubated at room temperature for up to nine months. Mineral grains and sandstone were examined for evidence of bacterial colonization and diagenesis periodically throughout the experiments. Microbially colonized mineral surfaces were prepared for SEM examination by air drying and critical point drying. Enrichment techniques were used to isolate thirteen pure strains of bacteria from the mixed culture. Three strains were identified by fatty acid analysis.

Critical-point drying preserved bacterial morphology and details of the bacteria/mineral surface better than air drying, but air drying was found to be sufficient for reconnaissance observations. Bacteria colonized the surfaces of unconsolidated sediments and were concentrated in surface cracks. In experiments with rock, bacteria penetrated to the interior of 2 cm sandstone chips and colonized the surfaces of quartz, feldspar, and clay minerals. Isolated bacteria and microcolonies were the most commonly observed colonization pattern on sediment and rock surfaces. Pits and runnels in quartz overgrowth surfaces were observed beneath attached bacteria. These pits appear to be the result of quartz dissolution at the mineral/bacteria contact.

1. INTRODUCTION

Studies of how bacteria behave in, and affect the subsurface reservoir environment are important to the development of the technology of microbial enhanced oil recovery. The geochemical effects on minerals of bacterial metabolism within rock pores has been largely unrecognized and underappreciated by most geologists. Heterotrophic bacteria metabolize carbon substrates in order to gain the energy they need to reproduce. Organic matter, inorganic nutrients and minerals from the environment surrounding bacteria are actively transported into the cell via enzymatically catalyzed reactions, and metabolic by-products are discarded from the cell back to the environment (Brock, 1988). The cell spends energy to concentrate or dilute biochemicals and minerals at non-equilibrium concentrations for its own uses. Bacterial cells also have certain characteristics, such as electrical surface charge and the capability to concentrate ions at high concentrations on their surfaces, that increase their probability of affecting a rock-water interaction. In the subsurface sedimentary environment, bacterial activity occurs in the porespace microenvironment.

The purpose of this study is to investigate the hypothesis that bacterial metabolism of organic matter within pores alters the geochemical microenvironment and causes diagenetic changes in minerals and rocks. The approach of this research was to conduct experimental work with live bacteria in simulated subsurface sedimentary environments.

Despite the advances that have been made as the result of recent studies of the effect of bacteria on rocks, there are many significant unanswered questions. In many ways the subsurface represents a new environment to be explored, with a diverse biological flora of bacteria that inhabit the subsurface and that are geochemically active.

1.1 Bacterial Diagenesis

Many geologic investigations of ancient rocks suggest that bacteria have played an important role in rock formation and alteration. Kuznetsov et al., (1963), Ehrlilch (1981), Krumbein (1983), and Chilingarian and Wolf (1988) have reviewed in broad terms bacterial roles in a variety of diagenetic processes in both silicate and carbonate rocks.

The role of bacteria in the precipitation of calcium carbonate has been reported from many field studies of ancient rock. Oppenheimer (1961) reported that marine bacteria formed spherical aragonite grains on the Bahamian Bank. Ashirov and Sazonova (1962) studied a calcite layer beneath the Pokrovka oil reservoir in the Volga-Ural Basin and determined that it was formed as the result of the reduction of calcium sulfate by sulfate reducing bacteria. Folk and Chafetz (1980,1983) and Chafetz and Folk (1984) reported that bacteria formed low magnesium, microporus calcite travertines and pisoliths in a variety of hot spring environments, and described five morphologies of bacterially precipitated calcite aggregates. Chafetz and Meredith (1983) reported a bacterial role in the formation of travertine pisoliths in Idaho hot springs. Folk, Chafetz, and Tiezzi (1985) further described the role of hot spring bacteria in the deposition of Italian travertines. Tiezzi and Folk (in prep.) described calcite cements from within stromatactis-type cavities of the Mississippian Lodgepole formation in Montana as cloudy with bacterial bodies. They attributed the precipitation of these calcite cements to the biochemical activity of live bacteria in the original reef. Folk (1986, 1987) proposed that gray colored carbonate nodules in the Portoro Limestone of Italy are the result of local bacterial diagenesis. Hiebert (1988) described fossil bacteria trapped in calcite cements of the Posidonienschiefer Shale and proposed that when the bacteria were alive, they performed a catalytic role in the precipitation of the entombing cements.

The role of a variety of bacteria that initiate carbonate precipitation has been experimentally studied in varying degrees of detail in the lab. Drew (1910) was an early worker who noted that denitrifying marine bacteria could precipitate carbonate in tropical and temperate seawater. Zobell (1946) reviewed processes of bacterial participation in carbonate precipitation and suggested that 1) bacterial production of ammonia during metabolism caused a pH rise, 2) that sulfate reduction was necessary , and 3) that only certain bacteria were capable of intiating precipitation events. LaLou (1957) experimentally produced a carbonate scum on the surface of a mud-enriched seawater aquarium and attributed its formation to bacterial alteration of pH, CO_2 and H_2S. Oppenheimer (1961) suggested that bacteria formed spherical aragonite bodies in Bahama Bank muds. Berner (1969, 1971) sytematically investigated carbonate during bacterial degradation of marine organisms. He found that the type of organic matter available to bacteria determined pore water conditions. He concluded that in restricted pore water conditions, the kinetics of base versus acid production during metabolism determined whether carbonate precipitated of dissolved. Sulfate reduction was reported to be an essential process for carbonate precipitation. McCallum and Guhathakurta (1970) noted bacterial bodies forming the nucleus of aragonite needles precipitated from nutrient-enriched seawater solutions. Krumbein (1974) found that marine heterotrophic bacteria which precipitate aragonite, Mg-calcite and monohydro-calcite under both aerobic and anaerobic conditions were first coated, and then encapsulated with carbonate. Krumbein noted that the actual cells were destroyed by later crystal growth. Morita (1980) and del Moral *et al.*, (1987) also reported bacterial bodies incorporated in carbonate precipitated from nutrient enriched seawater in laboratory experiments.

The alteration of silicate minerals by microorganisms has been reported via several mechanisms. Eno and Reuzer (1955) experimented with *Aspergillus niger* (a fungus) and found that potassium was released from biotite, muscovite and microcline because of metabolically increased acidity of the soil environment. Webley, Duff, and Mitchel (1960) and Duff et al., (1963) made studies of silica degrading soil bacteria. They correlated silicate degradation with bacterial production of organic acids, especially 2-ketogluconic acid. Savotsin (1972) studied non-heterotrophic silicate degrading bacteria from tropical soils and reported that 26% of total potassium was released from orthoclase in 40 days of growth, an order of magnitude faster rate than weathering with carbonic acid alone. In laboratory experiments with three strains of carbohydrate-utilizing bacteria, Rades-Rohkohl et al., (1978) presented scanning electron micrographs of bacterially colonized synthetic quartz crystals covered with microdivots complete with bacterial bodies within each, but felt chemical data did not support an interpretation that bacteria dissolved quartz in her experiments. She reported total dissolved SiO_2 in solution decreased for six days, then increased to saturation in six days but did not result in statistically significant changes. Eckhardt (1979) found that organic acids produced by heterotrophic yeasts released 62% K, 33% Mg, 29% Fe, 59% Al and 6% Si from biotite over a 12 week growth experiment.

Erlich (1981) reviewed silicate dissolution by glucose-utilizing heterotrophic soil bacteria. He reported that the weathering and dissolution of silicate minerals was caused by chelation of cations by organic acids and bases and did not comment on any precipitation effects. Krumbein and Jens (1981) reviewed indirect physical evidence of bacterial dissolution of silicates, and bacterial precipitation of silicates, but not mechanisms of dissolution. He proposed that carboxylic acids are important in silicate weathering as chelating agents and concluded that microcrystalline quartz can be mobilized under natural conditions by a variety of microorganisms. However, he left the question of direct or indirect bacterial dissolution open for future studies.

Bennett and Siegal (1987) reported dissolution of quartz at near-neutral pH by organic acid complexation in a shallow, crude-oil contaminated sand aquifer in Bemidji, MN. They proposed that the source of organic acids was the microbial metabolism of crude oil. Bennett et al. (1988) suggested that biodegradation of the organic component of a silica complex could release the silica and create supersaturated conditions. This would favor quartz overgrowths, precipitation of cherts and the replacement of carbonate cements by silica during diagenesis. Ferris, Fyfe, and Beveridge (1988) proposed a geochemical mechanism for the fossilization of individual bacteria based on experiments with *Bacillus subtilis* grown in a solution saturated with iron and colloidal silica. Berthelin (1988) reviewed in detail the environmental constraints and chemical mechanisms of microbial weathering in the soil zone.

1.2 Extent of Subsurface Bacterial Colonization

Attention to bacterial alteration of minerals has been focused on the surface and soil zone environments. Until recently, bacteria were thought to be absent from the deep subsurface (Ghiorse and Wilson, 1988). In fact, the subsurface is an excellent environment for bacterial life (Balkwill 1989, Fredrickson et al., 1989). At low temperatures (<80°C), interconnected porosity, abundant surface area, and the presence of dissolved organic matter in the interstitial pore water allow bacteria to grow in and and move through a sedimentary rock environment. The small size of bacteria, their colloidal properties, and ability to adapt to extreme environmental conditions allow them to flourish in many subsurface environments. The results of recent drilling and aseptic sampling work has extended the depth to which indigenous bacteria are known to live in the subsurface to 237 M (Carl Fliermans, pers. com. 1989). It is likely that as sampling for indigenous microbes proceeds deeper, the range of documented bacterial activity in the subsurface will increase. Bacteria introduced into oil wells for MEOR are known to flourish at depths of several thousand feet, to bottom-hole temperatures of about 80°C.

1.3 A Method of Studying Bacteria/Rock Interactions

The study of subsurface bacteria in -situ is impossible except in the rare core especially drilled for that purpose. Therefore, a laboratory system that models the subsurface is an appropriate method with which to study bacteria/rock interactions.

A microcosm ("little world") is an experimental system that models a complex natural environment with both biologic and physical components (Ehrlich, 1981). The design of microcosms may be tailored to the purpose of studying a specific question. Some have been designed to model aquifer systems and to study the effects of changing environmental conditions on bacterial communities (Wilson and Noonan, 1984). Another use of microcosms is to define the processes that effect the transport and fate of specific environmental pollutants in subsurface conditions (Bouwer and McCarty, 1984). Bacterial transport in microcosms by interstitial fluid flow has been reported by Yen (1986) and Jang and Findley (1982) and others. Motile bacteria may move on thier own with a velocity of up to 0.3 cm/minute in aqueous solution (for example *Pseudomonas aeurginosa*, Brock, 1988, p.87).

While microcosms have been utilized for studies of bacteria and the fate of organic compounds, there has been little published research to date on the diagenetic effect that various bacteria might have on the rock substrates included in microcosm experiments. In our work, microcosms were designed to simply model a subsurface sedimentary environment. The goal of the microcosm experiments was to produce bacterially colonized and altered geologic materials for analysis.

In nature, bacteria live in mixed communities in which the various bacteria symbiotically interact with each other. However, most experimental work with bacteria has been carried out with pure strains of bacteria. In this study, however, the important observations are the net effect of natural populations of bacteria on mineral surfaces, therefore the pure strain approach was rejected, and mixed colonies were used.

1.3.1 SEM and Sample Preparation

Scanning Electron Microscopy (SEM) is well suited to the study of bacteria/mineral interactions. SEM has ample depth of field and resolution to image micron size bacteria on rugged microtopographic surfaces (Weise and Rheinheimer 1978). The SEM images only the mineral surface, which is the crucial factor in bacteria/rock interactions.

In most cases typical geological sample preparation techniques destroy the delicate structure of bacterial tissue by heat, desiccation, and chemical additives. Desiccation is a highly destructive process to cell wall morphology. Surface morphology of cells can be preserved with the critical-point drying (CPD) technique (Costerton 1980, and Newbury 1981). Since the focus of this research is to study the bacteria-rock interaction, microbiological sample preparation techniques were adapted that preserved bacterial tissue on the mineral and rock surfaces of interest.

2. OBJECTIVES

The objectives were to develop and refine methods to study bacterial diagenesis and to make preliminary observations on bacterial colonization patterns and alteration of minerals.

3. MATERIALS

Quartz, calcite and sandstone were selected and prepared as the geologic components of microcosms. Rounded quartz sand from a high energy shallow marine environment (from Mobile Bay courtesy of Prof. R.L. Folk) was wet seived, and the 1-2 phi fraction (0.25-0.5 mm) saved for further preparation. The quartz was cleaned by continuous washing with distilled water, surface detritus was dislodged by 5 minutes of sonic dissagregation, and washed again with distilled water until no visible debris was contained in the rinse water. The sand was then steam autoclaved and stored in sealed sterile glass flasks until used in experiments. Calcite crystals with fresh smooth surfaces were prepared by crushing a large single crystal of iceland spar and cleaned in the same manner as the quartz grains. Sandstone was selected from core drilled from the Cretaceous Shannon formation, Bar Margin facies, Natrona County Wyoming. The rock was broken into pieces approximately 2 cm in diameter, steam autoclaved, allowed to dry in a covered beaker, and stored in sealed sterile bottles until used in experiments.

The cleaned mineral grains and rock were characterized in the SEM as controls for comparison with bacterially altered material. The samples were air dried, mounted on aluminum stubs with double sticky tape and sputter gold coated for 40 seconds at 40 milliamperes.

The fluid component of each experiment was slightly nutrient-enriched filtered seawater. A light "sweet" crude oil from the Shannon reservoir of Teapot Dome Field in Natrona County, Wyoming was used as the organic nutrient for all the bacterial growth experiments. The oil comes from the same formation as the sandstone used in the microcosm experiments.

The microorganisms used in experiments were a mixed community of naturally occurring soil and subsurface bacteria that utilize hydrocarbon as energy and carbon source. The culture was obtained courtesy of Prof. C. H. Oppenheimer. All bacterial growth experiments were inoculated from fresh log-phase culture and all experiments were carried out with the complete mixed culture.

4 METHODS

Isolation of strains was by standard enrichment techniques (Brock 1988, Pope 1989). Five of the mixed community isolates were analyzed by gas chromatography (by Dr. M. Sinclair, Northfield, VT) for total fatty acid composition of their cell walls.

Microcosms were constructed of 500 ml sterile beakers that contained components of mineral or rock, fluid, organic matter, and bacteria as described in the Materials section. The beakers were covered, kept at room temperature and atmospheric pressure. The bacterial inoculum was adjusted so that the initial concentration of bacteria in the microcosm fluid was 1×10^6 cells/ml. A control for each microcosm was constructed with all the same components except for the bacterial inoculum. Mineral and rock material from controls were sampled and analyzed in the same manner as from the test microcosms.

Most experiments were conducted as batch growth experiments in which a finite amount of organic matter was introduced at the start of the experiment. This method represents a closed system with respect to nutrient supply and accumulation of metabolic by-products. Fluid composition, bacterial culture, and type of organic matter were held constant throughout all growth experiments. Amounts of aeration, agitation of the mineral sediment, and mineral and rock substrate were varied throughout many experiments. The experiments were observed over a three to nine month period.

Two experiments and a control were maintained as semi-open systems in which the organic matter and media fluids were partially replaced on a daily basis. The basic design and components were similar to the batch system except that once every 24 hours, 250 ml of media was removed from the microcosm and replaced with fresh media plus 2.5 ml crude oil. In this system, nutrients were replaced, and by-products and cells removed. Minerals or rock fragments were sampled from the microcosms at periodic intervals with sterile instruments and prepared as described below for SEM analysis.

Air drying and critical point drying (CPD) techniques for SEM sample preparation were compared for their success in preserving bacterial morphology in relation to the mineral surface.

Observations were made with a Philips Model 515 SEM (Cell Research Institute, UT) and a JOEL JSM T330A Scanning Microscope (Dept. of Geological Sciences, UT). Most SEM work was carried out at 20KeV accelerating voltage, but many machine settings were investigated.

5 RESULTS

5.1 Isolation and Characterization of Bacterial Strains

Thirteen strains of bacteria were isolated ,yet many more were clearly present in the mixed culture. An example of the scanning electron micrograph morphological type catalog of bacterial isolates is presented (Figures 1, and 2).

5.2 Comparison of SEM Sample Preparation Techniques

Air drying samples resulted in poor preservation of bacterial bodies. The bodies appear flattened or otherwise distorted and spread-out across mineral surfaces. In some instances, three-dimensional morphologies were preserved but surface detail and normal shape was obliterated. Critical point drying of samples resulted in excellent cell morphology preservation. This method allowed detailed study of the bacteria/rock attachment and diagenetic effects at the cell/mineral surface.

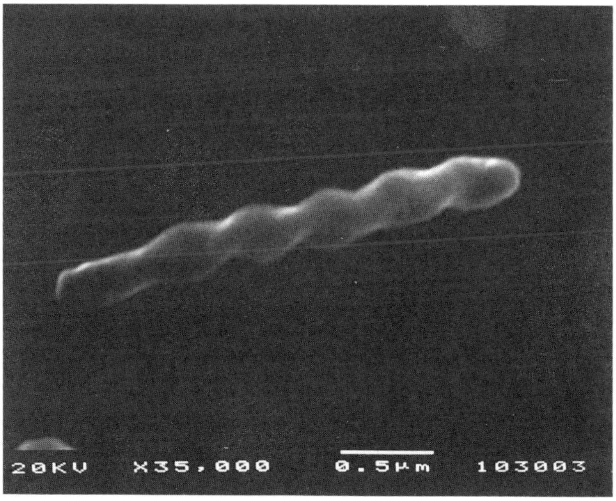

Figure 1. Scanning electron micrograph of bacteria strain LYBYR (as yet unidentified), illustrating its distinctive morphology.

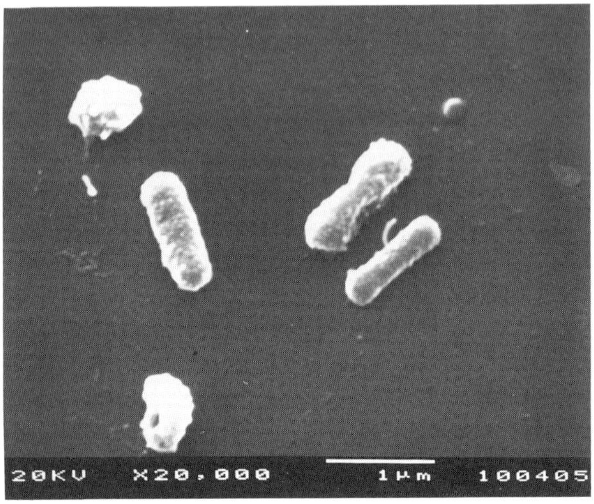

Figure 2. Scanning electron micrograph of bacteria strain FE (Bacillus sp.)

5.3 Microcosm Methodology

Within three days of inoculation, bacteria had colonized some surfaces of unconsolidated quartz sand and penetrated to the interior of 2 cm sandstone chips and colonized internal pore surfaces. On quartz sand grains, bacteria were observed to be concentrated in cracks and crevices, while sparse colonization was observed on exposed smooth surfaces (Figures 3 and 4). Bacterial colonization of internal pore surfaces of sandstone chips showed a different colonization pattern. A variety of bacterial types were observed to colonize smooth quartz overgrowth surfaces, and clay surfaces with nearly equal density (Figures 5 and 6).

Bacteria were most commonly observed colonizing rock and mineral surfaces as single isolated cells. Microcolonies were also commonly observed. On sand grains microcolonies were observed in cracks and in depressions of the surface. Within the pores of sandstone, microcolonies occurred randomly, even on the sharp terminations of quartz overgrowths. Most bacteria showed no sign of polymer adhesion, flagella, or attachment filament and may have adhered directly or by cryptic filaments at the bacteria/mineral contact (Marshall 1976, Rades-Rohkohl 1978, and Marshall 1980).

5.4 Diagenetic Effects of Bacteria

On smooth quartz surfaces, several types of bacteria (spheres, curved rods, and some rods) were observed to be attached in "micro-divots" and runnels which ranged from 0.1- 0.5 microns deep, and from between 1-3 microns in length and width. The margins of the divots are not sharply defined, rather they gently grade from the deepest directly beneath the bacteria outwards into the plane of the mineral surface. The bacterial divots were most commonly observed on the quartz overgrowths of the Shannon sandstone samples from the semi-open

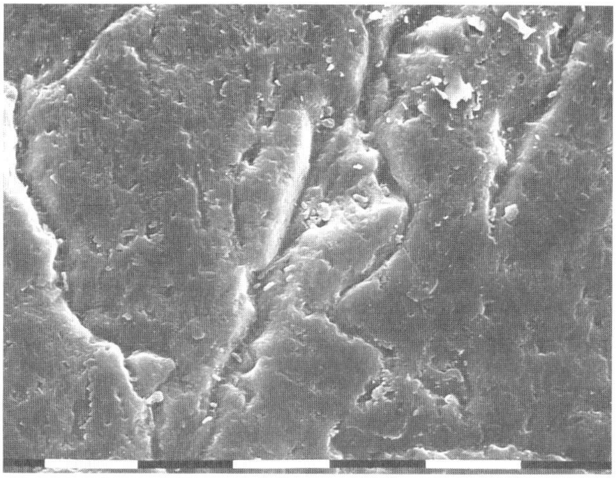

Figure 3. Bacteria concentrated in cracks on the surface of an unconsolidated quartz grain from a microcosm experiment. Scale bar = 10 microns.

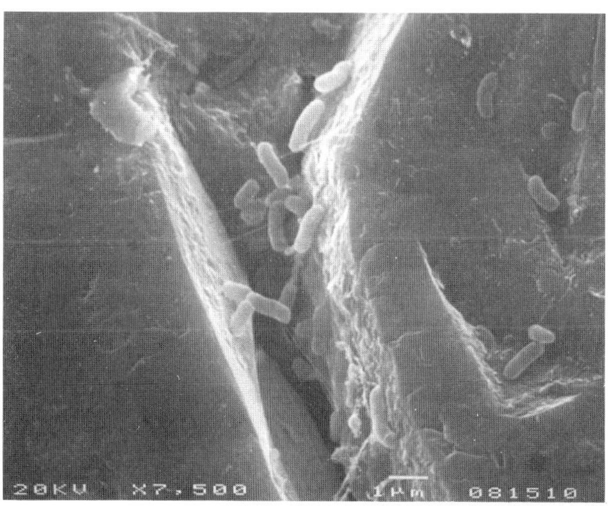

Figure 4. Bacteria concentrated in cracks and pits on the surface of a quartz sand grain from a microcosm experiment.

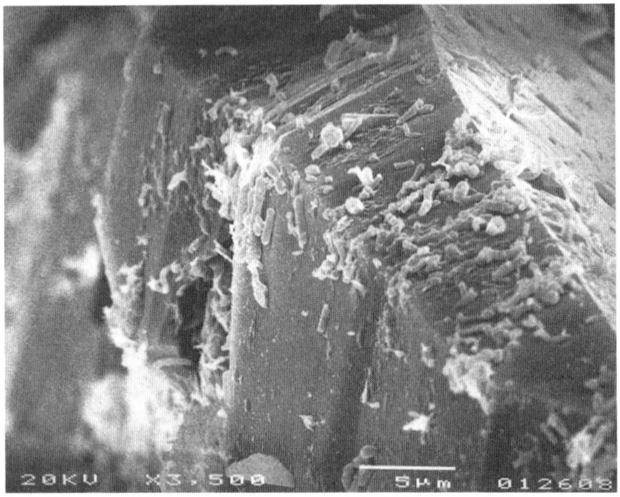

Figure 5. Mixed community of hydrocarbon-utilizing bacteria colonizing the surface of a quartz overgrowth within an internal pore of the Shannon Sandstone after three days in a microcosm experiment.

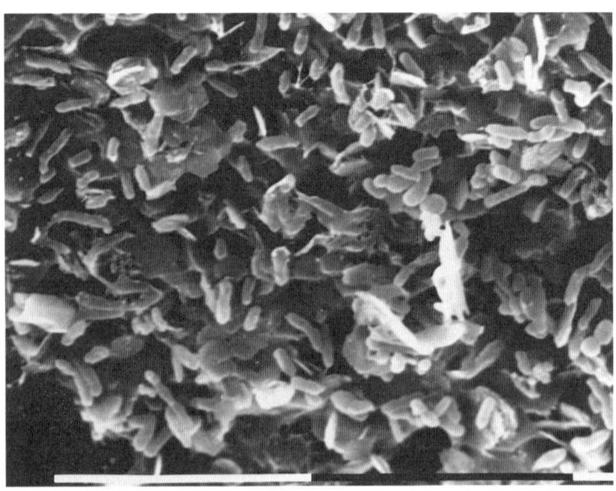

Figure 6. Mixed community of hydrocarbon-utilizing bacteria colonizing the surface of a clay-rich surface within an internal pore of the Shannon Sandstone after three days in a microcosm experiment. Scale bar = 10 microns.

microcosm model. Not all types of bacteria were observed in divots. There appears to be a high correlation between the occurrence of micro-divots and attached bacterial bodies. Few divots have yet been observed without an attached bacteria.

The divots are not apparent in top-down views of bacteria on mineral surfaces because of the lack of divot edges. The divots are visable only in side views (profiles) at an angle that allows observation of the bacteria/mineral surface contact. The divots are not an effect of the SEM. Stereo pair micrographs of bacteria in divots confirm that a real depression is present (Figures 7 and 8). In addition, bacteria not in divots were documented attached next to bacteria in divots, thus substantiating that the divot is not an artifact.

6. DISCUSSION

6.1 Bacterial Colonization on Mineral Surfaces

The hydrocarbon-utilizing bacteria in these experiments colonize mineral and rock surfaces as individual cells and as microcolonies. Much of the literature that describes growth of bacteria in experimental porous media reports that bacteria eventually form a continuous biofilm over the surface they colonize (Characklis and Cooksey 1983, Costerton et. al., 1985), yet this was not observed. The microcolonies observed in this work are unlikely to represent examples of stunted or nutrient-limited growth, since post-attachment growth on many mineral surfaces is indicated by dividing cells. One suggestion for these anomalous observations is that most experimental work reported in the literature was performed with carbohydrate-utilizing bacteria. Hydrocarbon-utilizing bacteria may not produce the copius extracelluar slime (often composed of polysacharide) as do carbohydrate-utilizing bacteria and may not produce surface conditions that initiate biofilm formation.

The observation of maximum bacteria concentration in the cracks and crevasses of quartz sand grains collected from the microcosm experiments is consistent with Weise and Rheinheimer (1978) who found a correlation between site and density of bacterial colonization and rough spots on the microtopography of a sand grain from a high energy marine environment. In contrast, bacterial distribution on pore surfaces within sandstone was not controlled by surface microtopography. It is unclear what accounts for this difference. Several hypotheses may be suggested: 1) different types of bacteria preferentially select adhesion surfaces based on chemical or mineralogical characteristics other than roughness, such as elemental composition, degree or type of organic matter sorbed on the mineral surface, or electrical characteristics of the surface, or 2) micro-scale fluid flow past the mineral surface could sort the bacteria into pockets of low-velocity flow, depositing the bacteria in cracks and depressions. Data from this work does not support one hypothesis over the other.

Different types of bacteria may preferentially colonize different types of minerals. Several hypotheses can be suggested for what controls this phenomena: 1) bacteria are attracted to minerals whose chemical composition includes nutrients essential to growth, 2) rough surface microtopography may offer a good attachment site, or represent a low velocity spot protected from fluid currents, or 3) organic matter adsorbed on certain minerals may attract motile chemotactic bacteria.

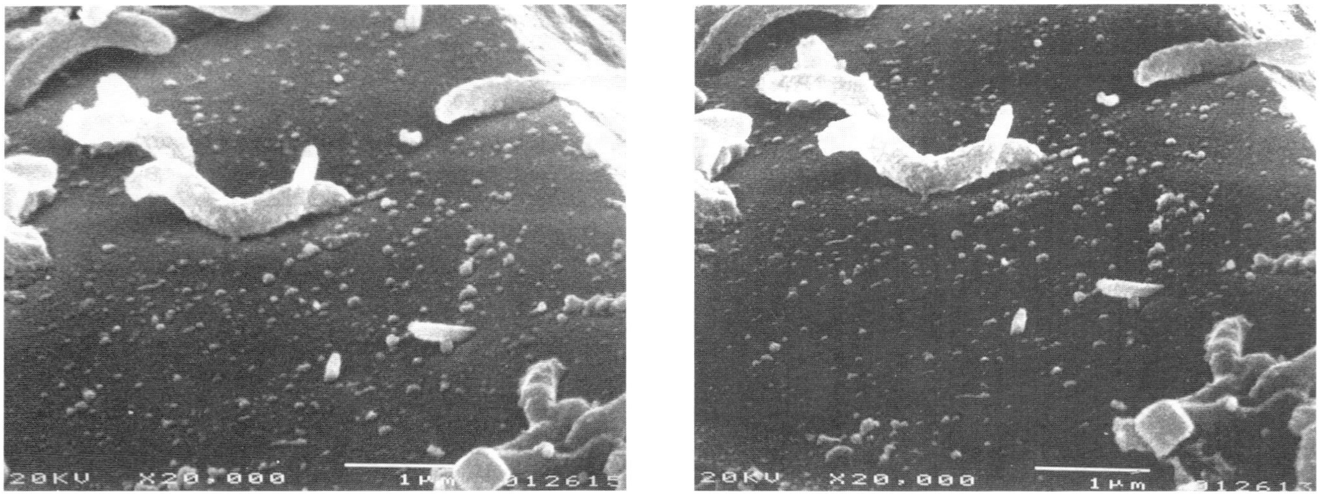

Figures 7 and 8. Stereo pair image of a curved rod-shaped bacterium resting in a microdivot. Stereo pair images confirmed that the microdivots are real depressions, not photographic illusions.

6.2 Bacterial Diagenesis

The observation of hydrocarbon metabolizing bacteria producing microdivots on quartz surfaces is of geological significance. The bacteria could be forming the divots by several mechanisms, including: 1) by producing organic acids that complex and mobilize Si from the quartz surface, 2) by incorporating Si directly into their cell material, or 3) by a combination of the above. Bennett and Siegal (1987) proposed a bacterial mechanism for quartz dissolution in a crude oil contaminated aquifer. They presented evidence of an increase in the solubility of quartz brought about by bacterially produced dissolved organic acids and proposed that quartz was complexed and mobilized by the organic acids in neutral pH conditions. They also presented scanning electron micrographs of etch pits, slightly larger than the microdivots observed on microcosm quartz grains, but without bacterial bodies.

Organic acids are the primary intermediate byproduct of the microbial metabolism of crude oil. These acids are released into the environment as metabolic byproducts as well as being retained within the cell (Brock, 1988). The production and release of fatty acids is a common net effect of the wide variety of bacterial metabolism pathways of hydrocarbon. Bacterial metabolism of crude oil may represent an alternative to thermal cracking of hydrocarbons as a source of organic acids in the subsurface . Bacterially generated fatty acids in the subsurface represent a mechanism for the dissolution, transport and, perhaps, precipitation of quartz (Bennett et al., 1988).

7. CONCLUSIONS

1. Microcosm and SEM methods are suitable for study of bacterial diagenesis. Critical-point drying preserved bacterial morphology and details of the bacteria/mineral surface better than air drying, but air drying was found to be sufficient for reconnaissance observations.

2. Bacterial colonization patterns of sediment and rock vary. Colonization of sediment surfaces was found to be concentrated in cracks. Bacteria colonizing internal pore surfaces of sedimentary rock appeared to be more homogeneously distributed. Bacteria may preferentially colonize certain minerals over others.

3. Certain bacteria may cause quartz dissolution at the bacteria/mineral contact.

8. ACKNOWLEDGEMENTS

This research was supported by a Subcontract Grant (Department of Energy, # 19-88BC 14205.00) from Alpha Environmental, Inc., Austin, Texas.

9. REFERENCES

Ashirov, K. B., and Sazonova, I. V., 1962, Biogenic sealing of oil deposits in carbonate reservoirs: (translated from) Microbiologiya, v. 31, no. 4, pp. 680-683.

Balkwill, D. L., 1989. Numbers, diversity, and morphological characteristics of aerobic, chemoheterotrophic bacteria in deep subsurface sediments from a site in South Carolina, Geomicrobiology, vol. 7, no. 1/2, p. 33-52.

Barker, W. W., 1985, Bacterial trace fossils in Eocene kaolin, Proceedings of the 43rd annual meeting of the electron microscopy society of America.

Bennett, P. C., and Siegal, D. I., 1987. Increased solubility of quartz in water due to complexing by organic compounds, Nature, vol. 36, no. 6114, pp. 684-686.

_______, Melcer, M. E., Siegal, D. I., and Hassett, J. P.,1988. The dissolution of quartz in dilute aqueous solutions of organic acids at 25° C, Geochimica et Cosmochimica Acta, vol. 52, pp. 1521-1530.

Berner, Robert A., 1969, Chemical changes affecting dissolved calcium during the bacterial decomposition of fish and clams in sea water: Marine Geology, v. 7, pp. 253-274.

_________, 1971, Bacterial processes effecting the precipitation of calcium carbonate in sediments: Johns Hopkins University Studies in Geology, no.19, pp. 247-251.

Berthelin, J., 1988, Microbial weathering processes in natural weathering, in Physical and Chemical Weathering in Geochemical Cycles, A. Lerman and M. Meybeck (eds.), New York :Kluwer Academic Publishers, p. 33-59.

Bouwer, Edward J., and McCarty, Perry L., 1984, Modeling of trace organics, biotransformation in the subsurface, Groundwater, vol. 22, no. 4, p. 433-440.

Brock, Thomas D., and Madigan, Michael, T.,1988, Biology of Microorganisms, Fifth Edition. Englewood Cliffs, New Jersey, pp. 835.

Chafetz, H. S., and Meredith, J. C., 1983, Recent travertine pisolites (pisoids) from southeastern Idaho, USA *in* T.M. Peryt (ed.), Coated Grains: New York, Springer-Verlag, pp. 450-455.

_________., and Folk, Robert, L., 1984, Travertines: depositional morphology and the bacterially constructed constituents, Jour. Sed. Pet., vol. 54, no. 1, p.289-316.

Characklis, W. G., and Cooksey, K. E., 1983, Biofilms and microbial fouling, Advances in Applied Microbiology, vol. 29, p. 93-138.

Chiliingarian, G. W., and Wolf, K. H., 1988. Diagenesis I, Developments in Sedimentology #41, Amsterdam: Elsevier, 591 pp.

Costerton, J. W.,1980. Some techniques involved in study of adsorbtion of microorganisms to surfaces, p. 403-426. in: Adsorbtion of Microorganisms to Surfaces, Gabriel Bitton and Kevin C. Marshall eds. New York: John Wiley and Sons, pp.439.

_________, Marrie, T. J., and Cheng, K. J., 1985, Phenomena of bacterial adhesion, p. 3-43, in D. C. Savage and M. Fletcher eds., Bacterial Adhesion, London: Plenum Publishers.

Danielli, H. M. C., and Edington, M. A., 1983, Bacterial calcification in limestone caves, Geomicrobiology, vol. 3, no. 1, p. 1-15.

del Moral, A., Roldan, E., Navarro, J., Montilova-Sanchez, M., and Ramos-Cormenzana, A., 1987, Formation of calcium carbonate crystals by moderately halophilic bacteria, Geomicrobiology Journal, v. 5, pp. 79-87.

Drever, James I., 1988, The Geochemistry of Natural Waters, 2nd Ed., Englewood Cliffs, New Jersey: Prentice Hall, 437 pp.

Drew, G. H., 1910, The action of some denitrifying bacteria in tropical and temperate seas, and the bacterial precipitation of calcium carbonate: J. Mar. Biol. Sci. Assoc. U. K., v. 9, pp. 142-155.

Duff, R. B., Webley, D.M., and Scott, R. O., 1963, Solubilization of minerals and related materials by 2-ketogluconic acid producing bacteria, Soil Science, vol. 95, p. 105-114.

Eckhardt, F. E. W.,1979, Uber die einwirkung heterotropher mikroorganismen auf die zersetzung silikatisher minerale, Z. Pflanzenernaehr. Bedenkd. v. 142, p. 434-445.

Ehrlich, Henry Lutz, 1981. Geomicrobiology, New York: Marcel Dekker, Inc., 393 pp.

Eno, Charles, F., and Reuzer, Herbert W., 1955, Potassium availability from biotite, muscovite, greensand, and microcline as determined by growth of *Aspergillus niger*, Soil Science, vol. 80, p. 199-209.

Fairchild, T. R., Schopf, J. W., and Folk, R. L., 1973, Filamentous algal microfossils from the Caballos Novaculite, Devonian of Texas, Jour. Paleontology, vol. 47, no. 5, p. 946-952.

Ferris, F. G., Fyfe, W. S., and Beveridge, T. J., 1988, Metallic ion binding by *Bacillus subtilis*: implication for the fossilization of microorganisms, Geology, v. 16, pp. 149-152.

Folk, Robert L., 1986, White card and smoking gun: bacterio-diagenetic nodules in the Triassic Portoro Limestones of Liguria, Italy: Geol. Soc. Am. Abs. with Programs, v. 18, p. 604.

______, 1987, Detection of organic matter in thin-sections of carbonate rocks using a white card: Sedimentary Geology, v. 54, pp. 193-200.

______, and Chafetz, H. S., 1980, Quaternary travertines of Tivoli (Roma) Italy: bacterialy constructed carbonate rock: Geol. Soc. Am. Abs. with Prog., v. 12, p. 428.

____, and _____, 1983, Pisolites (pisoids) in Quaternary Travertines of Tivoli, Italy, *in* T.M.Peryt (ed.) Coated Grains: New York, Springer-Verlag, pp. 474-487.

____,_____, and Tiezzi, P. A., 1985, Bizzare forms of depositional and diagenetic calcite in hot-springs travertines, central Italy. *in* N. Schneidermann and P. M. Harris (eds.), Carbonate Cements: SEPM Spec. Pub. 36., pp. 349-369.

Fredrickson, J. K., Garland, T. R., Hicks, R. J., Thomas, J. M., Li, S. W., and McFadden, K. M., 1989. Lithotrophic and heterotrophic bacteria in deep subsurface sediments and their relation to sediment properties. Geomicrobiology, vol. 7, no. 1/2, p. 53-66.

Ghiorse, W. C. and Wilson, J. T., 1988, Microbial ecology of the terrestrial subsurface, Advances in Applied Microbiology, vol. 33, p. 107-172.

Hayat, M. A., 1981, Aldehydes, Chapter 3, in Fixation for Electron Microscopy, pp. 64-148, New York: Academic Press, pp. 501.

Hiebert, F. K., 1988, The role of bacteria in the deposition and diagenesis of the Posidonienschiefer, a Jurassic oil shale from southern Germany, Masters Thesis, The University of Texas at Austin, pp. 124.

Hirsch, P. and Raades-Rohkohl, E., 1983, Microbial diversity in a groundwater aquifer in N. Germany, Chapter 14 in Developments in Industrial Microbiology, v. 24, p. 183-200. Proceedings of the 39th General Meeting of the Society of Industrial Microbiology.

Jang, L. K., Sharma, M. M., Findley, J. E.,1982, An investigation of the transport of bacteria through porous media, pp.60-70. in: Microbial Enhancement of Oil Recovery.

Krumbein, Wolfgang E., 1974, On the precipitation of aragonite on the surface of marine bacteria, Naturwissenschaften v. 61, p. 167.

____________,1983, Microbial Geochemistry, Oxford: Blackwell Scientific Publications, 330 pp.

____________, and Jens, K.,1981, Biogenic rock varnishes of the Negev desert (Israel) an ecological study of iron and manganses transformation by Cyanobacteria and Fungi, Oecologia, vol. 50, p. 25-38.

Kuznetsov, Ivanov, M. V., and Lyalikova, N. N., 1963, Introduction to Geological Microbiology: New York, McGraw-Hill, 169 pp.

LaLou, C. 1957, Studies on bacterial precipitation of carbonates in seawater: Jour. Sed. Pet., v. 27, pp. 190-195.

McCallum, M. F., and Guhathakurta, K., 1970, The precipitation of calcium carbonate from seawaer by bacteria isolated from Bahama Bank sediments: J. Appl. Bact., v. 33, p. 649-655.

MacQuarrie, K. T. B., Sudicky, E.A., and Frind, E. O., 1990a, Simulation of biodegradable organic contaminants in groundwater, 1. numerical formulation in princepal directions. Water Resources Research, Vol 26, No. 2, p. 207-222.

MacQuarrie, K. T. B., and Sudicky, E.A., 1990b, Simulation of biodegradable organic contaminants in groundwater, 2. plume behavior in uniform and random flow fields. Water Resources Research, Vol 26, No. 2, p. 223-239.

Marshall, K.C.,1976, Interfaces in Microbial Ecology Cambridge: Harvard Univsersity Press, pp.156.

_________, 1980, Adsorbtion of microorganisms to soils and sediments in: Adsorption of Microorganisms to Surfaces, Gabriel Britton and Kevin C. Marshall eds.

May, Jeffrey A., and Perkins, Ronald, D., 1979, Endolithic infestation of carbonate substrates below the sediment-water interface, Jour. Sed. Pet., vol. 49, no. 2. p. 357-378.

Molz, F. J., Widdowson, M. A., and Benefield, L. D., 1986, Simulation of microbial growth dynamics coupled to nutrient and oxygen transport in porus media, Wat. Res. Res., vol. 22, No. 8, p. 1207-1216.

Morita, R.M. 1980, Calcite precipitation by marine bacteria: Geomicrobiology Journal, v. 2, pp. 63-82.

Newbury, D. E., 1981, Speciman Preparation, p. 211-262, in Practical SEM Electron and Ion Micorprobe Analysis, J. I. Goldstein and H. Yakowitz Eds., New York: Plenum Press, pp. 582.

Oppenheimer,C. H., 1961, Note on the formation of spherical aragonite bodies in the presence of bacteria from the Bahama Bank: Geochim. Cosmochim. Acta, v. 23, pp. 295-296.

Phelps, T. J., Raione, E. G., White, D. C., and Fliermans, C. B., 1989. Microbial activities in deep subsurface environments, Geomicrobiology, vol. 7, no. 1/2, p. 79-92.

Phelps, T. J., Ringleberg, D,. Hedrick, D., Davis, J., Fliermans, C. B., and White, D.C., 1988, Microbial biomass and activites associated with subsurface environments contaminated with chlorinated hydrocarbons, Geomicrobiology, vol. 6, p. 157-170.

Pope, L. M., 1989, General Microbiology Laboratory Handbook, MIC 119K, Austin, Texas: Kinkos Professor Publishing, pp.234.

Postek, M. T., Howard, K. S., Johnson, A. H., and McMichael, K. L., 1980, Speciman preparation, Chapter 5, p. 115-240, in Scanning Electron Microscopy, A Students Handbook, Ladd Industries, Inc., pp. 305.

Rades-Rohkohl, E., Franzle, O., and Hirsch, P., 1978, Behavior, activities and effects of bacteria on synthetic quartz monocrystal surfaces. Microbial Ecology, vol 4, pp. 189-205.

Savotsin, Peter, 1972, Microbial transformation of silicates, Zeitschrift fur Pflanzernahrung und Bodenkunde, vol. 132, p. 37-45.

Tiezzi, P. and Folk, R. L., 1988, Cloudy cements in bioherm cavities, Mississippian Lodgepole Limestone, Montana: product of bacterial acitivty during organic decay: Sedimentary Geology (in prep).

Weaver, Charles. E., 1989, Microbial Alteration in Clays, Muds and Shales, p. 137-142, Developments in Sedimentology 44, Amsterdam: Elsevier.

Webley, D. M., Duff, R. B., and Mitchell, W. A., 1960, A plate method for studying the breakdown of synthetic and natural silicates by soil bacteria. Nature, vol. 188, no. 4752, p. 766-767.

Weise, W. and Rheinheimer, G., 1978, Scanning electron microscopy and epifluorescence investigation of bacterial colonization of marine sand sediments, Microbial Ecology, vol. 4, pp. 175-188.

Wilson, J., and Noonan, M. J., 1984, Microbial acitivity in model aquifer systems, p. 117-134, in G. Bitton and C. P. Gerba eds., Groundwater Pollution Microbiology, New York: John Wiley and Sons, Inc., pp. 377.

Yen, T.F.,1986, Bacteria Transport Through Porus Media, Annual Report, December 31, 1985. DOE/BC/10508-36.

Zobell, C. E., 1946, Marine Microbiology, Waltham, MA: Chronica Botanica Co., pp. 240.

Reprinted from: *Microbial Enhancement of Oil Recovery – Recent Advances*, edited by E.C. Donaldson

CH. R–4

INVESTIGATIONS OF MICROBIAL MECHANISMS FOR OIL MOBILIZATION IN POROUS MEDIA

K. L. Chase, R. S. Bryant, T. E. Burchfield, K. M. Bertus and A. K. Stepp

ABSTRACT

Prior work at NIPER has identified key mechanisms that are important for improved oil mobilization by microbial formulations. Key mechanisms include wettability alteration, emulsification, oil solubilization, alteration in interfacial forces, lowering of mobility ratio, and permeability modification.

Certain microbial formulations appear to shift the wettability of Berea sandstone core samples toward a more water-wet condition. The microbial cells appear to be involved in this wettability alteration since no change was observed in samples tested with filtered microbial products with cells removed. Aggregation of the bacteria at the oil-water-rock interface may produce localized high concentrations of metabolic chemical products that result in oil mobilization.

Unsteady-state relative permeability tests were performed with microbial formulations in Berea sandstone cores. A decrease in relative permeability to water and an increase in relative permeability to oil was usually observed in microbial-flooded cores causing an apparent curve shift toward a more water-wet condition. Samples tested with the microbial formulation exhibited higher oil recovery before water breakthrough and lower residual oil saturations.

Additive flask tests and corefloods were performed to evaluate the effects of sodium bicarbonate on microbial metabolite production and oil mobilization. These tests indicate that sodium bicarbonate does appear to alter metabolic pathways of the microbial formulation used in these studies. Cores treated with a preflush of sodium bicarbonate showed increased oil recovery efficiency.

INTRODUCTION

Laboratory and field experiments have demonstrated the potential for mobilization of oil with microbes and microbial products, but the actual mechanisms involved in this process have not been fully identified. Mechanisms shown to be important in most enhanced oil recovery methods involve lowering the mobility ratio or increasing the capillary number. A significant increase in capillary number is achieved by lowering the interfacial tension between the displacing and displaced fluids. Another significant mechanism of oil mobilization is alteration of wettability. The displacement of a wetting fluid by a nonwetting one is less efficient than the displacement of a nonwetting by a wetting fluid. Wettability studies were conducted using the Amott (Amott, 1959), imbibition and forced displacement, and USBM (Donaldson, 1980) centrifuge methods. An advantage of the USBM wettability test over the Amott test is its improved sensitivity near neutral wettability. On the other hand the USBM test cannot determine whether a system has fractional or mixed wettability, while the Amott test sometimes can. In some fractional or mixed-wet systems, both water and oil will imbibe freely (Anderson, 1986). For a microbial flooding process, wettability alteration could be a result of adsorption of a metabolic product or microbial cells. Unsteady-state relative permeability tests were also conducted both with and without microbes present and curves were observed for differences in wettability.

Results of previous coreflood and micromodel studies have shown that oil recovery efficiency of a microbial formulation was considerably greater than that of a solution of metabolic products without any cells (Bryant et al, 1989). The results indicate that microbial cells play an important role in oil mobilization probably by producing high localized concentrations of chemical products at the oil-brine-rock interface.

Oil displacement experiments were conducted in 4-foot-long Berea sandstone cores and three-dimensional plots were used to evaluate the relative rates of transport of microbial cells and chemical products from the metabolism of microbial cells. The results showed that metabolic products transport through porous media at a faster rate than do microbes (Bryant et al, 1989).

Addition of sodium bicarbonate has been shown to improve oil recovery efficiency of microbial formulations (Bryant et al, 1989). Some chemical additives can stimulate microorganisms to produce a greater amount of metabolites, or alternately, produce a different metabolite by altering the cell's metabolic pathways (Russell and Grula, 1987). A patent by Clark (1986) describes the use of sacrificial agents as a preflush for microbial injection in a petroleum reservoir. The patent claims that sacrificial agents that are strongly absorbed on reservoir rock can facilitate the transport of microbial cells. Flask tests and corefloods were performed using sodium bicarbonate and Tronacarb™ (additive corefloods) to compare microbial growth and production of products with and without bicarbonate, and to determine if the bicarbonate is altering the amounts or types of microbial products or is acting as a sacrificial agent to prevent adsorption.

MATERIALS

Crude Oil and Brine

Crude oil samples were obtained from the Bartlesville Sand formation in Delaware-Childers field in northeastern Oklahoma. Delaware-Childers oil has a gravity of 31° API and a density of 0.87 g/cm^3. Brine with a concentration of 3% sodium chloride by weight was used for USBM wettability tests. Because of the possible presence of clays in the Berea sandstone cores, a concentration of 3% wt/vol potassium chloride was used for Amott and relative permeability tests. A brine containing 0.5% wt/vol sodium chloride was used for additive corefloods.

Microorganisms

Results from previous studies have shown that a combination of *Bacillus licheniformis* , NIPER 1 (ATCC No. 39307), and a *Clostridium* species, NIPER 6, was the most effective formulation for the recovery of residual oil. This microbial formulation was chosen for further additive studies. *Bacillus licheniformis* is a facultatively anaerobic, sporeforming rod that produces alcohols, biosurfactant and some organic acids when fermenting molasses. The anaerobic sporeforming *Clostridium* is a member of the butyric acid group that produce acetone, butanol, ethanol, isopropanol, butyric acid, acetic acid, propionic acid, carbon dioxide and hydrogen when fermenting sucrose. This particular *Clostridium* species also produces a biosurfactant.

Several combinations of microorganisms were used in the USBM wettability tests. NIPER Bac 1 is the combination of four microorganisms used in a microbially enhanced waterflood field project in the Mink Unit of Delaware-Childers field at Nowata, Oklahoma (Bryant, et al, 1990). NIPER BAC 1 is a combination of NIPER 1, NIPER 2, NIPER 3 and NIPER 4. NIPER 2 is a species of *Bacillus* that produces surfactant, acids, and some carbon dioxide. NIPER 3 is a species of *Clostridium* similar to NIPER 6 that produces carbon dioxide, ethanol, butyric acid and surfactant. NIPER 4 is a Gram-negative, facultatively anaerobic rod that produces carbon dioxide and acids when fermenting sucrose.

NIPER 1 alone was used for relative permeability and Amott wettability tests because USBM centrifuge wettability tests have shown NIPER 1 to be the most effective in alteration of the wettability index . NIPER 6 was eliminated from these tests because the abundant production of gas would interfere with the tests.

Nutrient

The molasses used in these experiments was obtained from Pacific Molasses Company in Oklahoma City, and its mineral content is as follows: total ash, 6.1%; calcium, 0.8%; phosphorous, 0.08%; magnesium, 0.35%; potassium, 2.4%; sulfur, 0.8%; and sodium, 0.2%. The amount of total suspended solids is 74%, of which 3% is total protein, 48% is total sugar (sucrose), and the remaining 23% is fiber. The concentration of molasses used in the experiments was 4% vol/vol in tap water with 0.1% wt/vol ammonium phosphate added to facilitate microbial metabolism.

For relative permeability and Amott wettability tests, 3% wt/vol potassium chloride was added to the nutrient solution. Microbes were cultivated in filtered molasses (0.45μ Millipore filter) solution with an antibiotic added to prohibit growth of the normal molasses flora.

For additive corefloods, NIPER 1 and 6 were grown in tryptic soy broth (TSB) for injection and followed by a slug of nutrient. Microbes were also cultivated in TSB for the imbibition phase of the USBM wettability centrifuge tests.

Core Samples

Blocks of Berea sandstone were obtained from Cleveland Quarries (Amherst, Ohio) and cut into cylindrical cores of 1.5 inches (3.8 cm) in diameter and 3 inches (7.6 cm) in length for Amott wettability and relative permeability tests. Core plugs for the USBM centrifuge tests were drilled 0.75 in (1.9 cm) in diameter and 1.2 in (3 cm) in length. Cores to be used for additive corefloods were drilled 1.5 in (3.8 cm) in diameter and 12 in (30.1 cm) in length. Larger 4-foot-long cylindrical cores that were 2 inches (5.1 cm) in diameter were also used for additive corefloods.

Core flood apparatus

Three-inch-long core samples for Amott wettability relative permeability tests were encased in rubber sleeves and confined in a hydrostatic coreholder at 350 psi, net confining pressure. The samples were waterflooded at constant pressure using nitrogen to drive a piston-type fluid reservoir.

A description of the experimental apparatus used for additive corefloods has been given previously (Bryant and Douglas, 1988). The fluid separators are piston devices used to inject microbial solutions and other fluids into the cores and were designed to prevent corrosive fluids from contacting the pumps. Cores were encased in rubber sleeves and placed inside stainless steel Hassler coreholders, and a hydrostatic net confining pressure of 250 psi (1723.7 kPa) was maintained throughout the test.

Core Preparation

All core samples were evacuated and then saturated with brine. Initial water saturation was established in USBM centrifuge samples by placing the samples in Delaware Childers crude oil and centrifuging at approximately 3000 RPM which is equivalent to a pressure gradient of approximately 10 psi. One set of the centrifuge plugs was stored for 7 days submersed in the crude to establish initial wettability conditions. Other sets of core plugs were left in the oil overnight.

Samples were prepared for Amott wettability and relative permeability tests by flushing with a refined mineral oil (45 cP, viscosity). The mineral oil was then displaced with crude oil by flushing. Permeability to oil at initial water saturation was determined using Darcy's law. All core samples to be used for wettability and relative permeability tests were aged with the crude oil in place to establish initial wettability conditions. The samples were stored submersed in the crude for a period of 7 days.

Cores to be used for additive corefloods were flushed with crude oil to establish an initial water saturation, then waterflooded with brine to until a residual oil saturation (S_{orwf}) was achieved, effectively simulating a flooded-out reservoir after waterflooding.

Gas Chromatography

Compositional analyses were performed using a Hewlett Packard 5980A gas chromatograph equipped with a flame ionization detector. A 6-ft (1.83 m) glass column packed with Poropak QS (800-100 mesh) was used for all analyses. A temperature program of 95 to 195°F (35 to 90.6°C) gave the best separation of compounds. Standards used were 0.1 or 1% by weight alcohols or fatty acids.

Surface Tension

Surface tensions were measured with a Fisher Instruments semi-automated ring tensiometer using the DuNuoy principle.

Viscosity

Viscosity measurements were performed using a Wells-Brookfield micro-viscometer.

METHODOLOGY

Flask Tests

Three flasks were filled with 500 ml of deionized water containing 4% by volume molasses and 0.1% by weight ammonium phosphate. The first flask was used as a control that contained no bicarbonate. Sodium bicarbonate was added to a second flask (1% by weight) and Tronacarb™ was added to a third flask. Tronacarb™ is a commercial product containing 96 wt % sodium bicarbonate, 2 wt % sodium carbonate, and 1 wt % sesquicarbonate. All three flasks were inoculated with NIPER 1 & 6 and incubated anaerobically at ambient conditions. Samples were taken from each flask at 24-hour intervals to analyze for standard microbial plate counts (Standard Methods, 1980), pH, surface tension or viscosity , and microbial products detectable by gas chromatography.

Additive Tests

The cores that had been waterflooded to residual oil saturation, S_{orwf}, were prepared for injection of an additive as a preflush to the microbial formulation. It was previously determined that a preflush strategy was more successful in oil mobilization when compared to injection of sodium bicarbonate as a mixture with the microbial formulation; however, better oil recovery was exhibited in synergistic tests when compared to controls without bicarbonate (Bryant et al, 1989).

For 12-inch corefloods, a 0.1 PV slug of 1% by weight sodium bicarbonate was used for the preflush additives, followed by 0.1 PV of microbial formulation then 0.2 PV of molasses. After a 3 day incubation period, each core was injected with a second 0.2 PV slug of nutrient and shut-in for a second 3-day incubation period. Two corefloods were performed in 4-ft long Berea sandstone cores using a different injection strategy for the microbes and molasses, and to compare the bicarbonate preflush oil recovery in the longer core. The microbial and molasses injection slugs were decreased to one-half of the amount normally injected, and the second shut-in period was eliminated.

Centrifuge USBM Wettability Tests

For these tests core samples were stored at initial water saturation and submersed in the crude oil for a period of 7 days. Other samples were stored for a period of 24 hours. Imbibition and drainage capillary pressure curves with brine (containing 3% by weight sodium chloride) and crude oil were generated by increasing rates of rotation to facilitate oil displacement with brine followed by brine displacement with oil. For other samples, imbibition and drainage capillary pressure curves were generated using a microbial formulation in TSB. The displacement curves of average saturation versus pressure were used to calculate the wettability index.

Unsteady-State Relative Permeability Tests

After the cores were aged in crude oil for 1 week, the crude oil was displaced dynamically with mineral oil. Permeability to mineral oil at initial water saturation was determined. The core was immediately waterflooded with either the 3% by weight potassium chloride brine or the microbial formulation. The cores were waterflooded at a constant pressure calculated from a theoretical flow rate of 0.1 cm^3/sec to overcome end effects. The brine (or microbial formulation) was injected at constant pressure, and water and oil volumes produced were measured as a function of time. These data were used to calculate relative permeability values for water and oil using the method of Johnson, Bossler and Naumann (1959).

Amott Wettability Tests

Following a 7-day period submersed in crude oil, the cores were placed in imbibition tubes. One core was placed in a tube full of brine solution (A1), and the other core was placed in a tube full of the microbial formulation (A4). The cores were left in the tubes and monitored daily until no more oil was produced. Both samples were stored in the tubes for a period of about 3 weeks.

Each sample was then removed from the wettability tube and loaded into a coreholder. The samples were flooded with either brine or microbial formulation to determine the additional volume of mobile oil that could be displaced dynamically. Sample no. A4 was flooded with the microbial formulation. The samples were then placed in imbibition tubes full of crude oil. Again, the samples were left in the tubes until equilibrium had been reached, then both samples were flushed with crude oil until no more brine (or microbial formulation) was produced. Residual water saturation was verified by a toluene distillation method. Results are presented in terms of a wettability index to both oil and water. A general outline of the Amott wettability procedure is presented in figure 1.

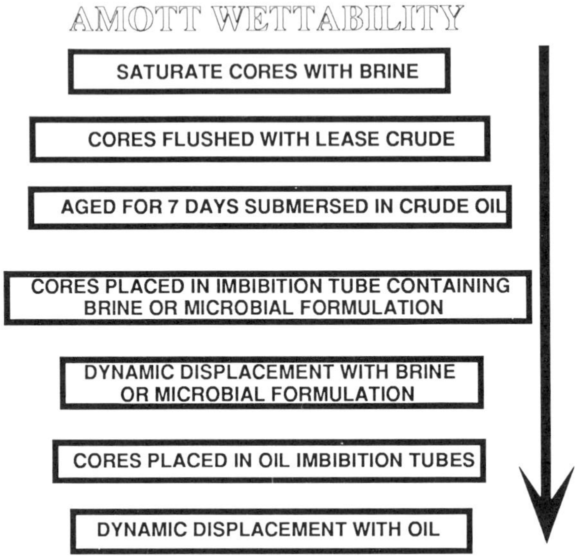

Fig. 1. Amott wettability procedure.

RESULTS AND DISCUSSION

Flask Tests and Additive Corefloods

Flask tests were performed to compare microbial counts and products, surface tension, and pH of NIPER 1 and 6 when fermenting molasses without additives (flask 1); with the addition of 1% sodium bicarbonate (flask 2) and with Tronacarb™ (flask 3). Results of flask test no. 1 are presented in table 1. A second flask test (flask experiment no. 2) was conducted with some minor modifications. Surface tensions were not measured because the tensiometer malfunctioned, and viscosity measurements were taken during the test to determine if addition of bicarbonate affected the viscosity of the microbial formulation. Results of the second flask test are presented in table 2.

TABLE 1 Flask experiment no. 1 with sodium bicarbonate and Tronacarb™

Time, hr	pH	Surface tension, mN/m	Aerobic, cfu/mL[1]	Anaerobic, cfu/mL	TGC[2], vol %
			Flask 1 Control - No bicarbonate		
0	6.7	44.0	3.0×10^3	3.0×10^3	0.000
24	6.4	47.5	2.4×10^7	2.1×10^7	0.133
48	5.0	45.2	3.6×10^7	4.2×10^7	0.166
72	4.8	45.8	2.9×10^7	4.8×10^7	0.103
96	4.7	47.0	3.4×10^7	2.9×10^7	0.129
120	4.6	46.5	1.9×10^7	2.9×10^7	0.122
144	4.4	47.0	2.0×10^7	2.6×10^7	0.132
168	4.2	45.8	1.7×10^7	2.9×10^7	0.218
			Flask 2 Sodium bicarbonate		
0	8.0	47.5	9.0×10^3	9.0×10^3	0.000
24	7.6	49.5	1.9×10^7	2.2×10^7	0.186
48	7.3	52.5	3.6×10^8	4.2×10^8	0.191
72	5.6	48.5	4.6×10^8	4.6×10^8	0.201
96	5.4	44.5	2.3×10^8	2.1×10^8	0.189
120	5.2	45.8	4.2×10^8	2.9×10^8	0.154
144	4.9	45.0	2.4×10^8	1.7×10^8	0.194
168	4.9	46.5	7.9×10^8	7.6×10^8	0.180
			Flask 3 Tronacarb™		
0	8.4	47.0	3.0×10^3	3.0×10^3	0.000
24	8.3	52.2	1.4×10^7	1.6×10^7	0.181
48	6.9	48.2	2.0×10^9	2.4×10^9	0.114
72	6.6	48.0	6.6×10^9	1.3×10^{10}	0.195
96	6.3	48.5	3.5×10^9	2.3×10^9	0.200
120	6.2	45.5	8.4×10^9	4.4×10^9	0.301
144	6.1	48.0	6.1×10^9	8.2×10^9	0.262
168	5.8	48.0	6.3×10^9	5.5×10^9	0.285

[1]cfu - colony forming units/mL.
[2]Total products detected by gas chromatography, vol %.

TABLE 2 Flask experiment no. 2 with sodium bicarbonate and Tronacarb™

Time, hr	pH	Viscosity, cP @ 77°F	Aerobic, cfu/mL[1]	Anaerobic, cfu/mL	TGC[2], vol %
Flask 1 Control - No bicarbonate					
0	6.7	1.30	7.7×10^3	1.0×10^0	0.053
24	6.4	1.31	5.6×10^7	3.7×10^7	0.052
48	5.0	1.25	8.6×10^7	5.6×10^7	0.137
72	5.0	1.36	1.7×10^8	9.0×10^7	0.125
96	4.8	1.30	9.1×10^7	1.3×10^8	0.139
120	4.5	1.24	5.7×10^7	5.3×10^7	0.192
144	4.3	1.48	2.1×10^7	6.7×10^7	0.173
168	4.2	1.25	4.0×10^5	9.9×10^6	0.184
Flask 2 Sodium bicarbonate					
0	8.0	1.30	5.8×10^3	1.0×10^0	0.056
24	7.8	1.23	3.6×10^7	3.6×10^7	0.042
48	6.4	1.10	8.9×10^7	1.3×10^8	0.048
72	6.4	1.34	3.4×10^7	1.0×10^8	0.140
96	6.6	1.19	5.3×10^8	1.9×10^8	0.122
120	6.4	1.35	7.6×10^8	8.6×10^7	0.143
144	6.2	1.34	7.5×10^7	1.2×10^8	0.138
168	6.4	1.27	7.5×10^7	8.8×10^7	0.142
Flask 3 Tronacarb™					
0	8.2	1.20	5.2×10^3	1.0×10^0	0.063
24	8.1	1.17	7.2×10^7	3.3×10^7	0.060
48	6.7	1.05	7.7×10^8	7.4×10^8	0.146
72	6.3	1.16	5.2×10^8	6.9×10^8	0.353
96	6.5	1.39	2.8×10^8	8.7×10^8	0.385
120	6.6	1.66	8.6×10^8	7.9×10^7	0.394
144	6.6	1.72	7.3×10^8	1.1×10^8	0.405
168	6.6	1.44	4.0×10^8	5.2×10^8	0.470

[1]cfu - colony forming units/mL.
[2]Total products detected by gas chromatography, vol %.

The data from both flask experiments (nos. 1 and 2) illustrate how consistently the pH drops in each flask from 24 to 48 hours (figure 2). This may be the optimal incubation time for production of short-chained fatty acids by NIPER 1 and NIPER 6. The Tronacarb™ stimulated microbial growth by a factor of 10 to 100 times in both flask tests (figure 3). The increased growth appeared to correlate with the production of metabolites detectable by gas chromatography (figure 4). Gas chromatography results for flask test 1 indicated that the total amount of microbial metabolites was greater initially in both flasks with sodium bicarbonate and Tronacarb™. In the Tronacarb™ flask, the amount of products increased over that of the bicarbonate after 120 hours. The primary metabolite responsible for this increase in flask test 1 was ethanol, although all flasks showed an increase in propionic and butyric acids with time.

In flask test 2, the Tronacarb™ flask had significantly higher total microbial product production. In this test an increase in both alcohols and fatty acids were responsible for the higher amount of detectable product. A slight increase in viscosity was also apparent in the Tronacarb™ flask. In the second test, the microbial counts and amount of detectable microbial products for the sodium bicarbonate flask were in the same range as those determinations for the control flask.

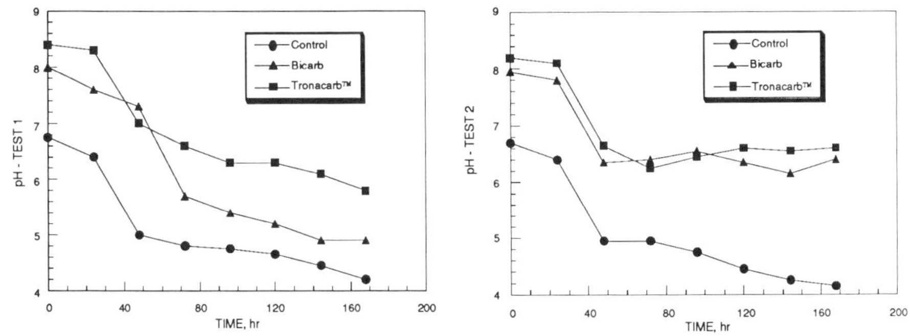

Fig. 2. pH as a function of time for microbial flask tests 1 and 2.

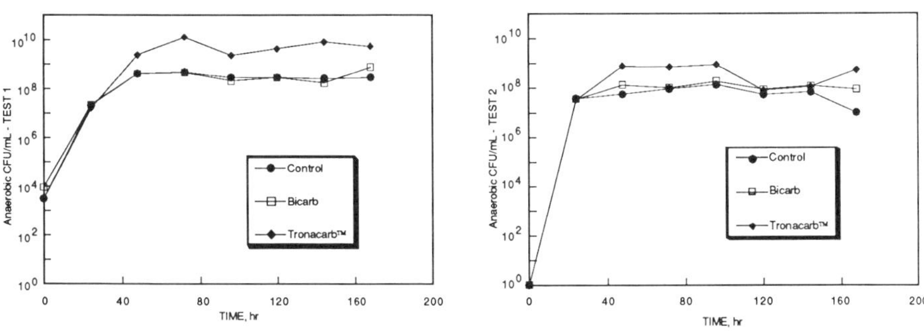

Fig. 3. Anaerobic Counts as a function of time for microbial flask tests 1 and 2.

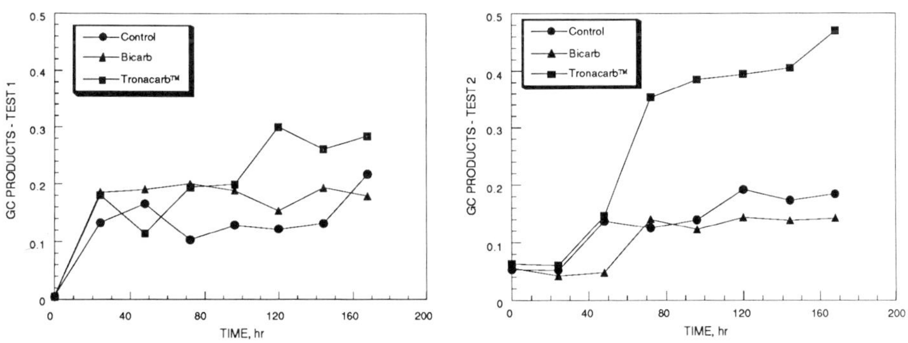

Fig. 4. Total GC products as a function of time for microbial flask tests 1 and 2.

Additive Corefloods

Table 3 and figure 5 show comparisons of oil production in 12-inch corefloods using sodium bicarbonate and Tronacarb™ as a preflush compared to a control core without preflush treatment. The microbial coreflooding results, along with the flask test results, indicate better oil mobilization occurred by using either sodium bicarbonate or Tronacarb™ as a preflush. This difference is even more pronounced at a brine throughput of one pore volume, which is probably more economically feasible.

TABLE 3 1-ft Berea sandstone cores with bicarbonate and Tronacarb™

Core	Total Injected, PV	S_{orwf}[1], %	S_{orcf}[2], % 1PV	S_{orcf}, % 2PV	ΔS_{orcf}, %[3] 1PV	ΔS_{orcf}, % 2PV	Er, %[4] 2 PV	Brine k, md
CONTROL	[5]0.5	35.3	28.1	24.0	7.2	11.3	32.0	925
BICARB ONATE	[6]0.6	33.4	21.2	19.1	12.2	14.3	42.8	830
TRONACARB™-1	[6]0.6	36.0	22.2	19.5	13.8	16.5	45.8	278
TRONACARB™-2	[6]0.6	40.3	30.5	28.5	9.8	11.8	29.2	828

[1]S_{orwf} = Oil saturation after waterflooding, percent pore volume
[2]S_{orcf} = Oil saturation after microbial treatment, percent pore volume
[3]$\Delta S_{orcf} = S_{orwf} - S_{orcf}$
[4]$Er\ \% = (S_{orwf} - S_{orcf})/S_{orwf} \times 100$
[5]Core was injected with 0.1 PV microbes, 0.2 PV molasses, shut-in 3 days, re-fed with 0.2 PV molasses, shut-in 3 days, and waterflooded.
[6]Core was injected with 0.1 PV of additive preflush (1%), then injected with 0.1 PV microbes, 0.2 PV molasses, shut-in 3 days, re-fed with 0.2 PV molasses, shut-in 3 days, and waterflooded.

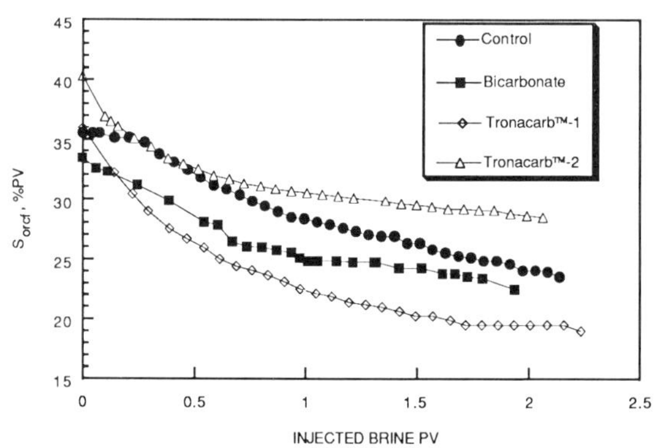

Fig. 5. Residual oil saturation vs time for 1 ft corefloods.

Results of the 4-ft long corefloods (table 4) demonstrated that the smaller volume of slug could recover oil, although not as effectively as a larger pore volume of microbial slug and the combination of two shut-in periods. The recovery efficiency for the core with a preflush of sodium bicarbonate was 26.6% higher than the core without preflush treatment.

TABLE 4 4-ft Berea sandstone coreflood experiments with and without bicarbonate preflush

Core	Total Injected, PV	S_{orwf}, %	S_{orcf}, % 1PV	S_{orcf}, % 2PV	ΔS_{orcf}, % 1PV	ΔS_{orcf}, % 2PV	Er, % 2 PV	Brine k, md
CONTROL	[1]0.25	32.9	30.1	29.5	2.8	3.4	10.3	385
BICARBONATE	[2]0.35	30.9	24.6	19.5	6.3	11.4	36.9	353

[1]Core was injected with 0.05 PV microorganisms, 0.1 PV 4% molasses, shut-in 3 days, re-fed 0.1 PV molasses, and waterflooded (no second shut-in period).

[2]Core was injected with a preflush of 0.1 PV of 1% sodium bicarbonate, then injected with 0.05 PV microorganisms, 0.1 PV 4% molasses, shut-in 3 days, refed 0.1 PV molasses, and waterflooded (no second shut-in period).

Centrifuge Wettability

Results of the USBM method centrifuge wettability tests are presented in tables 5 and 6. NIPER 1 used either alone or in combination with other microbes appears to shift the wettability index significantly in the positive direction toward a more water-wet condition. The core when treated with a filtered solution of products (cells removed) appeared to be in the same range of wettability index as the core used with the brine-oil system. The in situ microbial metabolism appears to be involved in wettability alteration since microbial products alone do not appear to alter wettability.

TABLE 5 USBM centrifuge wettability tests

3% Sodium Chloride Brine		0.5% Sodium Chloride Brine	
Sample No.[1]	Wettability Index	Sample No.[1]	Wettability Index
CP66	+0.153	CP53	-0.064
CP67	+0.145	CP54	-0.030
CP68	+0.267	CP55	-0.082
CP69	+0.196	CP56	+0.063
CP70	+0.223		
CP71	+0.222		
Average WI =	+0.201 ±0.046	Average WI =	-0.028 ±0.065

Delaware Childers oil and NIPER 1 & 6 filtered products in TSB:

Sample No.[2]	Wettability Index
CP60	+0.239
CP61	+0.117
CP62	+0.247
CP63	+0.176
CP64	+0.237
CP65	+0.100
Average WI	+0.186 ±0.065

TABLE 6 USBM centrifuge wettability tests

Delaware-Childers oil and different microbial solutions in TSB:

Sample No.[2]	Wettability Index	Microbial formulation
CP1	+0.315	Brine/Delaware Childers oil
CP2	+0.453	Brine/Delaware Childers oil
CP3	+0.739	NIPER Bac 1[3]
CP4	+0.992	NIPER Bac 1
CP23	+0.927	NIPER 1 and NIPER 3
CP24	+0.763	NIPER 1 and NIPER 3
CP9	+0.950	NIPER 1 - *B. licheniformis*
CP10	+0.959	NIPER 1
CP15	+0.137	NIPER 2 - *Bacillus* sp
CP16	+0.117	NIPER 2
CP11	+0.293	NIPER 3 -*Clostridium* sp
CP12	+0.138	NIPER 3
CP13	+0.283	NIPER 4 - Gram (-) rod
CP14	+0.262	NIPER 4
CP21	+0.355	NIPER 5 -*Clostridium* sp
CP22	+0.500	NIPER 5
CP19	+0.310	NIPER 6 - *Clostridium* sp
CP20	+0.314	NIPER 6

[1]The core was aged for 7 days with Delaware Childers oil at Swi.
[2]Core was aged overnight with Delaware Childers oil.
[3]NIPER Bac 1 is a consortium of 4 microbes used in field test, NIPER 1, NIPER 2, NIPER 3 and NIPER 4.

Amott Wettability

Results of the Amott wettability test are presented in table 7. The decrease in relative permeability to the microbial formulation at the end of the test shows a decrease in water mobility which would be seen in a trend to more water-wet conditions. The ratio of the permeability to water in the presence of residual oil to the permeability to oil in the presence of connate water (end-point k_{rw}) is sometimes used as an indicator of wettability--a value of less than 0.3 indicates water-wetness and a value near unity indicates oil-wetness (Raza, et al, 1968). The brine-oil core (A1) had a calculated end-point k_{rw} value of 0.355, whereas the sample tested with microbial formulation (A4) had a value of 0.220. The wettability index to water of 0.244 for control core A1 was very close to the USBM centrifuge wettability index values presented in table 4. The wettability index to water for the microbial sample was significantly higher; however, this core did imbibe some oil--the wettability index to oil was 0.162. The difference of 0.422 between the wettability index to the microbial formulation and the wettability index to oil is still twice the value of the wettability index to water for sample A1. The residual oil saturation for sample A4 was also 5.3% lower, which is consistent with the relative permeability results.

TABLE 7 Summary of Amott wettability test results

Sample Number System	A1 Brine-DC[1] Oil	A4 Microbial formulation-DC Oil
Permeability to air, md	325	271
Porosity, %	19.2	19.2
Immobile water saturation, % PV	21.7	22.9
Permeability to oil at initial water saturation, md	163	131
Water[2] imbibed, % PV	12.2	27.7
Oil displaced dynamically, % PV	31.1	19.7
Total oil recovered, % PV	43.3	47.4
Immobile oil saturation, % PV	35.0	29.7
Permeability to water[2] at immobile oil saturation, md	58	29
Endpoint k_{rw}[3], fraction	0.355	0.220
Oil imbibed, % PV	0	6.2
Water[2] displaced dynamically, % PV	41.6	32.6
Total water recovered, % PV	41.6	38.8
Wettability index [4] to water[2]	0.244	0.584
Wettability index [4] to oil	0	0.162

[1]DC Oil = Delaware Childers oil.
[2]Denotes microbial formulation for sample A4.
[3]k_{rw} =Effective permeability to water or microbial formulation (A4) at residual oil saturation/effective permeability to oil at initial water saturation.
[4]Wettability index = Fluid Imbibed/fluid imbibed + fluid displaced dynamically

Unsteady-State Relative Permeability Tests

Relative permeability measurements were conducted on adjacent samples from two sets of Berea sandstone that had different permeability ranges. Each set includes one sample with brine-oil relative permeability and another sample used for microbial formulation-oil relative permeability tests. The two samples from each group are presented together graphically for comparison studies. A summary of relative permeability test results is presented in table 8. Comparisons of the relative permeability fraction curves and ratio curves are presented in figures 6 and 7, respectively. A higher relative permeability to oil is apparent in both microbially flooded samples as compared to brine flooded samples. Sample R2 exhibited a significant decrease in relative permeability to water throughout the flood. This higher permeability core, flooded with a microbial formulation that had been incubated 24 hours, exhibited an increase in oil production and relative permeability to oil immediately and continuing throughout the flood. A slight shift to the right can be observed in the ratio curves of microbially flooded samples. These curves also indicate that sample R8 (flooded with a 120-hour culture) had an increase in oil production as compared with the brine sample toward the latter half of the flood.

Oil recovery before water breakthrough was derived from the fractional flow equations using the viscosity of the Delaware-Childers oil. The samples waterflooded with the microbial formulation had a higher oil recovery before water breakthrough. A greater difference was seen in relative permeability characteristics for the higher permeability samples using a microbial formulation that had been incubated for a period of 24 hours.

TABLE 8 Summary of unsteady-state relative permeability tests

			Initial Conditions	
Sample[1]	Permeability to air, md	Porosity, percent	Water Saturation, percent PV	Effective permeability to oil, md
[2]R1	981	22.6	30.5	650
[3]R2	1240	22.8	30.4	837
[2]R7	322	18.9	25.3	209
[4]R8	323	19.0	25.8	202

	Terminal Conditions		Oil Recovered		
Sample	Oil Saturation, % PV	Effective k to Water, md	Percent PV	Percent OIP[5]	Before Breakthrough[6]
R1	30.3	127	39.3	56.5	23.5
R2	24.7	107	44.9	64.4	31.6
R7	32.2	17	42.5	56.9	26.7
R8	28.2	19	46.0	62.0	28.2

[1]All cores were aged for 1 week with Delaware-Childers oil.
[2]Brine/oil (mineral oil) relative permeability.
[3]24-hour microbial solution/oil (mineral oil) relative permeability.
[4]5-day old microbial solution (optimal biosurfactant production)/oil (mineral oil) relative permeability.
[5]OIP = Oil in place.
[6]Calculated using oil viscosity of Delaware Childers oil.

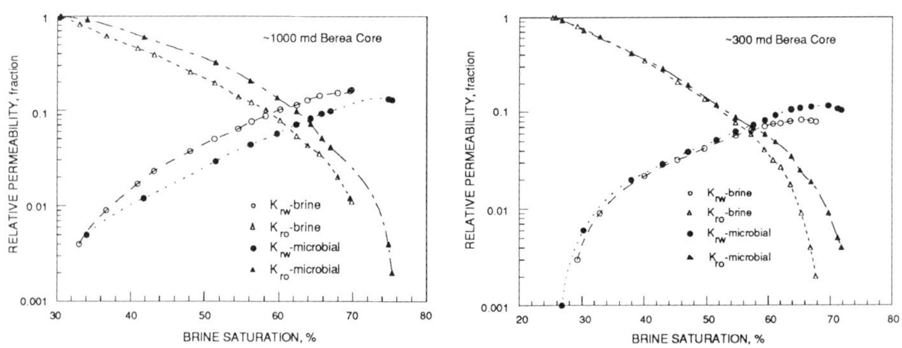

Fig. 6. Unsteady-state relative permeability fraction curves for different permeability cores.

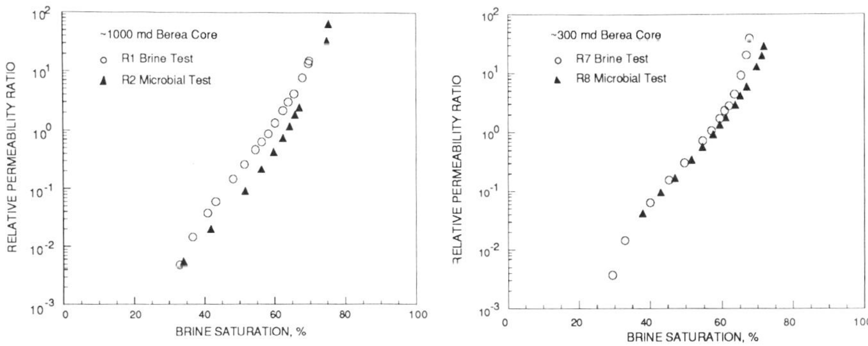

Fig. 7. Unsteady-state relative permeability ratio curves for different permeability cores.

CONCLUSIONS

Additive flask tests indicate that sodium bicarbonate and Tronacarb™ may alter microbial metabolic pathways. Since the additive corefloods showed about the same increase in oil recovery with either sodium bicarbonate or Tronacarb™ as a preflush, the increase in microbial cell and microbial product production, which was observed in flask tests by Tronacarb™, does not appear to be the sole mechanism of improved oil mobilization.

Microbial formulations containing NIPER 1 appear to shift the wettability index of Berea core samples from just slightly water-wet to a more positive value. The microbial cells appear to be involved in this wettability alteration since no change was observed in samples tested with filtered microbial products (cells removed). In situ microbial metabolism appears to be involved in wettability alteration. Results of the Amott wettability test indicate that the wettability index to water for the microbially treated core was twice that of the value for the control core, indicating a more water-wet condition.

A significant decrease was seen in k_{rw} curves (a decrease in water mobility) for two out of three samples containing microbial formulation. Both relative permeability test samples showed an increase in relative permeability to oil; although, one sample containing microbial formulation had a 20% higher initial effective permeability to oil. However, it has been shown that cores with varying permeabilities but the same rock type (pore geometry) have almost identical relative permeability characteristics (Morgan and Gordon, 1970). A slight shift to the right was observed at the brine saturation at which oil and water relative permeabilities are equal (crossover). Samples tested with the microbial formulation had higher oil recovery before water breakthrough and lower residual oil saturations. Steady-state water-oil relative permeability tests may provide more meaningful data since Delaware-Childers crude oil, which has a low viscosity of 7.5 cP, could be used instead of refined mineral oil.

Although all wettability tests with microbial formulations exhibited a change in the direction of a more water-wet condition, changes in the range of neutral wettability are very hard to distinguish. Berea sandstone is very water-wet, and any alteration toward a more oil-wet condition is difficult to attain. Tests performed on a core that is definitely in the range of oil-wet, would be helpful in further evaluating the effects of microbial systems on wettability.

REFERENCES

American Public Health Association, #907. Standard Plate Count, In: Standard Methods for the Examination of Water and Wastewater. 15th edition, 1980.

Amott, E. Observations Relating to the Wettability of Porous Rock, *Trans.*, AIME **216**, pp 156-62, 1959.

Anderson, W. G. Wettability Literature Survey--Part 2: Wettability Measurement, Journal of Petroleum Technology, November, 1986.

Bryant, R. S. and J. Douglas. Evaluation of Microbial Systems in Porous Media for Enhanced Oil Recovery. SPE Reservoir Engineering, SPE paper 16284, May, 1988.

Bryant, R. S., T. E. Burchfield, K. L. Chase, K. M. Bertus, and A. K. Stepp. Optimization of Microbial Formulations for Oil Recovery: Mechanisms of Oil Mobilization, Transport of Microbes and Metabolites, and Effects of Additives. SPE paper 19686. Presented at the 64th Annual Technical Conference and Exhibition of the Society for Petroleum Engineers, San Antonio, TX, October 8-11, 1989.

Bryant, R. S., T. E. Burchfield, D. M. Dennis, and D. O. Hitzman. Microbial-Enhanced Waterflooding: Mink Unit Project, SPE/DOE paper 17341, SPE Reservoir Engineering, February, 1990.

Clark, J. B. Oil Recovery Process, U.S. Patent No. 4,610,302, September 9, 1986.

Donaldson, E. C. Equipment and Procedures for Fluid Flow and Wettability Tests of Geological Materials, Bartlesville Energy Technology Center, report DOE/BETC/IC 79/5, U.S. DOE, May, 1980.

Johnson, E. F., Bossler, D. P., and Naumann, V. O. Calculation of Relative Permeability from Displacement Experiments, *Trans*, AIME, January, 1959.

Morgan, J. T. and D. T. Gordon. Influence of Pore Geometry on Water-Oil Relative Permeability. Journal of Petroleum Technology, October 1970.

Raza, S. H., L. E. Treiber, and D. L. Archer. Wettability of Reservoir Rocks and Its Evaluation. Producers Monthly, April, 1968.

Russell, H. H. and Grula, M. M. Nutritional Aspects of MEOR, paper presented at the 1987 U.S. DOE Symposium on Applications of Microorganisms to Petroleum Technology, Bartlesville, August 12-13, DOE Conf-870858.

Reprinted from: *Microbial Enhancement of Oil Recovery - Recent Advances*,
edited by E.C. Donaldson
Elsevier Science Publishers B.V., Amsterdam - Printed in The Netherlands

CH. R–5

MECHANISMS OF OIL DISPLACEMENT BY MICROORGANISMS

S.A. KIANIPEY AND E.C. DONALDSON
School of Petroleum & Geological Engineering, University of Oklahoma

ABSTRACT

Mechanisms of oil displacement at the oil-brine-sand interfaces by bacteria were investigated by microscopic observations and capillary pressure changes using unconsolidated, thin, reservoir flow cells. Three genera of bacteria, Bacillus, Pseudomonas, and Clostridium, were injected into water-wet and oil-wet cells of unconsolidated sand, saturated with brine and crude oil. The flow cells were placed under a microscope for visual and photographic observations. The flow cells were connected to a manometer to examine the complete capillary pressure hysteresis loop before and after introduction of bacteria.

The thin reservoir flow cells were first saturated with brine, and then displaced to irreducible water saturation with crude oil. Oil and brine displacements were then made to determine the complete capillary pressure relationship. Next, the bacteria and nutrients were introduced and incubated in the flow cell for 24-48 hours. Microscopic observations were recorded photographically. After incubation, oil and brine displacements were again made to determine the capillary pressure hysteresis loop.

During incubation, some emulsification of the oil at the oil-brine-sand interfaces were observed. Biogenic gas production caused pressure increases in the flow cells. Final measurements of the capillary pressure hysteresis loop showed decreases in residual oil saturation of 9-24% and changes in wettability to more water-wet systems.

This work shows the potential of microbial enhanced oil recovery, and investigates some of the mechanisms of oil displacement by three genera of bacteria. These considerations are important with regard to the technical feasibility of microbial enhancement of oil recovery.

INTRODUCTION

Two major concerns, namely the ever-increasing demand for petroleum-derived energy and the rapid depletion of discovered oil fields, have directed several researchers to seek out an alternative oil recovery technique to somehow mobilize the residual oil remaining after waterflooding. Although its development has not been completely researched and some of its technical difficulties have yet to be resolved, the utilization of different bacterial species to recover additional residual oil from a number of reservoir models and low-producing oil fields has been successfully demonstrated (1-13). Except for the proposed use of indigenous microbes to selectively plug high permeability zones (14-17) and investigation of the transport of nutrients through sandstone cores (14, 18), very limited work has been initiated in the use of injected microorganisms to mobilize residual oil. Some of the proposed mechanisms responsible for oil displacement by the in situ growth of bacteria include biogenic gas production to increase the reservoir pressure (2,15,19), microbial production of surface active agents (6,10,12,13), and production of organic acids and co-surfactants (19,20).

The intent of this research was to explore and investigate some of the mechanisms responsible for oil displacement in unconsolidated, thin reservoir flow cells by three species of microorganisms. The species of the bacteria used were Bacillus licheniformis strain JF-2, Pseudomonas aeruginosa, and Clostridium acetobutylicum. Capillary hysteresis loops were developed, which were used to evaluate and identify the critical factors contributing to the displacement of oil in the flow cells (21-23). Visual and photographical observations of the flow cells, during the incubation of bacteria at oil-brine-sand interfaces, showed production of biogenic gas and release of oil droplets with possible emulsification of oil droplets. Thus, the microorganisms apparently released and mobilized oil at the interfaces by gas and solvent production. The decrease in residual oil saturation of the flow cells was due to in situ growth of microorganisms.

EXPERIMENTAL APPARATUS AND PROCEDURES

Capillary Pressure Apparatus

The schematic diagram of the experimental equipment is shown in Fig. 1. The equipment consists of a burette, an unconsolidated thin reservoir flow cell, optical and photographic apparatus, a manometer, an overhead reservoir, a drainage vessel, and valves. The optical and photographic apparatus used in these experiments were an Olympus Zoom-Stereo microscope, model Sr-Tr 001 and an Olympus automatic exposure system with a built-in microcomputer, model PM-10A.

Construction of Unconsolidated Thin Reservoir Flow Cells

A schematic diagram of an observation flow cell is shown in Fig. 2. Several sterilized clear glass plates (7.5 cm x 7.5 cm x .03 cm) were prepared. A Five Minute Clear Epoxy was applied to opposing corners of one side of each glass plate, to roughen the glass surface to avoid breakthrough of fluids. Epoxy-coated Vollumoid gaskets, 1 cm wide and 2 mm thick, were cut and placed at the edges between two glass plates. The gaskets and one of the two hypodermic needles (18 gage, 4 cm long) were cemented to the face of a glass plate. Then, the inner edges of each gasket were coated with epoxy to avoid fluid absorbance by the gaskets. The second glass plate was then cemented on top of the assembly.

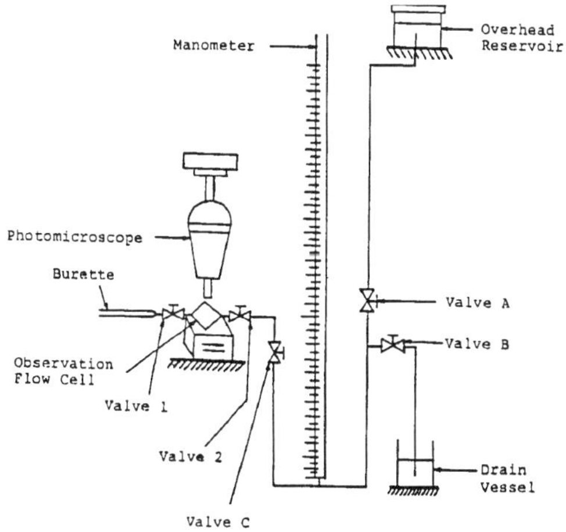

Figure 1. Schematic diagram of Capillary Pressure apparatus.

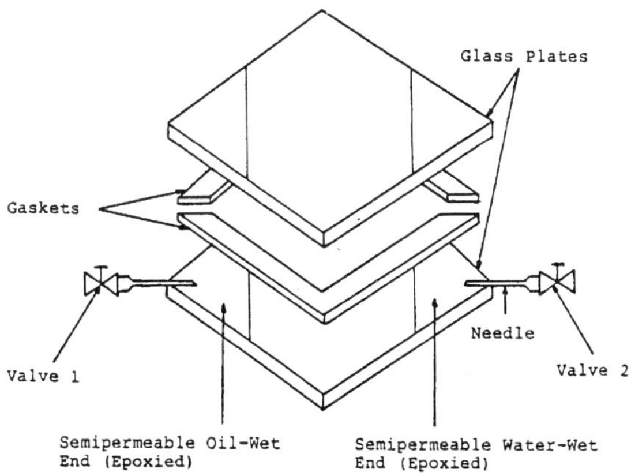

Figure 2. Schematic diagram of unconsolidated thin reservoir flow cell.

The fine sand of 80-90 mesh and matrix Ottawa sand of 30-40 mesh (average diameter .417 mm) were prepared. The sand grains were water-wet in nature, but wettability of sand grains can be altered by adsorption of polar resins from crude oil (28). To insure that sand grains were either strongly water-wet or oil-wet, the 80-90 mesh fine sand and 30-40 mesh Ottawa sand used in these experiments were treated with 10% HCl and silicone solution, respectively.

A few grains of water-wet Ottawa sand were first poured into the cell. Then, fine water-wet sand was placed in the cell to the level indicated in Fig. 2. Next, water-wet or oil-wet Ottawa quartz sand (depending on the type of system made) was used as matrix sand. the last one-third of the flow cell was then filled with oil-wet fine sand followed by a few grains of oil-wet 30-40 mesh sands. Then, the second hypodermic needle was inserted in the groove and cemented. Three-way nylon valves were attached to each of the two needles to control the amount of fluid entering the flow cells. the bulk volume of the constructed flow cells ranged from 4.2 to 4.8 ml.

Fluid Properties

A light crude oil (44.5°API) and 5% NaCl + 0.1% sodium sulfite (Na_2SO_3) brine solution were used for all experiments. Sodium sulfite was used to remove dissolved atmospheric oxygen from stored brine solution to discourage indigenous aerobic microbial growth. The crude oil was washed with distilled water three times in a separatory funnel before it was used (to remove sludge acquired during production). The brine used in this study was filtered through 0.22 um filter membrane to reduce the amount of particulate matter in the brine.

Species of Microorganisms

Three isolated species of microorganisms, Bacillus licheniformis JF-2, Pseudomonas aeruginosa, and Clostridium acetobutylicum were provided by the University of Oklahoma, Department of Botany and Microbiology. These microorganisms were used anaerobically for the experiments.

The biosurfactant-producing Bacillus licheniformis JF-2 (13) was grown in medium F (Table 1). Pseudomonas aeruginosa was used because it has been reported to produce a biosurfactant type compound which lowers the oil-water interfacial tension (9,10). For this research, this organism was grown in medium F.

Clostridium acetobutylicum was grown in enriched Thioglycollate (THIO) medium (24). This species produces very large amounts of CO_2 (19,20). Acetate and butyrate may be produced as well as acetine and butanol

Table 1. Medium F used for the cultivation of Bacillus licheniformis JF-2 and Pseudomonas aeruginosa[14].

Components	Amount Used (g/liter)
KH_2PO_4	2.70
K_2HPO_4	14.00
NaCl	30.00
$(NH_4)_2SO_4$	1.00
Glucose	20.00
$MgSO_4$	0.50
Yeast extract	1.00
Trypticase	1.00
$NaNO_3$	1.80
Sodium thioglycollate	1.80
Trace Metal solution (ml)	10.00

(19,20,25). This microorganism was used in this study due to the reasons stated above.

Saturation of Flow Cells

The flow cells were first flushed with CO_2 for 15 minutes to remove the air from the pore spaces. The water-wet matrix sand was then vacuum saturated vertically with brine. The crude oil was then injected through the oil-wet fine sand matrix. The volume of fluids injected into the flow cells during the saturation were measured to determine the total and matrix sand pore volume of the flow cells.

Procedure

The brine level in the manometer (Fig. 1) was set at the hydrostatic datum (zero level) and all the valves were closed. A 2 ml burette (0.01 ml calibration) was filled with crude oil and attached to the oil-wet end of the flow cell at valve 1. Valve 2, at the water-wet end of flow cell, was connected to the manometer. The brine level in the manometer could be changed to either increase or decrease the pressure in the flow cell. At the same time, changes of fluid saturation monitored by recording the burette readings. The brine level of the manometer was decreased by 1 cm intervals causing oil to enter the flow cell and displace the brine to irreducible brine saturation (brine desaturation). The brine level was lowered by another 1 cm interval only when there was no change in the burette readings for at least one hour.

Next, the brine level in the manometer was raised by 1 cm intervals causing brine to enter the flow cell and displace oil to residual oil saturation. Next, the brine level in the manometer was reset to zero. The desaturation of brine by oil was repeated to complete the capillary pressure hysteresis loop. Valves 1, 2, and C were then closed and, the brine level in the manometer was reset to zero.

Microscopic and photographic observations of the flow cell were made during desaturation and saturation, to observe the path of the injected fluid front through the porous matrix. A 35 mm camera was used to photograph any event which appeared to be significant.

After the capillary pressure relationship of the flow cell was established, a mixture of nutrient media and microorganisms were prepared. A plate count of the inoculum was made to determine the concentration of cells in the medium. Then, the solution containing the microorganisms was injected into the water-wet end of the flow cell until residual oil saturation was obtained. Next, valves 1 and 2 were closed and the flow cell was disconnected and incubated at 37°C for 24 to 48 hours. The flow cell was periodically checked under a microscope with observations beginning after 18 hours of incubation to visually observe and photograph any changes at the oil-brine-sand interfaces.

After the incubation period was completed, the flow cell was connected to the manometer and pressure increase due to the growth of the microorganisms was recorded.

The capillary pressure hysteresis loop after treatment by microorganisms was once more determined (as outlined above) and compared to the capillary pressure relationship determined before bacteria were introduced. Visual and photographic observations of important events were made. After completion of the experiment, the microbial culture in the flow cell was examined to determine if the culture had become contaminated with indigenous bacteria.

Properties of Porous Matrix

Calculation of the permeability of the porous matrix was not possible due to variations in the cross-sectional area of the cell, and the presence of the semi-permeable ends in the flow cells. The pore volume of the flow cell containing the matrix sand was determined during the saturation process by measuring the amount of brine required for complete saturation and by weighing the flow cell before and after saturation.

The wettability of the matrix sand was measured using the method of Donaldson et al (26). This method uses the areas under the capillary pressure curves to measure the wettability of the sand as follows:

$$W = \log(A_1/A_2)$$

where A_1 and A_2 are the areas under the oil displacement curve and the water displacement curve, respectively. When W is positive, the system is water-wet; negative values and values near zero represent oil-wet and intermediately-wet systems, respectively. An Apple IIe computer, general plotting program, was used to determine areas A_1 and A_2.

RESULTS

Bacillus

Experiment 1 was performed with a water-wet flow cell. The pore volume was 1.4 ml. The drainage-imbibition capillary pressure curve as a function of water saturation is shown in Fig. 3. The residual oil saturation before injection of bacteria, at the end of the first brine flood (imbibition curve), was 0.20. The wettability number obtained from Fig. 3, before introduction of bacteria, was 0.247, indicating a water-wet system.

When the first capillary hysteresis loop was completed, 0.7 ml of medium F and Bacillus licheniformis strain JF-2 was injected into the flow cell. The bacterial count of the mixture was 1.6×10^6 cells/ml of medium. The flow cell was then incubated for 24 hours at 37°C.

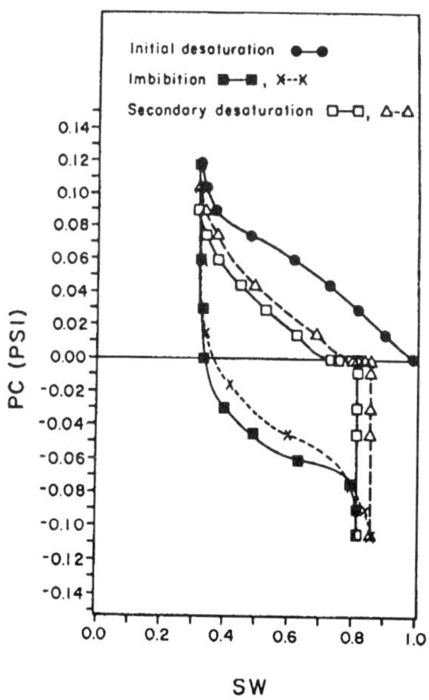

Figure 3. Capillary hysteresis loops before (solid line) and after injection of microorganisms (dashed line) for experiment #1 in water-wet matrix.

The flow cell was periodically checked under a microscope. Biogenic gas production was first observed at the inlet after 18 hours of incubation and increased slowly throughout the flow cell as time progressed. Figs. 4 and 5 show the gas production due to the in situ growth of organisms. The produced gas accumulated at the top face of the flow cell and little gas was observed on the underside of the cell. The pressure of the flow cell was increased by 0.1 cm of manometer scale (0.0014 psi) due to the biogenic gas production. During incubation, some emulsification of the oil took place with formation of small oil droplets throughout the flow cell, Fig. 6. These oil droplets were observed on both faces of the flow cell and were observed only after injection of microorganisms. The oil droplets varied in size, ranging from 0.0041 mm to 0.08 mm. It is assumed that emulsification at the interface was due to production of biosurfactant by the microorganisms, however, no definitive tests for this were developed.

Capillary pressure hysteresis loop was then obtained (Fig. 3). The residual oil saturation decreased from 0.20 to 0.16. The relative change in residual oil saturation was a decrease of 20%. The wettability number increased from .247 to .285. Visual observation indicated that the areal sweep efficiency of the brine flood improved after the injection of bacteria.

Experiment 2 was performed with an oil-wet flow cell. The pore volume was 1.25 ml. The same procedure explained, previously, was carried out. The drainage-imbibition capillary pressure curve as a function of water saturation is shown in Fig. 7. The residual oil saturation before the injection of bacteria was 0.184. The wettability number determined from Fig. 7, before the injection of microorganisms, was -0.269 indicating an oil-wet system.

When the pre-microbial injection capillary hysteresis loop was complete, 0.6 ml of medium F containing 9.5×10^6 cells/ml of _Bacillus licheniformis_ strain JF-2 was injected and incubated for 24 hours at 37°C. Visual observation of the flow cell during incubation indicated similar results in terms of gas production and formation of oil droplets as experiment 1. The capillary hysteresis loops are shown in Fig. 7. The residual oil saturation decreased from 0.184 to 0.14 (-24%). The wettability number increased from -0.269 to -0.10, showing that the system became more water-wet after treatment with the microorganisms. Visual observation indicated qualitatively that the sweep efficiency of the water flood after microbial treatment were improved.

Pseudomonas

Experiment 3 was performed with a water-wet flow cell, using _P. aeruginosa_. Results of this experiment is presented in Fig. 8. The pore volume of matrix sand was 1.4 ml. The residual oil saturation was determined

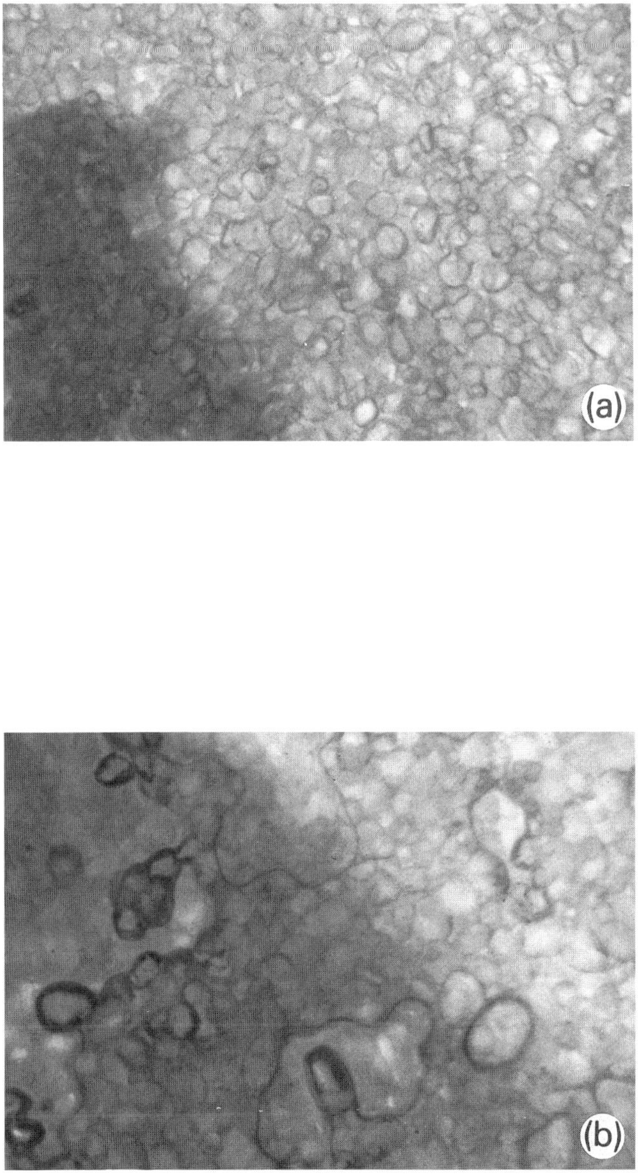

Figure 4. Biogenic gas production at the inlet; (a) before injection of microorganisms; (b) after injection of B.licheniformis strain JF-2.

Figure 5. Biogenic gas production in matrix sand; (a) before injection of microorganisms; (b) after injection of B.licheniformis strain JF-2.

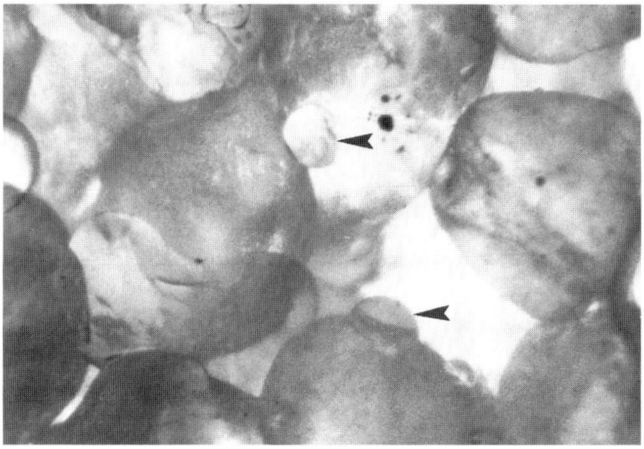

Figure 6. Emulsification of oil (oil droplets) at the oil-brine-sand interfaces due to biosurfactant production; after injection of B.licheniformis strain JF-2.

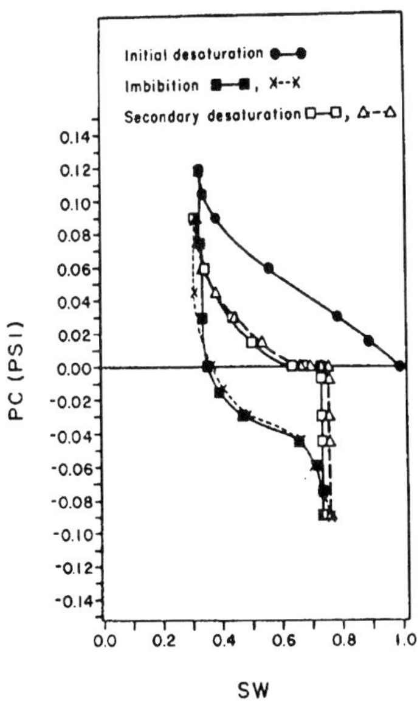

Figure 7. Capillary hysteresis loops before (solid line) and after injection of microorganisms (dashed line) for experiment #2 in oil-wet matrix.

(Fig. 8) to be 0.157 and the wettability number, before the injection of bacteria, was 0.53.

A 0.7 ml volume of medium F containing $7.2x10^5$ cells/ml of P. aeruginosa was injected into the flow cell. The flow cell was then incubated for 48 hours at 37°C. There was no indication of gas production during the incubation period; however, production of oil droplets began after 39 hours. The accumulation of oil droplets was principally concentrated at the inlet in the matrix sand.

The residual oil saturation, after injecting the bacteria and determination of the capillary hysteresis loop, decreased form 0.157 to 0.142 (-9.5%). The wettability number also changed very little from 0.53 to 0.481 toward a slightly more oil-wet system.

An oil-wet flow cell was used in experiment 4. Results of this experiment is shown in Fig. 9. The pore volume of the flow cell containing matrix sand was 1.25 ml. The residual oil saturation and wettability number were 0.264 and -0.16, respectively.

A 0.5 ml volume of medium F containing $2.8x10^6$ cells/ml organisms was injected into the flow cell. The results of this experiment were similar to those of experiment 3. Gas production was not observed but small oil droplets formed at the inlet. The residual oil saturation changed from 0.264 to 0.24 (-9.1%). There was no observed change of wettability.

Clostridium

A water-wet flow cell was used for experiment 5. The matrix sand pore volume was 1.35 ml. The capillary pressure versus saturation relationship for this experiment is shown in Fig. 10. The residual oil saturation and wettability number were 0.163 and 0.281, respectively.

A 0.7 ml volume of enriched Thioglycollate (THIO) medium containing $4.5x10^6$ cells/ml of Clostridium acetobutylicum was prepared and injected into the flow cell. The flow cell was incubated for 40 hours at 37°C. Gas production began after 34 hours of incubation and progressed with time. The produced gas was observed throughout the flow cell and it was much greater than the gas production of experiments 1 and 2. The pressure of the flow cell was increased by 0.5 cm of manometer scale (.0075 psi) due to biogenic gas production. Also, oil droplets were observed at the oil-brine-sand interfaces throughout the flow cell.

The capillary hysteresis loops are shown in Fig. 10. The residual oil saturation decreased from 0.163 to 0.137 (-16%). The wettability number increased from 0.281 to 0.332.

An oil-wet flow cell with a pore volume of 1.6 ml was used next for

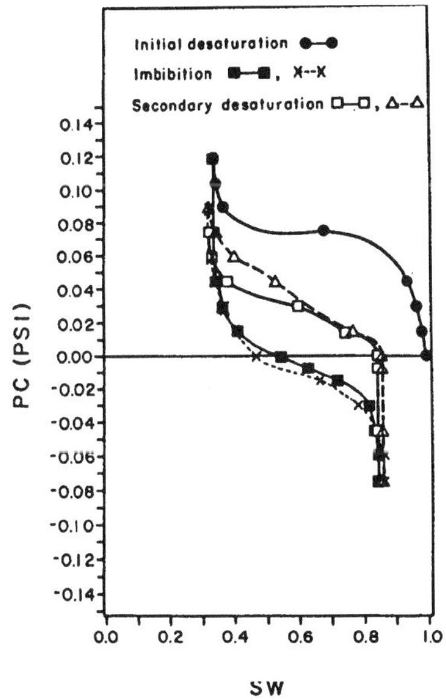

Figure 8. Capillary hysteresis loops before (solid line) and after injection of microorganisms (dashed line) for experiment #3 in water-wet matrix.

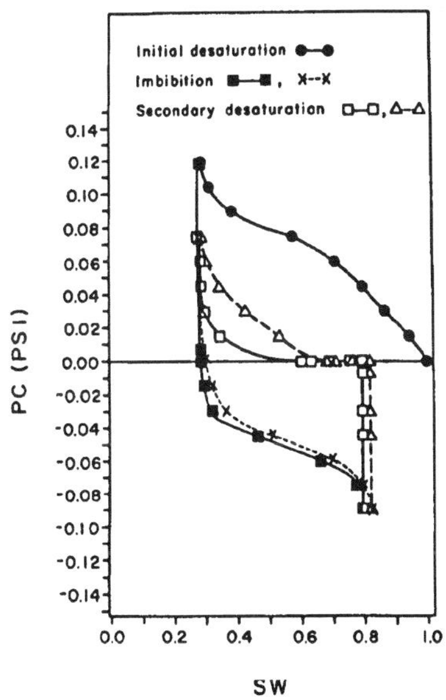

Figure 9. Capillary hysteresis loops before (solid line) and after injection of microorganisms (dashed line) for experiment #4 in oil-wet matrix.

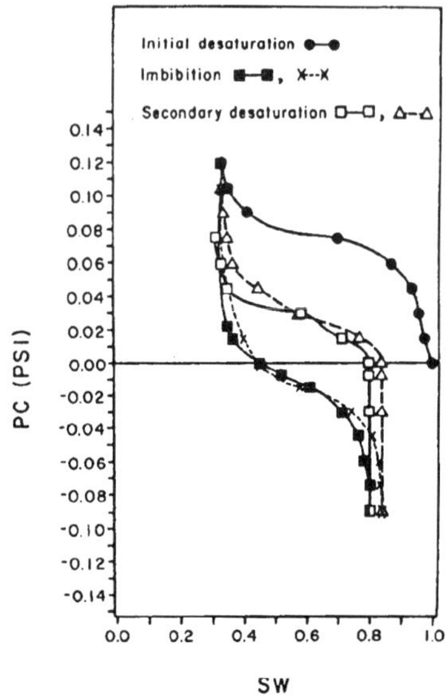

Figure 10. Capillary hysteresis loops before (solid line) and after injection of microorganisms (dashed line) for experiment #5 in water-wet matrix.

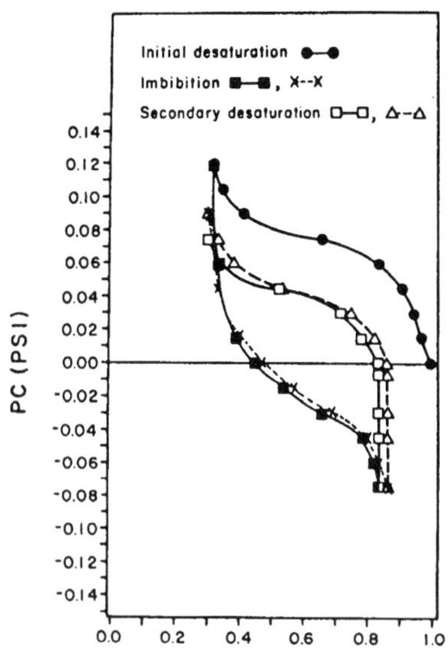

Figure 11. Capillary hysteresis loops before (solid line) and after injection of microorganisms (dashed line) for experiment #6 in oil-wet matrix.

injection of C. acetobutylicum. The capillary pressure/brine saturation relationships before and after microbial treatment are shown in Fig. 11. The residual oil saturation and wettability number before microbial treatment were 0.206 and -0.97, respectively.

A 0.8 ml volume of enriched Thioglycollate (THIO) medium containing $2.6x10^6$ cells/ml organisms was injected and incubated for 40 hours at 37°C. Gas and oil droplet productions were similar to those of experiment 5. The residual oil saturation decreased from 0.206 to 0.181 (-12.1%). The wettability number was increased from -0.97 to -0.378.

Microscopic analysis of the effluent at the end of all of the experiments showed that the organisms had the same morphology as those that were injected. This indicates that the flow cells were not contaminated by other indigenous bacteria during experiments. Residual oil saturations and wettability numbers for all the experiments are shown in Table 2.

Table 2. Residual oil saturations and wettability numbers for the experiments.

	$Swor_b$	$Swor_a$	RDR	W_b	W_a	RCW
Exp. 1_w	.20	.16	20%	.247	.285	15%
Exp. 2_o	.184	.14	24%	-.269	-.10	63%
Exp. 3_w	.157	.142	9.5%	.53	.481	9%
Exp. 4_o	.264	.24	9.1%	-.16	-.16	0%
Exp. 5_w	.163	.137	16%	.281	.332	18%
Exp. 6_o	.206	.181	12.1%	-.97	-.378	61%

w : Water wet system

o : Oil wet system

$Swor_b$: Residual oil saturation before injection of bacteria

$Swor_a$: Residual oil saturation after injection of bacteria

RD : Relative decrease in residual oil saturation

W_b : Wettability number before injection of bacteria

W_a : Wettability number after injection of bacteria

RCW : Relative change in wettability number

DISCUSSION OF RESULTS

The observed decrease in residual oil saturation in these experiments is believed to be a direct result of the in situ growth and metabolism of microorganisms.

Bacillus licheniformis strain JF-2 anaerobically produces biosurfactant and solvent in the form of an alcohol (13). Therefore, bacterial biosurfactant probably caused emulsification of the oil in experiments 1 and 2, forming oil droplets at the oil-brine-sand interfaces. It also has been shown by other researchers (27,28) that presence of non-ionic surfactants causes the emulsification of oil, forming small spheres. The biosurfactant produced by the bacteria reduced the interfacial tension between the oil and brine. the reduced interfacial tension reduces the capillary forces that trap residual oil in the porous media. The reduction of capillary forces free the oil droplets and allows the oil to be displaced and recovered(3).

Gas production by strain JF-2 was observed in experiments 1 and 2. The produced gas was undoubtedly an important factor in the oil recovery. The produced gas by this strain of microorganism is probably N_2 and CO_2, since these gases are predominantly produced by anaerobic bacteria in the presence of high nitrate concentrations. The gas production increased the pressure in the flow cell by the introduction of a third flowing phase, which further increased the mobility of the oil.

The wettability changes toward more water-wet systems probably contributed to reduction of the residual oil saturation, since oil in a water-wet system is less apt to cling to sand grain surfaces.

Visual observation indicated that the areal microscopic sweep efficiency improved after injection of microorganisms. The sweep efficiency improvements were mostly observed at the opposing corners of flow cell which was previously unswept and contained a larger oil saturation.

Production of gas, biosurfactant, other metabolic products, and the change of wettability contributed to a reduction of the residual oil saturation.

Production of metabolic products by Pseudomonas aeruginosa (9,10) in experiments 3 and 4 were less effective for oil production than the products produced by the other two species of bacteria that were used. The fact that this bacteria had negligible gas production, and no significant increase in the wettability of the sand matrix, also contributed to a lesser decrease in residual oil saturation.

Many references cite the advantageous use of gas-, and acid-, and/or solvent-producing bacteria, such as Clostridium acetobutylicum, in the microbial enhanced oil recovery process (19,20,25). By using this microorganism in experiments 5 and 6, it was clearly shown that this bacteria produces significant amounts of biogenic gas in the form of carbon dioxide. The gas produced increases the flow cell pressure significantly. Also, we believe that the carbon dioxide produced due to in situ growth of bacteria reacts with the brine in the flow cell, forming a weak carbonic acid. The carbonic acid and other produced metabolic products of the c. acetobutylicum, such as low molecular weight acids and solvents in the form of ketone and alcohols (19,20), together cause a removal of sludge and inorganic fine particles from the surface of the sand grains in the flow cell. As a result, the wettability of the system, especially the oil-wet sand, is notably increased to a more water-wet system.

The above factors, plus the fact that this bacteria also produced limited amounts of biosurfactant as indicated by the results, result in the favorable decrease in residual oil saturation.

Microscopic analysis of the effluent at the end of all of the experiments showed that the organisms had the same morphology as those that were injected. This indicates that the flow cells were not contaminated by other indigenous bacteria and lead us to believe that the decrease in residual oil saturations in these experiments was the result of the in situ growth and metabolism of the injected microorganisms.

This work showed the potential of microbial enhanced oil recovery, and further revealed some of the mechanisms of oil displacement by three species of bacteria.

CONCLUSIONS

1. The in situ growth and metabolism of injected bacteria decreased the residual oil saturation in the unconsolidated thin reservoir flow cells by 9-24%.

2. The indicated oil recovery increase is due to biosurfactant, biogas, and other products produced by the bacteria.

3. The in situ growth of the bacteria increased the water wettability of the matrix sand.

4. The in situ microbial growth mobilized residual oil by emulsifying the oil, wettability reversal, gas pressurization, and possibly by other mechanisms.

5. Of the three species of microorganisms used in these experiments, Bacillus licheniformis strain JF-2 showed the best results in the reduction of residual oil saturation and sweep efficiency observed qualitatively in the flow cells.

ACKNOWLEDGMENT

This paper was first presented as SPE Paper 15601 at the 61st Annual

Technical Conference of the Society of Petroleum Engineers, New Orleans, LA, October 5-8, 1986.

REFERENCES

1. Jang, L.K., Sharma, M.M., Findlay, J.E., Chang, P.W., and Yen, T.F.: "An Investigation of the Transport of Bacteria Through Porous Media", In, E.C. Donaldson and J.B. Clark, (editors) Proceedings of the 1982 International Conference on Microbial Enhancement of Oil Recovery, Nat. Tech. Inf. Service, Springfield, VA, CONF-8205140 (1983) 60-70.
2. Hitzman, D.O.: "Petroleum Microbiology and the History of Its Role in Enhanced Oil Recovery", In, E.C. Donaldson and J.B. Clark, (editors) Proceedings of the 1982 International Conference on Microbial Enhancement of Oil Recovery, Nat. Tech. Inf. Service, Springfield, VA CONF-8205140 (1983) 162-218.
3. VanPoolen, H.K. and Associate: Fundamentals of Enhanced Oil Recovery, PennWell Publishing Co., Tulsa, OK (1980) 58-111.
4. Gale, W.N. and Sandvik, E.I.: "Tertiary Surfactant Flooding: Petroleum Sulfonate Composition-Efficiency Studied", Soc. Pet. Eng. J. (Aug. 1973).
5. Wilson, S.B. and Zajic, J.E.: "Co-surfactant Production by an Anaerobic Salt Tolerant Species of Bacillus", In, J.E. Zajic and E.C. Donaldson (editors), Microbes and Oil Recovery, Volume 1, International Bioresources Journal, Bioresources Publications, El Paso, TX (1985) 321-329.
6. Singer, M.E.: "Microbial Biosurfactants", In, J.E. Zajic and E.C. Donaldson (editors), Microbes and Oil Recovery, Volume 1, International Bioresources Journal, Bioresources Publications, El Paso, TX (1985) 19-38.
7. Finnerty, W.K., and Singer, M.E.: "Microbiology and Enhanced Oil Recovery", Oil and Gas J., (1984) 4: 7-12.
8. Cooper, D.G., MacDonald, C.K., Duff, S.J.B., and Kosaric, N.: "Enhanced Production of Surfactin from Bacillus subtilis by Continuous Product Removal and Metalcation Additions", Appl. Environ. Microbial. (1981) 42: 408-412.
9. Gurerra-Santos, L., Kappeli, O., and Fiechter, A.: "Pseudomonas aeruginosa Biosurfactant Production in Continuous Culture with Glucose as Carbon Source", Appl. Environ. Microbial. (1984) 48: 301-305.
10. LaRiviere, J.W.M.: "The Production of Surface Active Compounds by Microorganisms and Its Possible Significance in Oil Recovery. I. Some General Observations on the Change of Surface Tension in Microbial Cultures", Antoine Van Leeuwenhock J. Microbial. Serol. (1955) 21: 1-8.
11. Cooper, D.G., Zajic, J.E., Gerson, D.F., and Manninen, K.I.: "Isolation and Identification of Biosurfactants Produced During Anaerobic Growth of Clostridium pasteurianum", J. Ferment. Technol. (1980) 58: 83-86.
12. Jenneman, G.E., McInerney, M.J., Knapp, R.M., Clark, J.B., Feero, J.M., Revus, D.E., and Menzie, D.E.: "A Halotolerant, Biosurfactant Producing Bacillus Species Potentially Useful for Enhanced Oil Recovery", Develop. Indust. Microbial. (1983) 24: 485-492.
13. Javaheri, M., Jenneman, G.E., McInerney, M.J., and Knapp, R.M.: "Anaerobic Production of A Bio-Surfactant by Bacillus Licheniformis JF-2", appl. Environ. Microbial. (1985) 50: 698-700.
14. Jenneman, G.E., Knapp, R.M., McInerney, M.J., Menzie, D.E., and Revus, D.E.: "Experimental Studies of In Situ Microbial Enhanced Oil Recovery", Soc. Pet. Engr. J. (Feb. 1984) 33-37.
15. Raiders, R.A., Freeman, D.C., Jenneman, G.E., Knapp, R.M., McInerney, M.J., and Menzie, D.E.: "The Use of Microorganisms to Increase the Recovery of Oil from Cores", Paper SPE 14336, presented at 60th Annual Technical Conference and Exhibition of SPE in Las Vegas, NV, Sept. 22-25, 1985.

16. Jenneman, G.E., McInerney, M.F., and Knapp, R.M.: "Microbial Penetration Through Nutrient Saturated Berea Sandstone", Appl. Environ. Microbial. (Aug. 1985) 50: 383-391.
17. Torbati, H.M., Donaldson, E.C., Jenneman, G.E., Knapp, R.M., McInerney, M.J., Menzie, D.E.: "Depth of Microbial Plugging and Its Effect on Pore Size Distribution in Berea Sandstone Cores", In, J.E. Zajic and E.C. Donaldson (editors) Microbes and Oil Recovery, Volume 1, International Bioresources Journal, Bioresources Publications, El Paso, TX (1985) 213-225.
18. Jang, L.K., Sharma, M.M. and Yen, T.F.: "The Transport of Bacteria in Porous Media and Its Significance in Microbial Enhanced Oil Recovery", Paper SPE 12770, presented at the 1984 California Regional Mtg., Long Beach, Ca, April 11-13.
19. Grula, E.A., Russell, H.H., Bryant, D., Kenaga, M. and Hart, M.: "Isolation and Screening of Clostridia for Possible Use in Microbially Enhanced Oil Recovery", In E.C. Donaldson and J.B. Clark, (editors) Proceedings of the 1982 International Conf. on Microbial Enhancement of Oil Recovery, Nat. Tech. Inf. Service, Springfield, VA, CONF-8205140 (1983) 43-47.
20. Donaldson, E.C.: "Use of Capillary Pressure Curves for Analysis of Production Well", Paper SPE 13809, presented at the 1985 Production Operation Symposium, Okla. City, OK, March 10-12.
21. Leverett, M.C.: "Capillary Behavior in Porous Solids", Trans. AIME (1940) 152-169.
22. Collins, R.E.: Flow of Fluids Through Porous Materials, The Petroleum Publishing Co., Tulsa, OK (1976) 23-33.
23. Ahsan, S.S.: "A Microscopic Study of Fluid Flow in Porous Media (Capillary Pressure Aspects)", University of Oklahoma Masters Thesis, 1985.
24. Dowell, V.R., Jr., Lombard, G.L., Thompson, F.S., and Armfield, A.Y.: Media for Isolation, Characterization, and Identification of Obligately Anaerobic Bacteria, U.S. Dept. of Health, Education and Welfare, Atlanta, GA (1977) 38.
25. Yarbrough, H.F. and Coty, V.F.: "Microbially Enhanced Oil Recovery from the Upper Cretaceous Nacatock Formations, Union County, AR", In, E.C. Donaldson and J.B. Clark, (editors) Proceedings of the 1982 International Conf. on Microbial Enhancement of Oil Recovery, Nat. Tech. Inf. Service, Springfield, VA, CONF-8205140 (1983) 149-153.
26. Donaldson, E.C., Thomas, R.D., and Lorenz, P.B.: "Review of Petroleum Oil Saturation and Its Determination", Soc. Pet. Eng. J., (December 1969).
27. Donaldson, E.C., Thomas, R.D.: "Microscopic Observations of Oil Displacement in Water-Wet and Oil-Wet System", Paper SPE 3555, presented at the 1971 SPE Annual Fall Mtg., New Orleans, LA, Oct. 3-6.
28. Donaldson, E.C., and Crocker, M.E.: "Characterization of the Crude Oil Polar Compound Extract", DOE/BETC/RI 80-5, NTIS, Springfield, VA (1980) 1-27.

Reprinted from: *Microbial Enhancement of Oil Recovery - Recent Advances*, edited by E.C. Donaldson

CH. R–6

ISOLATION AND CHARACTERIZATION OF ANAEROBIC HALOPHILIC BACTERIA FROM OIL RESERVOIR BRINES

D. GEVERTZ, J.R. PATEREK, M.E. DAVEY and W.A. WOOD
Salk Institute Biotechnology/Industrial Associates, Inc. (SIBIA), P.O. Box 85200, San Diego, CA 92186.

ABSTRACT

Three metabolic types of obligately anaerobic, halophilic bacteria were isolated from oil reservoir brines and characterized to better understand the ecology of an oil reservoir. These included a methanogenic bacterium, a sulfate-reducing bacterium, and a fermentative bacterium. All species were isolated by plating from enrichment cultures containing substrates specific for each metabolic type. The methanogenic bacterium was isolated by enrichment with trimethylamine. Methane production occurred only with trimethylamine compounds or methanol as substrates. Sodium, magnesium, and potassium were all required for growth. This organism appears to be a member of the genus *Methanohalophilus*, based on substrate utilization and general growth characteristics. The sulfate-reducing bacterium was isolated by enrichment with a lactate-sulfate medium containing 3% NaCl. This isolate utilized lactate as an electron donor for sulfate reduction and, in addition, contained desulfovirdin, typical of the genus *Desulfovibrio*. In contrast to the methanogen, only sodium was required for growth. Finally, the sheathed bacterium was isolated by enrichment with glucose. Sodium was also required by this isolate for growth. This organism is a fermentative bacterium surrounded by a sheath, and is capable of reducing elemental sulfur to hydrogen sulfide. These are all characteristics of the genus *Thermotoga*.

INTRODUCTION

Conventional waterflooding techniques used in secondary oil recovery leave a large percentage of the oil still in place (ca. 60-70%) (Mungan and Johansen, 1978). As waterflooding progresses, the ratio of oil to water in produced water decreases because water preferentially sweeps zones of highest permeability in the reservoir, leaving oil behind in the low permeability zones. Enhanced oil recovery (EOR) techniques were developed for use after waterflooding in the final, or tertiary phase of oil recovery (Morrow and Heller, 1985). Some of these processes employ steam, polymers, or surfactants to help recover residual oil. Microbial enhanced oil recovery (MEOR) uses bacteria and their metabolic products to achieve this goal. MEOR can be carried out by several means (Brown *et al.*, 1986; Finnerty and Singer, 1983). Microorganisms can be grown in fermentors and their products, such as emulsifiers, solvents, polymers or surfactants, can be injected into a reservoir to aid in oil recovery. Secondly, indigenous or injected cells can be stimulated by the injection of nutrients, generally in the form of carbohydrate, to grow and produce products *in situ*. Alternately, bacteria capable of using components present in crude oil are injected, limiting the need for nutrient supplementation.

The ability of bacteria to survive and grow under reservoir conditions is a prerequisite for a successful MEOR process. A multitude of extreme environments may be present in oil reservoirs with conditions such as anaerobiosis, high pressure, high temperature and high salinity, a broad spectrum of pH values, low nutrient concentrations, and the presence of potentially toxic compounds. A review of reservoir conditions in the top ten oil-producing states revealed that approximately one-fourth of reservoirs have acceptable conditions for potential MEOR applications (Clark *et al.*, 1981). The criteria for selection of suitable conditions included temperatures below 75°C, pH values near neutrality, and salt concentrations of less than ten percent. These environmental conditions have led to the suggestion that halotolerant, thermophilic bacteria capable of anaerobic growth are good candidates for an MEOR process (McInerney, 1983).

The indigenous bacteria present in an oil reservoir are an important element in an MEOR process for at least two reasons. First, they may be stimulated directly by the addition of nutrients and, second, they may interact with exogenous cells injected into the formation. Therefore, it is essential to understand the microflora of the reservoir so as to predict how this population will respond to perturbation by nutrients and/or other bacteria. Some of the organisms that have been identified from oil reservoir brines include sulfate-reducing bacteria (Zobell, 1958), pseudomonads (Ilzuka and Komagata, 1964), methanogens (Belyaev *et al.*, 1983), and *Bacillus* species (Davis, 1967; Pfiffner *et al.*, 1985).

In this study, we have isolated and characterized three nonsporulating anaerobic, halophilic bacteria from oil field brines. This information adds to our understanding of the microbiology of the oil reservoir. These organisms belong to the genera *Methanohalophilus*, *Desulfovibrio*, and, possibly, *Thermotoga*, and are believed to be part of the natural population present at this site. Other species belonging to these genera have been isolated from marine and hypersaline environments worldwide.

MATERIALS AND METHODS

Enrichment cultures

Field brine was collected from the North Burbank unit in northeastern Oklahoma in sterile evacuated bottles under anaerobic conditions. Brine enrichments were made by the addition of field brine to concentrated nutrient solutions for minimal dilution of the brine. All enrichments were set up under anaerobic conditions in stoppered 100 ml serum bottles and incubated at 35-45°C.

Anaerobic techniques

Modified Hungate technique (Hungate, 1969), or an anaerobic chamber (Coy Laboratory Products, Ann Arbor, Michigan) was used for all media preparation as well as for isolation and cultivation of bacteria.

Media

The medium used for primary isolation of methanogenic bacteria was made in a field brine base and contained the following (in g/l): NH_4Cl, 0.5; yeast extract, 2.0; trypticase peptone, 2.0; casamino acids, 2.0; K_2HPO_4, 0.2; cysteine·HCl, 0.25; sodium bicarbonate, 2.0; sodium carbonate, 2.0; and trimethylamine·HCl, 3.0. The medium was adjusted to pH 6.8-7.0 with KOH. Sodium sulfide, 0.25 g/l, was used as a reductant. The gas phase was nitrogen/carbon dioxide (80:20). "BBM" medium for isolation of fermentative bacteria contained the following (in g/l): NaCl, 66.6; $CaCl_2$, 15.3; $MgCl_2 \cdot 2H_2O$, 5.1; KNO_3, 1.0; yeast extract, 0.5; NH_4SO_4, 0.13; sucrose, 4.0; PIPES [piperazine-N,N'-bis-(2-ethanesulfonic acid)] sodium salt monohydrate, 8.5; α-glycerol phosphate, 0.4; and trace element solution, 1.0 ml (Widdel and Pfennig, 1984). The medium was adjusted to pH 6.8 with NaOH. Cysteine·HCl, 0.2 g/l, was used as a reductant. Medium designated "MMS" was used to test substrate utilization and antibiotic sensitivity of fermentative isolates (Huber et al., 1986). Postgate's Medium C, with minor modifications and containing 3% NaCl, was used for the isolation of sulfate-reducing bacteria; sulfate-free growth was tested using Postgate's Medium D (Postgate, 1984a). Oxidation of short-chain fatty acids, palmitate, and benzoate coupled to the reduction of sulfate was tested using a modification of the media by Fowler et al.(1986). Hydrogen utilization was tested with the medium of Badziong and Thauer (1978). All media contained 1 mg/liter of resazurin as a redox indicator.

Growth studies

Growth studies were performed in Balch tubes or serum vials containing 5-10 ml of medium. Each tube was inoculated from a liquid culture of the isolated organism. Growth was monitored by increase in absorbance at 550-600 nm. In the case of the methanogen, growth was also monitored microscopically by fluorescence at 420 nm and by production of methane.

Microscopy

Routine observations on isolates were made with an Olympus BH-2 phase microscope equipped with a fluorescence attachment. Electron micrographs were taken with a Philips 300 or CM-10 transmission electron microscope. Thin sections were prepared by fixing cells in 3% glutaraldehyde and 1% paraformaldehyde, followed by osmication and embedding. Negatively stained samples were prepared by allowing the cells to attach to a freshly ionized carbon formbar grid, followed by a brief wash in double distilled water, and staining with 2% uranyl acetate.

Chemicals

Vancomycin, streptomycin, chloramphenicol, rifampicin, Coenzyme M, and bromoethanesulfonic acid (BES) were purchased from Sigma Chemical Company (St. Louis, MO). Clindamycin and cephalothin were purchased from Fluka Chemical Co. (Ronkonkoma, NY).

Gas chromatography

Methane production was monitored using a Hewlett Packard gas chromatograph model 5890 equipped with a thermo-conductivity detector. A carbosieve S11 column was used for separation of the gases. Helium was used as a carrier gas. The oven temperature was held at 35°C for seven minutes and then increased at a rate of 32°C per minute to a final temperature of 225°C, after 12 minutes. The injector temperature was 110°C, and the detector 230°C.

RESULTS

Description of site and composition of reservoir brine

Reservoir brine was collected from the tank battery of Tract 5 in the North Burbank unit operated by Phillips Petroleum Company (Bartlesville, OK). This field is a sandstone formation located in northeastern Oklahoma with an average temperature of 45°C and a depth of 3,000 feet. Brine was collected after it passed through the skimmer tank, just prior to reinjection. The brine had a total dissolved solids content of approximately 10% and a pH of 6.2. Chemical analysis showed that the major cations in the brine were: sodium, 3.0%; calcium, 0.6%; magnesium, 0.12%; and barium, 0.07%. The major anion was chloride, at a concentration of 6.5%.

Preparation of brine enrichments

Enrichments were prepared by adding 100 ml of reservoir brine to serum bottles containing nutrients either as dry solids or concentrated solutions. Various types and combinations of nutrients were utilized in order to examine the microbial population under differing conditions. The addition of nutrient generally resulted in an increase in cell number from 10^3 cells per ml of brine to 10^8 cells per ml. Dominant members of the enrichment cultures were isolated by plating on media of the same general composition as that used for enrichment. Pure cultures of organisms were obtained by successive transfers on solid medium or in roll tubes. A methanogenic bacterium, a sulfate-reducing bacterium, and a fermentative isolate were subsequently identified by microbiological and biochemical characterization.

Isolation and characterization of microorganisms

Methanohalophilus species. The methanogenic isolate (designated BT-5), is a nonmotile, irregular coccus 0.8-1.2 μm in diameter, as measured by phase microscopy. An electron micrograph of this organism, shown in Figure 1, revealed cells of irregular size and shape. The cells stain gram negative and have a tendency to form clumps. Colonies in roll tubes were raised, circular and smooth, with a light brown to brownish-green color.

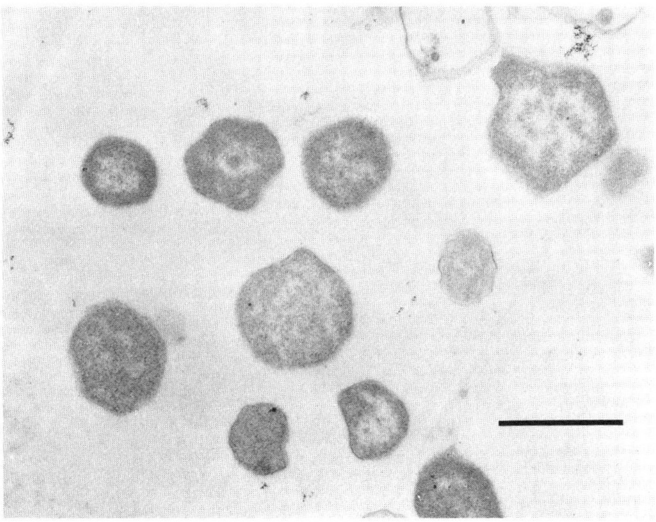

Fig. 1. Thin section of the methanogenic isolate BT-5. Bar, 1 μm.

Some of the distinguishing characteristics of isolate BT-5 are shown in Table I, along with those of other type species. Methanol and methylamines were used by this organism as substrates for methanogenesis. The traditional substrates for methane formation, hydrogen and carbon dioxide, acetate, and formate, were not utilized by this isolate. Bromoethanesulfonic acid (BES), a metabolic inhibitor (Gunsalus *et al.*, 1978), and tetracycline both inhibited growth. This organism was insensitive to penicillin, D-cycloserine, erythromycin, clindamycin, and cephalothin. These observations are consistent with the archeobacterial pattern of sensitivity to antibiotics. Lysis of cells occurred in distilled water, sodium dodecyl sulfate, or Triton X-100.

Sodium, potassium and calcium ions were all needed for growth. The sodium chloride range for this organism was 1.5-23%, with the optimum occurring at 11.6-14.5%. The temperature range for growth was 20-60°C, with an optimum at 40°C. The optimum pH for this isolate was 6.8. Isolate BT-5 has characteristics similar to *M. mahii* obtained from the Great Salt Lake (Paterek and Smith, 1988) as shown in Table I.

TABLE I

Comparison of BT-5 Isolate to Members of the Genus Methanohalophilus

Characteristic	BT-5	*M. mahii*[a]	*M. zhilinae*[b]
Gram reaction	negative	negative	negative
Morphology	coccus	coccus	coccus
Substrates:			
methylamines, methanol	+	+	+
H_2/CO_2, formate acetate	-	-	-
pH optimum	6.8	7.5	9.2
Temperature range	20-60°C	20-50°C	ND
Optimum temp.	40°C	35°C	45°C
Salt range (NaCl)	1.5-23%	1.5-23%	1.2-12.2%
Optimum salt conc.	11.6-14.5%	11.6%	4.1%
Inhibitors			
BES	+	+	ND
tetracycline	+	+	+
others		cephalothin	chlormaphenicol

BES, bromoethanesulfonic acid
ND, not determined
+, positive response or reaction
-, negative response or reaction

[a] Paterek and Smith, 1985; 1988.
[b] Mathrani *et al.*, 1988.

Thermotoga-like species. The fermentative isolate, designated T5, was morphologically distinct because the cell was surrounded by a sheath or "toga" structure (Huber et al., 1986). Isolate T5 was variable in length (5-20 μm) and 1.0 μm in diameter, as measured by phase microscopy. The electron micrograph in Figure 2 shows the sheath-like structure surrounding the cell. Isolate T5 exhibited diverse morphologies with regard to cell size and amount of sheath when grown at varying salt concentrations. Cells appear rod-shaped, either singly or in short chains, and stain gram negative. This isolate formed smooth, raised, white colonies on agar plates.

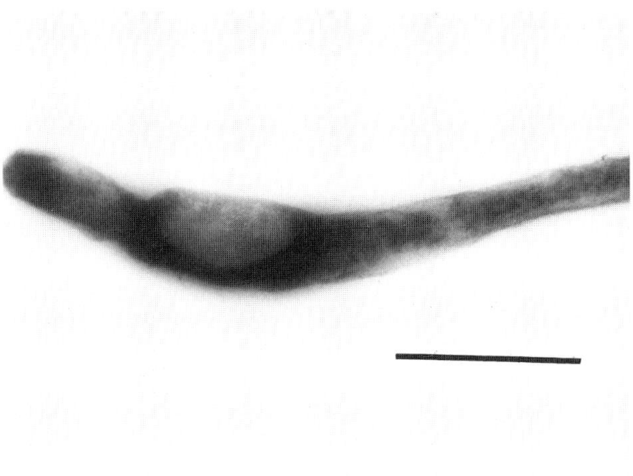

Fig. 2. Negative stain of the fermentative isolate T-5. Bar, 1 μm.

Table II compares the physiological characteristics of isolate T5 with members of the genus Thermotoga. Different carbohydrates were utilized as an energy source by this organism. Approximately 0.1% yeast extract was required for growth. Elemental sulfur was facultatively reduced to hydrogen sulfide. Growth was inhibited by rifampicin at a concentration of 100 μg/ml.

Sodium chloride was needed for growth, in the range of 0.25-10%, with an optimum of 2.5%. The temperature range for growth was 30-55°C, with an optimum of 50°C. This isolate possesses most of the characteristics of the genus Thermotoga listed in Table II.

TABLE II

Comparison of T5 Isolate with Members of the Genus *Thermotoga*

Characteristic	T5	*T. maritima*[a]	*T. neapolitana*[b,c]	*T. thermarum*[c]
Gram reaction	negative	negative	negative	negative
Sheath-like structure	present	present	present	present
Motility	+	+	+/-	+
Carbohydrates fermented	+	+	+	+
Obligate anaerobe	+	+	+	+
Elemental sulfur reduced	+	+	+	-
Temperature range	30-55°C	55-90°C	55-90°C	55-84°C
Optimum temperature	50°C	80°C	80°C	80°C
Salt range (NaCl)	0.25-10%	0.24-3.75%	ND	0.2-0.55%
Optimum salt conc.	2.5%	2.7%	ND	0.35%
Sensitivity to rifampicin (100 μg/ml)	+	-	-	+

ND, not determined
+, positive response or reaction
-, negative response or reaction

[a] Huber *et al.*, 1986.
[b] Jannasch *et al.*, 1988.
[c] Windberger *et al.*, 1989.

Desulfovibrio species. The sulfate-reducing bacterium, designated SSR, is a motile, curved rod, 2-5 μm in length and 1 μm in diameter, as measured by phase microscopy. An electron micrograph of this organism, shown in Figure 3, revealed comma-shaped cells with polar flagella. These organisms occur either singly or in pairs and stain gram negative. Raised, black colonies were formed on agar plates containing ferrous iron.

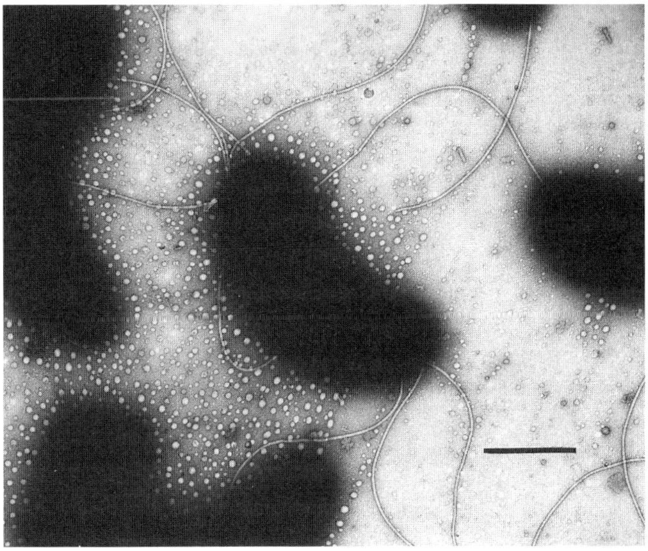

Fig. 3. Negative stain of the sulfate-reducing isolate SSR. Bar, 1 μm.

Table III lists the characteristics of isolate SSR, along with those of other halophilic species of Desulfovibrio. Hydrogen, formate, and lactate were utilized as electron donors for sulfate reduction. Acetate, ethanol, propionate, butyrate, palmitate, and benzoate were not used as electron donors, and no growth was observed with these substrates. Growth in sulfate-free media occurred with pyruvate and fumarate. An increase in the viscosity of the culture over time was apparent in media containing yeast extract, which is typical of salt-requiring strains of Desulfovibrio (Ochynski and Postgate, 1963). The pigment desulfovirdin was present in this organism, as evidenced by fluorescence of concentrated cells under ultraviolet light after the addition of 1 N sodium hydroxide (Postgate, 1984a).

Sodium chloride, in the range of 0.5-15%, was required for the growth of this organism. The optimum sodium chloride concentration ranged from 2-6%. Potassium chloride could not substitute for sodium chloride, indicating that isolate SSR is not a strain of D. salexigens (Postgate, 1984b). The temperature range for growth was 25-45°C, with an optimum of 35-40°C.

TABLE III

Comparison of SSR Isolate to Other Halophilic Strains of *Desulfovibrio*

Characteristic	SSR	*D. desulfuricans*[a,b] (*aestuarii*)	*D. salexigens*[a,b]
Gram reaction	negative	negative	negative
Morphology	vibrio	vibrio	vibrio
Desulfovirdin	+	+	+
Electron donors for sulfate reduction:			
hydrogen, formate, lactate	+	+	+
ethanol	-	+	+
acetate, butyrate, palmitate	-	-	-
benzoate	-	-	ND
Temperature range	25-45°C	25-45°C	25-45°C
Temperature optimum	40°C	30-36°C	30-36°C
Optimum salt conc. (NaCl)	2-6%	2.5%	2.5-5%
Ion requirement	Na^+	ND	Cl^-

ND, not determined
+, positive response or reaction
-, negative response or reaction

[a] Postgate, 1984b.
[b] Widdel, 1988.

DISCUSSION

Three physiological types of bacteria were isolated from reservoir brines. The methanogenic isolate, BT-5, was obtained by enrichment with trimethylamine, a typical substrate for halophilic methanogens. Halophilic methanogens comprise a unique group because they do not use the traditional substrates for methanogenesis, *i.e.*, hydrogen/carbon dioxide, acetate, and formate. Instead, these methanogens generate methane from methanol or N-methyl compounds (Mathrani *et al.*, 1988; Paterek and Smith, 1988). The only recognized genus of halophilic methanogens, *Methanohalophilus*, contains two members. Table I compares the characteristics of *M. zhilinae* (Mathrani *et al.*, 1988), *M. mahii* (Paterek and Smith, 1988), and isolate BT-5. *M. zhilinae* was isolated from an alkaline hypersaline lake, and is distinguished from the other two organisms by its pH optimum of 9.2. Isolate BT-5 closely resembles M. mahii, isolated from the Great Salt Lake. Methanogenesis has been studied in a number of hypersaline lakes (Giani et al., 1984; Oremland et al., 1982; Paterek and Smith, 1985) and resulted in the isolation of several halophilic species.

The fermentative sheathed isolate from brine closely resembles members of the genus Thermotoga. T. maritima, the first described member of this genus, was isolated from geothermally heated areas of marine sediments in Italy and the Azores (Huber et al., 1986). Cells are characterized by the presence of a sheath-like structure and an extreme temperature requirement, the highest reported for eubacteria. Two additional species, T. neapolitana (Jannasch et al., 1988) and T. thermarum (Windberger et al., 1989) differ from T. maritima in physiological properties, but still retain the sheath-like structure and a high temperature requirement. Isolate T5 resembles these species both morphologically and physiologically, although it is capable of growth at mesophilic temperatures. Another possibility is that this isolate belongs to the genus Thermosipho, a new genus within the Thermotogales (Huber et al., 1989). One distinguishing characteristic of Thermosipho is growth at lower temperatures than Thermotoga. Further characterization of the T5 isolate by genetic techniques, such as DNA homology to other species and 16S ribosomal RNA sequencing, will establish its relationship to these genera. Reservoir brines would constitute a new environment for isolation of Thermotoga or Thermosipho.

Sulfate-reducing bacteria have long been associated with oil deposits (Davis, 1967; Zobell, 1958). Their presence can have a serious economic impact on the oil industry, causing problems such as corrosion of pipes, plugging of injection wells and souring of fuel gas and oil (Cord-Ruwisch et al., 1987; Iverson and Olsen, 1984). The sulfate-reducing bacterium described here is a member of the genus Desulfovibrio and, based on its general characteristics, is probably a salt-requiring strain of D. desulfuricans (Postgate, 1984b). Sulfate reduction has been reported to occur in several hypersaline environments such as the Dead Sea (Truper, 1969), Big Soda Lake (Smith and Oremland, 1987) and solar salt ponds (Klug et al., 1985).

Carbohydrates, such as molasses, have traditionally been used as substrates for MEOR processes; initiation of growth depends on metabolism of carbohydrate by fermentative bacteria. The Thermotoga-like isolate, T5, is capable of using carbohydrates directly, and is undoubtedly one of a number of fermentative bacteria present in reservoir brines. Metabolism of carbohydrate will result in the production of fermentation end products and other types of metabolic by-products. One class of by-products produced in response to hypersaline conditions are osmoregulants, compounds used to maintain internal osmotic balance (Le Rudulier and Bouillard, 1983). Osmoregulants such as betaine and choline can, in turn, be broken down to methylamines, which serve as substrates for methanogenesis (King, 1984a). The end products of carbohydrate metabolism, for example, hydrogen and acetate, can be utilized as electron donors for sulfate reduction, giving rise to the growth of sulfate-reducing bacteria (Widdel, 1988). In many ecosystems, methanogenesis and sulfate reduction are competing processes for the same substrates; namely, hydrogen and acetate (Nedwell and Banat, 1981; Oremland and Taylor, 1978; Schönheit *et al.*, 1982). Hypersaline environments present a situation in which different substrates give rise to methane and hydrogen sulfide and, consequently, these metabolic types are not competitive (King, 1984b; Oremland *et al.*, 1982). Surprisingly, sulfide is generally not produced in our carbohydrate enrichments, perhaps because of limited sulfate in the brine, exhaustion of other nutrients by fermentative bacteria, competition for substrates with other organisms, or the buildup of toxic end products. These interactions illustrate the type of community metabolism which will likely take place in an MEOR process upon the addition of nutrients, and underscores the need for understanding the microbiology of the oil reservoir before designing a MEOR process. Patterns of how the indigenous population responds to nutrient stimulation can then be established, so as to predict and control their behavior.

The reservoir conditions at this site dictate that bacteria must be able to tolerate, survive, and grow anaerobically at salinities of up to 10% and at temperatures of approximately 45°C. The three organisms discussed here are capable of proliferating under these conditions. These organisms are readily isolated from reservoir brine at this site and have been identified in brines from other locations in the Burbank field. We have isolated another *Thermotoga*-like organism from reservoir brine collected from a site in Texas as well. This suggests that these organisms were selected by, or adapted to, this extreme environment. Since these organisms are easily cultured on synthetic media in the laboratory, it raises questions about the source and concentration of nutrients present in the reservoir environment. Based on the composition of this brine, nutrients are most likely present at low concentrations. One form of adaptation to such nutrient deprivation is the formation of ultramicrobacteria, which are metabolically quiescent (Novitsky and Morita, 1978; Roszak and Coldwell, 1987). The ability to survive under these conditions would indicate that the organisms best suited for MEOR are those already present in the reservoir. Screening programs have been devised based on this concept, *e.g.*, examining bacteria obtained from brine and evaluating their ability to respond to nutrients and to produce exopolymers, surfactants and other products beneficial to MEOR (Grula *et al.*, 1982; Lazar, 1983; Pfiffner et al., 1986). A successful MEOR process will ultimately depend on a sound understanding of microbial physiology and ecology.

ACKNOWLEDGEMENTS

We acknowledge Michael McCaffery and Lance Washington at the University of California at San Diego for preparing electron micrographs of our isolates. This work was funded by Phillips Petroleum Company, Bartlesville, Oklahoma.

REFERENCES

Badziong, W. and Thauer, R.K., 1978. Growth yields and growth rates of Desulfovibrio vulgaris (Marburg) growing on hydrogen plus sulfate and hydrogen plus thiosulfate as the sole energy sources. Arch. Microbiol., 117: 209-214.

Belyaev, S.S., Wolkin, R., Zenealy, W.R., DeNiro, M.J., Epstein, S. and Zeikus, J.G., 1983. Methanogenic bacteria from the Bondyuzhskoe oil field: general characterization and analysis of stable-carbon isotopic fractionation. Appl. Environ. Microbiol., 45: 691-697.

Brown, M.J., Moses, V., Robinson, J.P. and Springham, D.G., 1986. Microbial enhanced oil recovery: progress and prospects. Crit. Rev. Biotech., 3: 159-197.

Clark, J.B., Munnecke, D.M. and Jenneman, G.E., 1981. In situ microbial enhancement of oil production. Devel. Indust. Microbiol. 22: 695-701.

Cord-Ruwisch, R., Kleinitz, W. and Widdel, F., 1987. Sulfate-Reducing bacteria and their activities in oil production. J. Pet. Tech., January: 97-106.

Davis, J.B., 1987. Microbes in injection waters. Petroleum Microbiology, Elsevier Publishing Co., New York, New York, pp. 449-470.

Finnerty, W.R., and Singer, M.E., 1983. Microbial enhancement of oil recovery. Biotechnology, 1: 47-54.

Fowler, V.J., Widdel, F., Pfenning, N., Woese, C.R. and Stackebrandt, E., 1986. Phylogenetic relationships of sulfate- and sulfur-reducing eubacteria. Sys. Appl. Microbiol., 8: 32-41.

Giani, D., Giani, L., Cohen, Y. and Krumbein, W.E., 1984. Methanogenesis in the hypersaline Solar Lake (Sinai). FEMS Microbiol. Lett., 25: 219-224.

Grula, E.A., Russell, H.H., Bryant, D., Kenaga, M. and Hart, M., 1983. Isolation and screening of Clostridia for possible use in microbial enhanced oil recovery. International Conference on the Microbial Enhancement of Oil Recovery. In: E.C. Donaldson and J.B. Clark (Editors), Proceedings, U.S. Department of Energy, Bartlesville Energy Technology Center, Bartlesville, Oklahoma, pp. 43-47.

Gunsalus, R.P., Romesser, J.A. and Wolfe, R.S., 1978. Preparation of Coenzyme M analogues and their activity in the methyl Coenzyme M reductase system of Methanobacterium thermoautotrophicum. Biochemistry, 17: 2374-2377.

Huber, R., Langworthy, T.A., König, H., Thomm, M., Woese, C.R., Sleytr, U.B. and Stetter, K.O, 1986. Thermotoga maritima sp. nov. represents a new genus of unique extremely thermophilic eubacteria growing up to 90°C. Arch. Microbiol., 144: 324-333.

Huber, R., Woese, C.R., Langworthy, T.A., Fricke, H. and Stetter, K.O., 1989. Thermosipho-Africanis new-genus new-species represents a new genus of thermophilic eubacteria within the Thermotogales. Sys. Appl. Microbiol., 12: 32-37.

Hungate, R.E., 1969. A roll tube method for the cultivation of strict anaerobes. Methods Microbiol., 3: 117-132.

Iizuka, H. and Komagata, K, 1964. Microbiological studies on petroleum and natural gas. II. Determination of pseudomonads isolated from oil-brines and related materials. J. Gen. Appl. Microbiol., 10: 223-231.
Iverson, W.P., and Olsen, G.J., 1984. Problems related to the sulfate-reducing bacteria in the petroleum industry. In: R.M. Atlas (Editor), Petroleum Microbiology, MacMillan Publishing Co., New York, New York, pp. 619-641.
Jannasch, H.W., Huber, R., Belkin, S and Stetter, K.O., 1988. Thermotoga neapolitana sp. nov. of the extremely thermophilic, eubacterial genus Thermotoga. Arch. Microbiol., 150: 103-104.
King, G.M., 1984a. Metabolism of trimethylamine, choline and glycine betaine by sulfate-reducing and methanogenic bacteria in marine sediments. Appl. Environ. Microbiol., 48: 719-725.
King, G.M., 1984b. Utilization of hydrogen, acetate and "non-competitive" substrates by methanogenic bacteria in marine sediments. Geomicrobiol. J., 3: 275-306.
Klug, M., Boston, P., Francois, R., Gyure, R., Javor, B., Tribble, G. and Vairavamurthy, A., 1985. Sulfur reduction in sediments of marine and evaporite environments. In: D. Sagan (Editor), The Global Sulfur Cycle, NASA Technical Memorandum 87570, July (1985), pp. 128-157.
Lazar, I., 1983. Some characteristics of the bacterial inoculum used for oil release from reservoirs. In: J.E. Zajic, D.C. Cooper, T.R. Jack and N. Kosaric (Editors), Microbial Enhanced Oil Recovery, PennWell Publishing Co., Tulsa, Oklahoma, pp. 73-82.
Le Rudulier, D. and Bouillard, L., 1983. Glycine betaine, an osmotic effector in Klebsiella pneumonia and other members of the Enterobacteriaceae. Appl. Environ. Microbiol., 46: 152-159.
Mathrani, I.M., Boone, D.R., Mah, R.A., Fox, G.E. and Lau, P.P., 1988. Methanohalophilus zhilinae sp. nov., alkaliphilic, halophilic, methylotrophic, methanogen; Int. J. Sys. Bacteriol., 38: 139-142.
McInerney, M.J., 1983. Physiological types of microorganisms useful for enhanced oil recovery. In: E.C. Donaldson and J.B. Clark (Editors), Proceedings of the 1982 International Symposium on Microbial Enhanced Oil Recovery. Department of Energy, Bartlesville, Oklahoma, pp. 38-42.
Morrow, N.R. and Heller, J.P., 1985. Fundamentals of enhanced oil recovery. In: E.C. Donaldson, G.V. Chilingarian and T.F. Yen (Editors), Enhanced Oil Recovery, 1. Fundamentals and Analysis, Elsevier, Amsterdam, The Netherlands, pp. 47-74.
Mungan, N., and Johansen, R.T., 1978. Miscible Fluid Displacement. In: D.C. Bond (Editor), Secondary and Tertiary Oil Recovery Processes, The Interstate Oil Compact Commission, Oklahoma City, Oklahoma, pp. 40-66.
Nedwell, D.R. and Banat, I.M., 1981. Hydrogen as an electron donor for sulfate-reducing bacteria in slurries of salt marsh sediment. Microb. Ecol., 7: 305-313.
Novitsky, J.A. and Morita, R.Y., 1978. Possible strategy for the survival of marine bacteria under starvation conditions. Mar. Biol., 48: 289-295.
Ochynski, F.W. and Postgate, J.R., 1963. Some biological differences between fresh water and salt water strains of sulphate-reducing bacteria. In: C.H. Oppenheimer (Editor), Marine Microbiology, C.C. Thomas, Springfield, Illinois, pp. 426-441.
Oremland, R.S., Marsh, L.M. and DesMarais, D.J., 1982. Methanogenesis in Big Soda Lake: an alkaline, moderately hypersaline desert lake. Appl. Environ. Microbiol., 43: 462-468.
Oremland, R.S., Marsh, L.M. and Polcin, S., 1982. Methane production and simultaneous sulphate reduction in anoxic, salt marsh sediments. Nature, 296: 143-145.
Oremland, R.S. and Taylor, B.F., 1978. Sulfate reduction and methanogenesis in marine sediments. Geochim. Cosmochim. Acta, 42: 209-214.
Paterek, J.R. and Smith, P.H., 1985. Isolation and characterization of a halophilic methanogen from Great Salt Lake. Appl. Environ. Microbiol., 50: 877-881.

Paterek, J.R. and Smith, P.H., 1988. Methanohalophilus mahii [gen. nov., sp. nov.] a methylotrophic halophilic methanogen. Int. J. Sys. Bacteriol., 38: 122-123

Pfiffner, S.M., Jenneman, G.E., Walker, Jr., G.B., Javaheri, M., McInerney, M.J. and Knapp, R.M., 1985. Isolation and characterization of halotolerant, thermophilic, anaerobic, biopolymer-producing and biosurfactant-producing bacteria. In: E.J. Zajic and E.C. Donaldson (Editors), Microbes and Oil Recovery (Vol. 1), International Bioresources Journal, pp. 285-294.

Pfiffner, S.M., McInerney, M.J., Jenneman, G.E. and Knapp, R.M., 1986. Isolation of halotolerant, thermotolerant, facultative polymer-producing bacteria and characterization of exopolymer. Appl. Environ. Microbiol., 51: 1224-1229.

Postgate, J.R., 1984a. The Sulfate-Reducing Bacteria, Second Edition, Cambridge, University Press, Great Britain, pp. 1-50.

Postgate, J.R., 1984b. Genus Desulfovibrio. In: N.R. Kreig and J.G. Holt (Editors), Bergey's manual of systematic bacteriology, Vol. 1, Williams and Wilkins, Baltimore, Maryland, pp. 666-672.

Roszak, D.B. and Coldwell, R.R., 1987. Survival strategies of bacteria in the natural environment. Microbiol. Revs., 51: 365-379.

Schönheit, P., Kristjansson, J.K., and Thauer, R.K., 1982. Kinetic mechanism for the ability of sulfate reducers to out-compete methanogens for acetate. Arch. Microbiol., 132: 285-288.

Smith, R.L., and Oremland, R.S., 1987. Big Soda Lake (Nevada). 2. Pelagic Sulfate Reduction. Limnol. Oceanogr., 32: 794-803

Truper, H.G., 1969. Bacterial sulfate reduction in the Red Sea hot brines. In: E.T. Degens and D.A. Ross (Editors), Hot Brines and Recent Heavy Metal Deposits in the Red Sea, Springer-Verlag, New York, NY, pp. 263-271.

Widdel, F., and Pfennig, N., 1984. Dissimilatory sulfate- or sulfur-reducing bacteria. In: N.R. Krieg and J.G. Holt (Editors), Bergey's Manual of Systematic Bacteriology, Volume 1, Williams and Wilkins, Baltimore Maryland, pp. 663-679.

Widdel, F., 1988. Microbiology and ecology of sulfate- and sulfur-reducing bacteria. In: A.B.J. Zehnder (Editor), Biology of Anaerobic Microorganisms, John Wiley and Sons, Inc., New York, New York, pp. 469-585.

Windberger, E., Huber, R., Tricone, A., Fricke, H., and Stetter, K.O., 1989. Thermotoga thermarum sp. nov. and Thermotoga neapolitana occurring in African continental solfataric springs. Arch. Microbiol., 151: 506-512.

Zobell, C.E., 1958. Ecology of sulfate-reducing bacteria. Producers Monthly, 22: 12-29.

Reprinted from: *Microbial Enhancement of Oil Recovery - Recent Advances*, edited by E.C. Donaldson

CH. R–7

ISOLATION AND CHARACTERIZATION OF NOVEL HALOPHILIC ANAEROBIC BACTERIA FROM OIL FIELD BRINES

V. K. Bhupathiraju*, P. K. Sharma*, M. J. McInerney*, R. M. Knapp**, K. Fowler* and W.Jenkins*.

Department of Botany and Microbiology* and the School of Petroleum and Geological Engineering**, University of Oklahoma, Norman, OK 73019.

ABSTRACT

The development of a succesful microbially enhanced oil recovery process requires a thorough understanding of the factors affecting microbial activity in the oil reservoir. A study has, therefore been initiated to evaluate the ecological and physiological factors governing the microbial activity in the oil field brines collected from the South East Vassar Vertz Sand Unit, Payne County, Oklahoma. The brines were highliy saline containing 11 to 19% NaCl and 1 to 2% of calcium and magnesium ions. Diverse populations of anaerobic heterotrophic bacteria were present at densities of 2.4 x 10^3 to 4.6 x 10^3 MPN/ml. Of the several strains isolated, five strictly anaerobic and obligately halophilic strains were selected for further characterization. None of the isolates grew at salt concentrations less than 3% (w/v). All strains grew in a mineral salts medium containing glucose, yeast extract and casamino acids in the presence of NaCl concentrations of up to 20% (w/v), but the optimum concentration for growth was around 12%. The doubling time of the isolates ranged between 7 and 11 hours. Addition of nitrate to the growth medium did not stimulate growth of any of the isolates. However the addition of sulfite as the external electron acceptor stimulated the growth yield of the strain TTL-30. All of the isolates metabolized a variety of carbohydrate substrates but not several of the tested aromatic compounds and amino acids. Acetate, butanol and propanol were detected as the fermentative products on glucose. The change of gas phase from N_2/CO_2 to H_2/CO_2 stimulated the growth of strains TA and SA, but completely inhibited the growth of the strain TTL-30. Methane was never produced by any of the five strains showing that these strains are not methanogens. The ability to use a fermentable carbohydrate make these strains suitable candidates for microbial selective plugging process in hypersaline environments.

INTRODUCTION

Microbially enhanced oil recovery (MEOR) process requires that microorganism either indigenous or injected, anaerobically metabolize added nutrients and produce useful products such as gases, biosurfactants, polymers etc. under the environmental conditions that exist in the reservoir (Clark et al., 1981; McInerney and Westlake, 1990). The key requisite for initiating such a process is a thorough understanding of the factors affecting the microbial activity in the oil reservoir so that compatible organisms and nutrients can be selected. An important challenge for MEOR processes is determining the composition of the fluids. The fluid phase represents the most important part of the reservoir in that, it is in this phase that the nutrients required for growth and metabolism will be transported. Information about the fluid composition can be used to establish baseline data, modify experimental procedures if needed, and predict the success of the pilot. The composition of water phase in the reservoir can influence the microorganisms ability to metabolize certain carbon sources, the displacement of oil by waterflooding enhanced oil recovery processes, and the property of clay.

The role of indigenous bacteria in in-situ enhanced oil recovery has been loosely defined. In one previous field experiment, indigenous bacteria numerically dominated those of the injected strains of Clostridium acetobutylicum (Yarborough and Coty,1983). In laboratory tests where pure cultures were injected into cores, indigenous cultures that grew under the given laboratory conditions were found in high numbers in core effluent (Bryant and Douglas, 1987). In core experiments using exclusively indigenous bacteria, sweep efficiency improvements and oil recoveries were comparable to experiments performed with injected pure cultures (Raiders et al.,1989). Thus injection of large numbers of bacteria into the reservoir may not be needed.

Our research over the past several years has focused on the use of indigenous microorganisms and their metabolic products to improve volumetric sweep efficiency. The procedure involves the addition of a nutrient mixture to stimulate the growth of microorganisms in highly permeable water channels. The growth of microorganisms in these regions blocks the pores and immobilizes the chanelling water which in turn improves the volumetric sweep efficiency and recovers additional oil. A candidate reservoir has been identified for the proposed pilot test. The reservoir is the South East Vassar Vertz Sand Unit (SEVVSU) reservoir, Payne County, Oklahoma. A study has been initiated to evaluate the ecological and physiological factors governing the microbial activity in the reservoir.

Several authors have reported that levels of NaCl above 1% inhibit the ability of bacteria to grow in sandstones (Smith and Collins, 1985). Salt concentrations above 3% have been cited as the cause for failure in some MEOR studies (Grula et al., 1986). However bacteria have been isolated that can produce useful products such as biosurfactants and polymers in brines with NaCl concentrations of upto 8% (Javaheri et al., 1985; Pfiffner et al., 1986), and have been shown to selectively plug sandstone systems and recover additional oil (Raiders et al., 1989). Halophilic bacteria have been isolated from salt lakes, salt plains and salt-treated foods. Few cases of anaerobic halophilic bacteria in oilfield brines have been reported, although several halotolerant species have been described (Rodriguiez-Valera, 1988). Thus , it is not known whether MEOR processes will be effective in highly saline reservoirs where salt concentrations may exceed 10% NaCl. The purpose of the present study was to evaluate the ecological and physiological factors affecting microbial activity in highly saline brines from SEVVSU and to determine which nutrient source would best stimulate the growth and metabolism of the indigenous microorganisms. Also, we report the isolation and characterstics of five strictly anaerobic and obligately halophilic bacteria.

2 METHODS AND MATERIALS

2.1 Sampling procedure

Brine samples were obtained from several wells in the South East Vassar Vertz Sand Unit in Payne County, Oklahoma. Samples were collected from the waterflood brine source well, the injection manifold, a well that has not been waterflooded and several wells in the proposed test pattern. Samples were coproduced brine and oil taken from the well-head of producing wells. All samples were collected in sterilized bottles. Samples were analyzed immediately for hydrogen sulfide and within four hours for bacterial numbers.

2.2 Chemical analyses of brines

Salinity, alkalinity and hardness measurements were performed with the aid of Hach chemical analysis kits (Hach Chemical Co, 3rd eds.1975). Sulfate analysis of the brines were performed by barium precipitation (Hach Chemical Co, 3rd eds.1975) and results were confirmed by high pressure liquid chromatography (Montgomery, 1987). Hydrogen sulfide was determined colorimetrically (Tanner, 1989). Known amounts of sulfate and sulfide were added to brine samples to test the accuracy and reliability of the analysis.

2.3 Bacteriological analyses

Certain physisologically important groups of bacteria were enumerated using standard plate count and most probable number techniques. The numbers of aerobic and anaerobic heterotrophic bcteria were determined using plate count agar (Difco) with and without 10% NaCl added. Three plates were inoculated at each dilution. The plates were incubated at room temperature for 21 days for aerobic counts and 35 days for anaerobic counts. The anaerobic plates were incubated in the anaerobic chamber. Aerobic and anaerobic heterotrophic bacteria were also enumerated using a three-tube most probable number technique using plate count broth with 15% NaCl. Anaerobic molasses-nitrate using bacteria and sulfate-reducing bacteria were also enumerated by a three-tube most probable number method using a molasses-nitrate medium (Knapp et al., 1989) and API-RST medium (Tanner et al., 1985) respectively. Each medium contained 15% NaCl and was incubated at 35°C until no further growth was observed. The methods for preparation and use of anaerobic media were those of Bryant (Bryant, 1972) as modified (Balch and Wolfe, 1976).

Tests to determine the best nutrient source for the stimulation of indigenous microbial populations were done by amending the brines with various nutrient mixtures. Brine samples were asceptically dispensed into sterile serum tubes which were closed with rubber bungs and crimp sealed. Where indicated sterile solutions of, molasses, sucrose, glucose, and fructose were added to give a final concentration of 0.2% (w/v except for molasses which was added on a v/v basis). Ammonium nitrate and sodium sulfate were added to give a final concentration of 0.3% (w/v). Sodium lactate and sodium thiosulfate were added to give a final concentration of 0.1% (v/v) and 0.2% (w/v), respectively. Nitrogen (as ammonium chloride) and phosphate (as sodium phosphate) were added at concentrations of 0.025% (w/v) and 0.07% w/v), respectively. Unamended brine samples served as controls. All manipulations were done anaerobically. Tubes were incubated at 35° C and routinely checked microscopically and turbidimetrically for growth.

2.4 Isolation and culture methods

Several anaerobic halophilic bacteria were isolated from the brine samples. Anaerobic heterotrophic halophiles were isolated by serially diluting brine samples and plating them on plate count agar with 15% NaCl. Plates were incubated anaerobically at room temperature. Isolated colonies were restreaked several times to obtain pure cultures. Anaerobic nitrate-using halophiles were enriched in molasses nitrate medium with 15% NaCl. These enrichments were serially diluted and agar roll tubes with molasses nitrate medium with 15% NaCl were inoculated. Several isolated colonies were selected and transferred to similar medium. Seven anaerobic heterotrophic halophiles and thirty anaerobic nitrate-using halophiles were isolated. Of these, four anaerobic heterotrophic halophilic strains designated TA, WA, SA, QB, and one molasses-using halophilic strain designated TTL-30 were selected for further studies. Molasses nitrate medium with 15% NaCl was modified by deleting molasses and adding glucose (5%,w/v),casamino acids (1%,w/v) and yeast extract(1%,w/v) to grow and maintain all the halophiles.

2.5 Growth and metabolic characterstics

Growth was followed by measuring the increase in absorbance at 660 nm. All experiments were done in duplicates and the results represent mean values. Volatile acids and alcohols were analyzed with a Varian 3400 series gas chromatograph, operating at 120°C for acids and 80°C for alcohols, with nitrogen as the carrier gas, using a Carbopack c/o. 3% Carbowax 20m/0.1% H_3PO_4, 30"x1/4"ODx4mm ID glass column.

The ability of the isolated strains to utilize different substrates were tested by replacing yeast extract, casamino acids, and glucose with the test compound, at 10mM (unless otherwise indicated) and comparing the growth (A 660nm) and the amount of acetate produced, with that obtained in controls. The effect of gas phase on the growth of the isolates was tested by changing the gas phase from 80%N_2:20%CO_2 to 80%H_2:20%CO_2. The effect of external electron acceptors on the growth of the isolates was tested in medium with glucose as an electron donor. Additions included elemental sulfur, sulfate, thiosulfate, sulfite, nitrate, and nitrite at a concentration of 0.3%(w/v).

3 RESULTS

3.1 Chemical analyses of brines

Since a thorough understanding of the fluids in the reservoir is essential for initiating a microbially enhanced oil recovery process, chemical analyses were performed to provide a detailed discription of the dominant chemical species in the brines. The chemical composition of the brines collected throughout the reservoir are listed in Table 1.

TABLE 1

Chemical analyses of brines

Sample source[a]	Major components							
	Total dry residue (mg/ml)	Total hardness as $CaCO_2$ (mg/ml)	Calcium hardness as $CaCO_2$ (mg/ml)	NaCl salinity (mg/ml)	Total alkalinity (mg/l)	Sulfate (mg/l)	Sulfide (mg/l)	pH
Source Well	129	36.8	25.0	116	38	2430	<0.1	7.0
Pump manifold	165	48.6	36.4	149	50	944	0.3	6.7
Well 5-1	163	45.4	32.8	149	24	1320	2.0	7.2
Well 7-2	170	28.2	19.8	154	104	780	1.0	6.7
Well 1A-7	197	61.6	41.8	182	42	< 1	<0.1	6.9

[a] The location of the sample sites in the oil reservoir have been described elsewhere (Knapp et al.,1988).

The brines from South East Vassar Vertz Sand Unit were highly saline with NaCl salinities ranging from 11% in the Source Well, to 15% in the water flooded portion of the reservoir, and 19% in well 1A-7 which has not yet been waterflooded. The brines also contained high levels of divalent cations, particularily calcium as indicated by the hardness data, ranging from 2 to 4%. Moderate levels of alkalinity were also detected in most of the brines indicating carbonates as the important anions. Detectable levels of iron and nitrate were also found; however, little or no phosphate or nitrite was detected (data not shown). The pH of the brines was close to neutrality. The sulfate concentration (2430 mg/l) was high in the brine from the source well when compared to brines from the waterflooded portion of the reservoir. In contrast, little or no sulfate was detected in brines from portions of the reservoir not affected by water flooding indicating that the source of sulfate is from the aquifer that supplies the make-up water for the waterflood. Sulfide was detected in those wells which had been waterflooded, indicating a potential for anaerobic corrosion problems.

3.2 Bacteriological analyses

Because of the high salinities of the brines, it was important to determine whether the brines had indigenous microbial populations which could be useful for enhanced oil recovery. The number of

TABLE 2

Populations of heterotrophic bacteria in brine samples

Sample source [a]	Cell concentration (cells/ml)			
	Aerobic		Anaerobic	
	10% NaCl[b]	15% NaCl[c]	10% NaCl[b]	15% NaCl[c]
Pump manifold	$2.4 * 10^4$	NA[d]	NA[d]	NA[d]
Well 5-1	$1.1 * 10^4$	$3.2 * 10^4$	$1.1 * 10^2$	$2.5 * 10^3$
Well 7-2	$1.1 * 10^4$	$3.4 * 10^4$	$4.6 * 10^1$	$3.5 * 10^2$
Wel 7-5	$2.4 * 10^4$	$7.6 * 10^4$	$2.4 * 10^1$	$3.6 * 10^3$

[a] The location of the sample sites in the reservoir have been described elsewhere (Knapp et al.1988).

[b] Cell numbers were determined using the plate count agar with 10% NaCl and incubated for 21 days for aerobic counts and 35 days for anaerobic counts.

[c] Most probable number of organisms estimated using plate count broth with 15% NaCl and incubated at 35°C for 35 days.

[d] Not available.

heterotrophic bacteria was estimated using plate count agar and broth. When this medium was used without the addition of NaCl, many colonies were found on plates inoculated with the first dilution, but none were found on any of the subsequent dilutions. This was observed with both aerobic and anaerobic incubations which indicated that some components present in the brine were required, and as the brine was diluted, the component became limiting, which prevented growth at higher dilutions. The simplest explanation was that the bacteria present in the reservoir are halophiles and required high salt concentrations for growth. The addition of NaCl in amounts greater than 10% allowed the enumeration of these bacteria in the reservoir.

The numbers of heterotrophic bacteria detected in the reservoir are listed in Table 2. Aerobic heterotrophic bacteria were more numerous than anaerobic heterotrophic bacteria. Higher counts (both aerobically and anaerobically) were observed when NaCl concentration was increased from 10% to 15%, indicating the halophilic nature of the organisms. Both aerobic and anaerobic heterotrophic bacteria were consistently found in the waterflooded portion of the reservoir.

The numbers of anaerobic, molasses nitrate-using bacteria and sulfate-reducing bacteria are listed in Table 3. Anaerobic molasses-nitrate using bacteria were present in all the brines which were analyzed and in greater concentrations than the anaerobic heterotrophic bacteria. Thus there are numerous organisms in the reservoir brines whose growth can be stimulated by the addition of nutrients. No sulfate-reducing bacteria were detected in brine from well 1A-7, however significant numbers were detected in the waterflooded portion of the reservoir which is consistent with levels of sulfate detected in the brines and indicating the potential for corrosion problems.

TABLE 3

Numbers of sulfate-reducing bacteria and molasses nitrate using bacteria in brine samples

Sample source[a]	Most probable number[b] (per ml)	
	Sulfate reducers	Nitrate users
Pump manifold	$1.5 * 10^{3}$	$1.1 * 10^{3}$
Well 5-1	$2.4 * 10^{2}$	$2.4 * 10^{3}$
Well 7-2	$1.5 * 10^{2}$	$4.6 * 10^{3}$
Well 7-5	$1.1 * 10^{2}$	$4.6 * 10^{3}$
well 1A-7	ND[c]	$2.7 * 10^{3}$

[a] The location of the sample sites in the reservoir have been described elsewhere (Knapp et al., 1988).

[b] Numbers of bacteria estimated using the three tube most probable number method. Samples were incubated at 35°C for 35 days.

[c] Not detected.

TABLE 4

Growth in brine samples using different nutrient additions

Additions	Growth[b]				
	Samples sites[c]				
	Well 1A-1	Well 1A-9	Well 7-2	Well 8-2	Well 8-3
Molasses	++	++	++	++	++
Molasses + Nitrate	+++	+++	+++	+++	+++
Sucrose	-	++	-	+	+
Sucrose + Nitrate	-	-	-	++	+
Sucrose + Phosphate +Nitrogen + Nitrate	-	-	-	-	-
Glucose	-	+++	-	-	+
Glucose+ Nitrate	-	-	-	-	++
Glucose + Phosphate +Nitrogen + Nitrate	-	-	-	-	-
Fructose	-	-	-	-	-
Fructose+ Nitrate	-	-	-	-	-
Fructose+ Phosphate +Nitrogen + Nitrate	-	-	-	-	-
Lactate	-	-	-	-	+
Lactate + sulfate	-	-	-	-	-
Lactate + thiosulfate	-	-	-	-	-
Acetate	-	-	-	-	++
Acetate + sulfate	-	-	-	-	-
Acetate + thiosulfate	-	-	-	-	+

[a] Concentration (each in %w/v): 0.2, Molasses; 0.2, Sucrose; 0.2, Fructose; 0.2, Glucose; 0.1, Acetate; 0.1, Lactate; 0.3, Nitrate; 0.3, Thiosulfate; 0.2, Sulfate; 0.025, Nitrogen as ammonium chloride; 0.07, Phosphate.

[b] Growth: +++ OD > 0.3; ++ OD 0.1-0.3; + OD 0-0.1; - No turbidity. Tubes were checked microscopically to verify the presence of microbial cells.

[c] The location of the sample sites in the reservoir have been described elsewhere (Knapp et al.,1988).

Since these studies showed the presence of indigenous microbial populations it was important to determine which nutrient source would best stimulate microbial growth. The brines from different portions of the reservoir were amended with various nutrient mixtures (Table 4). Results from these experiments show that molasses alone supported good growth of the indigenous populations. However, the addition of molasses together with ammonium nitrate resulted in the largest increases in the growth of the indigenous microbial populations. In some brines, the addition of either glucose or sucrose with or without ammonium nitrate stimulated growth. The addition of a phosphate source and an additional nitrogen source to brines ammended with individual sugars did not stimulate microbial growth. Thus, a complex nutrient source is required for optimal growth of the indigenous populations.

3.3 Isolation, growth and metabolic properties of the halophilic isolates

Several anaerobic halophilic bacteria were isolated from the brine samples. Seven anaerobic heterotrophic halophiles and thirty anaerobic molasses nitrate using bacteria were isolated in pure culture. Of these four anaerobic heterotrophs designated TA, WA, SA, QB, and one anaerobic-nitrate using halophile designated TTL-30 respectively, were further characterized. The characterstics of the isolates are listed in Table 5.

TABLE 5

Characterstics of the halophilic isolates

Parameters	Strain TA	Strain WA	Strain QB	Strain SA	Strain TTL-30
Morphology	Rod	Cocci	Cocci	Cocci	Rod
Gram stain	Negative	Positive	Negative	Positive	Positive
Doubling time (h)	7.25	7.25	10	11	10
Temperature range(°C)	NA[a]	NA[a]	NA[a]	35	30
NaCl conc. range (% w/v)	9-18	9-18	6-22	6-22	9-22
Optimum salt conc. for growth (%w/v)	12	12	6	12	15
Motility	Negative	positive	positive	positive	negative

[a] NA=Not available

All the strains were obligately anaerobic and required high NaCl concentrations for growth (Figure1). No growth was observed in medium with less than 3% NaCl (w/v) concentration. Strain TA was a gram negative, non-motile, rod-shaped bacterium usually occuring in pairs, having an optimal NaCl requirment of 12% w/v. Strains WA, QB, and SA were all motile cocci. These strains had similar cellular characterstics excepting that the WA occurred mostly in chains while QB and SA occurred either singly or in pairs. These coccoid strains differed in their gram reaction, the ability to grow at different salt concentrations, the optimal NaCl concentration requirment for growth (Figure

1) and the doubling time. Strains QB and SA were the most versatile with respect to their ability to grow at a wide NaCl concentration range (6-22% [w/v]) and had an optimum salt requirment of 6% and 12% w/v respectively. Strain TTL-30 was a non-motile, gram positive, rod shaped bacterium with an abilty to grow at a fairly wide NaCl concentration (9-22% [w/v]) with an optimum concentration of 15% NaCl (w/v).

All of the strains utilised a variety of substrates such as glucose, maltose, fructose, mannose, yeast extract, peptone, galactose, arabinose, sucrose, trypticase, and inulin. Strain TTL-30 was the most versatile in its ability to utilize other substrates such as pyrogallic acid, and glycerol in addition to the above (Table 6). Strain TA utilized succinate in addition to the substrates mentioned earlier. Substrates tested but not utilized by any of the isolates include xylulose, mannitol, gallic acid, resorcinol, fumaric acid, pyruvic acid, dihydrobenzoic acid, citric acid, crotonic acid, ascorbic acid and methanol. Acetate, butanol and propanol were detected as the end products of fermentation on glucose, except TTL-30 did not produce butanol. Other gaseous products have not been analyzed as yet.

TABLE 6

Substrates examined as energy sources for growth of the strain TTL-30

Substrate [a] (10mM)	Growth [b] (A 660nm)	Acetate [c] (mM)
Glucose	0.25	15.0
Fructose	0.05	9.8
Sucrose	0.06	9.7
Galactose	0.10	3.3
Mannose	0.15	12.0
Maltose	0.20	15.3
Trypticase	0.13	15.8
Yeast extract (1%)	0.26	16.1
Fucose	0.10	NA
Glycerol	0.07	5.26
Pyrogallic acid	0.10	1.0

[a] Substrate was added at 10 mM (unless otherwise mentioned) to mineral salts medium described in methods and materials.

[b] Values indicated have been corrected for carry over in control.

[c] Values indicated have been determined after 10 days of incubation at 35°C and have been corrected for carry over in control.

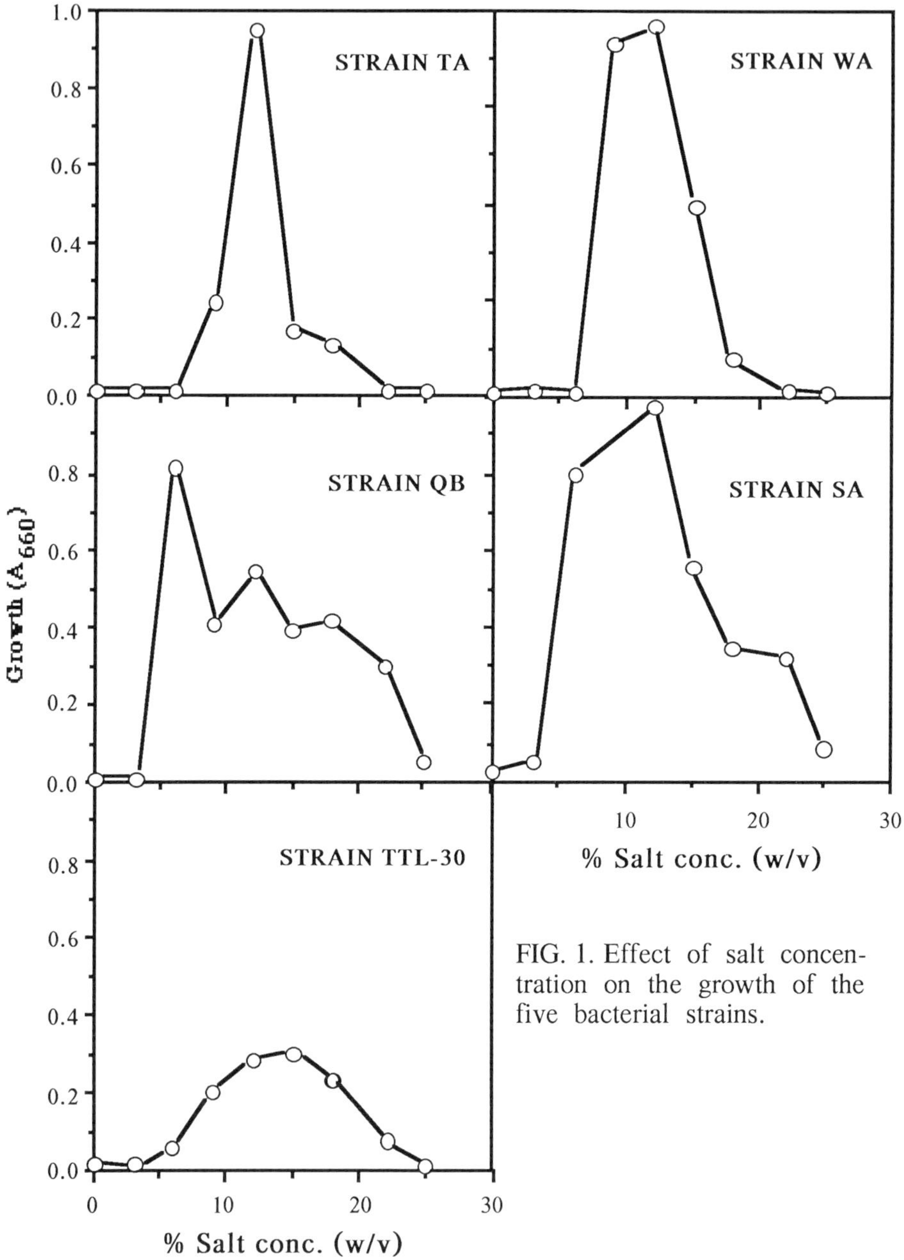

FIG. 1. Effect of salt concentration on the growth of the five bacterial strains.

The effect of added external electron acceptors and the gas phase on the growth of the isolates was also tested (data not shown). No external electron acceptor seemed to stimulate the growth of the isolates except for sulfite which stimulated the growth of the strain TTL-30. The change of gas phase from N_2/CO_2 to H_2/CO_2 stimulated the growth of strains TA and SA but completely inhibited the growth of the strain TTL-30, and did not have any affect on the growth of strains WA and QB. Methane was not produced by any of the strains indicating that they are not methanogens.

4 DISCUSSION

The characterization of the microbial ecology of the reservoir is a critical first step to the development of a successful microbially enhanced oil recovery process. The brines obtained from the SEVVSU reservoir contained diverse populations of bacteria which can grow in highly saline environments. Our studies showed that the brines did contain organisms which grow readily upon simple nutrient additions. A molasses nitrate mixture best stimulated the growth of the microbial populations. Large amounts of gas and, at times, polymers and emulsifiers were found in the brines amended with molasses and nitrate. Thus indigenous organisms are present which produce useful products for oil recovery. These organisms have been naturally selected to grow in this harsh environment, avoiding the need for extensive strain development. Also, since they already exist in the reservoir problems encountered in injecting cells into the formation are eliminated. Moreover, the organisms grew only when large concentrations of NaCl were added to the culture medium, indicating that they are halophiles. Thus, they will not survive outside of the reservoir and their accidental release into the environment will not cause any environmental or public health problems.

Our studies indicate that sulfate reduction will be a problem in this field throughout its remaining life. The sulfate content of the source well stimulated sulfate-reducing activity in the waterflooded portion of the reservoir. Studies show that if sulfate reduction is evident, the metabolism of sulfate reducing bacteria can be inhibited by adding high levels of nitrate to the medium (Jenneman et al.,1986). In experiments where high levels of nitrate were added, hydrogen sulfide was not evident and the numbers of sulfate-reducing bacteria were reduced (Knapp et al., 1989). Since the nitrate additions inhibited sulfide production in our brine samples, nitrate additions will effectively control sulfate-reducing activity in the pilot project.

Brines from most mid continent United States oil reservoirs contain high concentrations of NaCl, thus tolerance to NaCl is one of the most important characterstics needed for organisms to be useful in MEOR. Bacteria are known which grow at high NaCl concentrations (approximately 32%) (Kushner,1978). However it would be preferable to have organisms which can grow over wide ranges of salinities. Several halotolerant anaerobic bacteria have recently been described (Rodriques-Valera, 1988), however relatively little is known about obligately anaerobic, moderately or extremely halophilic bacteria (Zeikus et al., 1983; and Oren et al., 1983). We have isolated five anaerobic obligately halophilic bacteria from oil field brines capable of growing in highly saline environments. Based on the morphological characterstics, gram reaction, ability to grow at a wide concentration range of NaCl , optimal salt requirement for growth , ability to utilize a variety of substrates, and fermentative products produced from glucose, these strains seem to be very different from previously reported anaerobic halophiles such as *Clostridium lortetii* (Oren, 1983), *Halobacteroides halobius* (Oren et al., 1984), *Haloanaerobium prevalens* (Zeikus et al., 1983) and *Halobacteroides* acetoethylicus (Rengpipat et al., 1988). The ability to grow at high salt concentration ranges and use a fermentable carbohydrate substrate make these strains suitable candidates for microbially selective plugging processes in hypersaline environments.

5 REFRENCES

Balch, W. E., and Wolfe, R. S., 1976. " New Approach to the Cultivation of Methanogenic Bacteria : 2-Mercaptoethanelsulfonic acid (HS-CoM)-dependant Growth Of Methanobacterium ruminatium in a pressurized Atmosphere," Appl. Environ. Microbiol. 32: 781- 791.

Bryant, M. P., 1972. " Commentary on the Hungate Technique for Culture of Anaerobic Bacteria," Amer. J. Chem. Nutr. 25: 1324-1328.

Bryant, R. S., and Douglas, J., 1987. " Evaluation of Microbial Systems in Porous Media for Enhanced Oil Recovery ", in Symp. on Oil field Chemistry, Society of Petroleum Engineers, San Antanio.

Clark, J.B., Munnecke, D.M., and Jenneman, G. E., 1981. In Situ Microbial Enhancement of Oil production., Dev. Ind. Microbial. 22:695-701.

Finnerty, W. R. and Singer, M. E., 1983. Microbial Enhancement of Oil Recovery. Bio/technology 1: 47-54.

Grula, E. A., Russel, H. H., and Grula, M. M. 1986. Field Trials in Central Oklahoma Using Clostridium Strains for Microbially Enhanced Oil Recovery, In, Donaldson, E. C., and Zajic, . E., (eds.) Microbes and Oil Recovery, Volume I, Bioresources Publications, El Paso, Texas, pp 1-18.

Hach Chemical Co. 1985. Hach Water and Wastewater Analysis Procedure Manual, 3rd Edition, Hach Chemical Co., Ames, IA.

Javaheri, M., Jenneman, G. E., and McInerney, M. J., and Knapp, R. M., 1985. " Anaerobic Production of a Biosurfactant by Bacillus licheniformis strain JF-2 ". Appl. Environ. Microbiol. 698-700.

Jenneman, G. E., McInerney, M. J., and Knapp, R. M., 1986. " Effect of Nitrate on Biogenic Sulfide Production, " Appl. Environ. Microbiol. 51: 1205-1211.

Knapp, R. M., McInerney, M. J., and Menzie, D. E., Raiders, R.A., 1989. "Microbial Field Pilot Study," Final report , DOE/BC/14084-6, National Technical Informational Services, Springfield, VA.

Kushner, D. J., 1978. Life in High Salt and Solute Concentrations: Halophilic Bacteria, in D. J Kushner (eds.) Microbial Life in Extreme environments, Academic Press,New York, pp.317-368.

Montgomery, A. J., 1987. Oxidation of Hydrogen Sulfide by a Sulfide Resistant Strain of Thiobacillus denitrificans in Mixed Culture, M.S. Thesis, University of Oklahoma, Norman .

McInerney, M. J., and Westlake, D. W. S., 1990. " Microbially Enhanced Oil Recovery " in H.L. Ehrlich and C. L. Brierley (eds.) Microbial Mineral Recovery, McGraw - Hill Publishing Company, New York, pp 409-445

Oren, A., 1983. Clostridium lortetii sp. nov. a halophilic obligately anaerobic bacterium producing endospore with attached gas vacuoles. Arch. Microbiol. 136: 42-48.

Oren, A., Weisburg, W. G., Kessel, M., Woese, C. R., 1984. Halobacteroides halobius gen. nov., sp. nov., a moderately halophilic obligatory anaerobic bacterium from the bottom sediments of the dead sea. System. Appl. Microbiol. 5: 58-70.

Pfiffner, S. M., McInerney, M. J., Jenneman, G.E, and Knapp, R. M. , 1986. " Isolation of Halo--tolerant, Thermotolerant, Facultative Polymer-Producing Bacteria and Characterization of the exopolymer. Appl. Environ. Microbiol. 51:1224-1229.

Raiders, R.A., Knapp, R.M., McInerney, M.J., and Menzie, D.E., April.1989 ."Microbial selective Plugging and Enhanced Oil Recovery,"J. Indust. Microbiol., 4: 215-230.

Rengpipat, S., Langworthy, T. A., and Zeikus, G. J., 1988. Halobacteroides acetoethylicus sp.nov. a New Obligately Anaerobic Halophile Isolated from Deep Subsurface Hypersaline Environments. System. Appl. Microbiol. 11: 28-35.

Rodriguiez-Valera, F., 1988. Halophilic bacteria, Volume I, CRC Press, Boca Ranton, FL, pp1-29.

Smith, R. J. and Collins, A. G., 1985. " Microbial Enhanced Oil Recovery: A Review," Zajic, J.E., and Donaldson, E.C, (eds.) Microbes and Oil Recovery, Volume I, Bioresources El Paso, Texas, pp 1-18.

Tanner, R. S., Haack, T.K., Semet, R. F., and Greenley, D. E., 1985. " A Mild Steel Tubular Flow System for Biofilm Monitoring, " Proceedings of the United Kingdom National Academy of Corrosion Engineers Corossion '85, Harrowgate, Uk 259-269.

Tanner, R. S., 1989. Monitoring Sulfate-Reducing Bacteria: Comparision of Enumeration Media. J. Micro. Methods. 10: 83-90.

Yarborough, H. F., and Coty, V. F., 1983. " Microbially Enhanced Oil Recovery from the Upper Cretaceous Macatoch Formation, Union County, Arkansas, " In, Clark, J. B., and Donaldson, E. C., (eds). Proceedings of the 1982 International Conferance on Microbial Enhancement of Oil Recovery, Conf. 8205140, National Technical Information Service, Springfield, VA, pp 149-153.

Zeikus, J. G., Hegge, P. W., Thompsoon, T. E., Phelps, T. J., Langworthy, T. A., 1983. Isolation and description of Haloanaerobium prevalens gen. nov., sp. nov., an obligately anaerobic halophile common to Great Salt Lake Sediments. Curr. Microbiol. 9:225-234.

Reprinted from: *Microbial Enhancement of Oil Recovery - Recent Advances*, edited by E.C. Donaldson

CH. R-8

ANAEROBIC MICROBIAL DEGRADATION OF AROMATIC HYDROCARBONS

D. GRBIC-GALIC
Environmental Engineering and Science Program, Department of Civil Engineering, Stanford University, Stanford, California 94305-4020 (U.S.A.)

ABSTRACT

In the last decade, it has become widely recognized that nonoxygenated aromatic hydrocarbons undergo microbial transformations under anoxic conditions. The compounds which are transformed include mononuclear and polynuclear homocyclic, nitrogen heterocyclic, and sulfur heterocyclic aromatic hydrocarbons. The conditions conducive to the transformations range from denitrifying and iron(III)-reducing, to sulfate-reducing and fermentative/methanogenic. Anaerobic or facultatively anaerobic bacteria capable of degrading aromatic hydrocarbons have been enriched and, in some cases, isolated from various environments, both pristine and contaminated with pollutant mixtures containing aromatic hydrocarbons: anaerobic sludge, soil, freshwater sediments, estuarine sediments, marine sediments, and ground water aquifers. At this point, it is maybe premature to conclude that such bacteria are ubiquitous and that they can all degrade a very broad range of hydrocarbons; however, they are being detected in numerous environmental samples, and their activity can be demonstrated *in-situ*, e.g. in ground water aquifers contaminated by petroleum or creosote. Once the nutritional requirements, physiology and ecology of these microorganisms are well understood, they will represent a considerable potential to be tested and used in microbially-enhanced oil recovery.

INTRODUCTION

Utilization of indigenous microorganisms capable of transforming oil constituents and consuming them as carbon and energy sources, appears to be an attractive possibility for mobilization of oil. Anaerobic microorganisms which do not require introduction of oxygen as an electron acceptor, offer the greatest utilization potential provided they can successfully degrade selected oil components at reasonably high rates. Such anaerobes may be denitrifiers, sulfate reducers, iron(III) reducers, or fermenters coupled in food chain with methanogens. The latter physiological groups are by far the most attractive because they do not require an exogenous electron acceptor other than carbon dioxide (for methanogens), and this gas can be supplied in abundance through metabolism of fermentative bacteria. An active, indigenous fermentative/methanogenic community could represent a self-sustained oil recovery machinery which would not depend on external additions, and would thrive on constituents of oil as sources of carbon, energy, and reducing power for biosynthetic reactions. Gases produced by such microorganisms (carbon dioxide and methane) could become a significant factor in oil mobilization.

It has been known for quite a while that oxygenated aromatic compounds (alcohols, aldehydes, acids, amino acids, phenols) are susceptible to anaerobic microbial transformations in the presence of an external electron acceptor (nitrate, sulfate), or fermentatively. The transformations of simple oxygenated aromatics, such as benzoic acid, are based on reductive reactions which convert the aromatic ring into an alicyclic ring; this was first recognized by Evans (1977). The alicyclic moiety carrying an oxygenated substituent is prone to hydrolytic ring cleavage, whereby the original aromatic substrate is finally converted to an aliphatic acid and subsequently degraded to end products (carbon dioxide and [depending on the conditions] methane). The current knowledge on anaerobic degradation of oxygenated aromatics has been summarized in several excellent review papers, for example Young (1984), Berry et al. (1987a), and Evans and Fuchs (1988).

Nonoxygenated aromatic hydrocarbons are stable resonant structures which were believed, until recently, to be completely recalcitrant in the absence of oxygen. Under aerobic conditions, microorganisms attack these compounds using mono- and di-oxygenases, powerful enzymes which split the oxygen molecule and insert atoms of oxygen onto the aromatic structure. Anaerobically, the use of moleculer oxygen as a reactant is prohibited; and yet, if the transformation of nonoxygenated homocyclic aromatic hydrocarbons under anaerobic conditions were initially reductive as is the case with oxygenated aromatics, the product would be a completely reduced alicyclic ring which is even stabler and more difficult to degrade than an aromatic ring. Therefore, the knowledge on anaerobic transformations of oxygenated aromatic compounds did not support the possibility of anaerobic microbial attack on homocyclic aromatic hydrocarbons. However, there were data available (summarized by Chapman, 1979) which indicated that heterocyclic aromatic hydrocarbons with one or more nitrogen atoms in the ring structure (e.g., indole, purine, pyrimidine) could undergo hydroxylation both in the presence and in the absence of oxygen. Molecular oxygen was not necessary for this transformation, since the oxygen functionality was derived from water. The first documentation of a microbial transformation of this kind dated from 1952 (Whiteley). This information was useful as a starting point in the attempt to understand the feasibility of anaerobic degradation of homocyclic aromatic hydrocarbons.

2 METHANOGENIC CONDITIONS

2.1 Homocyclic Aromatic Hydrocarbons

Interestingly enough, the first observation suggesting the possibility of anaerobic microbial transformation of nonoxygenated homocyclic aromatic hydrocarbons was the result of an experiment under methanogenic conditions. Methanogenesis is the ultimate form of anaerobic microbial metabolism, and occurs concurrently with fermentation at redox potentials (E_h) which are usually lower than -200 mV (measured at pH 7); no exogenous electron acceptor other than carbon dioxide is required (Thauer et al., 1977; Zehnder, 1978). This first observation was published by Ward et al. (1980). The authors experimented with laboratory microcosms containing saturated salt marsh and estuarine sediments contaminated by a catastrophic marine oil spill. They amended the microcosms with ^{14}C-labelled toluene, benzene, and naphthalene; after eight months of incubation, 2% of the label from benzene and up to 5% of the label from toluene were detected in gaseous products (carbon dioxide, methane), whereas naphthalene was not degraded. Since it was not likely that the compounds were reductively transformed (reduced rings are stabler under anaerobic conditions than aromatic rings), there was an exciting possibility that the initial transformation might have been oxidative. This possibility, however, was so unbelievable that it took several years before anybody decided to consider it seriously.

In 1983 and 1984, two groups of authors (Schwarzenbach et al., 1983; Reinhard et al., 1984) reported independent observations on selective removal of toluene and xylene isomers in anaerobic zones of landfill leachate-contaminated ground water aquifers. The selective removal could not be explained by sorption and dilution only. There was no concomitant production of saturated alicyclic rings. These observations pointed in the same direction as the findings of Ward et al. (1980). On the basis of these results, laboratory experiments were started at Stanford University, which resulted in publication of two papers elucidating the initial transformation of homocyclic aromatic hydrocarbons under methanogenic conditions (Vogel and Grbic-Galic, 1986; Grbic-Galic and Vogel, 1987). The authors used stable methanogenic consortia degrading ferulic acid, an aromatic lignin derivative; these consortia had been originally enriched from anaerobic sludge, through an intermediary step using lignocellulose as the growth substrate (Healy et al., 1980). The ferulate consortia were the basis from which stable methanogenic toluene-degrading and benzene-degrading consortia were established. These consortia degraded 3-30 mM of benzene or toluene as sole growth/energy substrates or dissolved in methanol, in 40-60 days of incubation at 35°C, in the dark. ^{14}C-Labelling of the aromatic hydrocarbons, gas chromatography (GC), gas chromatography/mass spectrometry (GC/MS), and ^{18}O-labelling of water in the prereduced defined liquid medium for microbial growth were used to follow the fate and transformation of toluene and benzene. It was concluded that the initial transformation reaction involved a hydroxylation (incorporation of ^{18}oxygen from [^{18}O]water into the aromatic molecule), which represents an oxidative reaction (redox state of the affected carbon atom becomes more positive), exactly opposite from the initial reductive transformation in oxygenated aromatic compounds. This mechanism is shown in Figure 1. [^{18}O]Phenol was found as the early transformation intermediate from benzene, and [^{18}O]*p*-cresol, *o*-cresol, and benzyl alcohol as the earliest detectable intermediates from toluene. After 40-60 days of incubation at 35°C, about 50% of both [^{14}C]toluene and [^{14}C]benzene were converted to [^{14}C]carbon dioxide and methane, whereas the rest of the label remained in biomass and various transformation intermediates. Prolonged incubation (three to four months) resulted in stoichiometric conversion of toluene and benzene to carbon dioxide and methane.

The initial oxidative transformation of homocyclic aromatic hydrocarbons under methanogenic conditions may be catalyzed by an oxidoreductase similar to the one

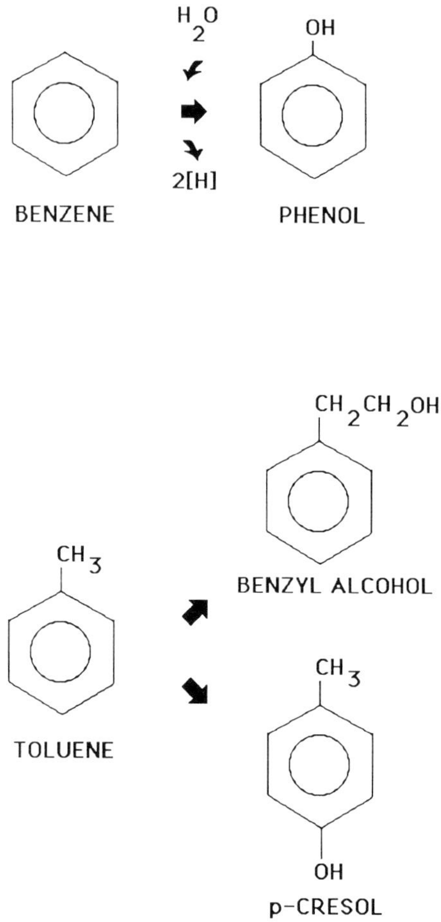

Fig. 1. Initial oxidative transformation of benzene and toluene under methanogenic conditions. [H] represents a proton and an electron bound to an unidentified carrier. Two of the three detected initial toluene oxidation products are shown: benzyl alcohol and *p*-cresol.

described by Hopper (1976, 1978; Hopper and Taylor, 1977) for anaerobic oxidation of *p*-cresol under denitrifying conditions. This hydroxylase was isolated and purified from a *Pseudomonas putida* and shown to consist of two flavoprotein subunits and two cytochrome *c* subunits. It does not require an external electron acceptor, and is presumably linked to an electron-transport system. In addition to iron-containing cytochromes, other transition-metal compounds, such as cobalt-containing corrinoids (Barker, 1967; Firth et al., 1969; Schrauzer and Sibert, 1969) are good candidates as possible catalysts in such anaerobic oxidations. Trifluoroacetates of Co(III) and

Fe(III), dissolved in trifluoroacetic acid, were shown to catalyze anaerobic oxidations of aromatic hydrocarbons under abiotic conditions (Kochi et al., 1973; Fukuzumi and Kochi, 1981; Schlesener et al., 1984).

The pathway of toluene degradation under methanogenic conditions has been elucidated (Grbic-Galic and Vogel, 1987; Grbic-Galic, 1990). After the initial oxidation reaction with water-derived oxygen, which creates either an alcohol (benzyl alcohol) or a phenol (*p*- or *o*-cresol), these compounds are gradually further oxidized to aromatic acids (benzoic acid, *p*-hydroxybenzoic acid, or *o*-hydroxybenzoic acid, respectively). The acids subsequently undergo typical transformation pathways for oxygenated aromatics as described before (Evans, 1977), including ring reduction, cleavage, and degradation of aliphatic intermediates (alcohols and acids) to gaseous products. Therefore, it can be concluded that anaerobic transformation of aromatic hydrocarbons follows the general aromatic transformation pathway, but has to be initiated through an oxidation. The reducing equivalents generated through such oxidations can be used in biosynthetic reactions, energy-yielding reactions such as methane production, and in further transformations of the aromatic intermediates in the pathway (ring reduction).

Current literature indicates that homocyclic monoaromatic hydrocarbons which are transformed under methanogenic conditions include benzene and toluene (Vogel and Grbic-Galic, 1986; Wilson et al., 1986, 1987; Grbic-Galic and Vogel, 1987; Grbic-Galic, 1989), *o*-, *m*- and *p*-xylene (Wilson et al., 1986, 1987; Edwards and Grbic-Galic, 1990; Grbic-Galic et al., 1990), and ethylbenzene (Wilson et al., 1987). Active microorganisms have been found in anaerobic sludge and in ground water aquifers contaminated by petroleum derivatives or creosote; the activity in ground water has been demonstrated not only in the laboratory, but also (indirectly) *in situ* (Wilson et al., 1987). The rates of transformation, as a rule, have been reported as slow (months necessary for the completion of degradation to occur). Our new investigations, however (Edwards and Grbic-Galic, 1990; Grbic-Galic et al., 1990) , indicate that some methanogenic communities may be considerably faster after the initial (long) acclimation period. We now have suspended enrichment cultures, obtained from saturated microcosms (originally from the creosote-contaminated aquifer material of Pensacola, Florida, ground water aquifer). Whereas the adaptation of the microcosms to degradation of toluene and *o*-xylene initially took 3.5 and 8 months at 35°C, respectively, the degradation of these compounds as sole substrates (5 mM) in active stabilized enrichments is completed in approximately ten days at the same temperature, in which period they are stoichiometrically converted to methane and carbon dioxide (Edwards and Grbic-Galic, 1990). The use of [^{14}C]toluene fully confirms these findings. The rate of the process can be influenced: if pH of the medium is lowered from 7.0 to 6.0, toluene degradation is completed in six rather than 10 days. The earliest transformation intermediate from toluene is *p*-cresol, and *p*-cresol seems to play a role in the *o*-xylene transformation pathway as well (Edwards, unpublished results). The addition of *p*-cresol to microcosms reduces the initial adaptation lag to three (toluene) or six months (*o*-xylene; Grbic-Galic et al., 1990).

Quite recently, we have isolated two fermentative bacteria from the complex methanogenic, aquifer-derived, toluene-degrading consortia, and they are currently being investigated (Edwards and Grbic-Galic, unpublished results). Furthermore, our newest findings indicate that propylbenzene, which has three carbon atoms in its side chain, is also transformed under methanogenic conditions, with aquifer solids from a jet-fuel contaminated aquifer (Traverse City, Michigan) used as the inoculum (Liang and Grbic-Galic, unpublished results). The list of homocyclic monoaromatic hydrocarbons which are degraded by complex methanogenic communities is presented in Fig. 2.

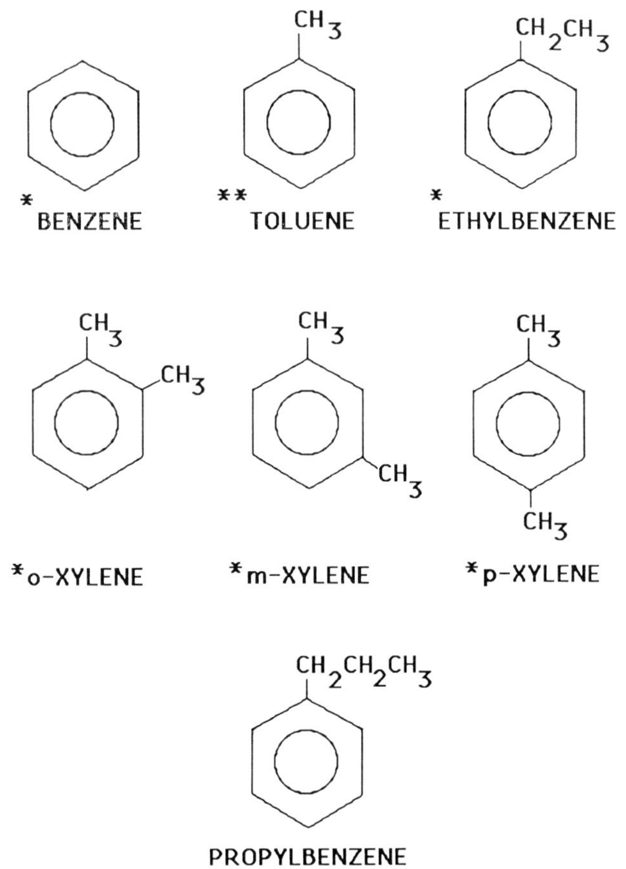

Fig. 2. Homocyclic monoaromatic hydrocarbons shown to be degraded under anaerobic conditions. All of the listed compounds have been demonstrated to undergo methanogenic degradation. *: compounds which are also degraded under denitrifying conditions. **: compounds which are also degraded under denitrifying and iron(III)-reducing conditions. There are some indications, but no certain proof, that ethylbenzene can be degraded not only methanogenically, but also with iron(III) as an electron acceptor.

Not only mononuclear, but also simple polynuclear homoaromatic hydrocarbons can undergo biodegradation under anaerobic conditions (Godsy et al., 1989; Grbic-Galic, 1989; Grbic-Galic et al., 1990). The compounds shown so far to be oxidized under methanogenic conditions include indene, naphthalene, and acenaphthene (Fig. 3); the active microorganisms have been derived either from anaerobic sludge, or from creosote-contaminated ground water aquifer solids. Contrary to the experiments with mononuclear aromatics, we have never been able to detect the earliest oxidation intermediates from polynuclear aromatic hydrocarbons. However, we have identified oxidized intermediates further down the transformation line, such as 2-ethylphenol and benzofuran from naphthalene (Grbic-Galic, 1989), or phthalic acid and *p*-cresol from acenaphthene. These intermediates indicate that the initial transformation must be again oxidative. The only source of oxygen available under the experimental conditions used was water. Numerous monoaromatic (alcohols, acids, phenols) and aliphatic intermediates (alcohols, acids) found in the naphthalene- and acenaphthene-degrading methanogenic systems suggest that these pathways also overlap with the degradation routes for oxygenated aromatic compounds, once the initial oxidation has occurred.

Saturated microcosms derived from the creosote-contaminated aquifer in Pensacola, Florida, completely degraded 10-20 mg/l of naphthalene or indene in one or two months, respectively (Grbic-Galic et al., 1990). The anaerobic sludge-derived

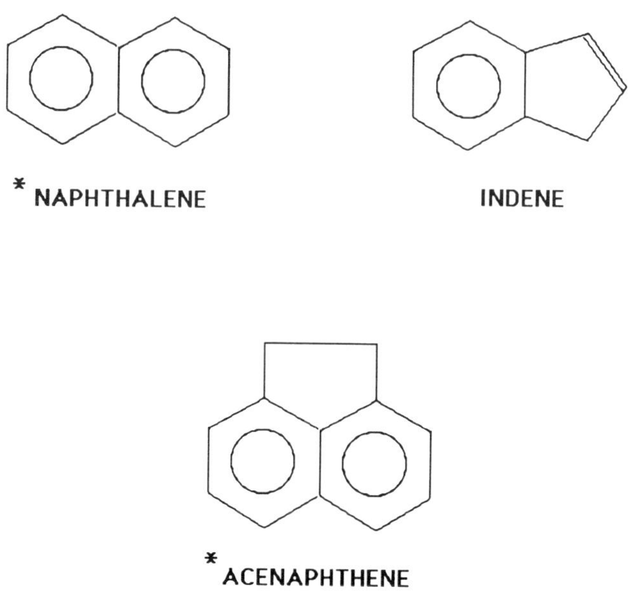

Fig. 3. Polynuclear homoaromatic hydrocarbons biodegraded under anaerobic conditions. All of the listed compounds have been shown to be degraded by complex methanogenic communities. *: compounds which are also degraded under denitrifying conditions.

consortia took a much longer time to degrade 1-2 mM naphthalene (Grbic-Galic, 1989) or acenaphthene. The naphthalene consortium, which was enriched through three consecutive steps (lignocellulose, ferulate, and finally benzene as substrates), converted naphthalene to stoichiometric amounts of methane and carbon dioxide in six months of incubation, following a month-long acclimation lag. The acenaphthene consortium, enriched through the same three consecutive steps and an additional, fourth step (naphthalene as the substrate), degraded acenaphthene to CH_4 and CO_2 (74% of the substrate converted to these gases) in seven months, which includes a 1.5-months long acclimation lag. The difference in rates may be due to the more efficient acclimation history of ground water bacteria in a creosote-contaminated site, where small polynuclears represent a significant pollutant fraction. The degradation of polynuclear aromatic hydrocarbons by complex methanogenic communities may be considered slow, but is significant. Under the conditions of an oil reservoir, several months may not be an overly long period of time.

2.2 Nonoxygenated Heterocyclic Aromatic Hydrocarbons

Heterocyclic pollutants which are prone to methanogenic microbial degradation include both nitrogen- and sulfur-heterocyclic compounds. These chemicals, which comprise small but significant fractions of oil and creosote, are frequently found as contaminants in ground water. Therefore, some of the earliest data on methanogenic degradation of nitrogen- and sulfur-heterocycles were obtained using aquifer-derived methanogenic microcosms (Godsy et al., 1987, 1989; Pereira et al., 1987; Godsy and Grbic-Galic, 1989; Kuhn and Suflita, 1989). Anaerobic sludge, soil, and freshwater sediment microorganisms, however, also degrade these compounds (Bak and Widdel, 1986; Pereira et al., 1988; Berry et al., 1987b; Madsen et al., 1988). As suggested earlier by Chapman (1979), heterocyclic compounds easily undergo an initial hydroxylation reaction, regardless of the conditions (aerobic or anaerobic), because the heteroatoms render the ring structure less stable. Pereira et al. (1988), using [^{18}O]water, proved that the oxygen functionality for oxidation of nitrogen heterocycles was derived from water. Fig. 4 shows this mechanism for quinoline (conversion to quinolinone), and Fig. 5 lists oil- and creosote-derived nitrogen and sulfur heterocycles which have been shown to be completely biodegraded methanogenically. The earliest transformation intermediates have been detected and identified for indole (oxindole; Berry et al., 1987), quinoline and isoquinoline (quinolinone and isoquinolinone, respectively; Pereira et al., 1988; Godsy et al., 1989). Such an intermediate was not identified for benzothiophene, but a long series of oxidized intermediates, including both homocyclic and sulfur heterocyclic mononuclear aromatic compounds (e.g., thiophene-2-ol, 4-hydroxybenzenesulfonic acid, alpha-hydroxyphenylacetic acid, benzyl alcohol), alicyclic compounds, and aliphatics (alcohols and acids), have been shown (Godsy and Grbic-Galic, 1989). Consequently, it is very likely that the initial transformation mechanism is the same for nitrogen and sulfur heterocycles.

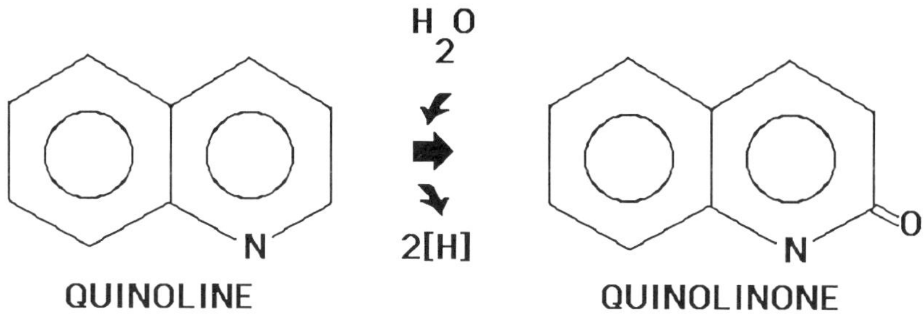

Fig. 4. Initial oxidative reaction of transformation typical for heterocyclic aromatic hydrocarbons which occurs under anaerobic, and also frequently under aerobic conditions (an alternative reaction under aerobic conditions is oxygenation, rather than hydroxylation, of the ring).

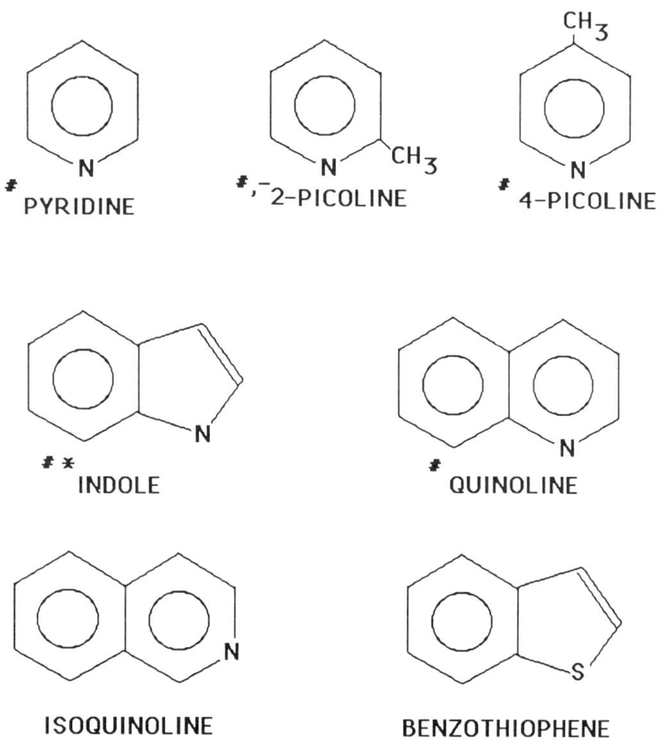

Fig. 5. Heterocyclic aromatic hydrocarbons demonstrated to be susceptible to anaerobic transformation. All of the compounds listed except 2-picoline have been shown to undergo decomposition to carbon dioxide and methane. *: compounds which are also degraded under denitrifying conditions. #: compounds which are also degraded under sulfate-reducing conditions. #,-: compounds degraded under sulfate-reducing, but not under methanogenic conditions.

Laboratory microcosms derived from a creosote-contaminated ground water aquifer in Pensacola, Florida, completely mineralized 40 mg/l indole in 25 days, 20 mg/l quinoline or isoquinoline in 55 or 60 days, and 10 mg/l of benzothiophene in 30 days, respectively (Grbic-Galic, 1989; Grbic-Galic et al., 1990). The rates reported for 0.2 mM indole in 9% anaerobic digested sludge at the same incubation temperature (22°C) were comparable (25 days), but for soil and freshwater sediments they were much slower (110-136 days; Madsen et al., 1988). In the sludge, an increase in microbial concentration, as well as an increase in temperature to 35°C, resulted in increased rates. The concentration of microorganisms in the Pensacola, Florida, aquifer solids was estimated at approximately 10^5 cells per gram of dry aquifer material (Godsy and Grbic-Galic, 1989). The concentration of bacteria in the experiments with soil and freshwater sediment (Madsen et al., 1988) was not reported, but it is known that the biomass levels in such habitats are at least two orders of magnitude higher than in ground water aquifers. Therefore, the difference in rates in aquifer microcosms versus sediment and soil microcosms may be due to previous acclimation of the microorganisms in the creosote-contaminated ground water aquifer to degradation of heterocyclic compounds (water-soluble fraction of creosote is significantly enriched in heterocycles). On the contrary, aquifer solid microcosms derived from a landfill leachate-contaminated aquifer in Norman, Oklahoma, took approximately eight months to degrade 0.2 mM pyridine and 4-picoline under methanogenic conditions (Kuhn and Suflita, 1989). It is possible that mononuclear heterocyclic compounds are more recalcitrant than binuclears under anaerobic conditions.

3 SULFATE AS AN ELECTRON ACCEPTOR: SULFATE REDUCTION

Sulfate reducers require sulfate as an exogenous electron acceptor for their anaerobic metabolism. This type of respiration occurs in various anaerobic habitats, usually after nitrate, an alternate electron acceptor, is depleted, and redox potential (E_h) drops to -200 mV at pH 7 (Stotzky, 1974; Postgate, 1984). Depending on pH, the range of E_h at which sulfate reduction by microorganisms occurs can be extended into more positive, as well as more negative regions. Sulfate is reduced to sulfide in several steps; if sulfate is unavailable, some sulfate-reducing bacteria can also use other oxidized sulfur species for respiration (e.g., sulfur or thiosulfate). In nature, methanogens and sulfate reducers frequently compete for common resources (e.g., hydrogen); since sulfate reducers are more successful, sulfate reduction will usually inhibit methanogenesis as long as significant concentrations of sulfate are present (Zehnder, 1978).

So far, only nitrogen heterocyclic hydrocarbons have been shown to undergo anaerobic degradation with sulfate as an electron acceptor. Bak and Widdel (1986) observed this activity in sulfate-rich marine sediments; the compounds transformed included pyridine, indole, and quinoline. A pure culture of a sulfate reducer, *Desulfobacterium indolicum* gen. nov., sp. nov., was isolated from sediment microcosms. This bacterium utilized a range of compounds, including indole and quinoline, as growth substrates, with

Airforce Office of Scientific Research grant AFOSR-88-0351, and by U.S. Department of Interior grant DOI-14-08-0001A-0357. The views and conclusions contained in this document are those of the author and should nt be interpreted as necessarily representing the official policies or endorsements, either express or implied, of the granting agencies or the U.S. Government.

8 REFERENCES

Bak, F. and Widdel, F., 1986. Anaerobic degradation of indolic compounds by sulfate-reducing enrichment cultures, and description of *Desulfobacterium indolicum* gen. nov., sp. nov. Arch. Microbiol., 146: 170-176.

Barker, H.A., 1967. Biochemical functions of corrinoid compounds. Biochem. J., 105: 1-15.

Berry, D.L., Francis, A.J. and Bollag, J. M., 1987a. Microbial metabolism of homocyclic and heterocyclic aromatic compounds under anaerobic conditions. Microbiol. Rev., 51: 43-59.

Berry, D.L., Madsen, E.L. and Bollag, J.-M., 1987b. Conversion of indole to oxindole under methanogenic conditions. Appl. Environ. Microbiol., 53: 180-182.

Bostrom, K., 1967. Some pH-controlling redox reactions in natural waters. In: R.E. Gould (Editor), Equilibrium Concepts in Natural Waters. American Chemical Society, Washington, D.C., pp. 286-311.

Chapman, P.J., 1979. Degradation mechanisms. In: P.H. Pritchard (Editor), Proceedings of the Workshop: Microbial Degradation of Pollutants in Marine Environments. U.S. EPA Report EPA-600/9-79-012, Gulf Breeze, Florida, pp. 28-66.

Edwards, E. and Grbic-Galic, D., 1990. Anaerobic biodegradation of homocyclic aromatic compounds. Abstr. Annu. Meet. Amer. Soc. Microbiol. American Society for Microbiology, Washington, D.C., Q-50, p. 196.

Engler, R.M., Antie, P.A. and Patrick, W.H., Jr., 1976. Effect of dissolved oxygen on redox potential and nitrate removal in flooded swamp and marsh soils. J. Environ. Qual., 5: 230-235.

Evans, W.C., 1977. Biochemistry of the bacterial catabolism of aromatic compounds in anaerobic environments. Nature, 270: 17-22.

Evans, W.C., and Fuchs, G., 1988. Anaerobic degradation of aromatic compounds. Ann. Rev. Microbiol., 42: 289-317.

Evans, P.J. and Young, L.Y., 1990. Toluene catabolism and its effect on *o*-xylene loss in denitrifying cultures. Abstr. Annu. Meet. Amer. Soc. Microbiol.. American Society for Microbiology, Washington, D.C., Q-46, p. 296.

Firth, R.A., Hill, H.A., Pratt, J.M., Thorp, R.G. and Williams, R.J., 1969. The chemistry of vitamin B12. Some further formation constants. J. Chem. Soc., A 1969: 381-386.

Fukuzumi, S. and Kochi, J.K., 1981. Electrophilic aromatic substitution: charge-transfer excited states and the nature of the activated complex. J. Am. Chem. Soc., 103: 7240-7252.

Godsy, E.M. and Grbic-Galic, D., 1989. Biodegradation pathways for benzothiophene in methanogenic microcosms. In: G.E. Mallard and S.E. Ragone (Editors), U.S. Geological Survey Toxic Substances Hydrology Program-Proceedings of the Technical Meeting. U.S. Geological Survey Water Resources Investigation Report 88-4220, Reston, Virginia, pp. 559-564.

Godsy, E.M., Goerlitz, D.F. and Grbic-Galic, D., 1987. Anaerobic biodegradation of creosote contaminants in natural and simulated ground water ecosystems. In: B.J. Franks (Editor), U.S. Geological Survey Toxic Waste-Ground Water Contamination Program: Proceedings of The Third Technical Meeting. U.S. Geological Survey Open-File Report 87-109, Tallahassee, Florida, pp. A17-A19.

Godsy, E.M., Goerlitz, D.F. and Grbic-Galic, D., 1989. Transport and degradation of water-soluble creosote-derived compounds. In: D.T. Allen, Y. Cohen, and I.R. Kaplan (Editors), Intermedia Pollutant Transport: Modeling and Field Measurements. Plenum Press, New York, pp. 213-236.

Grbic-Galic, D., 1989. Microbial degradation of homocyclic and heterocyclic aromatic hydrocarbons under anaerobic conditions. Dev. Industr. Microbiol., 30: 237-253.

Grbic-Galic, D., 1990. Anaerobic microbial transformation of nonoxygenated aromatic and alicyclic compounds in soil, subsurface, and freshwater sediments. In: J.-M. Bollag and G. Stotzky (Editors), Soil Biochemistry, vol. 6. Marcel Dekker, New York, pp. 117-189.

Grbic-Galic, D. and Vogel, T.M., 1987. Transformation of toluene and benzene by mixed methanogenic cultures. Appl. Environ. Microbiol., 53: 254-260.

Grbic-Galic, D., Henry, S.M., Godsy, E.M., Edwards, E. and Mayer, K.P., 1990. Anaerobic degradation of aromatic hydrocarbons and aerobic degradation of trichloroethylene by subsurface microorganisms. Proceedings of the American Chemical Society Meeting 1990, (in print), 46 pp.

Healy, J.B., Jr., Young, L.Y. and Reinhard, M., 1980. Methanogenic decomposition of ferulic acid, a model lignin derivative. Appl. Environ. Microbiol., 39: 436-444.

Hopper, D.J., 1976. The hydroxylation of *p*-cresol and its conversion to *p*-hydroxybenzaldehyde in *Pseudomonas putida*. Biochem. Biophys. Res. Commun., 69: 462-468.

Hopper, D.J., 1978. Incorporation of [^{18}O]water in the formation of *p*-hydroxybenzyl alcohol by the *p*-cresol methylhydroxylase from *Pseudomonas putida*. Biochem. J., 175: 345-347.

Hopper, D.J. and Taylor, D.G., 1977. The purification and properties of *p*-cresol-(acceptor) oxidoreductase (hydroxylating), a flavocytochrome from *Pseudomonas putida*. Biochem. J., 167: 155-162.

Knowles, R., 1982. Denitrification. Microbiol. Rev., 46: 43-70.

Kochi, J.K., Tang, R.T. and Bernath, T., 1973. Mechanisms of aromatic substitution. Role of cation-radicals in the oxidative substitution of arenes by cobalt(III). J. Am. Chem. Soc., 95: 7114-7123.

Kuhn, E.P. and Suflita, J.M., 1989. Microbial degradation of nitrogen, oxygen and sulfur heterocyclic compounds under anaerobic conditions: studies with aquifer samples. Environ. Toxicol. Chem., 8: 1149-1158.

Kuhn, E.P., Colberg, P.J., Schnoor, J.L., Wanner, O., Zehnder, A.J.B. and Schwarzenbach, R.P., 1985. Microbial transformation of substituted benzenes during infiltration of river water to ground water: laboratory column studies. Environ. Sci. Technol., 19: 961-968.

Kuhn, E.P., Zeyer, J., Eicher, P. and Schwarzenbach, R.P., 1988. Anaerobic degradation of alkylated benzenes in denitrifying laboratory aquifer columns. Appl. Environ. Microbiol., 54: 490-496.

Lovley, D.R. and Lonergan, D.J., 1990. Anaerobic oxidation of toluene, phenol, and *p*-cresol by the dissimilatory iron-reducing organism, GS-15. Appl. Environ. Microbiol., 56: in print.

Lovley, D.R., Baedecker, M.J., Lonergan, D.J., Cozzarelli, I.M., Phillips, E.J.P. and Siegel, D.I., 1989. Oxidation of aromatic contaminants coupled to microbial iron reduction. Nature, 339: 297-300.

Madsen, E.L., Francis, A.J. and Bollag, J.-M., 1988. Environmental factors affecting indole metabolism under anaerobic conditions. Appl. Environ. Microbiol., 54: 74-78.

Major, D.W., Mayfield, C.I. and Barker, J.F., 1988. Biotransformation of benzene by denitrification in aquifer sand. Ground Water, 26: 8-14.

McCabe, C., Sassen, R. and Saffer, B., 1987. Occurrence of secondary magnetite within biodegraded oil. Geology, 15: 7-10.

Mihelcic, J.R. and Luthy, R.G., 1988a. Degradation of polycyclic aromatic hydrocarbon compounds under various redox conditions in soil-water systems. Appl. Environ. Microbiol., 54: 1182-1187.

Mihelcic, J.R. and Luthy, R.G., 1988b. Microbial degradation of acenaphthene and naphthalene under denitrification conditions in soil-water systems. Appl. Environ. Microbiol., 54: 1188-1198.

Mikesell, M.D., Kukor, J.J. and Olsen, R.H., 1990. Isolation of hydrocarbon-degrading denitrifying bacteria from contaminated aquifer material. Abstr. Annu. Meet. Amer. Soc. Microbiol. American Society for Microbiology, Washington, D.C., Q59, p. 298.

Nazina, T.N. and Razonova, Y.P., 1980. Ecological conditions for the distribution of methane producing bacteria in oil bearing strata of Apsheron. Mikrobiologiya (Moscow), 49: 123-129.

Pereira, W.E., Rostad, C.E., Leiker, T.J., Updegraff, D.M. and Bennett, J.L., 1988. Microbial hydroxylation of quinoline in contaminated ground water: evidence for incorporation of the oxygen atom of water. Appl. Environ. Microbiol., 54: 827-829.

Pereira, W.E., Rostad, C.E., Updegraff, D.M. and Bennett, J.L., 1987. Fate and movement of azaarenes in an aquifer contaminated by wood-treatment chemicals. Environ. Toxicol. Chem., 6: 163-176.

Ponnamperuma, F.N., 1972. The chemistry of submerged soils. Adv. Agron., 24: 29-96.

Postgate, J.R., 1984. The Sulphate-Reducing Bacteria. Cambridge University Press, New York.

Reinhard, M., Goodman, N.L. and Barker, J.F., 1984. Occurrence and distribution of organic chemicals in two landfill leachate plumes. Environ. Sci. Technol., 18: 953-961.

Schlesener, C.J., Amatore, C. and Kochi, J.K., 1984. Kinetics and mechanism of aromatic oxidative substitutions via electron transfer. Application of Marcus theory to organic processes in the endergonic region. J. Am. Chem. Soc., 106: 3567-3577.

Schrauzer, G.N. and Sibert, J.W., 1969. Electron transfer reactions catalyzed by vitamin B12 and related compounds: the reduction of dyes and riboflavin by thiols. Arch. Biochem. Biophys., 130: 257-266.

Schwarzenbach, R.P., Giger, W., Hoehn, E. and Schneider, J.K., 1983. Behavior of organic compounds during infiltration of river water to ground water. Field studies. Environ. Sci. Technol., 17: 472-479.

Stotzky, G., 1974. Activity, ecology, and population dynamics of microorganisms in soil. In: A. Laskin and H. Lechevalier (Editors), Microbial Ecology. CRC Press, Inc., Cleveland, Ohio, pp. 57-135.

Streitwieser, A., Jr. and Heathcock, C.H., 1976. Introduction to Organic Chemistry. Macmillan Publ. Co., New York.

Stumm, W. and Morgan, J.J., 1981. Aquatic Chemistry. John Wiley & Sons, New York.

Thauer, R.K., Jungermann, K. and Decker, K., 1977. Energy conservation in chemotrophic anaerobic bacteria. Bacteriol. Rev., 41: 100-180.

Vogel, T.M. and Grbic-Galic, D., 1986. Incorporation of oxygen from water into toluene and benzene during anaerobic fermentative transformation. Appl. Environ. Microbiol., 52: 200-202.

Ward, D.M., Atlas, R.M., Boehm, P.D. and Calder, J.A., 1980. Microbial biodegradation and chemical evolution of oil from the Amoco spill. AMBIO, J. Human Environ. Res. Manag., Royal Swedish Acad. Sci., 9: 277-283.

Whiteley, H.R., 1952. The fermentation of purines by *Micrococcus aerogenes*. J. Bacteriol., 63: 163-175.

Wilson, B.H., Bledsoe, B. and Kampbell, D., 1987. Biological processes occurring at an aviation gasoline spill site. In: R.C. Averett and D.M. McKnight (Editors), Chemical Quality of Water and the Hydrologic Cycle. Lewis Publ., Inc., Chelsea, Michigan, pp. 125-137.

Wilson, B.H., Smith, G.B. and Rees, J.F., 1986. Biotransformation of selected alkylbenzenes and halogenated aliphatic hydrocarbons in methanogenic aquifer material: a microcosm study. Environ. Sci. Technol., 20: 997-1002.

Young, L.Y., 1984. Anaerobic degradation of aromatic compounds. In: D.T. Gibson (Editor), Microbial Degradation of Organic Compounds. Marcel Dekker, Inc., New York, pp. 487-523.

Zehnder, A.J.B., 1978. Ecology of methane formation. In: R. Mitchell (Editor), Water Pollution Microbiology, vol. II. John Wiley & Sons, New York, pp. 349-376.

Zeyer, J., Eicher, P., Dolfing, J. and Scwarzenbach, R.P., 1990. Anaerobic degradation of aromatic hydrocarbons. In: D. Kamely, A. Chakrabarty, and G.S. Omenn (Editors), Biotechnology and Biodegradation. Gulf Publ. Co., Houston, pp. 33-40.

Zeyer, J., Kuhn, E.P. and Schwarzenbach, R.P., 1986. Rapid microbial mineralization of toluene and 1,3-dimethylbenzene in the absence of molecular oxygen. Appl. Environ. Microbiol., 52: 944-947.

Reprinted from: *Microbial Enhancement of Oil Recovery - Recent Advances*, edited by E.C. Donaldson

CH. R–9

STRICT ANAEROBIC BACTERIA AND THEIR POSSIBLE CONTRIBUTION TO THE ENHANCEMENT OF OIL RECOVERY

Sergey S. Belyaev, Irene A. Charakchian and Veronika G. Kuznetsova
Institute of Microbiology, USSR Academy of Sciences, Moscow 117811, U.S.S.R.

ABSTRACT

Several strict anaerobic bacteria belonging to different phylogenetic taxons were isolated from the Tatar and Siberian oil-fields under different physico-chemical conditions. Data on their morphology and physiology were presented. All isolated strains were capable of producing oil-releasing compounds: biopolymers, organic acids or gases. Methanogenic bacteria were shown to produce polysaccharides. The polysaccharide of Methanococcoides euhalobius was partly purified and characterized. An active process of modern acetogenesis from CO_2 + H_2 was revealed in oilfields that were studied. Some acetogenic strains capable of producing volatile fatty acids from CO_2 and H_2 were isolated from stratal waters of the Tatar and Siberian oil fields.

INTRODUCTION

Strict anaerobic bacteria are an important component of any deep subsurface niche and, in particular, of the communities forming in the oil deposits of many strata. They are identified in anoxigenic subsurface waters of different gradients of mineralization up to brines (150-280 g/l) and in core samples taken at different depths. Strict anaerobic bacteria of different phylogenetic taxons are not only revealed, but they appear to function actively in a wide range of environmental conditions. Using modern biogeochemical methods, the processes of methanogenesis and sulfate reduction were shown to take place in both fresh and mineralized subsurface waters, at different depths and at different temperatures. Few data are available about the process of acetogenesis occurring in nature and almost nothing about the activity of this process in the deep subsurface. But, recently we managed to determine that the process of acetate formation from CO_2 + H_2 occurs actively in the formation waters of the oil deposits of Tataria. Significant rates of the processes were registrated within a broad mineralization range up to 200 g/l (Borzenkov et al., 1989).

It is known that anaerobic bacteria have a high bio-degradation potential. They are involved in interspecies hydrogen transfer and in regulating hydrogen flows they influence the complete community in the formation. Undoubtedly, this kind of organism can influence processes of oil recovery to some extent. In this report we shall not mention sulfate-reducing bacteria whose active development in oilfields is undesirable. Our topic will be some methanogenic and acetogenic bacteria which have been isolated from oilfields of Tataria and western Siberia. The physico-chemical conditions in the oil reservoirs of Tataria and western Siberia are very different. The objects of our field investigations in Tataria consist of two oil deposits (the Bondushskoe and Romashkinskoe) which also differ from each other. The depth of the oil bearing horizons in Tataria ranges from 1500 to 1700 m and the temperature of the formation varies from 30° to 40°C. Formation waters of the Bondushskoe oil field are chlorine-calcium-type Devonian brines with a total mineralization up to 300 g/l. Formation waters of the Romashkinskoe oil field belong to a chlorine-sodium type with low mineralization varying from 2.5 to 40 g/l. Oxygen O_2 and H_2S were not found in the formation waters of the Tatar oilfields. The oxidative-reductive condition is slightly reduced. The injection of fresh water in water-flooded oilfields sometimes leads to a decrease in mineralization of the formation waters.

The object of our field investigation in western Siberia is the Mychpay oil field in the Nizhnevartovsk region. The depth of the hydrodynamically connected oil-bearing layers is from 1700 to 1850 m. The temperature ranges from 60° to 80°C. Formation waters of the Mychpay oilfield belong to a hydrocarbon type with a mineralization of about 20 g/l (sulfates are absent). The Mychpay oilfield has been depleted by water flooding for more than 10 years. At present collected waste waters, with a mineralization of about 15 g/l are injected into the stratum.

METHODS

The samples of formation liquids were obtained from producing and injecting wells in sterile vessels under a blanket of N_2 without contact with air.

The enrichment cultures of methanogenic bacteria were obtained using PBBM media for slightly mineralized waters (Kinealy and Zeikus, 1981) and GF media for highly mineralized waters (Charakchian et al., 1989). The enrichment culture of acetogenic bacteria was obtained using the mineral media described by Braun et al. with some modifications (Braun et al., 1979). A mixture of CO_2 and H_2 was used as the growth substrate. Possible processes of sulfate reduction and methanogenesis were inhibited using sodium molydate (1 mM) and 2-bromethane sulfonic acid (10 mM), respectively. For preliminary purification the method of serial dilutions was used. For methanogenic bacteria the dilutions were made with addition of the antibiotics penicillin (2 mg/ml) and vancomycin (100 μg/ml). For acetogenic bacteria the dilutions were made using the inhibitors of sulfate reduction and methanogenesis. Pure cultures were obtained from a single colony formed on solid media in an anaerobic chamber (Scholzen Technic, FRG). The anaerobic techniques of Hungate and Balch were used for media preparation and isolation of cultures (Balch et al., 1979).

Methane and acetate were determined using a Pyc Unicum 4550 gas chromatograph with a flame ionization detector. The total carbohydrate content was determined using the phenol-sulfuric acid method (Dubois et al., 1956). To obtain the polysaccharide samples, the cultural medium and lysed cell fractions were treated several times with chloroform (N-butanol mixture, 5:1) until the protein components were completely removed. Then the deproteinized supernatants were dialysed and lyophilized. The composition of the polysaccharide was determined using the carbohydrate analyzer LC-200 made by Biotronic, Germany. The viscosity was determined using a glass capillar viscometer at 20°C.

RESULTS AND DISCUSSION

Thermophilic methanogenic bacteria (TM 337) were isolated from Siberian oilfields, characterized by high formation temperature and low mineralization of formation waters. The cells of this organism consist of long thin rods, often forming filaments. The rods were 0.3 - 0.4 μm in diameter and 5.0 - 7.0 μm in length with filaments were 12 - 14 μm in length. The cells of a strain of TM 337 were Gram negative. Ultrathin sections revealed an electron-dense cell wall sacculi composed of pseudomurein typical for the order Methanobacterialis. The temperature range for growth was from 20° to 80°C, with an optimum temperature of 65°C. The optimum pH for growth was 7.0 - 7.4. Generation time under the optimal conditions was 12.9 ± 0.5 hours. Isolated TM 337 could use only CO_2 and H_2 as substrates for growth and methanogenesis (it had no special growth requirements). The organism showed a considerable tolerance to NaCl; at a medium mineralization higher than 40 g/l the intensity of methanogenesis was reduced, however, the organism could grow and produce methane in a solution of NaCl up to 100 g/l (Fig. 1).

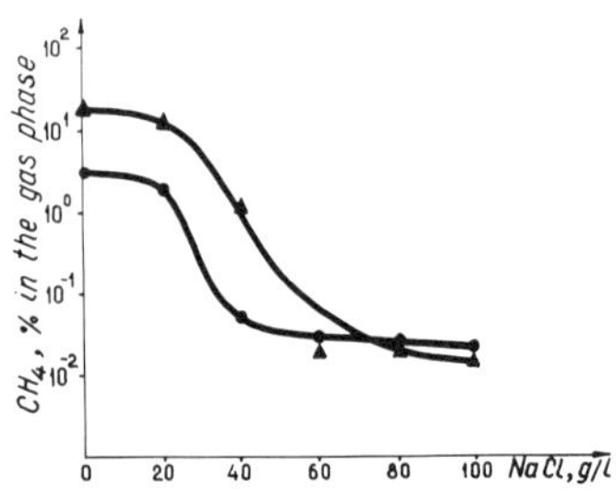

Fig. 1. Methane concentration in the gas phase at different NaCl content.

—▲— Methanobacterium str.TM 337.

—●— Methanobacterium str.41.

No definite antigenic relationship occurred between strain TM 337 and the known thermophilic methanogenic strains Methanobacterium thermoautotrophicum Δ H, Methanobacterium thermoformicicum, Methanobacterium bryantii MoH and Methanothrix thermophilum. Weak fluorescence (+1) was observed in the reaction with antiserum to Methanobacterium thermoaggregans. The fluorescence was so weak, however, that it was not possible to identify our isolate as this strain. The immunological data mentioned above show that the isolate TM 337 is probably a new thermophilic representative of the Methanobacterium genus.

Another rod-shaped methanogenic bacterium (labeled No. 41) was isolated from the low mineralized waters of the Romashkinskoe oil field (Tataria). The cells of this organism were 0.6 ± 0.1 μm in diameter and 3.5 - 4.5 μm in length with some filaments ranging between 7 - 14 μm. The cells stained Gram positive and from electron micrographs the cell envelopes were typical in organization of Methanobacteriales. The organism could grow at temperatures ranging from 20° to 60°C with an optimal temperature of 37°C. The optimal pH value was 7.2 - 7.4. The generation time under optimal conditions was 23.4 ± 0.3 hours. Growth and methanogenesis were supported by CO_2 and H_2. Strain 41 also showed some tolerance to NaCl. As in the case of TM 337 the methanogenesis was inhibited at NaCl concentrations greater than 20 g/l, but the organism grew and produced methane up to NaCl concentrations of 120 g/l. Therefore, both of the rod-shaped methanogenic organisms were halotolerant.

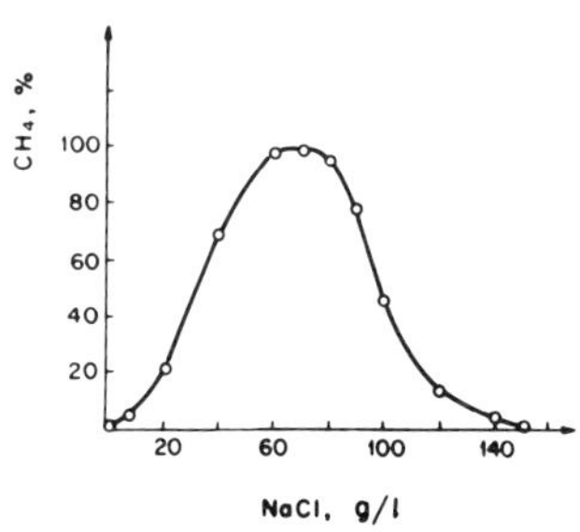

Fig. 2. Development of Methanococcoides euhalobius on the media with different NaCl concentrations.

The halophilic methanogenic bacterium Methanococcoides euhalobius was isolated from formation waters of the Bondushskoe oilfield (Tataria). These waters have a salinity of 140 g/l and contain high Ca^{2+} and Mg^{2+} concentrations (11.2 and 2.4 g/l, respectively) (Obraztsova et al., 1987). It is mesophilic and a moderate halophile. The cells of this organism were small irregular cocci varying in size from 1 - 3 μm with an optimal growth temperature of 37°C. The optimal pH value was from 6.8 - 7.2. Growth and methanogenesis of the organism was possible in NaCl concentrations ranging between 30 - 120 g/l. The optimal NaCl concentrations of the medium was 60 - 70 g/l (Fig. 2). The organism needed high amounts of Ca^{2+} and Mg^{2+}, 68 and 12 mM, respectively, or 10 and 2.49 g/l $CaCl_2.2H_2O$ and $MgCl_2.6H_2O$ as optimum values. Mcc. euhalobius could not grow in the absence of Ca^{2+} in the medium and Mg^{2+}, and at the optimal concentrations could not replace Ca^{2+}.

Methanol and methylamines were the substrates used for growth and methanogenesis. The generation time determined for trimethylamine as the most preferable substrate was 13.7 hours. The organism required biotin for growth.

Under laboratory conditions all of the above-mentioned methanogenic cultures, isolated from different oilfields, produced polysaccharides. The polysaccharide production by different methanogenic organisms is shown in Fig. 3.

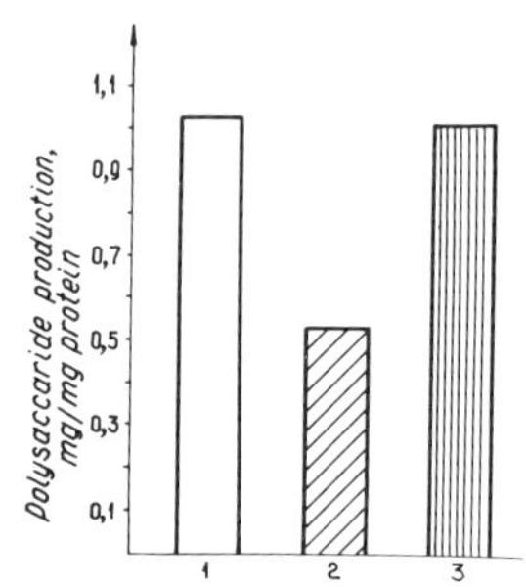

Fig. 3. Polysaccharide production by methanogenic bacteria.

1. Methanobacterium str. 41.
2. Methanococcoides euhalobius.
3. Methanobacterium str. TM 337.

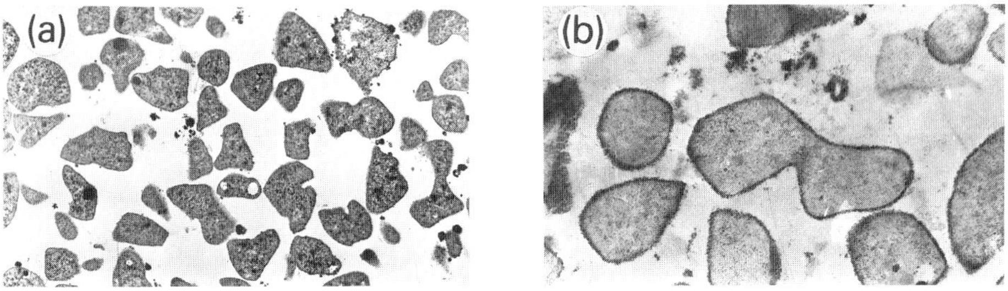

Fig. 4. The cytochemical staining of cells of Mcc. euhalobius for polysaccharide.

a. Cells without staining x6000.

b. Cells stained with alceanic blue x15000.

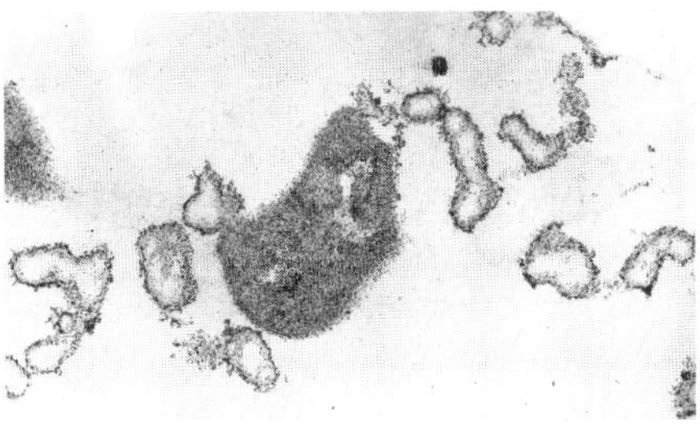

Fig. 5. The excretion of the cell matter to the medium.

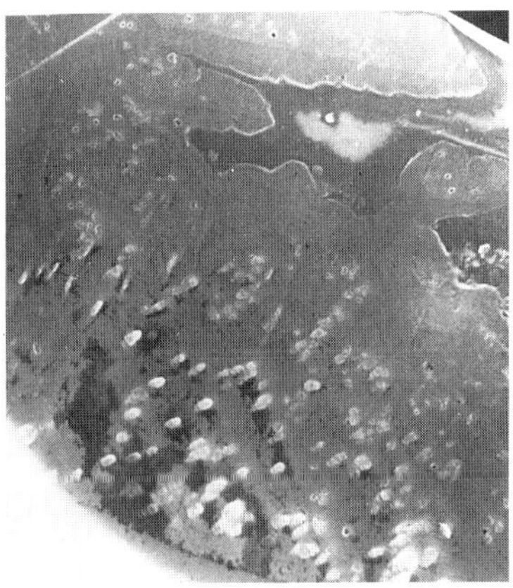

Fig. 6. Biofilm and the cells attached to it.

The production of polysaccharide in the course of the development of Methanococcoides euhalobius was studied in more detail. The cytochemical reactions for polysaccharide with alceanic blue revealed a polysaccharide component within the cell wall. The outer layer of the cell wall had a tendency to slice off and cytochemical staining revealed the polysaccharide component not only in the cell envelope but also in the growth medium (Fig. 4). Electron micrographs showed the excretion of cell matter to the medium (Fig. 5). The excreted matter formed a film which was clearly seen on photographs obtained with the scanning electron microscope (Fig. 6). The cells were attached to this film. In natural environments, the polysaccharide film probably fixes the cells connecting them to one population. We obtained the specimens of polysaccharide for analysis from the cultural liquid and the cells showed that their compositions were identical with glucose and mannose in the ratio of 1.5:1,0.

Some properties of the polysaccharide obtained were studied. The polysaccharide was shown to increase the viscosity of brine. Thus, at 0.08% of concentration of polysaccharide, the viscosity of a solution of 15% NaCl and 0.5% $CaCl_2$ increased from 1.005 ± 0.005 to 14.113 ± 0.005 mPs. The well-known polysaccharide xantan, in the concentration range from 0.01 - 0.08% precipitated from a solution of the same mineralization; but Xantan at the 0.05% concentration in freshwater increased the viscosity up to 10 - 15 cP, or by one order of magnitude more. A polysaccharide of Mcc.euhalobiius had chelating properties. It binds Ca^{2+} ions from $CaCl_2$ solutions beginning from 10^{-3}M and higher. In the salinity range of 40 - 120 g/l (the growth range for Mcc.euhalobius) the polysaccharide content in the medium was approximately constant: 0.5 - 0.7 mg/mg protein. The accumulation of polysaccharide in the medium containing 40 g/l NaCl was a little higher than 0.8 mg/mg protein. This is probably caused by adaptation of the cells to low salinity and protection of the cell's integrety at decreased salinity compared to optimal NaCl concentrations. Snowers and Gunsalus (1988) published data showing that cells of Methanosarcina thermophila formed thick polysaccharide envelopes in a medium with low NaCl concentration, but that the same cells lost this envelope in a marine medium.

The content of the polysaccharide bound to the cells of Mcc.euhalobius in a salinity range of 40 - 80 g/l does not depend on salinity and is equal to 0.1 ± 0.02 mg/mg protein. At a NaCl concentration of 120 g/l the bound polysaccharide content decreased sharply to 0.03 mg/mg protein. The culture growth is also inhibited to some extent (30 - 40%) at this NaCl concentration. The bound polysaccharide content was dependent on Ca^{2+} concentration in the medium (Fig. 7). We should note that the optimal $CaCl_2$ concentrations for bound polysaccharide accumulations do not coincide with the optimal growth $CaCl_2$ concentration (10 g/l). At optimal growth concentration thc polysaccharide content in thc medium and the bound polysaccharide content decreased.

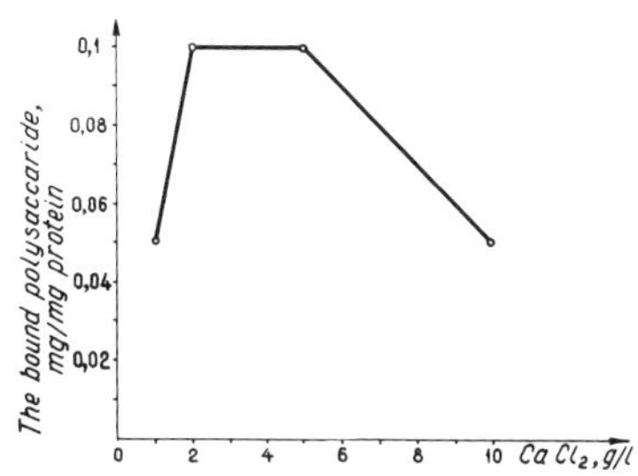

Fig. 7. The dependence of the bound polysaccharide content in the cell fraction of Mcc. euhalobius on the $CaCl_2$ concentration.

SUMMARY

Some bacterial strains producing volatile fatty acids from CO_2 + H_2 were isolated from formation waters of the Tatarian and Siberian oilfields. Two examples from the isolated strains form only acetate (70-100 mg/l) and one strain (No. 4175) forms acetate (80 - 90 mg/l) and propionate (50 - 60 mg/l) (Table 1).

The rated of modern acetogenesis in oil reservoir waters were shown to range from 4.6 - 70 μm acetate/l per day. The number of acetogenic bacteria using H_2 + CO_2 in stratal waters of the Tatarian oil fields was 10^4 - 10^7 cell/l. Therefore, there are a number of reasons to conclude that acetogenic bacteria using hydrogen as energetic source play a significant role in the organic matter turnover, and that the acids they produce, influence the collector solving the carbonates.

TABLE 1

The production of volatile acids from CO_2 + H_2 by the isolated acetogenic strains

	4172 str	4175 str	1236 str
acetic acid (mg/l)	70-80	70-90	90-120
propionic acid (mg/l)	0	50-60	0

The products of strict anaerobic bacteria (methanogenes and acetogenes) isolated from oilfields are active oil-releasing compounds (i.e., gases, organic acids and biopolymers). The ability of the methanogenic halophilic organism Mcc. euhalobius to produce polysaccharide under laboratory conditions is of special interest. Polysaccharide produced by this anaerobic organism resists high salinity combined with high Ca^{2+} and Mg^{2+} concentrations. Moreover, unlike the widely used xantan, carbohydrates do not serve as a substrate for its production, but instead methanol and methylated amines are the substrates. It is possible that the activity of strict anaerobic bacteria in the biosynthesis of oil releasing products is preserved under formation conditions as well.

REFERENCES

Balch, W.E., Fox, G.E., Magrum, L.I., Woese, C.R. and Wolfe, R. S., 1979. Methanogenes: reevaluation of a unique biological group. Microbiol. Rev., 43: 260-296.

Borzenkov, I.A., Charakchian, I.A. and Belyaev, S.S. Acetogenesis in stratal waters of oil deposits. Abstr. 9th Int. Symp. on Environmental Biogeochemistry 4-8 Sept., 1989. Moscow, USSR, p. 189.

Brawn, M., Schoberth, S. and Gottschalk, G., 1979. Enumeration ofbacteria forming acetate from H_2 and CO_2 in anaerobic habitats. Arch. Microbiol., 120: 201 - 204.

Charakchian, I.A., Mityushina, L.L., Kuznetsova, V.G. and Belyaev, S.S., 1989. Methanococcoides euhalobius morphology and fine structure and the role of calcium in morphogenesis. Mikrobiologia, 58: 825 830.

Dubois, M., Gilles, K.A., Hamilton, I.K., Rebers, P.A. and Smith, P., 1956. Calorimetric method for determination of sugars and related substances. Anal. Chemistry 28: 350 - 356.

Obraztsova, A.Ya., Shipin, O.V., Bezrukova, L.V. and Belyaev, S.S., 1987. Characteristics of the coccoid methylotroophic methanogenic organism Methanococcoides euhalobius sp. nov.

Reprinted from: *Microbial Enhancement of Oil Recovery – Recent Advances*,
edited by E.C. Donaldson

CH. R–10

A METHOD TO DETERMINE THE NUMBER OF HYDROCARBON DEGRADING BACTERIA IN MICROBIAL ENHANCED OIL RECOVERY

Lichaa, Ada, and Oppenheimer, Carl. Alpha Environmental Inc.
P.O. Box 90218 Austin, Texas 78709.

ABSTRACT

A method is described that measures the concentration of hydrocarbon degrading bacteria (HDB) in samples of fluid collected from oil wells treated with MEOR bacteria. Total HDB populations were enumerated by a serial dilution method and direct counted by optical microscopy. Examples from three field studies indicate that hydrocarbon degrading bacterial populations can be monitored through time and that changes in these populations in an MEOR treated reservoir may provide a measure of bacterial transport.

INTRODUCTION

The changes in bacterial populations in treatment and observation wells before and after MEOR applications is an indication of the condition of the system, and may provide a measure of the pattern of transport of the bacteria in the subsurface geological environment. To measure these changes, it is essential to have a simple, reliable and fast technique to be able to enumerate hydrocarbon degraders in a large number of samples. The method described here has been used to enumerate the HDB population of over 1000 wells treated for MEOR or paraffin control purposes. Bacterial populations were measured before treatment to establish the background levels of indigenous hydrocarbon degraders, and after treatment to evaluate the success of the MEOR application. To treat an oil well Alpha adds variable amounts of hydrocarbon degrading bacteria, a biocatalyst, inorganic nutrients and water all at concentrations related to the well characteristics (Oppenheimer and Hiebert, 1989). Field experience to date lead Alpha to believe that the biocatalyst provides a condition for aerobic activity . We also believe that the bacterial products control the crystalline nature of paraffins converting them to a thixotroph. The oil in place may be released by the following mechanism: Hydrocarbon-degrading microorganisms transported by formation brines attach to rock and oil surfaces. As the microbes metabolize oil, they produce acids (which act as surfactants), gases (primarily CO_2), and new cells. The net effect of the metabolic by-products is to chemically and physically release oil droplets from rock surfaces and pore spaces.

The enumeration technique described here is a modification of the most probable number method (de Mann, 1975; Standard Methods for the Examination of Water and Wastewater, 1985). A dilution technique was used (Oppenheimer, C., Miget, R., and Kator, H.,1979) to enumerate hydrocarbon degraders based on replicate dilution flasks in which the bacterial concentration was determined by changes in the appearance of the oil or water in the dilution flasks compared to control flasks. In Alpha's Laboratory the same dilution and observation technique was used. In addition, the numbers of bacteria were determined by microscopy. In the present study, this technique was used to determine microbial concentrations of three MEOR projects: (1) a water flood injection in The West Geff field of the Aux Vasas formation in Fairfield, Illinois; (2) a single well injection in the LaVernia oil field of the Navarro formation, Wilson County, Texas, and (3) a single well injection in the Shannon Sandstone Reservoir of the Teapot Dome Field, Naval Petroleum Reserve#3, Natrona County, Wyoming.

2 MATERIALS

2.1 Equipment

A Nikon Biological Optiphot differential interference contrast "NT" microscope was used to enumerate the hydrocarbon degrading bacteria (HDB). The Nomarski differential interference contrast (DIC) allows direct observation of the details of unstained bacteria and the presence of oil water interfaces does not distort the cell morphology. A 40x dry objective was used to eliminate the problems associated with the use of immersion oil between the cover slip and the objective.

2.2 Glassware

125 ml Erlenmeyer flasks were used for sample dilutions, and sterile disposable 5 ml pipettes were used for the dilution transfers. All washed glassware was rinsed with ethyl alcohol followed with double distilled water. Our experience shows the adverse effect of surfactants used in routine washing. The alcohol removes this potential problem and uniform results are obtained routinely. Microscopic glass slides with covered glass were used to examine few drops of each dilution sample for the microscopic reading.

3 METHODS

3.1 Nutrient Seawater Media (NSM)

The media used for analysis of the hydrocarbon degrading bacteria consisted of: filtered seawater (30 ppt, if seawater is diluted, distilled and then deionized water may be added), 1 gm. ammonium sulfate/ liter, and 0.1 gm. of dipotassium phosphate/liter. The pH of the media is adjusted to 7.6 and 45 ml dispensed in 125 ml Erlenmeyer flasks covered with aluminum foil tops, and autoclaved.

3.2 Sampling and Handling

Samples of oil and water from treatment and observation wells were collected in sterile 8 oz. glass vials at the wellhead during normal pumping operations. The bottles were cooled, packed in ice and transported to the laboratory by express mail within twenty-four hours. Samples were analyzed for microbial activity within twenty-four hours of arrival.

The samples were brought to room temperature, and agitated in a vibratory disc for one minute to thoroughly mix the water and oil. The water was allowed to separate for a few minutes from the oil and an inoculum sample taken from the water phase. The time required for the separation to take place depended upon the viscosity of the oil and the properties of the water.

3.3 Measurement of Bacterial Concentration

Bacterial populations in the sample fluids were measured by the serial dilution method and direct microscopic count. All serial dilutions were prepared in duplicate. Flasks containing 45 ml of the nutrient seawater media were inoculated with 5 ml of the water phase of the oil well sample. After thoroughly mixing each dilution with a vibratory mixer, 5 ml aliquots from successive dilutions were transferred to continue the 1:10 dilution series to a point where it has been established that bacteria are eliminated. This point can be estimated, but continual experience can be used to establish the optimum dilution for any series. After the dilution was made, six drops of sterile crude oil (a light south Texas crude oil was routinely used having an API of approximately 30) were added to each dilution flask. The duplicate set was incubated aerobically and agitated mechanically for three to four days at room temperature. After incubation, several drops of sample were transferred to a glass microscope slide. Bacteria per field were counted. Ten fields of view which contain one or more bacteria per field was considered a positive reading for that dilution. This indicated a concentration of bacteria of approximately 10^5 cells per cubic centimeter of fluid. Concentration of bacteria in the sample was calculated from the last positive dilution flask.

4 RESULTS

Figure 1 shows the morphology of the bacteria and relationship between oil and water as seen in the microscope. The differential interference contrast allows the clear differentiation of bacteria in the

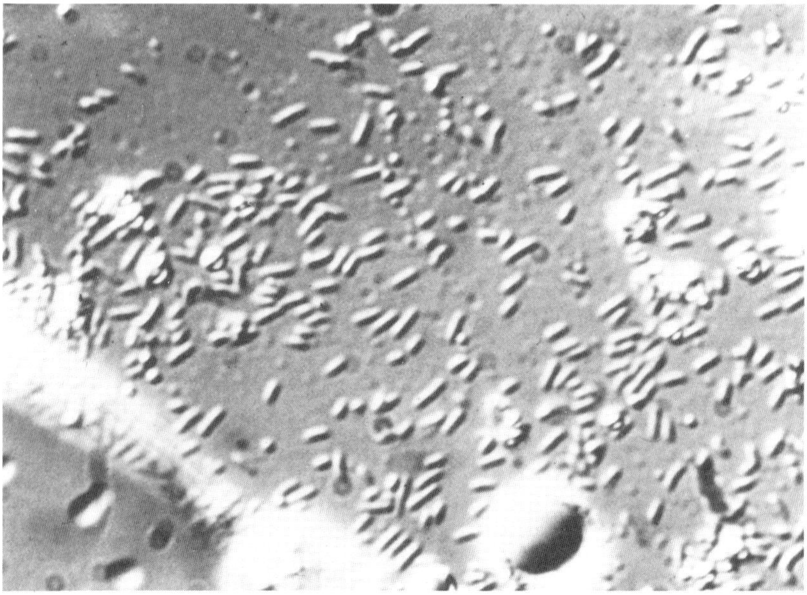

FIGURE 1. Emulsion of oil in nutrient seawater media. At the right side, some of the cells are driven out of the oil droplet into the water. In the reproduction of the photograph above, you will note that the bacterial phase contrast changes from light in oil to dark in water.

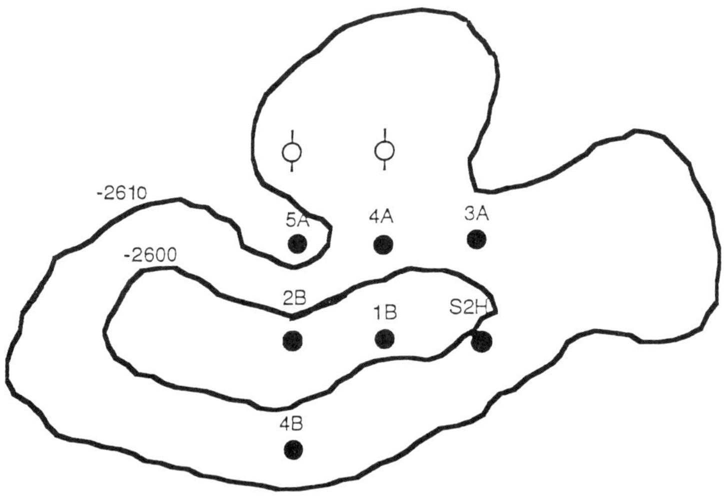

Figure 2. Pennington/Stevenson Lease, Wayne County, Illinois

C.I.=10 ft. well on 10 acre spacing

water and oil phases. The enumeration technique was used to monitor the activity levels of the wells, and the pattern of transport of the microorganisms in the treated field. Different fields varied in terms of persistence of activity and migration trends.

One of the treated fields, the West Geff field in Fairfield, Illinois, is located in the Aux Vasas Sandstone formation. The Pennington/Stevenson lease covers two thirds of the structure and consists of seven producing and two water injection wells (Figure 2). The treatment consisted of 100 bbls of formation water, 100 lbs of Alpha's bacteria (10^{12} cell/gm), 5 lbs of nutrients and 1 bbl of biocatalyst, which were introduced through the two water injection wells. Samples were taken from all seven producing wells before treatment (7/17/87), and starting two days after (8/7/87) up to a year (8/3/88) after treatment. No additional treatment was made. Results are presented in Table 1. The population of hydrocarbon degrading bacteria is reported in log number of bacteria per ml of reservoir fluid. Three production wells, 4B, S2H and 2B, located furthest from the injection wells, were the first to experience an increase in microbial population, within two weeks of treatment. Wells 5A,4A, and 3A, all adjacent to the injection wells, did not show increased microbial activity for up to a month after treatment Within six weeks of treatment however, all wells in the field showed a moderate level of microbial population increase. The increase in field-wide microbial population correlated well with an oil production increase during the first four months of post-treatment (Oppenheimer and Hiebert, 1989). At eight and a half months after treatment the HDB population for most of the wells was above pre-treatment levels. This concentration was maintained even a year after treatment.

The Lavernia field is located in the Navarro formation in Wilson County,Texas. The regional structure consists of a Northeast/Southwest striking monocline which dips slightly to the south-southeast. The La Vernia field trends NE/SW and is about one mile wide. Wells were treated with 26 bbls of formation water,1 bbl biocatalyst, 20 lbs bacteria (10^{12} cells/gm), and 3 lbs nutrients. The components were mixed, and delivered to the reservoir through the well annulus by gravity feed. The first field test was initiated in the Vorphal lease in June 1986. Well #7 was selected as the first treatment well (Figure 3). Population levels of hydrocarbon degrading microorganisms from samples collected before and after treatment of the Vorphal lease on a well by well basis are presented in Table 2. Little to no increase in microbial population and activity was observed in the wells that were down-dip from the treatment well. A significant increase in microbial population was observed in the wells to the north east of the treatment well (Vorphal #1A,9). An increase in oil production three weeks after

TABLE 1

Pennington/Stevenson Lease, Wayne County, IL
Population of hydrocarbon degrading bacteria in seven wells, reported in log number of bacteria per ml of formation water.

Date	Well 3A	Well 4A	Well 5A	Well 1B	Well 2B	Well 4B	Well S2H	
7/17/87	0	0	0	0	0	1	0	
8/7/87	0	1	0	0	1	2	0	Treatment
8/14/87	0	1	0	0	2	4	3	8/5/87
8/21/87	0	3	0	0	2	6	4	
9/4/87	0	4	2	1	0	5	5	
9/18/87	1	3	2	3	2	5	5	
12/4/87	0	0	1	0	ND	2	0	
4/25/88	2	2	1	2	2	0	0	
8/3/88	0	ND	1	1	2	2	0	

TABLE 2

Vorphal Lease, La Vernia field, La Vernia, Texas
Population of hydrocarbon degrading bacteria reported in log number of bacteria per ml of formation water.

Date	Well # 1A	Well # 7	Well # 8	Well # 9	Well #13	Well # 3	
May-86	0	0	0	0	0	0	
Jun-86	0	6	0	4	8	0	Well #7 treated
Jul-86	8	0	8	5	0	0	
Aug-86	8	3	1	3	4	0	
Sep-86	2	0	0	0	0	0	
Oct-86	0	0	0	0	0	1	
Nov-86	1	0	0	0	0	0	
Dec-86	0	0	0	0	0	0	
Jan-87	0	2	3	0	1	0	
Feb-87	0	0	0	0	0	0	
Mar-87	0	7	2	7	2	0	
Apr-87	0	0	0	0	0	0	
May-87	0	0	0	0	0	0	
Jun-87	0	7	2	7	2	0	4 wells treated

TABLE 3

Wiseman Lease, La Vernia Field, Texas
Population of hydrocarbon degrading bacteria in five wells reported in log number of bacteria per ml of formation water.

Date	Well#16	Well#17	Well#18	Well#19	Well#20
Aug-86	0	0	0	0	0
Oct-86	1	0	0	0	0
Nov-86	0	1	0	0	0
Dec-86	ND	1	2	0	3
Jan-87	7	4	4	3	5
Mar-87	7	6	6	4	8
Jun-87	0	5	4	3	2

Well #20 treated on
11/19/86

treatment on the Penshorn lease to the northeast of the Vorphal lease led to sampling of up-dip wells where high populations of bacteria were found (Oppenheimer and Hiebert,1989). Well #20 which is the deepest completion and furthest down-dip well on the Wiseman lease (Figure 3) was treated in December, 1986. Wells #16,17,18, and 19 which are located directly up-dip from well #20 were observed for signs of microbial activity. A month after treatment the microbial population had increased significantly even for well #16, which is located furthest from the injection well. Substantial increases in bacterial population were observed in all the wells for a period of six months (Table 3). None of the three leases were monitored beyond June, 1987

The Naval Petroleum Reserve #3 Teapot Dome Field, in Natrona County,Wyoming is located in the Shannon Sandstone formation. This is a highly fractured heterogeneous sandstone with intermingled bioturbated sands and laminated shales. Two wells #61-S-3 and 72-SX-3 (Figure 4), were selected as injection wells in the MEOR pilot area. The treatment composition consisted of: 115 bbls Shannon formation water, 5 bbls Alpha's biocatalyst, 150 lbs bacteria (10^{12} cells/gm) and 5 lbs of inorganic nutrients. The components were mixed, and delivered to the reservoir through the well annulus by gravity feed. The wells were treated on August 30, 1988. The two injection wells together with twenty-two surrounding observation wells were monitored for hydrocarbon degrading bacteria, as presented in Table 4. Forty two days after treatment, thirteen wells showed an increase in the level of the HDB population by two or more orders of magnitude above the natural pre-treatment background levels. Ninety days after treatment the HDB population has fallen to low or zero levels. Analysis made on 2/15/89 and 3/28/89 confirmed these findings. However, the last set of data suggests that some observation wells far from the injection wells showed a slight increase in HDB concentrations above pre-treatment levels, which indicates bacterial migration through the formation.

5 DISCUSSION

Field examples suggest that this enumeration method allows the monitoring of the HDB population in oil reservoirs. They also indicate that bacteria migrated from treatment wells to observation wells. Therefore, the enumeration of MEOR bacteria may be used as indicators of subsurface transport.

It is interesting to notice the difference in HDB population after treatment between the Teapot Dome field and the West Geff and La Vernia fields. Moderately high bacterial concentrations were found for the last two fields even a year after treatment, while they peaked up but

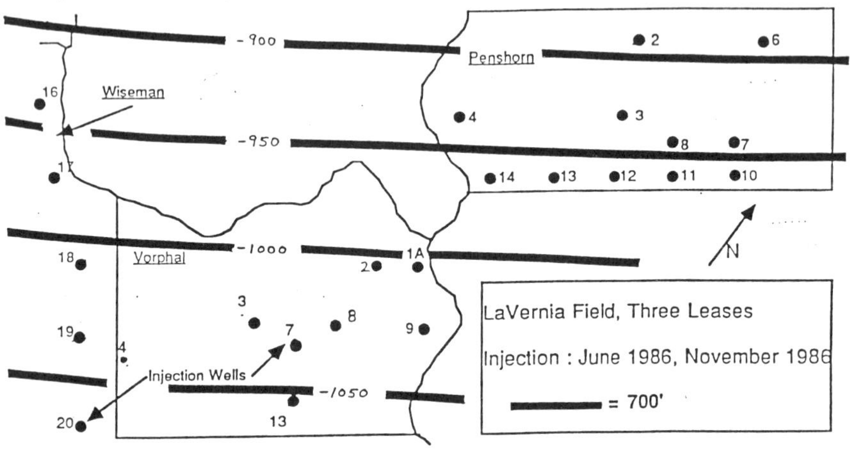

Figure 3. Structure-contour map on top of the Navarro Formation in the area of the LaVernia Field treatment sites. Contour interval = 50 ft. Injection wells are labeled. Most of the remaining wells served as "observation wells" for changes in microbial population and oil production.

TABLE 4

Shannon Reservoir, Teapot Dome, Wyoming
Population of hydrocarbon degrading bacteria before and after treatment of wells 61-SX-3 and 72-SX-3 on 8/30/88 reported in log # of bacteria per ml of well fluid

	Well #											
DATE	58	68	78	21	31	41	51	61	71	22	32	42
7/8/88	2,2	1,2,4	2,2	2,3,4	0,0	4,4	0,0	0,0	1,1	0,0	3,3	0,0
8/29/88	5,4	3,2	0,0	0,0	1,1	nd	3,3	0,0	0,0	0,0	4,4	0,0
8/30/88	TREATMENT											
9/23/88	2,2	5,5	2,2	0,0	3,3	2,2	2,2	5,5	0,0	0,0	5,5	0,0
10/10/88	3,3	0,0	2,4	1,1	2,2	1,1	4,5	4,4	0,0	4,3	6,8	3,3
12/6/88	0,0	0,0	0,0	0,0	1,1	1,2	0,0	1,1	0,0	0,0	2,2	0,0
2/15/89	3,3	1,1	0,0	1,1	1,0	1,1	0,0	1,0	1,1	1,1	2,1	0,0
3/28/89	3,3	1,1	1,2	0,0	0,0	1,1	3,4	1,1	2,1	2,3	2,2	2,1

	Well #											
DATE	52	62	72	82	33	53	63	73	34	44	64	45
7/8/88	3,3	0,0	0,0	nd	1,0,0	0,0	0,0	7,7,5	0,0	0,0	0,0	0,1,1
8/29/88	1,1	2,2	4,4	0,0	0,0	0,0	0,0	0,0	0,0	0,0	2,0	0,0
8/30/88												
9/23/88	0,1	3,2	8,7	3,2	0,0	0,0	1,1	0,0	nd	2,1	1,1	0,0
10/10/88	3,3	3,3	8,7	8,7	8,7	2,2	2,2	4,4	3,4	4,3	2,2	0,0
12/6/88	0,0	0,0	0,0	0,0	0,0	0,0	0,0	1,1	0,0	0,0	0,0	0,0
2/15/89	0,0	0,1	0,0	0,0	1,0	0,0	1,0	2,2	0,1	0,0	0,1	0,0
3/28/89	1,0	1,1	0,0	1,1	0,0	1,1	0,0	1,0	2,1	0,0	1,1	1,0

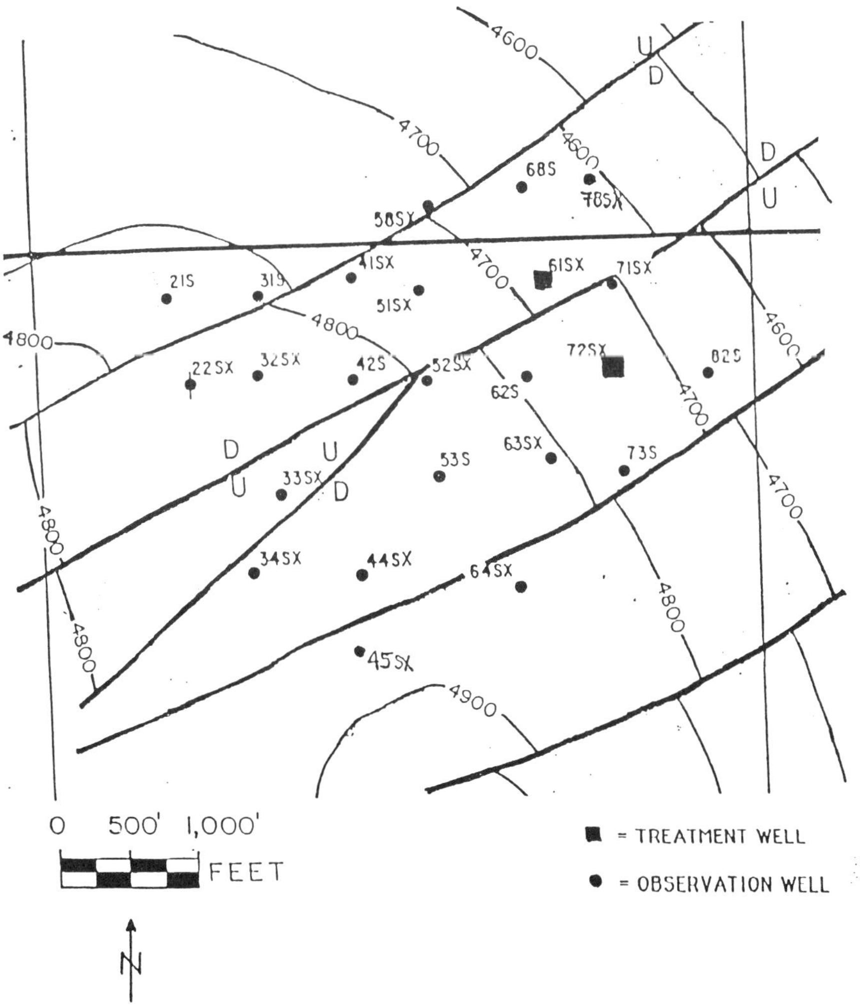

Figure 4. Alpha Environmental/DOE MEOR treatment area structure map on the top of the Upper Shannon Sandstone, Teapot Dome Field, N.P.R. no.3, Natrona County, Wyoming.

then decreased rapidly to low or pre-treatment levels for the Teapot Dome field. Every reservoir is a unique environment, and the population decline may represent the actual population down-phase of a natural population cycle. A nutrient or growth factor could have also been exhausted in the reservoir environment. As the cycle completes itself or the nutrients are replaced, the population should start to grow again. There is also a possibility of stimulation of indigenous HDB populations with nutrients, biocatalyst and/or the metabolic by-products of the MEOR bacteria. The added bacteria could have had a synergistic or antagonistic effect on the natural populations. We also know that many of the strains in a mixed community attach to mineral surfaces. Therefore the analysis of a fluid alone may give low population results.

6 CONCLUSIONS

The direct enumeration technique discussed above allows :

6.1 The measurement of hydrocarbon degrading bacterial populations in Oil Reservoirs. There is a distinct difference in microbial populations in treated and observation wells before and after MEOR treatment. The increases are significant, of two or more orders of magnitude.

6.2 The monitoring of MEOR bacterial transport in Oil Reservoir. The rapid increase in bacterial concentrations in observation wells indicates if and at what rate, they moved through the formation from the injection well.

7 RERERENCES

(1) de Mann, J.C., 1975. The probability of most probable numbers. J.Appl. Microbiol. 1, pp. 67-78.

(2) Oppenheimer, C.H., and Hiebert, F., 1989. Microbial Techniques for Paraffin Reduction in Producing Oil Wells. Final report. DOE contract No. FG19086BC 14014, DOE/BC/14014-9, (DE 89000741).

(3) Oppenheimer, C.H., Miget, R.J., and Kator, H.I., 1979. Ecological Relationships between Marine Microorganisms and Hydrocarbons in the O.E.I. study area, Louisiana Rice Univ. Studies, v. 65, pp.287-325.

(4) Standard Methods for the Examination of Water and Wastewater, 15th ed. 1985. American Public Health Association. pp 794-805.

Reprinted from: *Microbial Enhancement of Oil Recovery - Recent Advances*,
edited by E.C. Donaldson

CH. R–11

EFFECTS OF SODIUM CHLORIDE ON GROWTH AND METABOLISM OF TWO STRAINS OF *CLOSTRIDIUM*

M.M. Grula, H.H. Russell, S.M. Janloo and T. Conway
Department of Microbiology, Oklahoma State University, Stillwater, OK 74078

ABSTRACT

Sodium chloride will delay initiation of growth, decrease growth rate and ultimate yield of biomass and of metabolic products, of strains of fermentative clostridia isolated and selected for production of metabolites thought to be favorable for release of oils. Magnitude of the effect depends on concentration of NaCl, and various strains show marked diferences in sensitivity. For example, growth of one strain was completely inhibited by 0.5 M NaCl (3% W/V), whereas another strain, very similar otherwise, would grow to some extent in 1.1 M NaCl. Yields of metabolic products were decreased, and ratios of products formed could be altered by NaCl. Germination of spores (use of spores is favored for MEOR inocula) was much more sensitive to NaCl than growth of vegetative cells. Sodium chloride inhibits transport of nutrients into the cells, production of gas from sucrose by non-growing cells, and, at growth-inhibitory concentrations, completely destroys the mechanism providing the cell with energy. Added to growing cultures at that concentration, it stops growth very quickly, and causes loss of viability of 99% of the population. Efforts to produce strains with higher resistance to NaCl by "training" or "adaptation" were unsuccessful. Evidence was obtained that effects of NaCl occur mainly because the salt elevated the osmotic potential of the medium.

INTRODUCTION

Essential properties that a bacterial species must have in order to be successfully used in MEOR are:

1. Ability to grow abundantly *in situ*
2. Ability to produce oil-releasing metabolites *in situ*

A very desirable property would be the ability to sporulate *in situ*, with the spores remaining viable until conditions favored germination of spores, with consequent production of more oil-releasing metabolites. It is possible that germination of spores and metabolism of cells thereby produced could come about by refeeding. This is the rationale for the refeeding experiments of Russell (1988).

It is widely recognized that a limiting factor in the growth and metabolic activity of *Clostridium* species *in situ* is salinity (mainly but not entirely NaCl) in the water phase of the reservoir environment
(Russell, 1988).

The physiological properties of a bacterium that would be useful in MEOR should be such that it will grow abundantly and vigorously under the physicochemical conditions of the reservoir. McInerney (1988) defined such an organism as "a thermophilic anaerobic bacterium that grows in 5% NaCl at neutral pH . . . ". Other properties are also important; for example, the ability to be readily transported through porous rock. Spores (no net charge) are transported through cores more efficiently than vegetative cells (Jang et al., 1983). It is well known that vegetative cells of many kinds of bacteria will plug oil reservoirs (Updegraff, 1983). Since this plugging is not the result of a simple sieve-like action of the porous medium, but occurs because the net charge carried by vegetative cells is attracted to the opposite electric charge on the minerals of the medium, uncharged cells should pass through the medium unhindered, except by size and shape relative to pore size distribution, diameter of pore throats, etc.

Members of the genus *Clostridium* appear to be, *a priori*, bacteria that are most likely to meet all the requirements for a useful MEOR strain. Therefore, in the screening process whereby we isolated our strains from natural sources, conditions were established to favor the growth of *Clostridium* species (obligately anaerobic endospore formers). Isolates were chosen "to produce copious amounts of organic acids, alcohols, surfactant molecules, emulsifiers, and gases in a low cost medium such as molasses, in the presence of 5% NaCl at an upper temperature limit of 45°C" (Russell, 1988).

Two further requirements were placed on the strains isolated for potential use in MEOR, viz., non-pathogenicity, and no production of hydrogen sulfide. Any strain failing either of these tests was immediately discarded.

This method of isolation of pure cultures for potential use in MEOR, used by Russell, is a different approach from that of Lazar (1983) and other Eastern European workers. Lazar used mixed cultures, obtained as mixtures from specific sources, and adapted, as mixtures, to the conditions of the reservoir where they were to be used. According to Lazar (1983), such a mixed culture yielded an 18-32.5% oil release, as compared to 14.5% for a mixture of cultures originally isolated as pure cultures, and 3.5% for a pure culture.

In subsequent laboratory work on the properties of several of our isolates, it turned out that some of the most active in terms of growth and metabolite production were sensitive enough to salt not to have passed the initial screening. Since these strains were potentially useful because of high metabolite production, we felt it was worthwhile to study the nature of their salt-sensitivity, including the mode of action of salt. If this were understood, modification or lessening of the sensitivity by nutritional, physiological, or genetic means might be feasible.

2. MATERIALS AND METHODS

2.1 Cultures, their handling and maintenance

Strain SEG-3 was isolated from soil by Russell, using enrichment and screening procedures as described by him (1988). Two subcultures of SEG-3, the H and T strains, which had been maintained by different individuals, turned out to be different in certain properties, one of which was sensitivity to sodium chloride. We felt that it might provide insight to work with both strains. Other strains used in this work were also isolated from natural sources by Russell (1988).

Stock cultures were maintained in Hungate tubes containing approximately 7 g of sterilized soil. They were transfered every 6 months, and stored at ambient temperature. Working cultures were prepared by aseptically transferring a few mg soil to several tubes of TSB* sucrose. Cultures of SEG-3 strains could

*TSB = tryptic soy broth (Difco) or trypticase-soy broth (BBL).

lose their viability if kept in TSB-sucrose for over two months. Serial transfers in TSB-sucrose, if more than 4 or 5, resulted in loss of viability.

2.2 Preparation of inocula

Exponential cell inocula were prepared by transferring 0.1 ml of working culture to TSB-sucrose, and growing at 37°C, to an A_{540} (optical density or absorbance; 2-log G, where G is the percent transmittance at 540 μM) of 0.2-0.3.

Spore inocula were prepared by inoculating 10 ml glucose cooked meat medium with a loopful of exponential cells and incubating one week at 37°C. One milliliter of this broth was transferred to 100 ml of fresh broth. After 10 weeks at 37°, the culture showed 85-90% sporulation (direct microscopic observation). The entire culture was centrifuged, spores were washed and resuspended in a final volume of 10 ml sterile distilled water. This suspension was heat-shocked by adding 0.1 ml to 9.9 ml of sterile distilled water pre-equilibrated to 80°, and holding suspension at 80° for 20 min. Spores thus heat-shocked were used as an inoculum.

2.3 Composition and preparation of media

Cells were grown for many experiments on TSB with 1% sucrose. Neither SEG-3 strain would grow on TSB without a sugar added, although some other isolates would.

Often a chemically defined medium was used. It contained 0.1% NH_4Cl, 1.0% glucose or sucrose, 20 mg/100 ml $MgSO_4 \cdot 7H_2O$, and 0.05 M each of K_2HPO_4 and KH_2PO_4. In addition, it contained trace minerals at a final concentration of 4 x 10^{-3} of those given below (Grula, 1960).

Substance	Concentration (g liter^{-1})
H_3BO_3	0.5
$CaCO_3$	10.0
$CuSO_4 \cdot 5H_2O$	1.0
$FeSO_4(NH_4)_2 \cdot SO_4 \cdot 6H_2O$	50.0
KI	1.0
$MnSO_4 \cdot H_2O$	2.0
MoO_3	1.0
$ZnSO \cdot 7H_2O$	5.0

The medium referred to as the semi-defined medium is the same as the defined medium with 0.1% (W/V) yeast extract (Difco) added.

For some experiments a vitamin mixture was used to supplement the defined medium. Vitamins used, with final concentration in the medium, are given below.

Vitamin	Concentration (mg liter^{-1})
Biotin	2.0
Calcium pantothenate	400.00
Cyanocobalamin (B_{12})	1.6
Folic acid	2.0
Nicotinic acid	400.00
p-Amino benzoic acid	200.00
Pyridoxine·HCl	400.00
Riboflavin	200.00
Thiamine·HCl	400.00

For some experiments, the chemically defined medium of Andersch et al. (1982) for *Clostridium acetobutylicum* was used.

After preparation, the medium of choice was autoclaved at 121°C for 20 minutes. Sugars were autoclaved separately, and the vitamin stock solution was filter-sterilized; each was added aseptically after autoclaving.

All cultures were incubated at 37°C.

2.4 Measurement of gas

Gas was measured by the distance the plunger of a 30-50 ml syringe was displaced by gas pressure as the needle penetrated the septum of a Hungate tube. Volumes were converted to microliters (mm^3) by applying a conversion factor of 1000 mm^3 per cm^3.

2.5 Methods of growing anaerobic bacteria

A variety of methods were used (depending partly on the objectives of the experiment), from growing cells in a deep culture with a reducing agent (L-cysteine or thioglycollate) to the Forma anaerobic cabinet (Model 1024 glove cabinet). Techniques used the most were: 1) the pre-reduced, anaerobically sterilized roll tube technique of Hungate; and 2) the anaerobic cabinet or glove box. Occasionally, an anaerobic Gas-Pak jar was used.

In all manipulations of cultures, such as inoculations, washing of cells, or transferring cultures, all tubes, or flasks were flushed with nitrogen gas before use, and during the procedure, if several steps were required.

3. RESULTS

3.1 Effects of Sodium Chloride: Salt, above a certain minimum concentration, delays growth, with the length of the delay increasing with increasing NaCl concentration (Fig. 1). After growth is initiated, at low to moderate concentrations, rate of growth and final yield is less than, but may approach the values in a culture with no NaCl (Grula et al., 1990).

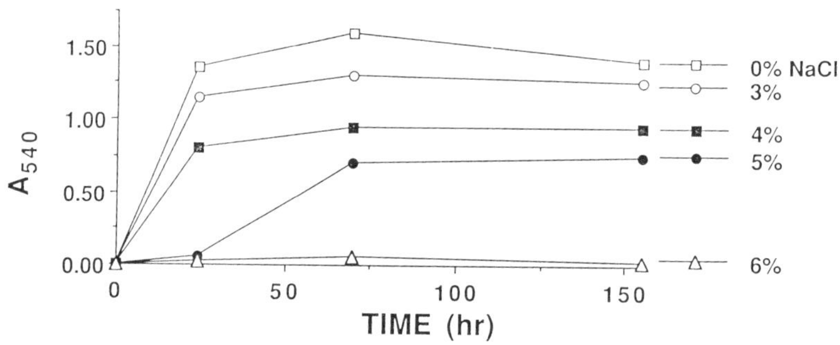

Fig. 1. Growth of Clostridium Strain HSEG-3 varying concentrations of Sodium Chloride. Medium: TSB-Sucrose.

At a certain concentration, which varies with the strain, growth is delayed indefinitely (in practice, up to 2 weeks or more). This is referred to as the minimum inhibitory concentration, or MIC. Among strains used in this work, the MIC varied from 3% W/V (0.51 M) to 7% or slightly higher (W/V), or 1.2M*.

At concentrations approaching the MIC, growth is never abundant, and metabolite production may be altered, as well as

---

* By oil field standards, these are not high concentrations.

simply decreased. For example, gas production may be completely stopped, even though some growth occurred (Table 1).

TABLE 1

Effect of sodium chloride on gas production by representative *Clostridium* strains at 37°C

Strain	Gas production - mm^3 at 48 h			
Carbon source:	Sucrose		Molasses	
% NaCl:	0.5	5.0	0.5	5.0
HS-4	1,330	NG*	9,290	NG*
HR-43	4,910	2,400	2,790	2,660
HR-9K	2,390	800	1,990	1,330
HR-14K	1,060	660	1,990	400
HR-27	5,310	acid only	1,330	acid only

*NG means that no growth occurred in 48 hr.

The value of the MIC may vary with the medium. With Strain TSEG-3 (a strain which produced copious amounts of gas from sugars, but is unusually salt-sensitive) in the chemically defined medium, a concentration of 1.5% completely prevented growth. In most complex media, however, the MIC of NaCl for this strain is 3.0%. Further data are given in Table 2.

TABLE 2

Effect of sodium chloride on growth and gas production by TSEG-3 in semidefined medium and in TSB-sucrose

NaCl %W/V	A540 4 hr[a]	5 hr	10 hr	24 hr	48 hr	Gas-mm^3 48 hr
TSB-sucrose						
0.0	0.01	0.05	1.35	>2	>2	15,500
1.0	0.0	0.02	0.0	1.62	>2	15,000
1.5	0.0	0.0	0.0	0.0	0.0	0
2.0	0.0	0.0	0.0	0.0	0.0	0
Semi-defined[b]						
0.0	0.18	1.77	>2	>2		11,500
1.0	0.045	1.45	1.92	>2		10,000
1.5	0.02	0.24	1.55	>2		8,000
2.0	0.0	0.0	1.53	>2		7,000
2.5	0.0	0.0	0.0	1.62		5,000

a Time after inoculation.
b Contains 0.1% W/V Difco yeast extract.

It appears from Table 2 that some substance present in yeast extract partially overcomes NaCl inhibition of growth and gas production of Strain TSEG-3. This conclusion is further supported by data in Table 3, which show a relationship between the concentration of yeast extract, and its ability to overcome NaCl growth inhibition.

TABLE 3

Effect of concentration of yeast extract in overcoming sodium chloride inhibition of Strain TSEG-3 in a semi-defined medium

0.1% Yeast Extract	NaCl %W/V	A540 24 hr[a]	A540 48 hr	A540 72 hr	Gas mm^3 72hr
	0	1.32	1.77	1.80	11,500
	2	0	1.27	1.57	6,500
	3	0	0	1.50	4,500
	4	0	0	0	0
0.5% Yeast Extract	0	>2	>2	>2	12,000
	2	0.85	>2	>2	9,000
	3	0	0	>2	7,500
	4	0	0	0	0

a Time after inoculation.

Data in Table 3 were obtained in a medium with no complex ingredients other than yeast extract. Tests of the following substances for ability to overcome NaCl inhibition were negative: raffinose, ferrous ion, manganous ion, and a mixture of B vitamins (see Materials and Methods).

Spore germination is more sensitive to NaCl than is growth of vegetative cells. This is shown by comparison of growth of a spore inoculum with an inoculum of actively growing (exponential phase) cells. Two strains, differing in sensitivity to NaCl, were used in this test. Data for Strain HSEG-3 are given in Table 4, and for Strain TSEG-3 in Table 5.

TABLE 4

Growth of exponential phase and spore inocula of HSEG-3 in the presence of graded concentrations of sodium chloride

NaCl %W/V	A540 24 hr[a]	42 hr	8 days
Exponential			
0	1.20	1.25	1.35
3	1.15	1.14	1.15
4	0.91	0.88	0.85
5	0.23	0.76	0.73
6	0.17	0.16	0.47
Spore			
0	1.00	1.10	1.00
3	0.00	0.48	0.56
4	0.00	0.32	0.48
5	0	0	0
6	0	0	0

a Time after inoculation.
Basal Medium: TSB-sucrose.

TABLE 5

Growth of exponential phase and spore inocula of TSEG-3 in the presence of graded concentrations of sodium chloride

NaCl %W/V	A540 24hr	48hr	8days
Exponential			
0	1.26	1.30	1.20
1	1.00	1.10	1.20
2	0.10	0.36	0.90
3	0.00	0.00	0.00
Spore			
0	1.10	1.00	0.95
1	0.00	0.00	0.00
2	0.00	0.00	0.00

In doing these tests, care was taken to insure that both spore and exponential cell inocula contained approximately the same number of viable units/ml. Conversion of a dormant spore to a metabolizing vegetative cell, capable of exponential growth, is a complex process involving triggering or initiation, germination (during which extensive proteolysis takes place) and outgrowth (Keynan and Halvorson, 1965). One or more component(s) of this process apparently is/are highly sensitive to sodium chloride in these strains.

Addition of a MIC of NaCl to an actively growing culture will stop growth almost immediately. After a relatively short time, viable count of cultures to which NaCl was added declines by about 2 logs. Lysis of cells takes place, resulting in a markedly lowered turbidity and release of highly viscous DNA. Some cells survive the NaCl shock, and are capable of growth upon transfer to a fresh medium, but they are not more NaCl-tolerant than the parent strain. Data are shown in Table 6.

TABLE 6

Decrease in viable count of strains HSEG-3 and TSEG-3 after addition of an inhibitory concentration of sodium chloride to exponentially growing cells

Time	Plate (Viable) Count cfu/ml	
	HSEG-3	TSEG-3
0[a]	4.7×10^{7}	6.0×10^{6}
45 min	1.0×10^{6}	2.8×10^{6}
135 min	9.0×10^{5}	8.0×10^{5}
3 hr	1.5×10^{4}	5.0×10^{4}

[a] Time of addition of NaCl. Final concentrations were 3% (w/v) - TSEG-3 and 6.0% (w/v) HSEG-3.

The fact that some cells do survive the toxic effects of NaCl, and are able to grow upon subculture suggests that the initial population contains some individual cells that are inherently more resistant to salt than the majority (about 99%) of cells. If this were true it should be possible to select a more NaCl-tolerant strain simply by continuous serial transfer in a medium with a sublethal concentration of NaCl. This approach was tried, using both strains, and with variation of most of the parameters that would be expected to influence favorably the selection of a salt-tolerant strain. After much effort, only a very minimal increase in salt tolerance was attained, which did not persist in a subculture. We decided to abandon further efforts in this direction. There is no proof that it cannot be done, but it will not be a simple task.

3.2 Mechanism of action of sodium chloride

(i) Background. It is possible that NaCl acts mainly or even exclusively, by raising the osmotic potential of the medium. It is well-known that microorganisms are profoundly affected by availability of water in their environment, including the concentration of solutes in the growth medium (Brown, 1976).

The osmotic potential of a medium is a function of the concentration* of all molecular species (both charged and uncharged) in the medium. If this concentration is higher than that in the interior of a cell, the cell will lose water, and undergo osmotic stress. Such stress may be manifested in a variety of ways. Conversely, if the medium has a lower concentration than the cell's interior, the cell will suffer hypoosmotic stress.

Both of these situations arise because of the existence of a membrane surrounding the cells which permits water to pass through freely, but is impermeable to the solute molecules. The osmotic potential of a medium is related to the thermodynamic activity (a) of the solvent (in this case, water) by the equation: $\pi = (RT/V)\ln a$, where R is the universal gas constant, T the absolute temperature, and V is the partial molar volume of the solvent. For an excellent discussion of the theory of osmotic potential, its relationship to water activity, availability of water to a cell, and effects of osmotic stress and hyper- or hypoosmotic shock on bacteria, see Csonka (1989).

Another possibility for the mode of action of NaCl on Clostridium species is a specific inhibition of a particular enzyme reaction or physiological function. Determination of the mode of action of NaCl is important as a basis for designing effective ways to relieve the inhibition, or to produce NaCl-resistant strains by recombinant DNA technologies (genetic engineering).

(ii) Observations which have a bearing on determination of the mode of action of sodium chloride. Tests of other compounds, both ionic and non-ionic, over the same range of osmolarities as NaCl showed that a very similar degree of growth inhibition

* More precisely, the activity.

occurred with these compounds. This was the case for KCl, $MgCl_2$, sorbitol, and α-methyl glucoside. The latter two compounds are non-ionic, and non-metabolizable by the *Clostridium* strains used in this study. Data are presented in Tables 7, 8, 9, 10, and 11.

TABLE 7

Comparison of the effects on strain HSEG-3 of solutions of NaCl and KCl of similar osmolarity

NaCl		A_{540}		Gas-mm^3
%(w/v)	Molarity	8 hr	24 hr	3 days
0		1.50	1.60	17,000
3	0.51	1.35	1.40	16,000
4	0.68	1.00	1.10	11,000
5	0.855	0.25	0.51	2,000
6	1.03	0.22	0.31	2,000
KCl		A_{540}		Gas-mm^3
%(w/v)	Molarity	9 hr	3 hr	3 days
0	0	1.50	1.60	17,000
3.9	0.52	0.95	0.90	4,000
4.9	0.66	0.25	0.32	3,000
5.9	0.79	0.20	0.33	3,000
6.9	0.93	0.20	0.22	3,000

TABLE 8

Comparison of the effects on strain TSEG-3 of solutions of NaCl and KCl of equal osmolarity

NaCl		A_{540}		Gas-mm^3
%(w/v)	Molarity	9 hr	3 days	3 days
0	0	0.64	1.32	35,000
1	0.17	0.33	1.30	24,000
2	0.34	0.22	1.10	21,000
3	0.51	0.00	0.00	0
KCl		A_{540}		Gas-mm^3
%(w/v)	Molarity	18 hr	48 hr	48 hr
0	0	1.00	1.40	30,000
1.3	0.17	0.12	1.51	13,000
2.5	0.34	0.06	0.14	2,000
3.9	0.52	0.00	0.00	0

TABLE 9

Effect of magnesium chloride on growth and gas production by Strains TSEG-3 and HSEG-3

$MgCl_2 \cdot 6H_2O$	A_{540}		Gas-mm^3	
(w/v)% NaCl Equivalent	TSEG-3	HSEG-3	TSEG-3	HSEG-3
0	1.30	1.45	25,000	16,000
1.0	1.25	----	20,000	
2.0	1.10	----	9,000	
3.0	0.00	1.20	0	14,000
4.0	0.00	0.95	0	5,000
5.0	0.00	0.42	0	5,000
6.0	0.00	0.35	0	4,000
7.0	0.00	0.08	0	0

Basal medium: TSB-sucrose.

All readings of A540 and gas volume were made at 24 hr. Solutions of $MgCl_2 \cdot 6H_2O$ were equivalent in osmolarity with solution of NaCl of the w/v percentages given above.

TABLE 10

Growth of Strain HSEG-3 in presence of concentrations of sorbitol and α-methyl glucoside equivalent in osmolarity to 2%, 4%, and 6% sodium chloride

NaCl Equivalent	A_{540}					
	Sorbitol			α-Methyl glucoside		
	6 hr	17 hr	22 hr	6 hr	17 hr	22 hr
0	0.70	1.20	1.20	0.70	1.10	1.20
2	0.22	0.86	0.90	0.14	0.95	1.00
4	0.09	0.55	0.62	0.12	0.10	0.09
6	0.09	0.06	0.06	0.05	0.05	0.05

Basal Medium: TSB-sucrose.

TABLE 11

Growth of Strain TSEG-3 in presence of concentrations of sorbitol and α-methyl glucoside equivalent in osmolarity to 1%, 2% and 3% sodium chloride

NaCl Equivalent	A540 Sorbitol 6 hr	Sorbitol 17 hr	Sorbitol 22 hr	α-Methyl glucoside 6 hr	α-Methyl glucoside 17 hr	α-Methyl glucoside 22 hr
0	0.18	1.08	1.20	0.18	1.08	1.20
1	0.22	0.90	1.08	0.22	0.80	0.98
2	0.09	0.05	0.05	0.08	0.34	0.64
3	0.07	0.05	0.03	0.06	0.04	0.03

Basal Medium: TSB-sucrose.

From Tables 7 - 11, it is evident that, using the criterion of inhibition of growth, KCl, $MgCl_2$, sorbitol and α-methyl glucoside are very similar quantitatively to NaCl. This suggests that a significant--possibly a major--component of NaCl inhibition is the osmotic potential of the medium. In the case of KCl, we do not know if the apparently greater toxicity is real, or, if so, what it means.

Additional evidence for a role of osmotic potential in growth and gas production inhibition of our *Clostridium* strains was obtained by the use of osmoprotectants. Osmoprotectants are compounds that mitigate or alleviate osmotic stress on sensitive organisms. They are non-specific in the sense that they relieve osmotic stress no matter what agent is causing it. Two osmoprotectants which have received considerable study are the amino acid proline, and glycine betaine, the trimethylated derivative of glycine. Both have been shown to overcome osmotic stress in a wide variety of bacteria, including representatives of the families Enterobacteriaceae, Pseudomonadaceae, Rhizobiaceae, and others (Walter et al., 1987). Glycine betaine is generally more effective than proline.

Not much has been done with the use of osmoprotectants with *Clostridium* species. These organisms, in so far as they have been studied, are quite susceptible to osmotic stress (Walter et al., 1987). Trials using osmoprotectants have been very limited, and results were inconclusive.

Our first trials of the ability of proline and glycine betaine to overcome NaCl growth inhibition of the *Clostridium* strains TSEG-3 and HSEG-3 were negative. Later tests, using much lower concentrations of these substances over a wide range, showed that both significantly overcome NaCl inhibition of growth and gas production (Grula et al., 1990). Data are presented in Table 12.

TABLE 12

Ability of proline and glycine betaine to overcome NaCl inhibition of growth and gas production by Strain TSEG-3 in a semi-defined medium

NaCl % (w/v)	Concentration Osmoprotectant	A540 8 hr	A540 20 hr	Gas-mm^3 24 h
0		1.28	1.63	9,000
1	0	0.80	1.38	6,000
	Proline 0.5 mM	0.87	1.45	8,000
	Proline 1.0 mM	0.84	1.48	8,500
	Glycine Betaine 5 μM	1.33	1.53	7,500
	Glycine Betaine 10 μM	1.25	1.50	9,000
1.25	0	0	1.18	5,800
	Proline 0.5 mM	0	1.44	7,000
	Proline 1.0 mM	0	1.39	6,000
	Glycine Betaine 5 μM	1.13	1.42	7,000
	Glycine Betaine 10 μM	1.06	1.33	7,000
1.50	0	0	0.85	0
	Proline 0.5 mM	0	1.43	6,500
	Proline 1.0 mM	0	0.80	6,000
	Glycine Betaine 5 μM	0.51	1.20	7,000
	Glycine Betaine 10 μM	0.43	1.30	8,000
1.75	0.01	0.65	0.65	0
	Proline 0.5 mM	0	1.24	6,500
	Proline 1.0 mM	0	0.22	4,500
	Glycine Betaine 5 μM	0.075	1.32	7,500
	Glycine Betaine 10 μM	0.43	1.42	7,500

It is clear that both compounds readily overcome NaCl inhibition of growth and gas production. In most cases, doubling of the concentration of osmoprotectant did not produce a greater effect. The highest concentration of salt used (1.75%) is very close to 0.3M; therefore, the molar ratios of NaCl to the osmoprotectants are ca. 600 (proline) and 6×10^4 (glycine betaine). This is strong evidence that osmotic stress plays an important role in NaCl toxicity for this strain of _Clostridium_.

Sodium chloride prevents the evolution of gas from sucrose or pyruvate by resting cells of _Clostridium_ (Strain TSEG-3). The significance of this for MEOR may be explained as follows.

The bacterium we have dealt the most with, and the most efficient gas producer, viz., Strain TSEG-3, strongly resembles _Clostridium acetobutylicum_, the species used in one of the most extensive field tests of MEOR carried out to date*.

We determined 44 cultural, biochemical, and nutritional characteristics of Strain TSEG-3; the only differences from _Clostridium acetobutylicum_ we found were the ability of TSEG-3 to ferment raffinose in 24-32 h, its lack of absolute requirements for biotin and _p-_aminobenzoic acid, and its failure to produce acetoin. In addition, our strain was highly sensitive to oxygen--more so than _Cl. acetobutylicum_, as described by McNeil and Kristiansen (1986).

Fermentative _Clostridium_ species belonging to the butyric-butanol group obtain energy largely from the fermentation of sugars. Hexoses, and disaccharide of hexoses (e.g., sucrose) which are fermented are converted to pyruvate via the Embden-Meyerhof pathway; after which pyruvate is converted to acetyl CoA, CO_2, and H_2 by a series of reactions known as the phosphoroclastic system. This series of reactions (one component of which is the release of H_2 by hydrogenase) is responsible for the fact that clostridia of this group have a high nutritional requirement for iron, even though they possess no heme pigments.

A key intermediate in the fermentation pathway of _Cl._

*This test was done in 1954-1955 in Arkansas by the Mobil Oil company. A summary of the results was published by Yarbrough and Coty (1982).

acetobutylicum is acetoacetyl CoA, which is formed by condensation of two molecules of acetyl CoA, and is a precursor of both acetone and n-butanol. In theory, 5 molecules of CO_2 are formed for every 2 molecules of glucose fermented.

Sodium chloride might directly block one or more of the enzyme reactions involved in gas production by butanol-forming anaerobes; this would indirectly inhibit growth by stopping essential energy-supplying reactions. To determine if NaCl was causing such a block, we tested the effect of NaCl on evolution of gas by resting (non-growing) cells from sucrose and pyruvate. (Pyruvate is the immediate precursor of CO_2).

Active non-growing cells of TSEG-3 were obtained by harvesting cells in mid-exponential phase, and keeping the cellular environment strictly anaerobic during all manipulations. Sodium chloride does inhibit gas production by these cells, from either sucrose or pyruvate, as shown in Table 13.

TABLE 13

Inhibition by NaCl of gas evolution from sucrose or pyruvate by resting cells of Clostridium Strain TSEG-3

Substrate	NaCl %w/v	Gas production at 14 hr mm^3
Pyruvate	None	6,000
Pyruvate	1%	3,000
Pyruvate	3%	0
Sucrose	None	11,500
Sucrose	1%	5,500
Sucrose	3%	0

All values are averages of quadruplicate determinations.

Reaction tubes (Hungate) were prepared with 7 ml phosphate buffer (0.05 M, pH 7.0), 1 ml cell suspension (200 ml culture concentrated to 20 ml suspension in phosphate buffer), 1 ml substrate (sucrose or pyruvate, 1%, and either 0, 1% or 3% NaCl. Growth temperature and reaction temperature 37°.

Sodium chloride at the MIC of 3% completely inhibited gas production, but this inhibition could occur by at least two distinctly different mechanisms:

1) a direct inhibition of one of enzymes of the

phosphoroclastic system; for example, pyruvate-ferredoxin oxido-reductase;

2) prevention of the uptake of the substrate, i.e., its transport across the cell membrane into the interior of the cell, where it is metabolized. It is known from previous work (Grula, E.A. et al., 1983) that NaCl, and KCl, at concentrations of 5% or 6%, will effectively stop the uptake of L-alanine, sucrose, or glucose by certain clostridia. Prevention of the transport of the substrate obviously would prevent formation of gas.

In order to decide between these alternatives, the first step should be to try the effect of NaCl on the isolated enzymes, free of the barrier of the cell membrane. All of our attempts to produce an active cell-free system were unsuccessful.

There is yet another possible mode of action of NaCl--it might be blocking one or more components of the basic energy-transducing system of the cell. If this were the case, transport, growth and metabolism of the cells would quickly cease, followed by loss of viability, and integrity of the cell (cell lysis). We have observed all of these effects after adding the MIC of NaCl to growing cultures of Clostridium cells.

Like other hypothetical modes of action of NaCl, this one (disruption of the cell's basic energy supply) could be direct or indirect.

Cells obtain directly utilizable energy by using external sources of energy (chemicals, such as glucose or ammonia, or light) to build up a protonic potential (protonmotive force, or PMF) across the cell membrane. The PMF usually consists of two components, viz., the $\Delta\Psi$, or difference in electrical potential between the inside and outside of the cell (interior more negative), and the ΔpH, or difference in pH between the inside and outside of the cell. The internal pH is usually higher than the external pH; the latter is very similar to the pH of the bulk medium. The PMF is defined as the sum of the ΔpH and the $\Delta\Psi$, or the total force tending to push a proton back into the cell. As protons traverse the membrane (mediated by a proton-translocating enzyme, on ATPase), they produce synthesis of ATP (oxidative phosphorylation).

Conversely, ATP synthesized (by a fermentation mechanism) inside the cell may expend its energy to move protons out of the cell, thereby setting up a ΔpH. The PMF provides energy directly for some cellular functions, such as motility, and many transport processes.

Most aerobic and facultative bacteria build up and maintain their PMF by aerobic respiration. Photosynthetic bacteris use the energy of light absorbed by their light-harvesting systems to generate a protonmotive force across the membrane. For a more in-depth discussion of energy transduction in microorganisms, see Gottschalk (1986) or Harold (1986).

Fermentative anaerobes constitute a group to which many species of Clostridium belong. These organisms do not possess a respiratory electron-transport system, and therefore must use ATP synthesized by fermentation to build up a ΔpH, a large portion of the PMF maintained by these bacteria.

If the PMF of a bacterial cell were disrupted, or dissipated by any substance, harm to the cell would be severe and rapid. Certain antibiotics which are known to specifically destroy the ΔpH, or the ΔΨ, of bacteria are extremely toxic (Jones, 1982). Since we had already shown that NaCl inhibits transport of several substrate molecules, and it is known that many forms of transport are directly dependent on the PMF, we thought it would be useful to determine the effect of NaCl on the PMF of our strains.

Measurement of the protonic potential in bacteria is not a simple task (Kashket, 1985). One method of determing the ΔpH in bacteria is by the use of a radioactively labelled weak acid as a probe. Such a compound will traverse the cell membrane in the unionized form, but not in the ionized. Thus the distribution of the radioactivity (found in the anion portion of the acid) between the inside of the cells, and the medium outside is determined by the pH inside and outside.

Using ^{14}C benzoic acid as a probe, and separating cells from medium by centrifuging through silicone oil to determine the ΔpH, we were able to get consistent results in measurements with Strain TSEG-3.

Repeated attempts, using both the flow dialysis method and the centrifugation method (Kashket, 1985), showed no ΔΨ, or potential difference between the interior and the exterior of these *Clostridium* cells. It is known that the $\Delta\Psi$ of several strictly anaerobic bacteria is very sensitive to O_2 (Kashket, 1985). In this review, Kashket concluded that ΔpH values are more reliable measurements than $\Delta\Psi$ since determinations of ΔpH by probe distribution methods and the independent technique of ^{31}P-NMR spectroscopy agree. In view of this, we decided to pursue only ΔpH measurements.

Representative values, obtained with TSEG-3 in 3 independent trials are given in Table 14.

TABLE 14

Values of ΔpH obtained in measurements with exponentially growing cells of Clostridium TSEG-3

ΔpH	Equivalent potential difference expressed in millivolts (mV)
1.74	102
1.85	109
1.78	105

We found that NaCl has a very destructive effect on the ΔpH of TSEG-3. Three percent NaCl, added to an exponential culture, completely destroyed the ΔpH within 5 min. A lower concentration (1.5%) produced changes as shown in Table 15.

TABLE 15

Reduction in ΔpH of *Clostridium* TSEG-3 by an intermediate concentration of NaCl

Time of sampling min.	Time in NaCl - min.	ΔpH in millivolts (mV)
5	0	72.9
10*	0	71.7
15	5	33.5
20	10	31.2
25	15	4.1

*Determinations of ΔpH of an exponentially growing culture were done 5 and 10 min. after removing a sample. Sodium chloride was added to sample 10 min. after it had been removed from the main culture.

Using the less NaCl-sensitive strain HSEG-3, we observed a much smaller effect on the ΔpH from a given concentration of salt. No significant effect occurred at 0.75% NaCl. AT 3.5%, the ΔpH, expressed in millivolts, was reduced from 64.1 to 50 in 15 minutes. Increasing the concentration to 4% resulted in a much greater reduction in the same time--from 60.2 to 7.1.

Potassium chloride, in preliminary trials, appeared to lower the ΔpH in a manner similar to NaCl, in an equimolar concentration. Four and one-half percent KCl (equivalent to 3.5% NaCl) reduced the ΔpH of HSEG-3 as given in Table 16.

TABLE 16

Reduction in ΔpH of HSEG-3 by 4.5% KCl

Time of Sampling-min.	Time in KCl - min.	ΔpH in millivolts (mV)
5	0	67.6
10*	0	68.8
15	5	62.9
20	10	61.7
25	15	61.7

*Time of addition of KCl.

The effect of 4.5% KCl is small. More data are needed to determine if this small decline in ΔpH is statistically significant. Data obtained so far indicate a relationship between the sensitivity of a given strain to NaCl growth inhibition, and the concentration of NaCl required to destroy the major energy-transducing mechanism of the *Clostridium* cell. Destruction of the ΔpH could be responsible for the failure of NaCl-treated cells to transport (take up) nutrients, and thus to produce gas.

4. DISCUSSION AND CONCLUSIONS

Some of our observation are consistent with the hypothesis and the ability that the reason why NaCl is toxic for fermentative *Clostridium* species is because it produces osmotic stress. Certainly this stress is a component of NaCl toxicity. On a more fundamental level, however, it is evident that loss of utilizable energy is responsible for many of the observed effects on the cells. We concluded this because certain chemicals which are known to cause disruption or destruction of the enrgy-transducing mechanism at well-defined sites, when added to Clostridium cultures, produce effects very much like those of NaCl. They do this at much lower concentration, on both a molar and a percentage basis, than NaCl; concentrations low enough to preclude osmotic stress.

We tested three such substances for their effects on growing cultures of TSEG-3 and HSEG-3. One was N,N^1 dicyclohexyl-carbodiimide (DCCD). This compound is known to inhibit proton translocation by binding with one subunit of the BF_O portion (membrane bound component of ATP phosphohydrolase-- thought to provide a gated channel through the membrane which allows protons to traverse the membrane). We found that DCCD at 200 μM lowered the ΔpH of TSEG-3 to zero within 5 min. (as 3% NaCl did).

Tests of the effects of DCCD on growth of the two strains revealed that, of the two, TSEG-3 was considerably more sensitive: 5 mg/ml permanently inhibited growth, whereas with strain HSEG-3, 5 mg/ml only delayed growth for 24 h.

If added to an exponentially growing culture, at a high enough concentration, growth was stopped, and lysis occurred. Some cells were able to recover, as they did with NaCl, as shown by renewed growth and gas production after incubation for 20-24 hr.

We also tried two ionophores (agents which facilitate the passage of ions across the membrane); one of which, valinomycin, collapses the $\Delta\Psi$, and the other, nigericin, collapses the ΔpH (Jones, 1982). With valinomycin, we found large differences in toxicity for the two strains; 200 μg/ml delayed growth of TSEG-3 for 18 hr, and gas production for 44 hr. This same concentration had no effect on HSEG-3.

Nigericin was much more toxic, on a molar as well as on a weight basis, than valinomycin. This is compatible with the hypothesis that the ΔpH is a quantitatively much larger contributor to the protonic potential in these bacteria than the $\Delta\Psi$. It was tested with only one strain, TSEG-3. A concentration of 1 μg/ml delayed growth but at 48 hr growth was basically the same as the control. Addition of an inhibitory level of nigericin to exponentially growing cells stopped growth immediately, and cell lysis ensued.

If the inhibitory effects of NaCl can be attributed to blocking of energy yielding reactions, as appears to be the case, by virtue of osmotic stress, the prospects for developing resistant strains are not good.

Yet there are distinct differences among strains in their sensitivity to agents that act solely, not be elevating the osmotic potential, but by destroying or collapsing the PMF of the cells found. These differences seem to show a pattern resembling their differences in sensitivity to NaCl. If the basis for these differences in sensitivity could be elucidated, it might be possible to make use of them, as a basis for isolation and cloning of one or more genes involved in resistance to NaCl.

It may be of interest to point out that the most sensitive strain was a much better gas producer than any of the less sensitive strains, i.e., for a given production of biomass, 2 to 3 times more gas was formed than with less sensitive strains. We can only speculate as to what this might mean. Possibly the T strain is more dependent on a PMF generated by ATP hydrolysis than the H strain or other less sensitive strains.

We have obtained some very preliminary evidence which indicates that some strains of <u>Clostridium</u> do possess a measurable $\Delta\Psi$. Assuming that this is confirmed, we propose an hypothesis as follows: Strains highly sensitive to NaCl, such as strain TSEG-3, depend on a PMF consisting solely of a ΔpH component. The ΔpH is established and maintained by proton extrusion energized by or coupled with ATP hydrolysis. ATP hydrolysis is catalyzed by the BF portion (soluble) of ATP phosphohydrolase, an enzyme very sensitive to NaCl. Less sensitive strains do not depend exclusively on ATP hydrolysis to maintain a proton-motive force, but are able to maintain a $\Delta\Psi$, enought to provide the cells with some utilizable energy even in the presence of NaCl.

Efficiency of gas production would basically reflect the very rapid metabolism of sugar necessary to maintain a protonic potential large enought to support growth. Such abundant gas production would, according to this hypothesis, be necessarily linked with high sensitivity to salt. Much more data will be needed to confirm (or refute) theis hypothesis. If true, a strain that combines rapid and copious gas production with a high salt tolerance would have to be very different from any we have seen to date.

5. ACKNOWLEDGEMENT

This work was supported by the Department of Energy under contract DE-AS19-80BC 10302.

6. REFERENCES

Andersch, W., Buhl, H., and Gottschalk, G., 1982. Acetone-butanol production by Clostridium acetobutylicum in an ammonium-limited chemostat at low pH values. Biotechnol. Lett., 4:29-32.

Brown, A. D., 1976. Microbial water stress. Bacteriol. Rev., 40:803- 846.

Csonka, L.N., 1989. Physiological and genetic responses of bacteria to osmotic stress. Microbiol. Rev., 53:121-147.

Jang, L.P., Chang, W., Findly, J.E., and Yen, T.F., 1983. Selection of bacteria with favorable transport properties through porous rock for the application of microbial enhanced oil recovery. Appl. Environ. Microbiol., 43:1066-1072.

Jones, C.W., 1982. Bacterial respiration and photosynthesis. Aspects of Microbiology, Vol. 5, American Society for Microbiology, Washington, D.C.

Keynan, A., and Halvorson, H., 1965. Transformation of a dormant spore into a vegetative cell. In: Spores III. L. L. Campbell and H.O. Halvorson (Editors). American Society for Microbiology, Ann Arbor, MI. (now Washington, D.C.), pp. 174-179.

Lazar, I., 1983. Some characteristics of the bacterial inoculum used for oil release from reservoirs. In: J.E. Zajic, D.C. Cooper, T.R. Jack and M.Kosaric (Editors). Microbial Enhanced Oil Recovery. Pennwell Publishing Co., Tulsa, OK, pp. 73-82.

McInerney, M.J., 1982. Physiological types of microorganisms useful for enhanced oil recovery. In: Proceedings of 1982 International Conference on Microbial Enhancement of Oil Recovery. E.C. Donaldson and J.B. Clark (Editors). Bartlesville Energy Technology Center, U.S. Department of Energy, Afton, OK.

McNeil, B., and B. Kristiansen, 1986. The acetone-butanol fermentation. In: Advances in Applied Microbiology, 31:61-92. A.I. Laskin (Editor), Academic Press, Inc. NY.

Russell, H.H., 1988. Isolation, screening, and field application of clostridial isolates to be used in microbial enhanced oil recovery (MEOR). Ph.D. dissertation, Oklahoma State University, Stillwater, OK 74078.

Updegraff, D.M., 1983. The effect of microorganisms on the permeability and porosity of petroleum reservoir rock. In: J.E. Zajic , D.C. Cooper, T.R. Jack and M. Kosaric (Editors). Microbial Enhanced Oil Recovery. Pennwell Publishing Co., Tulsa, OK., pp. 37-44.

Walter, R.P., J.G. Morris, and D.B. Kell, 1987. The role of osmotic stress and water activity in the inhibition of growth, glycolysis, and glucose phosphotransferase system of Clostridium pasteurianum. J. Gen. Microbiol. 133:254-266.

Yarbrough, H.F., and V.E. Coty, 1982. Microbially Enhanced Oil Recovery from the upper Cretaceous Nacotoch Formation, Union County, Arkansas. Proceedings, International Conference on Microbial Enhancement of Oil Recovery. Shangri-La, Afton, OK. pp. 149-153.

Reprinted from: *Microbial Enhancement of Oil Recovery - Recent Advances*, edited by E.C. Donaldson

CH. R-12

EFFECTS OF HIGH PRESSURE AND HIGH-PRESSURE GASES ON MICROORGANISMS

ROBERT E. MARQUIS
Departments of Microbiology/Immunology and of Dental Research, The University of Rochester, Rochester, NY 14642 USA

ABSTRACT

A general review of the interactions of pressure and temperature on microbial growth and survival is presented. In general, the pressures in petroleum reservoirs are not so extreme as to preclude microbial functions. It appears that adaptation for growth at either higher or lower temperatures is largely independent of pressure adaptation.

The general topic of hydrostatic pressure as an important environmental factor affecting the activities of microorganisms in deep petroleum reservoirs has been considered previously by the author (Marquis, 1983a, 1983b), and so this presentation will be primarily an updating on relevant aspects of high-pressure biology or barobiology. Molecular, cellular and organismal responses to high pressure have been considered in a recent proceedings volume (Jannasch et al., 1987).

High-pressure environments in the biosphere. The two major regions of the biosphere in which high pressures may have important ecological effects are in the deep ocean and the deep earth. It should be appreciated, however, that pressure is a major environmental factor for all living organisms, even those adapted to function best at about 1 atmosphere. Thus, for example, our enzymes and other functional components are designed specifically to function effectively at 1 atmosphere and have a pressure optimum for function as well as a temperature optimum, a pH optimum and an ionic-strength optimum. Denaturation occurs at high or low pressures much above or below the optimum for any particular enzyme. For most enzymes, high-pressure denaturation occurs in the kilobar range, and low-temperature denaturation occurs only at negative pressures. (1 atmosphere = 101.3 kPascals = 1.033 kg/cm^2 = 14.70 pounds/inch^2). However, for barophilic enzymes, denaturation could occur at positive pressures.

Pressure affects biological processes in relation to accompanying volume changes. Thus, pressure stimulates biochemical reactions or physiologic processes resulting in negative volume changes and retards those resulting in positive volume changes. It is important also to distinguish between effects of pressure on rates of reaction, for which the pertinent volume change is ΔV , or the volume of activation, and effects on equilibria, for which the pertinent volume change is ΔV, the volume of reaction, or the difference in the sum of the partial molar volumes of reactants and that of products. Thus, for example, ΔV can be used to predict the effect of pressure on the rate of an enzymatic reaction. In fact, the only readily available means to estimate ΔV is from the effect of pressure on the reaction rate. However, in many biological responses, ΔV is the pertinent volume change. For example, the effects of pressure on microtubular integrity have to do more with changes in the equilibrium between monomers and assembled structures than with changes in the rate at which assembly or disassembly occurs. ΔV can be estimated either dioatometrically or on the basis of pressure effects.

Pressure increases with depth in the ocean at a rate of about 1 atm for every ten meters depth due to the weight of the water column to a maximum of some 1,160 atm at the bottom of the Challenger Deep in the Pacific Ocean. Pressure increases with depth also in the earth at about the same rate but in a less continuous manner because of so-called overpressured zones and less common, underpressured zones. The maximum depth for life in the deep earth has not been determined. However, with the depth for maximal recovery of oil increasing in the range from 2,000 to 3,000 meters, the most pertinent pressures for enhanced recovery in producing wells are some 20 to 30 MPa. Undoubtedly, much higher pressures would have to be considered in attempts to define the depth limit of the subterranean biosphere.

In many regions of the earth, the limiting boundary is probably set more by high temperatures than by high pressures. The known upper temperature range for life is increasing steadily with isolation of more and more extremely thermophilic archaebacteria, and even some eubacteria, such as Thermotoga species. Still, the current upper range for growth is only approaching 120°C, and the general feeling is that the maximum is going to be below 150°C because of high-temperature instabilities of biological structures, including DNA and protein molecules. Some bacterial endospores are able to survive exposure to temperatures as high as 150°C in the moist environments required for growth but of course, they are not growing but exist in a dormant, dehydrated and mineralized state (Gerhardt and Marquis, 1989).

The environment in deep subterranean reservoirs is hostile to life not just because of high temperature and pressure but also because of other factors such as high salinity, high levels of toxic metal ions and extremes of pH. Pressure can act to sensitize microorganisms to these other factors, for example, to acid or alkali (Matsumura and Marquis, 1977). In contrast, it may act to enhance resistance, for example, to high salinity (Marquis, 1983b).

Despite many restrictions, bacteria may be highly active in deep oil reservoirs, and descriptions of their activities can be found in publications such as the book edited by R.M. Atlas (1984). The roles of bacteria in enhanced oil recovery have been considered specifically by Moses and Springham (1982) and in a more recent, short review by Moses (1987).

When oxygen is introduced into a reservoir with water during enhanced recovery operations, a succession of organisms develops. First, aerobic and facultative organisms, such as *Pseudomonas, Arthrobacter* and *Bacillus,* can carry out aerobic dissimilation of hydrocarbons to yield CO_2 and partially oxidized products such as organic acids. The latter are then substrates for anaerobic bacteria such as *Clostridium* and the organisms long considered to be the bane of the petroleum industry, the sulfate reducers. These organisms are obligately anaerobic and active mainly after oxygen depletion and reduction of the E_h value of the environment. Finally, methanogens can become active under anaerobic conditions and reduce CO_2 and CH_4. In the past few years, evidence has accumulated indicating that hydrocarbons can be degraded anaerobically under denitrifying or fermentative conditions as well as aerobically (Bertrand et al., 1989).

The organisms involved in all of these reservoir activities are from groups containing moderately barotolerant and baroduric members. Descriptors used to describe functional capabilities of organisms under pressure include: *barophobic* for organisms such as *Saccharomyces cerevisiae* which cannot grow under otherwise optimal conditions at pressures much above 20 MPa, *moderately barotolerant* for organisms such as *Escherichia coli* able to grow at pressure up to about 50 MPa, *baroduric* for organisms able to grow over a wide range of pressure up to 60 to 70 Mpa, *barophilic* for organisms that grow better at high pressure than at 0.1 MPa, and *obligately barophilic* for organisms such as the MT41 deep-sea bacterium isolated by Yayanos (1986) able to grow under high pressure but not at 0.1 MPa. As indicated, the maximum pressure for microbial growth has not been established. Some years ago, ZoBell (1958) reported growth of a sulfate reducer at the extreme conditions of 100 MPa and 104°C for organisms such as the MT41 deep-sea bacterium isolated by Yayanos (1986) able to grow under high pressure but not at 0.1 MPa. As indicated, the maximum pressure for microbial growth has not been established. Some years ago, ZoBell (1958) reported growth of a sulfate reducer at the extreme conditions of 100 MPa and 104°C, and there are reports of bacterial growth at pressures as high as 140 MPa. Thus, it seems that, although pressure in deep reservoirs would undoubtedly affect microbial activities in major ways, the pressures are not so extreme as to preclude microbial functions. However, other environmental factors in combination with pressure may be highly restrictive.

Growth of many bacteria isolated from the deep ocean is stimulated by pressure, and Yayanos (1986) has developed the view that most of the deep-sea bacteria have become so well adapted to their environment that they are characteristically barophilic. Similar adaptation could occur for organisms resident in the deep earth, but we currently have very little information on such adaptation. Microorganisms introduced into petroleum reservoirs, either intentionally or unintentionally, may not be adapted to function well under pressure. However, as indicated below, it should be possible to isolate laboratory adapted variants for use in operations to enhance oil recovery.

The hyperbaric regions of the biosphere most intensely studied during the past decade are those associated with hydrothermal vents at sites of new crust formation on the ocean floor. These vents were discovered in 1977 during exploration of a sea-floor spreading center near the Galapagos Islands (Corliss et al., 1979). Most of the deep ocean is sparsely populated except for specific niches in sediments, in association with settling particles or in the guts of marine animals. The major factors limiting populations in the deep ocean are nutrient sparasity, low temperatures close to 2°C and high pressures. However, the vent waters are heated and contain a source of energy for chemolithotropic CO_2 fixation in the form of sulfides or other reduced sulfur compounds. The vent community is diverse and includes *Riftis* tube worm, *Mytilis* clams and extremely thermophilic bacteria. The latter have stimulated one of the major debates of the decade in biological sciences since the publication of Baross and Deming (1983) in which the claim was made for growth of bacteria at temperatures as high as 250°C. The subsequent report of Trent et al. (1984) showed in a conclusive manner than what had been considered bacterial growth could be traced to artefact. Moreover, White (1984) and Bernhardt et al. (1984) have presented evidence that biomolecules such as nucleic acids and proteins are not stable at this high temperature. Even the endospore forms of bacteria are killed at 250°C, despite protection through dehydration and mineralization (Gerhardt and Marquis, 1990). Again, it seems that the maximum temperature for growth of any organisms is likely to be less than 150°C. The claim the life can occur at all temperature-pressure combinations at which water is liquid seems incorrect, or at very least not proven. Clearly, though, for growth to occur at temperatures much above 100°C, pressure is required to keep water in the liquid phase, and so growth at these high temperatures requires pressurization.

Pressure-temperature interactions. The interactions of pressure and temperature affecting microbial growth have been explored experimentally by a number of investigators, including the author for *Bacillus licheniformis*, *Enterococcus hirae* and *Bacillus stearothermophilus* (Marquis, 1983a, 1983b).

Yayanos and DeLong (1987) have used an ingenious apparatus for assessment of growth of bacteria, including marine barophilic strains, at multiple pressures and temperatures and have constructed three-dimensional plots with axes of pressure, temperature and growth rate. The hills and valleys above the plane of zero growth rate indicate in a clear way pressure temperature interactions. In general, increased pressure acts to increase the optimal temperature for growth, but at most, by only a few degrees. Moreover it appears that adaptation for growth at either higher or lower temperatures is largely independent of pressure adaptation. This independence is expected in terms of known ΔV and ΔH values for chemical bonds important for the maintenance of structure and function in biological molecules (Marquis, 1976). Clearly then, microbes cannot be grown at higher and higher temperatures simply by increasing growth pressure. In fact, temperatures above the optimum become even more restrictive for growth at higher pressures. Even for the thermophilic Archaebacterium *Methanococcus thermolithotrophicus*, Bernhardt et al. (1988) found that increased pressure could enhance growth rate but did not act to extend the temperature range for growth. The differences in adaptive mechanisms for growth at high temperature or high pressure are pluses in terms of molecular biology in that it should be possible to select pressure-sensitive mutants or variants that would be expected to differ fundamentally from temperature-sensitive mutants or variants.

Basic apparatus for use in high-pressure biology. Fig. 1 shows the standard apparatus used in the author's laboratory for growing cells under pressure. The culture to be pressurized is generally placed in a sealed vessel able to accommodate the relatively small changes in culture volume caused by pressure. The simplest sort of vessel is a tube stoppered with a neoprene rubber stopper or a serum cap. Care is taken to fill the tube totally to avoid air bubbles because the entrapped air would dissolve under pressure and could expose the culture to damaging levels of oxygen. Alternatively, cultures can be placed in syringes or sealed plastic bags, again with care to exclude air. For studies of oil displacement, rock cores or other vehicles for oil entrapment can be placed directly within chambers.

The sealed cultures are then placed in a stainless steel pressure vessel, which is filled with water, again with care to minimize inclusion of air bubbles, which would dissolve under pressure and reduce the pressure within the chamber. The pressure cylinder is of the type described by Johnson and Lewin (1946) and modified by Oppenheimer and ZoBell (1952). It is made of 300-series stainless steel, has a rubber O-ring seal and an HP-threaded closure at the top. The closure here is male for the cylinder and female for the cap. However, many commercial vessels have a female fitting for the cylinder with a male plug closure. Pressurization techniques and vessels in general have been described by Morita (1970) and by Marquis (1976).

The cylinder cap shown has a male fitting to connect through a needle-valve assembly to high-pressure (HP) tubing which is connected to an hydraulic pump with attached Bourdon gauge. The pump has a reservoir containing 50% (v/v) glycerol/water and an exit port and needle valve for decompression. Thus, the cylinder can be pressurized, but then after closure of the needle valve just above the cylinder cap, it can be disconnected from the pump and incubated under the desired conditions.

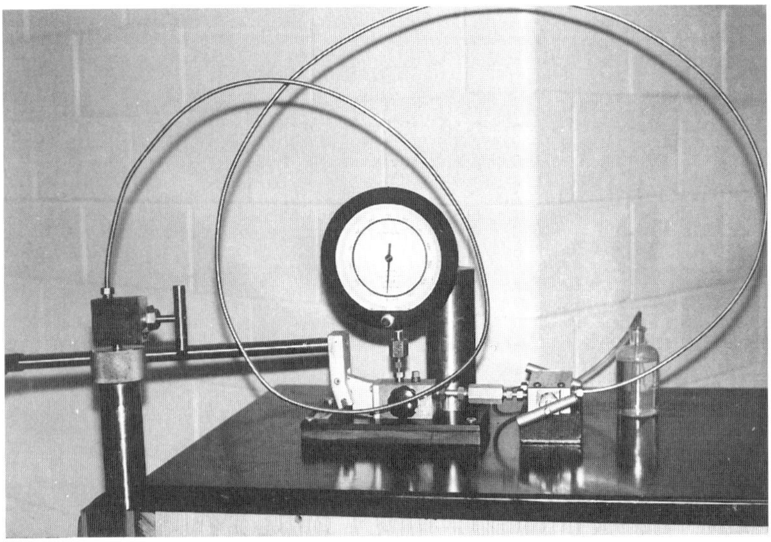

Fig. 1. Apparatus for growth of bacteria under pressure consisting of an hydraulic pump with a Bourdon pressure gauge, HP tubing, a needle-valve closure and a stainless-steel pressure chamber.

There are cylinders available commercially, for example, from Autoclave Engineers of Erie PA, with various modifications. Optical windows and electrical leads can be added to chambers to allow for acquisition of data without the need for decompression. In addition, cylinders can be constructed from nonmagnetic steel to allow for magnetic stirring powered from the outside. Yayanos and van Boxtel (1982) have designed a coupling device for quick high-pressure connections, and Jannasch et al. (1982) have described apparatus for isolation of deep-sea bacteria without decompression. Barnhardt et al. (1987) have described equipment for growing methanogens at high temperatures and pressures with gaseous substrates. Baross and Deming (1983) have described apparatus modified from a design of Yayanos (1969) for incubation of samples at very high temperatures under pressure in titanium syringes. Many other variants of pressure vessels and sampling apparatus have been or could be constructed. The need for specialized equipment has tended to slow development of barobiology. However, since the initial use of the Cailletet press in the latter part of the last century, relatively inexpensive pressure apparatus for biological work up to about 150 MPa has been available or can be easily constructed with the help of competent machine-shop personnel. Much of this equipment is directly usable for studies of microbial effects on oil recovery, and this volume has examples of the use of pressure equipment constructed specifically for petroleum microbiology.

Growth of pressurized cultures can be monitored without decompression by use of vessels with optical windows. Moreover, apparatus has been designed for taking samples without decompressing or for chemically fixing samples under pressure (Landau and Thibodeau, 1962). However, for most barobiological work, sampling involves decompression and recompression. Although, there has been debate regarding the damaging effects of this procedure, the general finding has been that it is not harmful. Certainly, in the author's laboratory, when we have incubated duplicate sets of cultures of bacteria, with one set repeatedly decompressed and recompressed and the other set let to grow without manipulation, no significant differences in growth rates or final extents of growth were detected. Even for organisms adapted for growth under pressure in the deep sea, Chastain and Yayanos (1990) did not find evidence of decompression damage apparent in electron micrographs. Also, microorganisms do not appear to suffer from the bends or caisson disease due to formation of intracellular bubbles (Hemmingsen and Hemmingsen, 1978). Care should be taken to compress and decompress cultures slowly to avoid excessive heating and cooling. However, the actual compressure/decompression does not appear to be harmful.

For organisms, including microorganisms, with gas-filled cavities, compression and decompression can present major problems. Compression of gas vacuolate bacteria results in collapse of the gas vacuoles, and it is possible to rupture these organisms by means of rapid decompressing (Hemmingsen and Hemmingsen, 1980). For an animal such as a human, the lungs must be kept inflated during pressurization, and generally normoxic mixtures of compressed gases with helium or helium and nitrogen are used in deep-diving procedures.

Adaptation to high pressure. As indicated above, it may be desirable to pressure adapt organisms introduced into deep reservoirs for enhanced recovery operations. Microorganisms resident in the deep ocean appear to be adapted genetically to function at high pressure and may even be obligately barophilic. Organisms indigenous to the deep earth would be expected to be similarly adapted. Recently, Bartlett et al. (1989) have described pressure-inducible synthesis of an outer-membrane protein of a deep-sea *Vibrio* organism. Regulation of synthesis was at the level of transcription. Microorganisms have been found (DeLong and Yayanos, 1986) also to undergo so-called homeoviscous adaptation to high pressure, and deep-sea bacteria may contain polyunsaturated fatty acids in their membranes. These fatty acids are uncommon for prokaryotes but presumably act to maintain membrane fluidity under pressure.

We have found that ordinary laboratory bacteria can become adapted to function better under pressure after prolonged culture at higher and higher pressures. We initially used the bacterium *Enterococcus hirae* ATCC 9790 for selecting barotolerant variants because of previous knowledge that one of the major effects of pressure on the organisms was to reduce acid tolerance. Also, under pressure, *E. hirae* becomes starved for ATP, and its barotolerance can be enhanced markedly by feeding it catabolites that are degraded rapidly. Therefore, to adapt the organism for growth at high pressures, we grew it in stab cultures in buffered medium at increasingly higher pressure during repeated subculturing. The variant designated APR-11 was isolated (Marquis and Bender, 1980). This variant was able to grow at pressures as great as 100 MPa, or some 20 MPa higher than the maximum for the parent strain under otherwise optimal conditions for barotolerance. The organisms was found to be defective in catabolite repression (Campbell et al., 1985) because of alterations in Enzyme II^{glu} of the phosphotransferase system (Marquis and Bender, 1987). The arginine deiminase system of the bacterium is generally under catabolite repression when glucose is being metabolized. However, in the APR-11 variant, glucose and arginine can be catabolized at the same time. The ammonia produced from arginine serves to neutralize the acid produced from glycolysis thereby reducing the acid stress to the organism. Thus, in a sense, the organism produces its own buffer. The APR-11 variant appeared to be present in populations grown at 0.1 MPa in very low numbers, and pressure acted only to select the variant and not as a mutagen. The APR-11 stain is genetically stable, even when cultured at 0.1 MPa. The important point to be made here is that barotolerant variants can be isolated, although the strategy for isolation may require detailed knowledge of the effects of pressure on the organism of interest.

Biological effects of high-pressure gases. High-pressure gases have effects of living organisms distinct from pressure effects. Perhaps the most studied of these effects is narcosis, which appears to depend on physicochemical interactions not involving covalent bonding of the narcotic agent. Thus, even a noble gas such as Xe can be used as a surgical anesthetic (Cullen and Gross, 1951). Hydrocarbon gases are narcotic. For example, acetylene and ethylene are somewhat more potent narcotically than N_2O or Xe, while methane is somewhat less potent (Macdonald, 1975). We found (Marquis, 1984) that methane is also inhibitory for microbial growth. CH_4 pressures for 50% reduction in growth rate ranged from about 2.5 MPa for *S. cerevisiae* to about 4.5 MPa for *E. hirae*. Growth inhibitory actions of gases are distinct from narcotic actions. We previously showed (Thom and Marquis, 1984) that there are two distinct classes of gases with respect to growth inhibition. Class-one gases such as He, N_2 and Ar do not inhibit growth. In fact, they act to reduce the inhibitory actions of hydrostatic pressure. Therefore, when cells were exposed to high pressures of these gases, growth inhibition was actually less than that caused by hydrostatic pressure alone applied without a gas phase. Class-two gases, such as CH_4, Kr, N_2O and Xe, are clearly inhibitory for growth, and sensitivity to them varies among organisms. All of the gases, including class-one gases, potentiate oxygen toxicity, and our current view is that they act by modifying free-radical metabolism in cells (Thom and Marquis, 1987). These gases also affect a variety of biological processes in addition to growth, for example, germination of bacterial spores (Enfors and Molin, 1978) and methane production by extreme thermophiles (Miller et al., 1988).

The inhibitory effects of CO_2 for microorganisms are well known in the food industry. The growth of a broad spectrum of food spoilage bacteria can be inhibited at CO_2 levels above about 5% in the gas phase at 0.1 MPa total pressure (Silliker et al., 1980). The inhibition is not due solely to acidification but occurs even in buffered media. The mechanisms for growth inhibition by CO_2 have not been clearly defined, although the gas, or rather, dissolved CO_2 can inhibit a number of enzymes, and bicarbonate is a weak acid that could upset the acid-base balance of cells.

Clearly, the gases associated with oil in deep reservoirs or gases produced in the reservoirs during enhanced recovery operations would be expected to have important effects on the microbes exposed to them. In deep oil reservoirs, the concentration of methane alone in associated waters may be some 90 mM or the equivalent of about 6 MPa (Buckley et al., 1958). Pressurization of wells with microbially produced gases, mainly CO_2, N_2 and H_2, is considered to be an important component of enhanced recovery. Dissolution of the gases in the petroleum phases is also considered to be important for mobilization. Of course, the gases would also dissolve in the aqueous phases where the producing organisms are and could moderate microbial activities desirable for enhanced recovery.

REFERENCES

Atlas, R.M., 1984. Petroleum Microbiology. MacMillan Publishing Co. New York.

Baross, J.A. and Deming, J.W., 1983. Growth of 'black smoker' bacteria at temperatures of at least 250°C. Nature 302:423-426.

Bartlett, D., Wright, M., Yayanos, A.A. and Silverman, M., 1989. Isolation of a gene regulated by hydrostatic pressure in a deep-sea bacterium. Nature 342:572-574.

Bernhardt, G., Ludemann, H-D., Jaenicke, R., Koenig, H. and Stetter, K.O., 1984. Biomolecules are unstable under "Black Smoker" conditions. Naturwissenschaften 71:583-586.

Bernhardt, G., Jaenicke, R. and Ludemann, H-D., 1987. High-pressure equipment for growing methanogenic microorganisms on gaseous substrates at high temperature. Appl. Environ. Microbiol. 53:1876-1879.

Barnhardt, G., Jaenicke, R., Ludemann H-D., Koenig, H. and Stetter, K.O., 1988. High pressure enhances the growth rate of the thermophilic arachaebacterium *Methanococcus thermolithotrophicus* without extending its temperature range. Appl. Environ. Microbiol. 54:1258-1261.

Bertrand, P., Caumetter, P., Mille, G., Gilewicz, M. and Denis, N., 1989. Anaerobic biodegradation of hydrocarbons. Sci. Prog. Oxf. 73:333-350.

Buckley, W.E., Hocott, C.R. and Taggert, M.S., Jr., 1958. Petroleum in the reservoir. In: L.G. Weeks (Editor), Habitat of Oil: A Symposium, Am. Assoc. Pet. Geol., Tulsa, Oklahoma: pp. 850-882.

Campbell, J. III, Bender, G.R. and Marquis, R.E., 1985. Barotolerant variant of *Streptococcus faecalis* with reduced sensitivity to glucose catabolite repression. Can J. Microbiol. 31:644-650.

Chastain, R.A. and Yayanos, A.A., 1990. The ultrastructure of deep-sea barophilic bacteria from hadal and abyssal depths. Abst. Ann Mtg. Amer. Soc. Microbiol. p. 259.

Corliss, J.B., Dymond, J., Gordon, L.I., Edmond, J.M., von Herzen, R.P., Ballard, R.D., Green, K., Williams, D., Bainbridge, A., Crane, K. and van Andel, T.H., 1979. Submarine thermal springs on the Galapagos Rift. Science 203:1073-1083.

Cullen, S.C. and Gross, E.G., 1951. The anesthetic properties of xenon in animals and human beings, with additional observations on krypton. Science 123:580-582.

DeLong, E.F. and Yayanos, A.A., 1986. Biochemical function and ecological significance of novel bacterial lipids in deep-sea prokaryotes. Appl. Environ. Microbiol. 51:730-737.

Enfors, S-O. and Molin, G., 1978. Mechanism of the inhibition of spore germination by inert gases and carbon dioxide. In: G. Chambliss and J.C. Vary (Editors), Spores VII, American Society for Microbiology, Washington, D.C., pp. 80-84.

Gerhardt, P. and Marquis, R.E., 1989. Spore thermoresistance mechanisms. In: I. Smith, R.A. Slepecky and P. Setlow (Editors), Regulation of Prokaryotic Development, American Society of Microbiology, Washington, D.C., pp. 43-63.

Hemmingsen, B.B. and Hemmingsen, E.A., 1978. Tolerance of bacteria to extreme gas supersaturation. Biochem. Biophys. Res. Commun. 85:1379-1384.

Hemmingsen, B.B. and Hemmingsen, E.A., 1980. Rupture of the cell envelope by induced intracellular gas phase expansion in gas vacuolate bacteria. J. Bacteriol. 143:841-846.

Jannasch, H.W., Wirsen, C.O. and Taylor, C.D., 1982. Deep-sea bacteria: Isolation in the absence of decompression. Science 216: 1315-1317.

Jannasch, H.W., Marquis, R.E. and Zimmerman, A.M., 1987. Current Perspectives in High Pressure Biology. Academic Press, London.

Johnson, F.H. and Lewin, I., 1946, The disinfection of *E. coli* in relation to temperature, hydrostatic pressure and quinine. J. Cel. Comp. Physoil. 28: 23-45.

Landau, J.V. and Thibodeau, I., 1962. The micromorphology of *Amoeba proteus* during pressure-induced changes in the sol-gel cycle. Exptl. Cell res. 27:591-594.

Macdonald, A.G., 1975. Physiological Aspects of Deep Sea Biology. Cambridge University Press.

Marquis, R.E., 1976. High-pressure microbial physiology. Adv. Microbial Physiol. 14:159-241.

Marquis, R.E., 1983a. Barotolerance and microbial enhancement of oil recovery. In: J.E. Zajic, D.C. Cooper, T.R. Jack and N. Kosaric (Editors), Microbial Enhanced Oil Recovery. PennWell Publishing Co., Tulsa, Oklahoma, pp. 8-13.

Marquis, R.E., 1983b. Barobiology of deep oil formations. In: E.C. Donaldson and J.B. Clark (Editors), International Conference on Microbial Enhancement of Oil Recovery. U.S. Department of Energy, Bartlesville, Oklahoma, pp. 124-128.

Marquis, R.E., 1984. Reversible actions of hydrostatic pressure and compressed gases on microorganisms. In: AS. Hust and A. Nasim (Editors), Repairable Lesions in Microorganisms. Academic Press, London, pp. 273-301.

Marquis, R.E. and Bender, G.R., 1980. Isolation of a variant of *Streptococcus faecalis* with enhanced barotolerance. Can. J. Microbiol. 26:371-376.

Marquis, R.E. and Bender, G.R., 1987. Barophysiology of prokaryotes and proton-translocating ATPases. In: H.W. Jannasche, R.E. Marquis and A.M. Zimmerman (Editors), Current Perspectives in High Pressure Biology. Academic Press, London, pp. 65-73.

Matsumura, P., and Marquis, R.E., 1977. Energetics of streptococcal growth inhibition by hydrostatic pressure. Appl. Environ. Microbiol. 33:885-892.

Miller, J.F., Shah, N.N., Nelson, C.M., Ludlow, J.M. and Clark, D.S., 1988. Pressure and temperature effects on growth and methane production of the extreme thermophile *Methanococcus jannaschii*. Appl. Environ. Microbiol. 54:3039-3042.

Morita, R.Y., 1970. Application of hydrostatic pressure to microbial cultures. In: J.R. Norris and D.W. Ribbons (editors), Methods in Microbiology, Volume 2. Academic Press, London, pp. 243-257.

Moses, V., 1987. Microbes and oil recovery. Microbiol. Sci. 4:306-309.

Moses, V. and Springham, D.G., 1982. Bacteria and the Enhancement of Oil Recovery. Applied Science Publishers, London.

Oppenheimer, C.H. and ZoBell, C.E., 1952. The growth and viability of sixty-three species of marine bacteria as influenced by hydrostatic pressure. J. Marine Res. 11: 10-18.

Silliker, J.H., Elliott, R.P., Baird-Parker, A.C., Bryan, F.L., Cristian, J.,H.B., Clark, D.W., Olson, J.C. and Roberts, T.A., 1980. Microbial Ecology of Foods, Volume I. Academic Press, New York.

Thom, S.R. and Marquis, R.E., 1984. Microbial growth modification by compressed gases and hydrostatic pressure. Appl. Environ. Microbiol. 47:780-787.

Thom, S.R. and Marquis, R.E., 1987. Free radical reactions and the inhibitory and lethal actions of high-pressure gases. Undersea Biomed. Res. 14:485-501.

Trent, J.D., Chastain, R.A. and Yayanos, A.A., 1984. Possible artifactual basis for apparent bacterial growth at 250°C. Nature 307:737-740.

White, R.H. 1984. Hydroltic stability of biomolecules at high temperatures and its implication for life at 250°C. Nature 310:430-432.

Yayanos, A.A., 1969. A technique for studying biological reaction rates at high pressure. Rev. Sci. Instrum. 40:961-963.

Yayanos, A.A., 1986. Evolutional and ecological implications of the properties of deep-sea barophilic bacteria. Proc. Natl. Acad. Sci. USA 83:9542-9546.

Yayanos, A.A. and van Boxtel, R., 1982. Coupling devise for quick high-pressure connections at 100 MPa. Rev. Sci. Instrum. 53:704-705.

Yayanos, A.A. and DeLong, E.F., 1987. Deep-sea bacterial fitness to environmental temperatures and pressures. In: H.W. Jannasch, R.E. Marquis and A.M. Zimmerman (Editors), Current Perspectives in High Pressure Biology. Academic Press, London, pp. 17-32.

ZoBell, C.E., 1958. Ecology of sulfate reducing bacteria. Prod. Month. 22:12-29.

Reprinted from: *Microbial Enhancement of Oil Recovery - Recent Advances*, edited by E.C. Donaldson

CH. R–13

PRODUCTION OF BIOSURFACTANT BY BACILLUS LICHENIFORMIS STRAIN JF-2

S.C. LIN, J.C. GOURSAUD, P.J. KRAMER, G. GEORGIOU and M.M. SHARMA*
Department of Chemical Engineering and *Department of Petroleum Engineering, University of Texas, Austin, TX 78712-1062

ABSTRACT

Bacillus licheniformis strain JF-2 produces a very effective surfactant under conditions typical of oil reservoirs. We have conducted a systematic investigation of the effects of environmental parameters (temperature, medium pH and NaCl) on : i) the production of the biosurfactant, ii) formation of fermentation end-products and iii) growth of *B. licheniformis* in batch cultures. The partially purified biosurfactant from JF-2 was shown to be the most active microbial surfactant found so far and gave an interfacial tension against decane of 0.016 mN/m. Optimal production of the surfactant was obtained in cultures grown in the presence of 5% NaCl at a temperature of 45°C and pH 7.0. The major fermentation end-products were lactic acid and acetic acid with smaller amount of formic acid and acetoin. Growth and biosurfactant formation were also observed in anaerobic cultures supplemented with a suitable electron acceptor such as $NaNO_3$.

1 INTRODUCTION

In a recent review (Sarkar et al., 1989), we have shown that *in-situ* microbial enhanced oil recovery offers some advantages compared to the conventional EOR processes. Although the exact mechanism by which the growth of microorganisms *in situ* facilitates the flow of oil is uncertain, the biosurfactant and biopolymer production and the selective plugging of high permeability zones (Brown et al. 1986) are expected to play critical roles in microbial enhanced oil recovery. It has been estimated (Lake, 1988) that the interfacial tension between oil and brine must be lowered to 10^{-2} to 10^{-3} mN/m before significant oil recovery can be achieved. Biosurfactants which are capable of reducing the interfacial tension to such low values have been reported (Cooper and Zajic, 1980; Brown et al, 1986). However, microorganisms producing these biosurfactants require oxygen for growth and are therefore unsuitable for *in situ* applications. Recently, Knapp and co-workers (Javaheri et al., 1985) isolated *Bacillus licheniformis* strain JF-2 from water injection brine. They were able to demonstrate that strain JF-2 grew and produced biosurfactant anaerobically in medium E supplemented with glucose and 0.1% $NaNO_3$. It was also shown that *B. licheniformis* JF-2 grew in medium with NaCl concentrations up to 10% and over a wide range of temperatures (up to 50°C) and pH values (4.6 to 9.0). Furthermore, growth was not inhibited by the presence of crude oil. The biosurfactant produced was not affected by the pH, temperature, or the NaCl and calcium concentrations typical of many reservoirs. These result indicate that *B. licheniformis* JF-2 is a good candidate for *in situ* microbial enhanced oil recovery processes. As a first step in understanding the kinetics of growth in porous media and the mechanisms of MEOR, we systematically investigated the effect of environmental parameters on the growth, biosurfactant production and formation of fermentation end-products. The accumulation of fermentation end-products can have an adverse effect on cell growth and thus, it may be a serious limitation of *in- situ* MEOR. On the other hand some microbially produced organic acids and alcohols can facilitate oil recovery by acting as co-surfactants.

2 MATERIALS AND METHODS

2.1. *Batch fermentations*

Batch fermentations were carried out in a 2-liter Multigen fermenter (New Brunswick Scientific Co., New Brunswick, NJ). The working volume was 1 liter. The fermenter was inoculated with 25 ml of fresh culture, which was grown in medium E : 0.1% $(NH_4)_2SO_4$, 0.025% $MgSO_4$, 1% glucose in 100mM phosphate buffer consisting of KH_2PO_4 and K_2HPO_4 supplemented with a 1% (v/v) trace salt solution. The trace salt solution contained the following (grams/l) : EDTA, 1.0; $MnSO_4$, 3.0; $FeSO_4$, 0.1; $CaCl_2$, 0.1; $CoCl_2$, 0.1; $ZnSO_4$, 0.1; $CuSO_4$, 0.01; $AlK(SO_4)_2$, 0.01; H_3BO_4, 0.01 and Na_2MoO_4, 0.01. Unless otherwise specified the growth medium also contained 5% NaCl. The medium pH was maintained constant by the addition of 1M NaOH or 1M HCl using an external pH controller. For anaerobic growth cultures were supplemented with 1% $NaNO_3$ as electron acceptor. Oxygen-free nitrogen (Linde National Specialty Gas Office, Somerest, NJ) was sparged throughout the fermentation to ensure anaerobic conditions.

Bacterial concentrations were monitored by taking samples and measuring the optical density at 600 nm with Ultraspec II (LKB Biochrom Ltd., Chambridge, England) periodically. The calibration curve for the bacterial dry weight as a function of the optical density was obtained as :

Dry weight of *B. licheniformis* JF-2 per ml of culture = 0.83 x OD at 600 nm

2.2. *Product analysis*

The production of biosurfactant was monitored by the decrease in the interfacial tension between the growth medium and decane by the spinning-drop technique (Cayias et al., 1975). Glucose concentration was determined by an enzymatic glucose analyzer (Yellow Spring Instrument Co., Inc., Yellow Springs, OH). A Waters HPLC instrument was employed for the analysis of fermentation end-products. A PRP-X300 column (7 μm, 250x4.1 mm I.D., Hamilton, Reno, NV), maintained at 50°C with a water bath was used for analysis. The flow rate of the mobile phase (1 mM sulfuric acid) was 1.5 ml/min. The elution from the column was monitored at 196 nm. Each sample was filtered with 0.45μm HV filter (Nihon Millipore Kogyo K.K., Yonezawa, Japan) prior to injection.The standard calibration curve between the peak areas and concentrations of each compound was constructed by injecting stardard solutions with known concentrations. The concentrations of the fermentation end-products were determined from the integration areas of the corresponding peaks.

2.3. *Biosurfactant characterization*

The surfactant was isolated from the growth medium by acid-precipitation following the removal of cells by centrifugation at 10,000 g for 10 minutes. For the acid-precipitation process, the fermentation medium was first acidified to pH 2.0 with concentrated HCl. The pellet, which was collected by centrifugation at 30,000 g for 45 minutes, was then extracted with methanol. Upon the evaporation of the methanol, the crude biosurfactant sample was obtained.

3 RESULTS

3.1. *Batch fermentations*

The optimal condition for the bacteria growth was found to be at a temperature of 45°C and pH 7.0 both aerobically and anaerobically (data not shown).A typical fermentation profile is shown in Figure 1. A lag phase of approximately 4-5 hours was observed before the beginning of exponential growth under aerobic conditions, while a 2-3 days lag phase was typical for anaerobic fermentations (data not shown). A maximum bacterial concentration of 1-1.5 g/l was obtained before growth ceased and the optical density began to decrease.

3.2. *Product analysis*

As shown in Figure 1, the interfacial tension of the medium reached the lowest value at the middle of the exponential phase and increased significantly during early stationary phase. The effects of growth temperature, medium pH and NaCl concentration on the interfacial tension are shown in Table 1, Table 2 and Figure 2, respectively. It was observed that both the aerobic and anaerobic cultures grown with a medium supplemented with 5% NaCl (wt/vol) at a temperature of 45°C and pH 7.0 gave the lowest interfacial tension values.

The major fermentation end-products were lactic acid and acetic acid with small amount of acetoin and formic acid. For the culture grown with a medium supplemented with 5% NaCl at 45°C and pH 7.0, the fractions of the carbon source utilized for the production of lactic acid and acetic acid were approximately 38% and 10%, respectively. About two percent of the carbon consumed was used for the production of acetoin and formic acid. The biomass yield for both the aerobic and anaerobic fermentations was in the range of 0.05 to 0.13 g/l of cell dry weight /g glucose consumed.

3.3. *Biosurfactant characterization*

About 0.1 g of crude biosurfactant per liter of fermentation broth was obtained. TLC analysis on cellulose plates followed by staining with rhodamine B showed that the major component of the

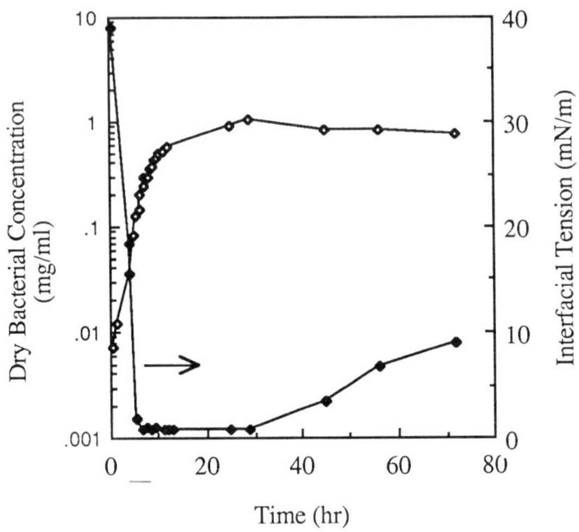

Fig. 1 Typical fermentation profile of *Bacillus licheniformis* JF-2 grown under aerobic conditions at constant pH.

TABLE 1

The effect of temperature on the growth and biosurfactant production of *Bacillus licheniformis* JF-2 in the presence of 5% NaCl at pH 7.0 under aerobic and anaerobic conditions

T (°C)	Anaerobic Fermentation		Anaerobic Fermentation	
	Minimal IFT* (mN/m)	Final Bacterial Concentration**	Minimal IFT (mN/m)	Final Bacterial Concentration
30	0.43	1.521	5.81	0.714
45	0.31	1.021	0.51	0.835
50	0.28	1.164	8.06	0.200

* IFT : The interfacial tension against decane.
** Bacterial concentration in mg/ml.

TABLE 2

The effect of medium pH on the growth and biosurfactant production of *Bacillus licheniformis* JF-2 in the presence of 5% NaCl at 45°C under aerobic conditions

Medium pH	Minimal IFT (mN/m)	Final Bacterial Concentration (mg/ml)
6.0	0.50	1.243
7.0	0.30	1.162
8.0	3.65	0.830

Reprinted from: *Microbial Enhancement of Oil Recovery - Recent Advances*, edited by E.C. Donaldson

CH. R–14

ENHANCED OIL RECOVERY AT SIMULATED RESERVOIR CONDITIONS

ERLE C. DONALDSON and TAWFIC OBEIDA
The School of Petroleum & Geological Engineering, University of Oklahoma, Norman, OK 73019

ABSTRACT

Experiments were conducted to elucidate the principal mechanism(s) of microbial enhanced oil recovery at simulated subsurface reservoir conditions up to 1500 meters depth. Two species of bacteria were used: Bacillus licheniformis JF-2 and Clostridium acetogutylicum.

Sandstone cores (70 cm^3 PV) were equilibrated to the desired simulated reservoir conditions, saturated with oil and brine, and flooded to residual oil saturation. The waterflood brine was displaced with a nutrient solution (3% cattle-feed-molasses in brine). The cores were then inoculated with bacteria (10^8 cells/ml) and shut-in. The pore pressure was monitored until there was no further increase in pore pressure. Finally, a pressure maintenance waterflood was conducted and the additional oil recovery was recorded (17-19%). Pore pressure increases up to 23 MPa (from an initial pressure of 17 MPa) due to biogenic gas was observed. The gas dissolved in the oil and brine establishing a solution gas drive mechanism. The MEOR recovery efficiency was directly related to the dissolved gas-oil ratio. The principal MEOR mechanism observed in this work was solution gas drive. All other possible mechanism (surface tension reduction, pH reduction, selective pore plugging, etc.) were shown to be minor compared to solution gas drive. The experiments suggested a procedure for field application of gas-generating bacteria.

INTRODUCTION

The combined total oil production by primary and secondary methods is generally estimated to be less than 40 percent of the original oil in place. The residual oil is estimated at about 50 billion cubic meters (320 billion barrels). Thus the target for all enhanced oil recovery methods is greater than the reserves that can be produced by conventional methods (GAO, 1986). Because of declining oil production and abandonment of old oilfields, the petroleum industry is challenged to find applicable cost-effective oil recovery technologies to increase the production. Several enhanced oil recovery technologies have been developed which have been locally successful, but no method which has general application to all petroleum reservoirs as emerged. The Bartlesville Office of the U.S. Department of Energy stimulated world-wide interest in the application of microbes for enhancement of oil recovery (MEOR) by sponsorship of conferences, basic research projects at universities, and field pilot studies beginning in 1980. After 10 years of research that indicates great potential for MEOR, the major petroleum companies have not made major investments in MEOR research although there are now some encouraging indications of interest.

A considerable amount of research has shown that both aerobic and anaerobic microbes are able to adapt to the severe subsurface conditions of temperature, pressure and salinity of petroleum reservoirs. A considerable flora of aerobic microbes occur around injection wells where aerobic conditions are maintained by dissolved oxygen. Anaerobic bacteria exists further away from the wellbore in the oxygen depleted regions (Ivanov and Belyaev, 1982; Moses and Springham, 1982).

The principal objectives of this work were to first examine the ability of two species of microbes to enhance oil recovery from sandstone cores at simulated reservoir conditions from ambient to a depth of 1500 meters (5000 feet): (1) Bacillus licheniformis (JF-2) and Clostridium acetobutylicum. Second, to determine the principal mechanism by which the microbes produced oil remaining in the core as residual oil after a normal waterflood. The microbes were indeed able to grow in the apparently adverse environment and they enhanced the recovery of oil by a solution gas-drive mechanism. The results indicate a specific technique for field applications.

BRIEF REVIEW OF BACKGROUND LITERATURE

Enhancement of Oil Recovery with Microbes

Beckman (1926) suggested that bacteria might be beneficial for the production of oil because products of their metabolism could assist in the release and transport of the oil in geologic structures. ZoBell (1946) patented a process for secondary recovery of petroleum using anaerobic bacteria such as Desulfovibrio hydrocarbonoclasticus. Later, in 1953 ZoBell used other fermenting bacteria to produce substantial amounts of organic acids and carbon dioxide to enhance oil recovery in laboratory experiments.

In 1957, Updegraff described the release of oil from cores saturated with a culture medium of Aerobacter aerogenes containing glucose and mineral salts. A non-microbial gas drive produced 13% additional oil (after secondary water drive), but biogenic gas, produced in situ, resulted in 33% additional oil production. Updegraff then proposed use of a fermentable carbon source, such as molasses, and anaerobic, or facultative anaerobic, gas-forming bacteria for enhancement of oil recovery.

Hitzman (1962) proposed the injection of bacterial spores and nutrients and tested this with oil saturated sand-packed columns. Spores of Clostridium roseum and molasses were pumped into the column. Germination of the spores and production of gas, and other metabolic products, led to about 30% oil recovery.

Karaskiewicz (1963) reported that anaerobic Desulfovibrio sp, Clostridium and other mesophilic bacteria, when grown in the presence of molasses and water, produce gases and surfactants. Using sandstone cores, he reported yields which were 30-140% greater than the yield from a normal waterflood when the period of fermentation was extended to several months.

Jenneman et al. (1982) and Jang et al. (1984) studied the feasibility of diminishing reservoir heterogeneity by growing bacteria in the high permeability zones of Berea sandstone cores. Jenneman showed that the addition of nutrients (carbon, nitrogen, and phosphate) in sufficient concentrations to support bacterial growth resulted in permeability reductions of 60-80%. In 1985 Torbati et al. reported work on the measurement of the pore size distributions of Berea sandstone cores before and after growth of bacteria in the cores. Their results showed that the bacteria were plugging the larger pores of the rock.

Enhancement of Oil Recovery with Acids and Surfactants

Mungan (1964, 1965) studied the roles of interfacial tension reduction and wettability changes on oil recovery. The interfacial tension of the surfactant solution was reduced to 1.1 and displacement tests were conduced using sandstone and alundum cores. Water-wet cores were prepared by treating the cores with 0.1 N hydrochloric acid and oil-wet cores were prepared by flowing a 1% solution of a silicone in heptane through the cores and then drying before use. Floods with 1.0 N HCl did not result in additional production of oil. Floods of water-wet cores with an amine produced approximately 6% additional oil (in addition to a normal waterflood). Amine floods of the oil-wet cores produced a maximum of 3% additional oil.

The small amount of additional recovery obtained with the acid flood was attributed to dispersions of particles that plugged high permeability zones. Floods with surfactants have been shown to be ineffective effective unless the interfacial tension is reduced to less than 0.1 mN/m. Floods with other surfactants also have been reported to be very poor (Donaldson et al., 1985).

Field Applications of MEOR

Johnson (1979) reported that he had inoculated at least 150 stripper wells (oil production < 0.3 m^3/day...2 bbl/day) with a mixture of *Bacillus* and *Clostridium.* Using a single well, "huff-and-puff" technique, he reported that the formation in the vicinity of the wellbores were cleaned by microbial production of gas, acids, and solvents resulting in a long-term increase of oil production.

Yarbrough and Coty (1982) reported on a field test in which a 2% solution of molasses was injected into a carbonate formation for a period of six months. *Clostridium acetobutylicum* was injected at regular intervals. After

70 days, fresh water began to appear at the production well and 10 days later fermentation products (carbon dioxide, acids, alcohols and acetone) began to appear at the production well. Then the production increased from 0.1-0.3 m^3/day (0.6-2.0 bbl/day).

Jaranyi et al. (review by Hitzman, 1982) used a mixture of anaerobic, thermophilic, bacteria that fermented molasses for a field trial in Hungary. Reduction of the produced crude oil viscosity from 42 to 20 cSt occurred and was attributed to the presence of dissolved fermentation gases, particularly carbon dioxide. Also some marginal enhancement of oil recovery was noted at several wells near the injection well. They also reported an increase of gas production (11% CO_2) from zero to 40 m^3/day (1,400 ft^3/day) four months after inoculation of the field.

In Poland, Karaskiewicz (review by Hitzman, 1982) conducted a number of single well ("huff and puff") field tests using mixtures of several bacteria (Clostridium,, Pseudomonas, Arthrobacter, etc) in 2.5% solutions of molasses. He reported increases ranging from 20-200% production from wells that were originally producing only 0.2 m^3/day (1.3 bbl/day). Although this represents a very small overall enhancement of oil recovery, he reported that the increased recovery was sustained for about 2 years after inoculation with bacteria.

Lazar and Constantinescu (1985) and Lazar (1986) reported on work conducted in Romania which involved: (1) examination of the microbial flora of petroleum reservoirs, (2) adaptation of the microbes to field conditions prior to injection, and (3) field testing of the adapted microbes. Several genera, ranging from 10^7-10^8 cells/ml, were identified in injection waters (Pseudomonas, Escherichia, Arthrobacter, Mycobacterium, Peptococcus, Micrococcus, Clostridium, and Bacillus). The results reported for treatment of seven reservoirs were: (1) a two-fold increase of oil production lasting from 1-5 years occurred in two of the seven reservoirs; (2) results from the other five reservoirs were inconclusive; and (3) more information on the microbial flora of the petroleum reservoirs would be helpful in design of the field applications.

Grula and Russell (1985) used Clostridium in field trials near Okmulgee, Oklahoma in a well completed at 530 meters (1750 feet) in a sandstone formation. After 12 m^3 (80 bbl) of 8.6% molasses and bacterial were injected, the well was closed for four weeks and the wellhead pressure and gas composition were monitored on a weekly basis. The oil production increased from 0.15-0.3 m^3/day (1-2 bbl/day) and was sustained past the 30 day period allotted for the test.

Effect of Environmental Conditions on Bacteria

Marquis et al. (1971) found that the maximum growth pressure for homofermentative (lactic acid) S. faecallis apparently depends on the source of adenosine triphosphate in the substrate. Cultures in a medium containing tryptone yeast extract and pyruvate showed no growth at pressures higher than 20 MPa (200 atm); however, growth occurred at 45 MPa when glucose, galactose, maltose or lactose were substituted for the pyruvate. Furthermore, if the medium was supplemented by addition of 50 mM of Mg^{++} or Ca^{++} ions. Thus this species of bacteria exhibits a wide range of barotolerance which depends on the composition of the growth medium.

Matsumura et al. (1974) found that calcium and magnesium cations enhance the acid resistance of many bacteria. The cations apparently enhance barotolerance by extending the effective pH range for growth under pressure. Almost all oilfield brines contain these cations which may prove to be an asset to MEOR.

Bubela (1983) found that an increase of pressure increased the optimum temperature, and rate of growth, of rod-shaped bacteria. At 100 kPa (1 atm), the optimum growth temperature was 50°C and the mean generation time was 17 hours; however, when the pressure was increased to 20 MPa the maximum growth rate temperature increased to 65°C and the mean generation time decreased to 12 hours. Bubela also observed that the morphology of the bacteria changed from rod-shaped to coccoidal at the higher pressure. Similar results were reported by Moses and Springham (1982). They reported that apparently some bacteria resist high temperature much better when subjected to high pressure. For example, a strain isolated from an oil well core exhibited an maximum growth temperature of 65°C at 100 kPa but would grow quite well when the pressure was increased to 40-60 MPa (400-600 atm).

Most bacteria which are commonly grown in cultures, or isolated from soil or fresh water, are inhibited by the presence of sodium chloride in excess of about one percent. Halotolerant microbes, however, are capable of growth at high salt concentrations. Bacillus subtilis and B. licheniformis are examples of organisms that depend on the presence of salt for growth and morphology. These rod-shaped microbes grow as normal cells in the presence of 0.8 M NaCl, but in the absence of NaCl they grow as abnormal strings of coccoid bodies with distorted cell walls. Rogers et al. (1970) found that at high temperatures, the rod mutants require NaCl for growth.

Ramsay et al. (1985) examined the effects of NaCl concentration and pressure on the growth and metabolism products of B. licheniformis. They found that at 0.1% NaCl, or greater, polysaccharide yield decreases from 0.24 to 0.08 g/l. Polysaccharide production, however, was not affected by an increase of pressure from 100 kPa-40 MPa. At NaCl concentrations between 0.05

and 2.0% carbon dioxide production increased by 350%. This indicates that polysaccharide production was suppresses and gas production was increased considerably by the presence of NaCl. Production of the metabolic products was not affected by the presence of crude oil.

The hydrogen ion concentration (pH) is a factor that affects the growth and reproduction of microbes. In MEOR processes, however, pH is not critical because most petroleum reservoir waters exhibit a range of pH from 6-8 which is suitable for microbial activity.

EQUIPMENT AND PROCEDURES

The principal objective of this work was to examine the ability of microbes to enhance the production of oil at simulated conditions of temperature and pressure up to a depth of 1500 meters (5000 feet), 55°C and 16.6 MPa respectively. An overburden pressure of 23 kPa/m (1 psi/ft) was used in each experiment. Brines with three different concentrations of NaCl were used in the experiments (2, 5 and 7.5% NaCl). The microbial nutrient solution used in all oil displacement tests was 3% cattle-feed molasses in the respective brine.

Cores 25 cm long and 3.8 cm in diameter (70 ml pore volume) are assembled by placing steel end-plates at the ends of the core and then wrapping this assembly in a Teflon sheet (to prevent interaction of the fluids in the core with the rubber sleeve); the assembly is then placed in a rubber sleeve (bicycle inner tube). This assembly was next inserted into heat-shrinkable tubing and heated in an oven to yield a tight, leak-proof, core assembly. The fitting on the end-plates are connected to tubing that penetrates the cap of a cylindrical pressure vessel (rated at 100 MPa 15,000 psi) and this assembly is then completed as described by Donaldson et al., 1980.

Water can be pumped into the annulus between the core assembly and the wall of the steel cylinder and raised to any desired pressure which simulates the overburden pressure. An adjustable back-pressure regulator is connected to the exit end of the core assembly. By setting this back-pressure regulator, fluid pressure in the core can be raised to any desired pressure to simulate equivalent subsurface hydrostatic pressures. When the heating mantle, designed to fit the steel cylinder, is set in place, the core holder is attached to the fluid flow lines. Liquid chromatographic pumps rated at 40 MPa (6000 psi) were used for fluid flow and a hand actuated pump was used to set the overburden pressure. Pressure transducers connected to recorders, and gauges, were used to monitor the overburden and fluid pressures and a thermocouple attached between the heating mantle and the wall of the steel cylinder is used to monitor the temperature. A preliminary test with a thermocouple placed at the center of a core showed that the center of the core will attain the same temperature registered between the heating mantle and the cylinder within 20-30 minutes.

All cores were cleaned with steam, dried at 120°C and then autoclaved to diminish the indigenous microbial flora to a negligible quantity, or perhaps to remove them completely. No pressure increase was observed when the cleaned cores were incubated for several weeks with cattle-feed molasses as the pore fluid. All fluids were evacuated to remove air, treated with 0.1% sodium sulfite to remove trace quantities of dissolved oxygen, filtered, autoclaved and stored in a refrigerator until used.

A steel microbial culture transfer vessel rated at 35 MPa was used to inject the cultures into the core at any given pore pressure. The vessel was equipped with a free moving piston, and inlet and outlet ports. This vessel was autoclaved and placed inside the glove box for culture transfer under anaerobic conditions (about 10^8 cells/ml in all cases).

The core was first evacuated, and flushed with carbon dioxide in nitrogen for 15 minutes to remove oxygen. The cores where then evacuated once more and the brine designated for use was admitted until it completely saturated the core. The designated overburden pressure and temperature were then applied and the crude oil (38°API) was injected into the core, displacing the water to its irreducible water saturation (all volumes entering an exiting the core from this point were measured). The oil volume in the core is labeled the original oil in place (OOIP), where OOIP = PV*(1-S_{iw}). S_{iw} was generally about 30% PV for these cores (Cottage Grove sandstone outcrop cores obtained about 3 km North of Bartlesville, Oklahoma).

A normal waterflood was then conducted, maintaining the integrity of the established simulated subsurface pore pressure. This waterflood produced the residual oil saturation which varied with respect to the environmental conditions that were used. When the residual oil saturation was established (oil could no longer be produced by water displacement), the brine flow rate and pressure were observed to determine the system permeability at residual oil saturation.

With the oil saturation at S_{or1}, one pore volume of brine containing 3% cattle-feed molasses was injected. The system was then incubated at the simulated reservoir conditions for about 2 weeks with careful observation of the pore pressure. If no pressure increase was observed, about 100 ml of the bacteria culture was transferred into the microbial transfer vessel. This vessel was then connected to the inlet port of the core assembly and the bacteria were injected into the core by applying gas pressure to the top of the free piston. Transfer of fluids into the core were made at pressures slightly greater than the pore pressure maintained by the back-pressure regulator which would open to expel fluid when the fluid pressure exceeded the set pressure rating.

After injection of the bacteria, the back-pressure regulator was set to a value several times higher than the established initial pore pressure in the core. The pore pressure was then constantly monitored with a recorder. In all of these experiments an increase of pore pressure was used as the indicator of microbial activity. When the pressure ceased to increase for a 24 hour period, it was assumed that microbial activity had stopped and at this point the system was once more water flooded with maintenance of the pore pressure and temperature. All produced fluids were measured and collected for analyses. Pressure maintenance during the water flood insured that the biogenic gasses would remain in solution and enhance production by solution gas drive and decrease of oil viscosity. This water flood resulted in additional oil production which was due solely to the microbial activity, and it established the second (lower) residual oil saturation (S_{or2}).

The effective permeability to brine was determined when the oil saturation was reduced to S_{or2} and the permeability reduction factor was calculated [PRF = (K at S_{or2})/(K at S_{or1})]. At this point the core was removed from the core holder, cleaned with toluene and steam, dried and saturated with water and oil, as described above to residual oil saturation S_{or1}.

The filtrate from the growth of bacteria in the brine solution at the simulated reservoir conditions was now used as a waterflooding agent to simulate a surfactant-acid solution flood. The pH and surface tension of this filtrate were measured and the solution was injected into the core at residual oil saturation S_{or1}. Any additional oil produced was carefully measured. This experiment was conducted with aqueous filtrates from both the *Bacillus* and *Clostridium* cultures to determine if the solution containing biosurfactants and acids would enhance oil recovery.

EXPERIMENTAL RESULTS

Preliminary Examination of Culture Characteristics

After the colonies of bacteria (*Bacillus licheniformis* JF2 and *Clostridium acetobutylicum*) were developed from stock cultures provided by the Department of Microbiology, the following tests were conducted:

(1) The Gram-stain test indicated that the cells were Gram positive,

(2) Streak plate tests indicated that they had the same morphology before and after use in the cores,

(3) Wet mount tests indicated that both cultures were composed of rod-shaped microbes that were motile,

(4) A viable plate count of the *Bacillus* inoculum just before use indicated that there were 10^8 cells/ml after a 24 hours incubation period at 37^oC at aerobic conditions,

(5) A direct count of the Clostridium cells indicated that the inoculum contained $9x10^7$ cells/ml after incubation for 18 hours at 37°C under anaerobic conditions.

(6) Tests for salinity tolerance showed that optimum growth of the Bacillus licheniformis JF-2 culture occurred in solutions containing 5% sodium chloride, (similar results also were reported by Jenneman et al., 1983). On the other hand, sodium chloride inhibited the growth of Clostridium acetobutylicum since its optimum growth occurred in fresh water (similar results were obtained by Buchanan and Gibbons, 1974).

Nine experiments were conducted with cores simulating depths up to 1500 meters (T°C - 25-55; P(kPa) = 7-16,500; Salinity = 2,5 and 7.5% NaCl). The initial values of pH and surface tension were 7 and $60*10^{-3}$N/m, respectively.

Core Tests with the Microbes

Core runs 1-6 were used to evaluate the performance of B. JF-2 with respect to MEOR at different reservoir conditions. Similarly core runs 7-9 were conducted to evaluate the performance of C. acetobutylicum. Oil recovery from waterflooding alone is reported as percent of the original oil in place (%OOIP). The additional oil recovery, obtained from MEOR, is reported both as %OOIP and as percent of the residual oil in place remaining after the waterflood (%ROIP). The residual oil recovery efficiency is $[E_{or} = 1.0 - S_{or2}/S_{or1}]$ (after Douglas, 1987):

Experiments with the cores involve three distinct states which are presented in Tables I, II and III. The tables list the core properties, experimental conditions and results for the six core runs using Bacillus and

TABLE I Experimental conditions and core properties for the MEOR core experiments using Bacillus and Clostridium

Run No.	Core No.	Core Porosity (%)	Core Pore Vol(ml)	Initial Pore P-gage(MPa)	Run T(°C)	NaCl (%)	Simulated Depth (meters)
			Results from Bacillus licheniformis JF-2				
1	CG1	25	71.0	0.0	25	5.0	0
2	GC2	24	70.9	5.5	35	5.0	500
3	CG3	24	70.4	11.0	45	5.0	1000
4	CG4	24	70.5	16.5	55	5.0	1500
5	CG5	25	72.0	16.5	55	7.5	1500
6	CG6	25	70.0	16.5	55	2.0	1500
			Results from Clostridium acetobutylicum				
7	CG7	24	70.0	5.5	35	2.0	500
8	CG8	24	70.0	11.0	45	2.0	1000
9	CG9	25	71.0	16.5	55	2.0	1500

the three core runs using Clostridium. The first stage of the work is to determine the petrophysical properties of the core and then established a residual oil saturation. For example, Run No. 1 was designed to simulate surface conditions, or the ambient conditions usually used for laboratory core flood experiments. The irreducible water saturation (S_{iw}) in Table II was 25.35 (S_o = 74.6%); as explained above, this was established by first saturating the core with the brine and then displacing the brine with the crude oil. The next step was to conduct a normal waterflood to measure the amount of oil recovery that could be obtained without the aid of bacteria and to establish the residual oil saturation (S_{or1} = 36.39% of the pore volume).

The second stage (Table II) was to conduct a waterflood and determine the amount of oil that could be produced without the aid of bacteria. For Run No. 1, the waterflood began with a water saturation in the core of 25.35% (S_{iw}) and ended with 63.61% (S_{wor}) which represents 38.26% production of the OOIP ($S_{wor} - S_{iw}$). At the end of the waterflood, the rate of water injection and the pressure drop across the core were used to calculate the permeability to water at the residual oil saturation existing at the end of the waterflood ($K_{w,Sor1}$).

TABLE II Water and oil saturations of the cores before and after waterflooding; and, the permeability to water at the end of the waterflood (water permeability at residual oil saturation)

Run No	S_{iw} (%)	S_{wor} (%)	S_{or1} ($1-S_{wor}$) (%)	OOIP Produced (%)	$K_{w,Sor1}$ (md)
		Results from Bacillus licheniformis JF-2			
1	25.35	63.61	36.39	38.26	6.2
2	41.57	82.63	17.37	41.06	23.6
3	42.05	81.15	18.85	39.10	10.9
4	31.92	65.76	34.24	33.84	17.7
5	30.21	69.10	30.90	38.89	28.6
6	36.43	79.27	20.73	42.84	23.0
		Results from Clostridium acetobutylicum			
7	31.86	68.14	31.96	58.30	32.8
8	28.30	71.70	28.30	59.10	31.3
9	34.40	65.60	34.40	54.80	30.3

The third stage (Table III) was the introduction of 3% cattle feed molasses followed by injection of the bacteria with the pore pressure maintained at the initial setting (Table I). After introduction of the bacteria, the back-pressure regulator was turned off which allowed the pore pressure to rise in the core to any value within the limits of the equipment to contain it. The increase of pressure with respect to time was monitored with a pressure recorder. When the pore pressure ceased to increase, indicating the end of

TABLE III Pore pressure at the end of the period of incubation; and, the core and brine conditions after the pressure maintenance waterflood that was run at the end of the period of incubation

P (MPa) At end of Incubation	S_{or2} (%)	E_{or} (%)	MEOR oil Recovery (%OOIP)	Gas Produced (cm^3)	$K_{w,Sor2}$ (md)	Brine pH	Surface Tension (10^{-3}N/m)
			Results from Bacillus licheniformis JF-2				
2.4	26.7	26.8	13.1	145	3.2	6.0	27.2
14.1	3.7	78.5	23.4	186	14.6	6.1	29.1
18.5	7.3	61.4	20.0	160	6.1	6.2	33.1
22.6	21.6	36.9	18.5	110	11.5	6.1	31.6
21.6	20.4	34.3	15.1	95	21.5	6.2	34.7
19.8	16.0	22.8	7.4	65	20.0	6.3	39.3
			Results from Clostridium acetobutylicum				
11.0	15.7	44.78	18.7	150	24.1	4.8	45.2
16.5	16.6	43.34	17.7	125	21.5	4.9	46.8
21.4	19.9	32.97	14.9	100	24.5	5.0	46.5

microbial activity, the back-pressure regulator was set to open at the maximum pore pressure established by the microbes (2.4 MPa pressure at the end of incubation for Run 1, Table III). A waterflood was then conducted, with maintenance of the pore-pressure, to determine the amount of additional oil recovery that could be obtained after incubation with bacterial using this procedure. For Run 1, the residual oil saturation was reduced from 36.39% (S_{or1}, Table II) to 26.7% (S_{or2}, Table III) which represents a residual oil recovery efficiency (E_{or}, Table III) of 26.8%. The percentage of the OOIP that was produced by microbial action, for Run 1, was 13.1%. The produced gas also was collected and measured at 101 kPa (1 atmosphere) and 25°C.

At the end of the waterflood, the injection rate and pressure drop across the core were used to calculate the permeability to water at the new residual oil saturation ($K_{w,Sor2}$). Finally, the pH and interfacial tension of the effluent brine from the core were recorded (6.0 and 27.2, respectively).

Biosurfactant-Bioacid Floods

Two additional experiments were conducted to determine if the aqueous solution from incubation of the bacteria in the cores (which contained surfactants and acids produced by the microbes) would recover additional oil if it was used as the waterflooding medium. Both runs were conducted at 55°C and 101 kPa (1 atm) pressure. The waterfloods were conducted at 101 kPa absolute pressure to avoid any contribution from biogenic gas.

Core GC10 (Table V) had a pore volume of 70.3 cm^3 and porosity of 24.0%; S_{iw} was 31.1% and the OOIP was 68.9% PV. Waterflooding produced 56.6% OOIP leaving a residual oil saturation, S_{or1} equal to 29.9%. The pH and surface tension of the brine were 7.0 and 60 $N*10^{-3}/m$, respectively. A flood conducted with filtered aqueous solution from the JF-2 incubation in a core (pH = 6, Surface Tension = 27) recovered 4.6% OOIP.

Core GC11 had a pore volume of 70.0 cm^3 and porosity of 24.0%; S_{iw} was 32.5% and the OOIP was 67.5% PV. Waterflooding (pH = 7, surface tension = 60) produced 60.7% OOIP leaving a residual oil saturation of 26.5%. A flood conducted with filtered aqueous solution from the Clostridium incubation in a core (pH = 4.8 and surface tension = 44.5) recovered 4.1% OOIP.

Solubility of Biogenic Gas

The biogenic gas was collected under water and analyzed by chromatography. The gas was a mixture of several constituents as shown in Table IV. A small amount of air was inadvertently introduced in collecting and handling the effluent gas. The overriding feature is that more than 75% of the gas was composed of carbon dioxide in all cases, and that JF-2 produced about 18% nitrogen but no hydrogen whereas Clostridium apparently did not produce nitrogen (the small amount in the gas was probably due to introduction of air: ratio of nitrogen to air = 3.7), but Clostridium produced instead about 12% hydrogen. The amount of methane was insignificant and probably came from the crude oil.

TABLE IV Analyses of the effluent gas from each of the core experiments

Run No.	Bacteria	H2 Mole%	N2 Mole%	O2 Mole%	CO2 Mole%	CH4 Mole%
1	JF-2	0.0	18.0	2.3	78.0	1.7
2	JF-2	0.0	17.6	3.1	77.4	1.9
3	JF-2	0.0	18.3	2.7	77.2	1.8
4	JF-2	0.0	18.1	2.9	77.3	1.7
5	JF-2	0.0	17.5	3.3	77.4	1.8
6	JF-2	0.0	18.5	3.5	76.0	2.0
7	Clos.	12.0	5.2	1.5	81.3	0.0
8	Clos.	11.4	6.2	2.1	78.8	1.5
9	Clos.	12.2	5.8	1.7	78.5	1.8

Because the cores were completely liquid-filled, all generated gas had to dissolve in the oil and water. In order to understand the role of the dissolved gas in MEOR, it was necessary to determine the amount of gas dissolved in each phase. From material balance:

$$nZ_i = n_L X_i + n_g Y_i \tag{1}$$

Substituting $K_i X_i$ for Y_i and making use of the fact that the mole fraction of component i in the oil and water phases (Z_i) is equal to 1.0, Eq. 2 is obtained:

$$\Sigma Z_i = \Sigma X_i[1.0 + n_g(K_i - 1)] = 1.0 \quad (2)$$

Equation 2 was solved by using a trial-and-error procedure: successive trial values of n_g were selected until the summation on the right hand side was equal to 1.0. The value of n_g that forces the summation to 1.0 is the correct value of n_g.

The following procedure was used to estimate the maximum solubility of the gas mixture (CO_2, H_2, and N_2) in oil and water at any given temperature and pressure:

(1) The gas-to-liquid equilibrium ratios (K_i) for CO_2 and N_2 in crude oil, at a given T and P, were estimated from equilibrium vaporization ratio charts (Jocoby and Rzasa, 1952),

(2) The mole fraction of each gas component dissolved in the oil phase (X_{io}) was calculated from $K_i = Y_i/X_{io}$,

(3) The maximum number of moles of each gas component dissolved in oil is then:

$$n_{io} = X_{io}n_o \quad (3)$$

Similarly, the maximum number of moles of each gas dissolved in water is:

$$n_{iw} = X_{iw}n_w \quad (4)$$

The moles of each component of collected biogas are:

$$n_{ci} = Y_i n_{gc} = n_{io} + n_{iw} \quad (5)$$

The maximum number of moles of gas that can be dissolved in the oil and water phases are:

$$n_{go} = \Sigma n_{io} \quad (6)$$

$$n_{gw} = \Sigma n_{iw} \quad (7)$$

If the total number of moles of collected biogenic gas is less than the maximum number of moles that can be dissolved in the two phases, then all of the collected gas was in solution. Otherwise, free gas will be present in the rock. This can be expressed as:

$$n_{gc} > [n_{go} + n_{gw}]_{max} \quad \text{(free gas present)} \quad (8)$$

The volume of gas dissolved in the oil phase and in the water phase can be calculated from the following equations (9 and 10):

$$V_{go} = [n_{go}/(n_{go} + n_{gw})]*V_{gc} \tag{9}$$

$$V_{gw} = V_{gc} - V_{go} \tag{10}$$

The amount of gas collect (at 101 kPa and 25°C), and the amount dissolved in the oil at reservoir conditions was used to calculate the gas/oil ratio (GOR). This GOR is the ratio of biogenic gas dissolved in the oil at the end of the incubation period to the volume of residual oil after waterflood. The results for all of the experiments are listed in Table V.

TABLE V List of the gas collected, the calculated amount of gas dissolved in the residual oil after waterflood, the gas/residual oil ratio and oil recovery (%OOIP) for each experiment

Run No.	Volume collected, V_{gc}, cm^3	Volume in oil, V_{go}, cm^3	Volume residual oil, V_{rol}, cm^3	GOR, cm^3/cm^3	Oil from MEOR, %OOIP
		Results from Bacillus licheniformis JF-2			
1	37.96	18.67	25.84	0.72	13.06
2	172.18	33.54	12.31	2.73	23.35
3	146.45	31.15	13.27	2.35	19.87
4	96.96	46.09	24.14	1.91	18.54
5	87.02	35.75	22.25	1.61	22.25
6	61.70	14.77	14.51	1.02	7.42
		Results from Clostridium acetobutylicum			
7	95.32	34.56	19.90	1.74	19.90
8	94.14	35.72	20.52	1.74	17.73
9	83.22	32.06	21.07	1.52	14.91
		Results from Biosurfactant-Bioacid Floods			
10	0	0	0	0	4.63
11	0	0	0	0	4.11

When the volume of oil recovered from MEOR (%OOIP) was plotted versus the gas/oil ratio, the volume of oil recovered was found to be a direct function of the GOR with a correlation coefficient (r_2) equal to 0.97 and standard deviation of 1.27 [$R_o = 4.3 + 7.1 * GOR$].

DISCUSSION OF EXPERIMENTAL RESULTS

The bacteria were able to grow vigorously at the simulated reservoir conditions as indicated by production of: (1) biosurfactants - JF-2 and _Clostridium_ lowered the interfacial tension an average of 32.5 and 46.2

$N*10^{-3}/m$, respectively; (2) organic acids - JF-2 reduced the pH from 7 to 6 and Clostridum from 7 to 5; and (4) biogenic gases as shown in Table IV. In fact, there is some indication that the high pore pressures increased the optimum growth temperatures of the bacteria by as much as $10^{o}C$. Bubela (1982) reported an increase of optimum growth temperature of $15^{o}C$ for rod-shaped bacteria when the pressure was increased from atmospheric to 20 MPa, and similar results were reported by Moses and Springham (1982).

Production of metabolic products by the bacteria did enhance oil recovery when the cores were waterflooded using a pressure-maintenance technique. Instead of bleeding off the pressure generated by the microbes and then conducting a waterflood, the maximum system pressure attained by bacteria treatments was retained during the waterflood. A back-pressure regulator was set to release at the system pressure and then the water was injected at the pressure required by the system. The result was very good recovery of residual oil; JF-2 and Clostridium produced averages of 52.8% and 40.4% of ROIP, respectively. The results indicate that a field MEOR project would be more successful if the excess pressure generated by microbial fermentation is maintained during the waterflood following the MEOR treatment of the field. This procedure will keep much of the biogenic gases in solution in the oil which will provide a solution gas drive caused by gradual release of gas as the oil migrates to the low pressure area of a production well.

In addition to solution gas drive, there also was probably some contribution to the increased production from lowering of the interfacial tension. Experiments 10 and 11 (using the filtrates from microbial fermentation for the waterflood) resulted in an average oil production increase of 10% which is probably due to reduced interfacial tension rather than reduced pH. The slight reduction of pH was insufficient to affect the minerals of the consolidated sandstone used for this work.

Some contribution to enhanced production also probably resulted from permeability reduction. Since bacteria preferentially grow in the larger pores, they reduce the rock permeability with biomass and metabolic products, as shown by Torbati et al., 1984. This reduction of permeability was indicated by the difference in the permeability to water at residual oil saturation 1, $K_{w,Sor1}$, (immediately after the preliminary waterflood) and residual oil saturation 2, $K_{w,Sor2}$, (permeability to brine after treatment with bacteria), Tables II and III. The reduction of permeability decreases the pore size heterogeneity and removes high permeability zones that provide channels for ineffective flow of injected water. These two effects lead to increased production of oil.

CONCLUSIONS

It is apparent that the major contribution to MEOR resulted from solution gas drive created by the biogenic gas. Apparently there also was some contribution from reduction of interfacial tension and permeability, but unfortunately it is not possible to separate the individual contributions of each mechanisms. Nevertheless, the work indicates that bacteria will enhanced oil recovery at simulated subsurface reservoir conditions. The system should be shut-in during the period of fermentation to allow buildup of the gas pressure and solution of the nascent gas in the oil phase. Furthermore, a pressure maintenance waterflood will enhance the additional oil production by stimulating the solution gas mechanism.

ACKNOWLEDGEMENTS

Dr. Michael McInerney, Department of Microbiology and his graduate students, for their assistance and advice on the growth of the microbes, and for providing the bacteria used in this work. Dr. Richard Mallinson, Department of Chemical Engineering for his assistance with the chromatographic analyses of the biogenic gases. To the Shell Companies Foundation for financial assistance.

NOMENCLATURE

cSt	centistokes
E_{or}	residual oil recovery efficiency
GOR	gas/residual oil ratio
K	permeability, md
K_i	ratio of the mole fraction of component i in the gas phase to the more fraction of component i in the liquid phase
n	moles in the mixture (liquid + gas)
n_{ci}	moles of the ith component of collected gas
n_{gi}	moles of collected gas component i
n_g	moles in the gas phase
n_{io}	moles of the ith component dissolved in the oil phase
n_{iw}	moles of ith component dissolved in water
n_L	moles in the liquid phase (oil and water)
n_o	moles of residual oil after waterflood
n_w	moles of water in the core after waterflood
PRF	permeability reduction factor
r_2	statistical correlation coefficient
R_o	oil recovery (%OOIP)
S_{or}	residual oil saturation
S_{or1}	residual oil saturation after waterflood

S_{or2} residual oil saturation after treatment with bacteria (after waterflood at the end of MEOR)

V_{gc} total volume of collected gas

V_{go} volume of dissolved gas in oil

V_{gw} volume of dissolved gas in water

V_{rol} volume of residual oil after waterflood, cm^3

X_i mole fraction of component i in the liquid phase

Y_i mole fraction of component i in the gas phase

Z_i mole fraction of component i in the water and oil phases

REFERENCES

Beckman, J.W., 1926. Industrial and Engineering Chemistry, News Ed., v. 4, Nov. 10, p. 3.

Bubela, B., 1982. Combined effects of temperature and other environmental stresses on MEOR. In: Donaldson, E.C. and Clark, J.B. (Editors), 1982. Proceedings of 1982 International Conference on Microbial Enhancement of Oil Recovery. CONF-8205140, NTIS, Springfield, VA 22161, pp. 118-123.

Buchanan, R.E. and Gibbons, N.E., 1974 (Editors). Bergey's Manual of Determinative Bacteriology. Williams and Wilkens, Baltimore, Maryland, 8th ed., pp. 560-563.

Donaldson, E.C., Chilingarian, G.V. and Yen, T.F. (Editors), 1985. Enhanced Oil Recovery, I - Fundamentals and Analyses. Elsevier Science Publishers, New York, Amsterdam, 357 pp.

Donaldson, E.C., Kendall, R.F., Pavelka, E.A. and Crocker, M.E., 1980. Equipment and procedures for fluid flow and wettability tests of geological materials. NTIS, Springfield, VA, DOE/BETC/IC-70/5, 40 pp.

GAO - General Accounting Office, 1986. Current and potential use of EOR. Briefing report to the Chairman, Subcommittee on Environmental, Energy and National Resources, House of Representatives. GAO/RCED-181BR, Washington, D.C.

Grula, E.A., Russell, H.H. and Grula, M.M., 1985. Field trials in Central Oklahoma using Clostridium strains for MEOR. In: Zajic, J. E. and Donaldson, E.C. (Editors). Microbes and Oil Recovery. Bioresources Publications, El Paso, TX, pp. 144-150.

Hitzman, D.O., 1982. Petroleum microbiology and the history of its role in enhanced oil recovery. Review in: Donaldson, E.C and Clark, J.B. (Editors), 1982. Proceedings of 1982 International Conference on Microbial Enhanced Oil Recovery. CONF-8205140, NTIS, Springfield, VA 22161, pp. 162-218.

Ivanov, M.V. and Belyaev, S.S., 1982. Microbial activity in waterflooded oilfields and its possible regulation. In: Donaldson,E.C. and Clark, J.B. (Editors), 1982. Proceedings of 1982 International Conference on Microbial Enhanced Oil Recovery. CONF-8205140, NTIS, Springfield, VA 22161, 219 pp.

Jacoby, R.H. and Rzasa, M.J., 1952. Equilibrium vaporization ratios for nitrogen, methane, carbon dioxide, ethane and hydrogen sulfide in absorber crude oils. Petroleum Trans. AIME, v. 195, pp. 99-110.

Jang, L.K., Sharma, M.M. and Yen, T.F., 1984. The transport of bacteria in porous media and its significance in MEOR. Soc. Pet. Engr. Paper 12770, California Regional Meeting.

Jenneman, G.E., Knapp, R.M., Menzie, D.E., McInerney, M.J., Revus, D.E., Clark, J.B. and Munnecke, D.M., 1982. Transport phenomena and plugging in Berea sandstone using microorganisms. In: Donaldson, E.C. and Clark, J.B. (Editors), 1982. Proceedings of 1982 International Conference on Microbial Enhanced Oil Recovery. CONF-8205140, NTIS, Springfield, VA 22161, 219 pp.

Jenneman, G.E. McInerney, M.J., Knapp, R.M., Clark, J.B., Ferro, J.M., Revus, D.E., and Menzie, D.E., 1983. A halotolerant, biosurfactant producing Bacillus species potentially useful for enhanced oil recovery. Develop. Industr. Microbiol. v. 24, pp. 485-492.

Johnson, A.C., 1979. Microbial oil release technique for EOR. Conf. on Microbiological Processes Useful in EOR. San Diego, CA, Aug. 29 - Sept. 1, DOE CONF-79087, NTIS, Springfield, VA 22161.

Karaskiewicz, J., 1963. Experiments in biological activation of petroleum reservoirs. Intern. Sci. Conf. Geochem., Microbiol Oil Chem. 3rd, Report 1, Geochem and Microbiol, pp. 566-578.

Lazar, I. and Constantinescu, P., 1985. Field trial results of MEOR. In: Zajic, J.F. and Donaldson, E.C. (Editors), 1985. Microbes and Oil Recovery. Bioresources Publications, El Paso, TX, pp. 122-143.

Lazar, I., 1986. Research on the microbiology of MEOR in Romania. Proc. 1st International MEOR Workshop, Apr. 1-3, DOE Report, DOE/BC/10852-1, pp. 124-151.

Matsumura, P. Keller, D.M. and Marquis, R.E., 1974. Restricted pH ranges and reduced yields for bacterial growth under pressure. Microb. Ecol., v. 1, pp. 176-189.

Marquis, R.E. and Zobell, C.E., 1971. Magnesium and calcium ions enhance barotolerance of streptococci. Arch. Microbiol., v. 79, pp. 80-92.

Moses, V. and Springham, D.G., 1982. Bacteria and Enhancement of Oil Recovery. Applied Science Publishers, London, 1982, 178 pp.

Mungan, N., 1964. Role of wettability and interfacial tension in waterflooding. Soc. Pet. Engr. J., pp. 115-122.

Mungan, N., 1965. Permeability reduction through changes in pH and salinity. J. Pet. Technol., pp. 1449-1453.

Rogers, H.J. McConnel, M. and Burdett, I.D., 1970. The isolation and characterization of mutants of Bacillus subtilis and lincheniformis with disturbed morphology and cell divisions. J. Gen. Microbiol., v. 61, pp. 155-171.

Torbati, H.M., Donaldson, E.C., Jenneman, G.E., Knapp, R.M., McInerney, M.J., and Menzie, D.E., 1985. Depth of microbial plugging and its effects on pore size distribution in Berea sandstone cores. In: Zajic, J.E. and Donaldson, E.C. (Editors). Microbes and Oil Recovery. Bioresources Publications, El Paso, TX, 370 pp.

Updegraff, D.M., 1957. Recovery of petroleum oil. U.S. Patent 2,807,570.

Yarbrough, H.F. and Coty, V.F., 1982. MEOR from the upper Cretaceous Nacatoch Formation, Union County, Ark. In: Donaldson, E.C. and Clark, J.B. (Editors), 1982. Proceedings of 1982 International Conference on Microbial Enhanced Oil Recovery. CONF-8205140, NTIS, Springfield, VA 22161, pp. 149-153.

Zoboll, C.E., 1946. Bacteriological Process for Treatment of Fluid-bearing Earth Formations. U.S. Patent No. 2,413,278, Dec 24.

Zobell, C.E., 1953. Recovery of Hydrocarbons. U.S. Patent No. 2,641,566, June 9.

Reprinted from: *Microbial Enhancement of Oil Recovery - Recent Advances,* edited by E.C. Donaldson

CH. R–15

HIGH-TEMPERATURE, SALT-TOLERANT ENZYMIC BREAKER OF XANTHAN GUM VISCOSITY

M.E. SLODKI AND M.C. CADMUS
Biopolymer Research Unit, U.S. Department of Agriculture, Agricultural Research Service, Northern Regional Research Center, 1815 N. University St., Peoria, IL 61604

ABSTRACT

Slodki, M.E. and Cadmus, M.C., 1990. High-temperature, salt-tolerant enzymic breaker of xanthan gum viscosity.

Soil enrichment growth on xanthan gum at 45°C in the presence of 3% NaCl afforded an extracellular xanthanase functional in 10% brines at 60° to 70°C. Even though the bacterial culture (NRRL B-14401) is a consortium from which no individual or combination of enzyme-producing bacteria were isolable, it has been reproducibly propagated from refrigerated or lyophilized broths. Structural analyses revealed three major products: pyruvated mannose; a tetrasaccharide bearing a terminal unsaturated glucuronic acid residue; and a pentasaccharide derived from unpyruvated side-chains. Both oligosaccharides were branched. The enzymes present were, consequently, a lyase that removed terminal pyruvated mannose from side-chains and a depolymerase that cleaved the cellulose-like backbone uniquely. A minor product of enzymolysis was a higher molecular-weight fragment that contained sequences of backbone glucosyls unsubstituted by side-chains. Such cellulosic sequences, which have not been taken into account in studies of the molecular behavior and interactions of xanthan gum, likely have practical consequences.

INTRODUCTION

Xanthan gum (Polysaccharide B-1459) from *Xanthomonas campestris* NRRL B-1459 was developed at the Northern Regional Research Center some three decades ago as part of a program on utilization of corn sugar for fermentative production of microbial polysaccharides. It was the first such product, and its food and industrial uses continue to grow. Major oil/gas field applications are in drilling fluids and as a mobility control agent in water-flooding operations. Low concentrations of the gum impart high viscosity to aqueous solutions. These solution are pseudoplastic; *i.e.*, display rapid shear thinning, viscosity regain immediately upon removal of shear forces, and high rheological yield. The latter property is a measure of the shear stress required for the onset of viscosity reduction, and consequently of the suspending power. In addition, the viscosity is remarkably stable to salts, divalent metal ions, temperature, extremes of pH, and to degradations either by chemicals or by mechanical shearing.

The above properties are advantageous for use in hydraulic fracturing fluids employed to stimulate flow of natural gas from subterranean formations of low porosity. In such fluids, the function of the gum is to suspend propping materials; e.g., sand, that prevent closure of the fracture and to allow their delivery deep into the fracture. When the proppage is in place, the viscosity of the carrier medium must be broken in order to allow gas to flow freely from the reservoir to the production well. Hypochlorite, currently used as a viscosity breaker, is difficult to control and potentially hazardous. An enzymic breaker would be a benign and potentially controllable agent.

2 STRUCTURE AND CONFORMATION OF XANTHAN GUM

Structure (1) depicts the average repeat unit of xanthan gum as determined by Jansson et al. (1975) and Melton et al. (1976):

```
         -4Glcβ-4Glcβ-4Glcβ-4Glcβ-
            3|        3|
        Ac-6Manα   Ac-6Manα
            2|        2|                       (1)
            GlcAβ     GlcAβ
            4|        4|
      Pyr-4,6Manβ     Manβ
```

The polysaccharide was determined to consist of a cellulose-like backbone of (1→4)-linked β-D-glucosyl (Glc) residues to which are appended, on alternate residues, trisaccharide side chains. In the commercial samples of the gum we used, as well as in those prepared in our laboratory, about half of the side-chain terminal β-D-mannosyl (Man) residues bear pyruvic acetal (Pyr) substituents (Sandford et al., 1977). It is not known whether these substituents occur in a regular manner. By contrast, the α-D-mannosyl residues (1→2)-linked by β-D-glucopyranosyluronic acid (GlcA) are fully 6-*O*-acetylated (Ac-6).

In solution, xanthan gum is denatured by heating and renatured by slow cooling in much the same way as DNA (Holzwarth, 1976; Holzwarth and Prestridge, 1977). Increasing the ionic strength raises the melting temperature. Although some hold that the transition reflects the concerted unwinding of an ordered single chain helix and simultaneous disruption of side-chain interaction with the backbone (Milas and Rinaudo, 1986), there is also good evidence for a double helical conformation (Lecourtier et al., 1986). Unfortunately, x-ray diffraction cannot distinguish between either possibility (Moorhouse et al., 1977). The work of Lecourtier et al. (1986), however, suggests that those reporting evidence for a single-stranded xanthan are indeed working with that species. In other words, the method of preparation and subsequent treatment influence the nature of the sample. At high ionic strengths, Lecourtier et al. observe a more compact double helical form. This structure would prevail in brines employed with hydraulic fracture fluids, and of necessity be the form attacked by salt-tolerant xanthanases. Fungal cellulase randomly cleaves the β-(1→4)-linked glucosyl main chain of xanthan in its unordered conformation, but does not attack the ordered helix (Rinaudo and Milas, 1980).

3 SALT-INTOLERANT AND -TOLERANT XANTHANASES

At the time of our initial studies on bacterial enzymes that degrade xanthan gum, the main objective was to find, through use of the soil enrichment culture technique, enzymes that degrade exocellular rhizobial polysaccharides. It was hoped that such enzymes would aid in determination of rhizobial polysaccharide structures and also generate oligosaccharides useful, as competitive inhibitors, for determining the specificities of interactions between legume lectins and the polysaccharides (Cadmus et al.,1987, 1990). Since a plentiful supply of xanthan gum was at hand, and its structure well-known, we decided to use it in order to gain experience with the enrichment technique as applied to polysaccharides. We were also mindful of the need for a salt-tolerant viscosity breaker, since fracture fluids are made up with brines in order to maintain the stability of subterranean clays. Consequently, enrichment cultures were carried out simultaneously in the presence and absence of 4% NaCl. We succeeded in isolating *Bacilli* (species not known) that excreted enzyme complexes having salt tolerance reflecting the enrichment conditions. One strain produced salt-sensitive enzyme(s), whereas the other excreted a salt-tolerant enzyme complex, but was difficult to grow except in the presence of other soil bacteria (Cadmus et al., 1982; Cadmus, 1983).

Earlier, Lesley (1961), Sutherland (1981, 1982) and Cripps et al. (1981) had reported xanthanase complexes that contained: 1) lyases that removed terminal pyruvated mannose with formation of unsaturated (*ene*-4,5) glucuronic acid residues and 2) hydrolases that cleaved the cellulosic backbone to give *linear* saturated and unsaturated oligosaccharides. As with our strain from the enrichment lacking salt, these enzymes were inactivated by moderate concentrations of salt. Our salt-tolerant strain, however, produced enzymes that retained one-third of maximum activity in 8% NaCl and that were stable up to 50°C in the presence of salt (Cadmus and Slodki, 1985).

Both enzyme preparations described by Cadmus et al. (1982) displayed activities not previously observed by others: hydrolysis of individual linkages in the side-chains of xanthan gum to yield D-mannose, pyruvated D-mannose, 6-*O*-acetyl-D-mannose and D-glucuronic acid. Neither unsaturated glucuronic acid nor D-glucose were detected, but a soluble, high-molecular-weight fragment was isolated that consisted primarily of (1→4)-linked D-glucosyl residues. These results were subsequently confirmed by Hou et al. (1986a) who also detected D-glucose and later isolated a depolymerase (Hou et al., 1986b). We recently observed the presence of this enzyme in our preparations after extended incubation with xanthan.

4 HIGH-TEMPERATURE, SALT-TOLERANT XANTHANASE

We had hoped that salt would stabilize the salt-tolerant xanthanase so that it would function beyond 50°C. Our goal was to obtain a xanthanase functional at 65-70°C, which is considered a moderate temperature range for gas-bearing formations. Incubations of soil enrichments at those temperatures were unsuccessful. Cell growth and viscosity reduction were obtained when, after progressively lower temperatures were used, the incubations were conducted at 45°C.

Unlike the earlier xanthanases, no single enzyme-producing organism, or pair of organisms, was isolable from the enrichment broth. The microbial consortium, however, is quite stable and may be stored on agar slants or in refrigerated or lyophilized culture broths (Cadmus et al., 1989; Cadmus and Slodki, 1989).

Fig. 1 shows a comparison between salt-tolerant (ST) and high-temperature (HT) xanthanases. In a medium of low ionic strength, the latter does not begin to lose activity until the temperature reaches 60°C.

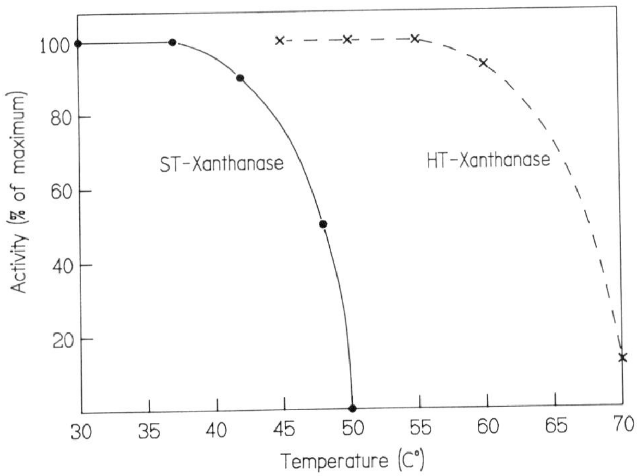

Fig. 1. Thermal stability of salt-tolerant (ST) and high-temperature (HT) xanthanases in a medium of low ionic strength (Cadmus et al., 1989). Activity was measured as the liberation of reducing power.

In a medium of high ionic strength (Fig. 2), 82% of the maximum activity of HT xanthanase remained at 65°C.

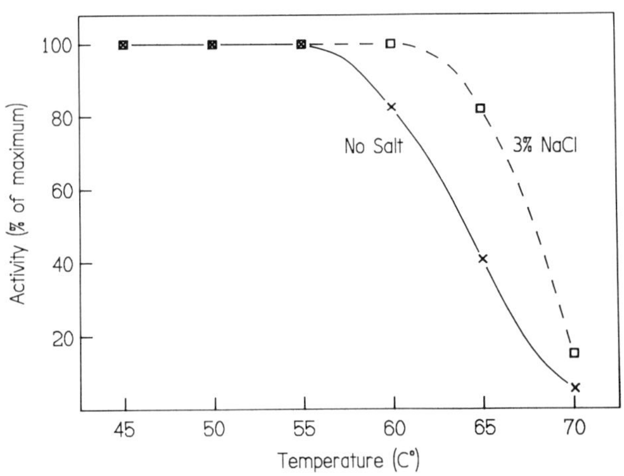

Fig. 2. Thermal stability of HT xanthanase in the absence and presence of salt (Cadmus et al., 1989). The incubation conditions and assay of activity were the same as in Fig. 1.

Fig. 3 illustrates the effect of pH on both the stability of HT xanthanase and on its activity in media of low ionic strength. Stability in the absence of substrate falls off sharply below pH 5, but 75% of the initial activity remains at pH 8.6. Enzymic activity is optimal at pH 5.8, with 85% of the maximal activity retained at pH 5.0 and 6.2.

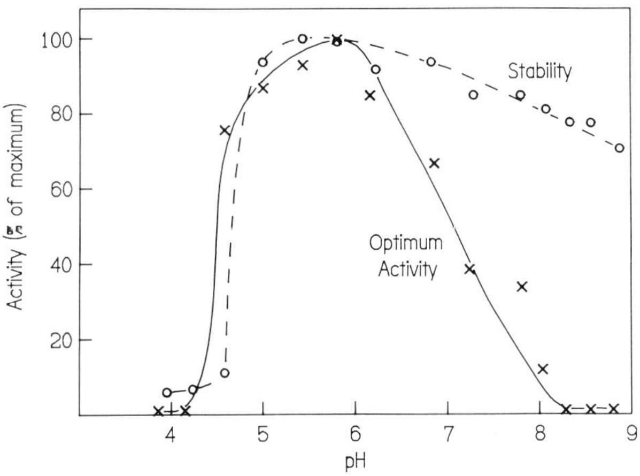

Fig. 3. Effect of pH on stability (-O-O-) and activity (-X-X-) of high-temperature xanthanase in media of low ionic strength (24 h, 25 C). The data are summarized in Cadmus et al. (1989). Acetate and phosphate buffers (0.05M) were used.

In Fig. 4 is shown the effect of added NaCl on HT xanthanase. Activity gradually diminishes until 40% remains in 10% salt. Calcium and magnesium chlorides at 1% concentrations have no effect.

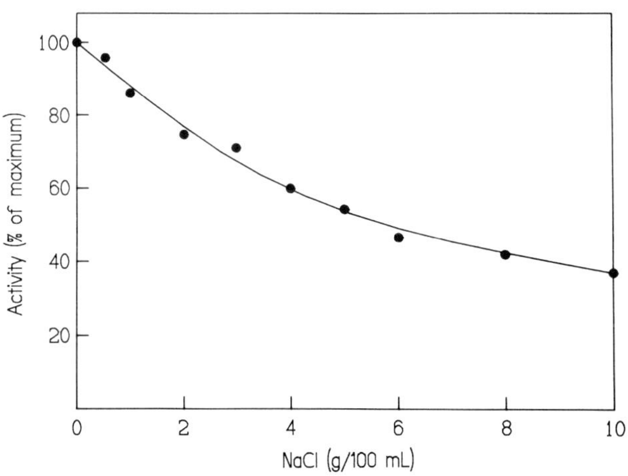

Fig. 4. Effect of added salt on activity of HT xanthanase incubated 6 h at 45°C (Cadmus et al., 1989).

The relationship between reducing power released and loss of viscosity (Fig. 5) appears reciprocal. Slow loss of viscosity suggests that the enzyme responsible is *exo*-acting.

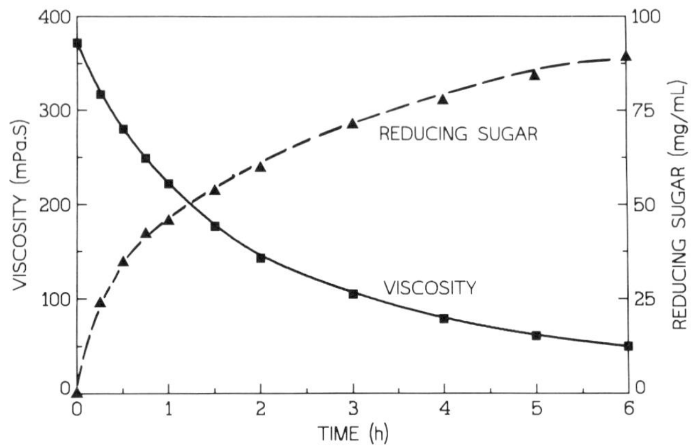

Fig. 5. Progress of HT xanthanase as a function of incubation time (45°C, 0.025M buffer, pH 6.0). Reducing sugar, ▲---▲; viscosity, ■---■ (Cadmus et al., 1989).

5 PRODUCTS OF XANTHAN DEGRADATION BY HT XANTHANASE

Cadmus et al. (1989) separated the products of xanthan enzymolysis by use of a dialysis apparatus equipped with a membrane of nominal 10,000 molecular weight (MW) cut-off. Low- and high-MW fractions were recovered in approximate 2:1 ratio and then analyzed by methylation analysis after further separation of the low MW components.

5.1 *Low MW products*

Pyruvated mannose and two *branched* oligosaccharides were the sole low MW products. The oligosaccharides, which arose from cleavage of backbone glucosyl linkages between residues bearing side-chains and succeeding glucoses, were a pentasaccharide (2) and an unsaturated tetrasaccharide:

Manβ-4GlcAβ-2(Ac6)Manα-(Glcβ-4)3Glc (2)

ene-4,5GlcAβ-2(Ac6)Manα-(Glcβ-4)3Glc (3)

Oligosaccharide (2) was derived from a backbone cellobiosyl bearing unpyruvated side-chains; (3) represented cellobiosyls bearing pyruvated side-chains, and was formed through removal of pyruvated mannose by action of a lyase. In accord with the previously determined structure of xanthan gum, the analyses detected pyruvated mannose and oligosaccharides (2) and (3) in equimolar amounts. A backbone-cleaving β-(1→4)-D-glucosidase apparently is the enzyme responsible for loss of viscosity, and differs in point of attack from the β-glucosidases present in other xanthanase preparations. With this finding, all possible glycosidic cleavages have been demonstrated for xanthan.

5.2 *High MW product; significance for xanthan properties*

Analysis of the fraction of the enzymic digest retained by the membrane filter revealed the presence of *O*-acetyl substituents, but no pyruvic acetal. Its UV spectrum indicated the presence of unsaturated uronic acid residues. The methylation analysis accorded with

```
Glcβ-4Glcβ-[4Glcβ-4Glcβ-4-Glcβ-4Glcβ]2-4Glcβ-4Glcβ
            3|              3|              3|
       Ac-6Manα        Ac-6Manα        Ac-6Manα
            2|              2|              2|                (4)
          GlcAβ ene-4,5-GlcAβ              GlcAβ
            4|                              4|
          Manα                             Manβ
```

structure (4), in which there is an excess of backbone glucosyls to which side-chains are not attached. An excess of glucose residues was also detected by Sutherland (1984) as a result of degradation of unordered xanthan by fungal cellulase. Hou et al. (1986b) detected residual heterogeneous high MW products following digestion of xanthan with their purified depolymerase. Incorporation of deuterium from $NaBD_4$ served to demonstrate the branched reducing end depicted in (4). The presumed sequence accords with a sequential depolymerization from the non-reducing end, in which the enzyme recognizes (*i.e.*, binds) a cellobiose unit that bears a side-chain and cleaves the anomeric linkage of the substituted glucosyl residue.

It is possible to calculate a number average molecular weight of ~4,600 from the amounts of labeled alditol and terminal glucosyl end groups. Retention by the membrane may, therefore, involve aggregated fragments that associate through celulose-like (extended ribbon) junction zones. Indeed, such cellulose-like sequences may be responsible for many aspects of the behavior of xanthan gum in solution. Among these are a tendency for aggregation at low ionic strength, synergistic increases in viscosity with plant galactomannans and difficulties encountered in demonstrating a double-stranded structure for native xanthan in contrast to apparent ease for fractions obtained from partially-degraded material (Pardossi and Brant, 1982; Sato et al., 1984a,b). The conformational change attendent to the melting out of the xanthan helical conformation and simultaneous disruption of side-chain hydrogen bonding to the backbone may be monitored calorimetrically or by n.m.r., optical rotation and viscosity. A relatively minor presence of cellulose-like junction zones might not be detected should their disruption require more drastic conditions. In other words, a double strand of native xanthan might not come apart fully* as judged by the above criteria.

Although the "nature of the ordered conformation is still under discussion" (Milas and Tinland, 1990), no consideration has been given, in the voluminous literature on conformation vs. properties, to the structural irregularities in xanthan gum and how they might affect its properties. For example, a recent molecular modeling study (Millane and Wang, 1990) of the

---

*Suggested by W. Burchard, University of Freiburg, GFR, 1984.

synergistic interaction of xanthan with β-(1→4)-D-mannose sequences in galactomannans concluded that a drastic conformational change of xanthan was required from a 5_1 helix to a 2_1 cellulose-like conformation. The structure (4) of the high-MW fragment indicates that a change of this sort need not be invoked, since regions having the latter conformation likely are present in native xanthan. It would be interesting to determine whether the high MW products of ST or HT xanthanases enhance the viscosities of galactomannans.

Possibly, a fully pyruvated preparation without cellulosic sequences may have been used by Lecourtier et al. (1986). In any event, it is clear that the HT xanthanase described here may be used to test such preparations or to screen other xanthan-producing organisms for the regularity of their polysaccharides.

Structure (4) also provides an inference about the biosynthesis of cellulosic sequences in xanthan gum. It is reasonable to assume that biosynthesis occurs by insertion of pentasaccharide subunits at the glucosyl or reducing end of the biopolymer chain. Degradation by the depolymerase proceeds from the non-reducing end. Given the low molecular weight of the fragment depicted in (4), it is apparent that the structural aberation must occur late in the biosynthesis of xanthan chains.

6 REFERENCES

Cadmus, M.C., Jackson, L.K., Burton, K.A., Plattner, R.D. and Slodki, M.E., 1982. Biodegradation of xanthan gum by *Bacillus* sp. Appl. Environ. Microbiol., 44: 5-11.

Cadmus, M.C., 1983. Salt-tolerant microbial xanthanase and method of producing same. U.S. Pat., 4,410,625.

Cadmus, M.C. and Slodki, M.E., 1985. Enzymic breakage of xanthan gum solution viscosity. Dev. Ind. Microbiol., 26: 281-289.

Cadmus, M.C., Jackson, L.K., Rutherford, W.M., Weisleder, D. and Slodki, M.E., 1988. Enzymic hydrolysis of rhizobial 4-*O*-methyl-D-glucurono-L-rhamnan. Carbohydr. Res., 184: 203-212.

Cadmus, M.C., Slodki, M.E. and Nicholson, J.J., 1989. High-temperature, salt-tolerant xanthanase. J. Ind. Microbiol., 4: 127-133.

Cadmus, M.C. and Slodki, M.E., 1989. Heat-stable, salt-tolerant xanthanase. U.S. Pat., 4,886,746.

Cadmus, M.C., Slodki, M.E., Nicholson, J.J., Kaneshiro, T., Weisleder, D. and Vucelic-Radovic, B., 1990. Enzymic and structural characterizations of an exopolysaccharide from an apparent *Rhizobium fredii* strain. Submitted for publication.

Cripps, R.E., Sommerville, H.J. and Holt, M.S., 1981. Xanthanase enzyme. Eur. Pat. Appl., 30, 393.

Holzwarth, G., 1976. Conformation of the extracellular polysaccharide of *Xanthomonas campestris*. Biochemistry, 15: 4333-4339.

Holzwarth, G. and Prestridge, E.B., 1977. Multi-stranded helix in xanthan polysaccharide. Science, 197: 757-759.

Hou, C.T., Barnabe, N. and Greany, K., 1986a. Biodegradation of xanthan by salt-tolerant aerobic microorganisms. J. Ind. Microbiol., 1: 31-37.

Hou, C.T., Barnabe, N. and Greany, K., 1986b. Purification and properties of a novel xanthan depolymerase from a salt-tolerant bacterial culture, HD1. Appl. Environ. Microbiol., 52: 37-44.

Jansson, P.-E., Kenne, L. and Lindberg, B., 1975. Structure of the extracellular polysaccharide from *Xanthomonas campestris*. Carbohydr. Res., 45: 275-282.

Lecourtier, J., Chauveteau, G. and Muller, G., 1986. Salt-induced extension and dissociation of a native double-stranded xanthan. Int. J. Biol. Macromol., 8: 306-310.

Lesley, S.M., 1961. Degradation of the polysaccharide of *Xanthomonas phaseoli* by an extracellular bacterial enzyme. Can. J. Microbiol., 7: 815-825.

Melton, L.D., Mindt, L., Rees, D.A. and Sanderson, G.R., 1976. Covalent structure of the extracellular polysaccharide from *Xanthomonas campestris*: evidence from partial hydrolysis studies. Carbohydr. Res., 46: 245-257.

Milas, M. and Rinaudo, M., 1986. Properties of xanthan gum in aqueous solutions: role of the conformational transition. Carbohydr. Res., 158: 191-204.

Milas, M. and Tinland, B., 1990. Behavior of xanthan in cadoxen. Carbohydr. Polym., 13: 47-56.

Millane, R.P. and Wang, B., 1990. A cellulose-like conformation accessible to the xanthan backbone and implications for xanthan synergism. Carbohydr. Polym., 13: 57-68.

Moorhouse, R., Walkinshaw, M.D. and Arnott, S., 1977. Xanthan gum - molecular conformation and interactions. In: P.A. Sandford and A. Laskin (Editors), Microbial Polysaccharides. American Chemical Society, Washington, D.C., pp. 90-102.

Pardossi, G. and Brant, D.A., 1982. Light scattering studies of a series of xanthan fractions in aqueous solutions. Macromolecules, 15: 874-879.

Rinaudo, M. and Milas, M., 1980. Enzymic hydrolysis of the bacterial polysaccharide xanthan by cellulase. Int. J. Biol. Macromol., 2: 45-48.

Sandford, P.A., Pittsley, J.E., Knutson, C.A., Watson, P.R., Cadmus, M.C. and Jeanes, A., 1977. Variation in *Xanthomonas campestris* NRRL B-1459: characterization of xanthan products differing in pyruvic acid content. In: P.A. Sandford and A. Laskin (Editors), Microbial Polysaccharides. American Chemical Society, Washington, D.C., pp. 192-210.

Sato, T., Norisuye, T. and Fujita, H., 1984a. Double-stranded helix of xanthan in dilute solution: evidence from light scattering. Polym. J. (Tokyo), 16: 341-350.

Sato, T., Kojima, S., Norisuye, T. and Fujita, H., 1984b. Double-stranded helix of xanthan in dilute solution: further evidence. Polym. J. (Tokyo), 16: 423-429.

Sutherland, I.W., 1981. *Xanthomonas* polysaccharides - improved methods for their comparison. Carbohydr. Polym., 1: 107-115.

Sutherland, I.W., 1982. An enzyme system hydrolyzing the polysaccharides of *Xanthomonas* species. J. Appl. Bacteriol., 53: 385-393.

Sutherland, I.W., 1984. Hydrolysis of unordered xanthan in solution by fungal cellulases. Carbohydr. Res., 131: 93-104.

Reprinted from: *Microbial Enhancement of Oil Recovery - Recent Advances*,
edited by E.C. Donaldson

CH. R–16

MICROBIAL ECOLOGY OF CORROSION AND RESERVOIR SOURING

D. W. S. WESTLAKE
Department of Microbiology, Faculty of Science, University of Alberta, Edmonton, AB, Canada. T6G 2E9

THE RESERVOIR AS A MICROBIAL HABITAT

An oil reservoir consists of a "rock" matrix containing mixtures of liquid and gaseous hydrocarbons. Two important parameters which determine the suceptibility of a reservoir to microbial penetration are the porosity and permeability of the "rock" matrix and the temperature of the reservoir. A review of the relationships of these parameters and microbial activity in reservoirs in relation to the use of microbes in enhanced oil recovery procedures was recently published (McInerny and Westlake, 1990). In general, other physical parameters such as pressure and salinity while influencing the types of bacteria which will be found can be considered to be secondary controlling physical factors relative to porosity, permeability and temperature.

Information generated in the last half of the 20th century clearly indicate that hydrocarbons as found in crude oils can support microbial growth if supplemented with a utilizable source of nitrogen, phosphate and oxygen. That is, certain aerobic bacteria but not anaerobes, can readily initiate growth using the low molecular weight components of the saturate and aromatic fractions of oil as a carbon and energy source. Such hydrocarbon-utilizing bacteria are ubquitous in the environment. Oil reservoirs, while containing energy rich hydrocarbons, however are usually deficient in oxygen, nitrogen and phosphate except at their peripheries which may be in contact with nutrient, oxygen and bacterial bearing aquifers. Under these conditions peripheral oil may show all of the characteristics of a bacterial degraded oil, that is, the loss of low molecular weight aromatic and saturate hydrocarbons. This hypothesis is supported, for example, by GC and isotope fractionation data obtained from a series of Mississippian oil reservoirs in the northwestern periphery of the Williston Basin field (Bailey et al., 1973). Such information supports the hypothesis that microbial activity in a reservoir originates by the introduction of bacteria, nutrients and oxygen via aquatic intrusions. Oil reservoirs not impacted by aquifers, or oil recovery techniques such as water flooding, should from a bacterial perspective, be a relatively stable environment.

MICROBIAL ECOLOGY OF PRODUCED OIL FIELD FLUIDS

Relatively large numbers of aerobic, facultative and anaerobic bacteria are frequently found in fluids produced from oil fields under going water-flooding techniques (Westlake et al., 1986). Aerobic organisms isolated included members of the genera - Aeromonas, Alcaligenes, Bacillus, Corynebacterium, Flavobacterium, Micrococcus and Pseudomonas while SRB (sulfate-reducing bacteria) and members of the genus Clostridium were readily detected under anaerobic conditions. The total number of isolates in each category varied greatly, but were usually highest in summer and fall samplings. A survey of 95 aerobic isolates selected at random indicated that two-thirds of them were able to grow anaerobically using ferric ions as electron acceptors. That is, these isolates are carrying out dissimilative iron reduction. These ferric iron-reducing isolates were initially classified as being members of the Pseudomonas genus (Obuekwe, 1980). These organisms could use sulfite, thiosulfate and elemental sulfur, but not sulfate, as an alternate electron acceptor to oxygen and were able to release ferrous ions from amorphous and crystalline iron compounds although at a very much reduced rate compared to that observed with ferric salts. Hydrogenase enzyme(s) are also present in these isolates.(personal communication (Dr. E. Laishley, Dept. of Biology, University of Calgary, Calgary, AB, Canada). The sulfide produced by the reduction of sulfite by iron-reducing bacteria shows both the normal and "reverse" isotope fractionation patterns which are diagnostic of dissimilatory sulfite reduction (Semple et al., 1987b).

When grown aerobically these iron-reducing organisms will utilize a wide range of carbon sources whereas anaerobically they will only grow on low molecular weight compounds such as lactic acid. The ability to grow, and then only at a very slow rate, on n-alkanes is found in only a few isolates. Hydrogenase activity is present although there is no cross reaction with DNA probes for desulfovibrio hydrogenases (Personal Communication, Dr. G. Vourdouw, Dept. of Biochemistry, University of Calgary, AB, Canada). Arnold et al. (1986) reported that Pseudomonas sp. #200 (Culture Collection, Department of Microbiology, University of Alberta) is capable of reducing ferric iron at a rate considerable higher than those previously reported for other bacteria. They also reported that the enzymes for this dissimilative pathway were induced when the oxygen tension was <0.01 atmospheres.

Iron-reducing organisms were divided into 4 physiological groups based on growth at 4C, salt tolerance and nature of sulfur compounds reduced and were classified as being strains of Alteromonas putrefaciens and not members of the Pseudomonas genus (Semple and Westlake, 1987a). [Group 1 isolates - grow at 4C and in media containing 4.5% NaCl; and reduce thiosulfate and sulfite: group 2 - isolates grow at 4C and in media containing 4.5% NaCl; and only reduce thiosulfate: group 3 - isolates grow in media containing 4.5% NaCl and reduce thiosulfate: group 4 - isolates grow in media containing 4.5% and 7.5% NaCl and reduce both thiosulfate and sulfite.] Studies on the DNA-relatedness of these groups (Semple et al. 1989) confirmed the existence of these 4 groups although some cross reactions were observed. The results of this study together with other information in the literature supported the re-classification of these organisms as being strains of Shewanella putrefaciens. These organisms also are found in dairy products, faeces, cutting oils. There is very little cross reaction between these isolates and common marine bacteria of the Shewanella, Alteromonas or Vibrio genera with the exception of a slight reaction with Vibrio harveyi (NCMB 1280). There is no cross reaction at the DNA level between these iron-reducing Shewanella putrefaciens isolates and SRB (Personnal Communication, Dr. G. Vourdouw, Dept. of Biochemistry, University of Calgary, Calgary, AB, Canada).

ROLE IN THE CORROSION AND RESERVOIR SOURING PROCESS

Studies (Obuekwe et al.,1981) showed that one of these iron-reducing isolates (Pseudomonas sp. #200) was able to colonize steel coupons by the production of a fibrous, exopolysaccharide material and in the process removed a surface coat from the steel coupon which was produced by incubation of the coupon in a sterile environment. Removal of this biofilm revealed that "pits" had been formed in the steel (Obuekwe, C. O., 1980; Westlake et al., 1986; Obuekwe et al., 1987a). Electrochemical studies (Obuekwe et al., 1981 b, c) showed that these iron-reducing isolates accelerate the anodic reaction of the electrochemical cell whereas the sulfate reducing bacteria act by cathodic depolarization. Obuekwe et al., (1987b) showed that the corrosion reaction was polarized when the available energy level was low but became readily depolarized when the energy level was increased. That is, the bacterial corrosion process is dependent on a suitable energy supply.

Thus considerable data is present which supports an active role for these iron-reducing bacteria in the corrosion process via: the removal of protective ferric oxide coatings on steel; the production of both sulfide and ferrous ions which will combine to produce corrosive ferrous sulfide (Obuekwe et al., 1987a); and possibly depolarization by the removal of proton and/or electron envelopes by their hydrogenase activity. In the absence of iron or under acidic conditions the sulfide-generating activities of these iron-reducing bacteria will add to the H_2S produced by SRB. Obuekwe et al., (1983) proposed that the interaction of facultative and anaerobic bacteria interact to produce a "cascade of sulfide generation" through a synergistic relationship where the intermediate product of reductive activities form the electron acceptor for subsequent reduction by other groups until sulfide is produced.

All of the above processes however are dependent upon the availability of suitable energy sources and in a reservoir the only energy source available is in the form of hydrocarbons. Under reservoir conditions the initiation of the bacterial utilization of hydrocarbons is dependent upon the activity of aerobic bacteria. It is only in those sections of reservoirs subjected to the incursion of aquatic systems (I.e.

aquifers or water flooding procedures) that conditions will be suitable for the initiation of bacterial growth. It also has to be recognized that under reservoir conditions that growth conditions will be sub-optimal. That is, carbon will not be oxidized to CO_2 by aerobic organisms but will be released as partially oxidized compounds. It is these compounds that permit the growth of facultative and anaerobic bacteria. This relationship was hypothesized to occur in reservoirs in eastern Europe (Kuznetsova and Gorlenko, 1965) where sulfate reduction was related to the activity of aerobic bacteria of the genus Pseudomonas. Jobson et al., (1979) in a laboratory study investigated the relationship between aerobic hydrocarbon-degrading bacteria, oil and the growth of SRB. As previously reported neither pure nor mixed cultures of SRB were able to grow on 4 different oils under anaerobic conditions. Such cultures however were able to grow, as determined by change in the chemical composition of the oil, on the residues from the aerobic degradation of these oils. Whether or not the SRB grew was determined in part by the source of the aerobic population used to degrade an oil, the chemical composition of the oil and whether the SRB were in pure or mixed culture. This study shows that there is a greater possibility for an association of aerobic hydrocarbon-degrading bacteria and the anaerobic SRB to take place under natural conditions than would be predicted from the pseudomonad relationship defined by Kuznetzova and Gorlenko (1965).

The information previously discussed is supportive of a synergistic relationship between aerobic petroleum-degrading bacteria and the anaerobic sulfide-generating bacteria as being a basis for the initiation and acceleration of the corrosion process and the reservoir souring observed in field undergoing water flooding recovery procedures. The incursion waters introduces the nutrients, O_2 and bacteria, which grow in the reservoir at sub-optimal rates and produce the low molecular weight, partially oxidized compounds used by anaerobes as carbon sources. The growth of anaerobes also should take place at sub-optimal rates because of the low substrate levels resulting from sub-optimal aerobic growth. It has been observed in chemostat studies that SRB when growing slowly excrete sulfite into the medium (Jobson, 1975) which is the most oxidized form of sulfur which can be reduced by members of the iron-reducing Shewanella putrefaciens. Sulfite is also the most oxidized form of sulfur reduced by the anaerobic clostridia which are frequently isolated from oil field fluids. The growth of non SRB organisms is not dependent upon the presence of sulfite. For example, clostridial species can produce reduced organic acids and alcohols as a means of disposing of the electrons generated from oxidative reactions and strains of Shewanella putrefaciens can use ferric ions as an alternate electron acceptor as well as being able to grow aerobically (i.e. they are facultative with respect to their relationship to oxygen).

Iron-reducing bacteria which show close a physiological relationship to the oil field isolates classified as S. putrefaciens are readily isolated from surface waters. These water sources are adjacent to oil fields which yielded the isolates used in these studies. The DNA relatedness of these surface water isolates and those previously isolated from oil field fluids is under investigation.

The result of this synergistic relationship between aerobes and anaerobes is the souring of reservoirs and the enhancement of corrosive conditions. The net effect being a decrease in the economic value of the reservoir through the loss of hydrocarbons and significantly increased operating costs through measures undertaken to control corrosive microbial activity.

The information reported here supports the view that it is only by studying how the diverse microbial types found in oil reservoirs interact will it be possible to effectively control and utilize microbial activities in the recovery and utilization of oil resources.

REFERENCES

Arnold, R. G., DiCristina, T. J. and Hoffman, M. R. 1986. Inhibitor studies of dissimilative Fe(III) reduction by Pseudomonas sp. strain 20 ("Pseudomonas ferrireductans"). Appl. Environ. Microbiol. 52: 281-289.

Bailey, N. S. L., Krouse, H. R., Evans, C. R. and Rogers, M. A. 1973, Alteration of crude oils by water and bacteria. Evidence from geochemical and isotope studies. Am. Assoc. Pet. Geol. Bull. 57: 1276-1290.

Jobson, A. M. 1975. Physiological characterization of a Desulfovibrio sp. isolated from crude oil. Ph. D. Thesis, University of Alberta, Edmonton, AB. Canada.

Jobson, A. M., Cook, F. D. and Westlake, D. W. S. 1979. Interaction of aerobic and anaerobic bacteria in petroleum biodegradation. Chem. Geol. 24: 355-365.

Kuznetsova, V. A. and Gorlenko, V.M. 1965. Effects of temperature on the development of microorganisms from flooded strata of the Romashinko oil field. Mikrobiologiya. 34: 274-278.

McInerney, M. J. and Westlake, D. W. S. 1990. Microbial enhanced oil recovery. In: Microbial Mineral Recovery. Eds. H. L. Erlich and C. L. Brierley. McGraw-Hill Pub. Co. pp. 409-445.

Obuekwe, C. O. 1980. Microbial corrosion of crude oil pipeline. Ph. D. Thesis. University of Alberta, Edmonton, AB. Canada.

Obuekwe, C. O., Westlake, D. W. S., Cook, F. D. and Costerton, J. W. 1981a. Surface changes in mild steel coupons from the action of corrosion-causing bacteria. Appl. Environ. Microbiol. 41: 766-774.

Obuekwe, C. O., Westlake, D. W. S., Plambeck, J. A. and Cook, F. D.. 1981b. Corrosion of mild steel in cultures of ferric iron reducing bacterium isolated from oil. I. Polarization characteristics. Corrosion. 37: 461-467

Obuekwe, C. O., Westlake, D. W. S., Plambeck, J. A. and Cook, F. D.. 1981c. Corrosion of mild steel in cultures of ferric iron reducing bacterium isolated from oil. II. Mechanism of anodic depolarization. Corrosion. 37: 632-637.

Obuekwe, C. O., Westlake, D. W. S. and Cook, F. D. 1983. Corrosion of Pembina crude oil pipeline. The origin and mode of formation of hydrogen sulfide. Eur. J. Appl. Microbol Biotechnol. 17: 173-177.

Obuekwe, C. O., Westlake, D. W. S. and Plambeck, J. A. 1987a. Bacterial corrosion of mild steel under the condition of simultaneous formation of ferrous and sulfide ions. Appl. Microbiol Biotechnol. 26: 294-298.

Obuekwe, C. O., Westlake, D. W. S. and Plambeck, J. A. 1987b. Evidence that available energy is a limiting factor in the bacterial corrosion of mild steel by a *Pseudomonas* sp. Can. J. Microbiol. 33: 272-275.

Semple, K. M. and Westlake, D. W. S. 1987a. Characterization of iron-reducing *Alteromonas putrefaciens* from oil field fluids. Can. J. Microbol. 33: 366-371.

Semple, K. M. and Westlake, D. W. S. 1987b. Sulfur isotope fractionation by strains of *Alteromonas putrefactiens* isolated from oil field fluids. Can. J. Microbiol. 33: 372-376.

Semple, K. M., Doran, J. L. and Westlake, D. W. S. 1989. DNA relatedness of oil field isolates of *Shewanella putrefaciens*. Can. J. Microbiol. 35: 925-931.

Westlake, D. W. S., Semple, K. M. and Obuekwe, C. O. 1986. Corrosion by ferric iron-reducing bacteria isolated from oil production systems. In: Biologically Induced Corrosion. Ed. S. C. Dexter. National Association of Corrosion Engineers. Houston, TX. U. S. A. pp. 193-200.

Reprinted from: *Microbial Enhancement of Oil Recovery - Recent Advances*, edited by E.C. Donaldson

CH. R–17

DEGRADATION OF GLUCOSE AND PRODUCTION OF H_2S BY A CONSORTIUM OF THERMOPHILIC BACTERIA UNDER SIMULATED RESERVOIR CONDITIONS

J.T.ROSNES[1], T.TORSVIK[1], T.LIEN[1] and A.GRAUE[2]

[1]Department of Microbiology and Plant Physiology, University of Bergen, Norway
[2]Department of Physics, University of Bergen, Norway

ABSTRACT

A consortium consisting of a thermophilic carbohydrate degrading bacterium and a thermophilic sulfate-reducing bacterium was enriched from oil-field production water originating from subsea hydrocarbon reservoirs in the Norwegian sector of the North Sea. The reservoir temperatures are in the range of 60-100 °C (140-212 °F) and the pressures are between 200-400 bars (2900-5800 psi). The consortium degraded sugars in a synthetic medium, containing sulfate, with hydrogen sulfide as the ultimate end product of the sulfate reduction. The degradation of glucose and the production of hydrogen sulfide have been determined both at ambient conditions and at elevated temperature and pressure conditions.

The sulfate-reducing bacteria form endospores and belong to the genus *Desulfotomaculum*. Two strains, T90A and T93B, were isolated after autoclaving mixed cultures at 120 °C (248 °F), allowing only sporeformers to survive. The strains grew autothrophically on H_2 + CO_2 and heterotrophically on fatty acids and on alcohols. Sulfate, sulfite and thiosulfate were used as electron acceptors. Pure cultures of the sugar fermenting bacterium have not been obtained.

Such consortia including biopolymer-degraders may have detrimental effects in sulfate-containing reservoirs when carbohydrates are added to the injection water either as a biopolymer to increase the viscosity, or as a substrate for bacteria in MEOR-processes. The effects of the bacterial carbohydrate degradation activity will be a reduced efficiency of the added biopolymer as a viscosity promotor and a high probability of reservoir souring.

INTRODUCTION

In oil reservoirs in the North Sea, injection of seawater is commonly used as a secondary oil recovery method. The injected seawater contains different kinds of bacteria, including fermentative and sulfate-reducing bacteria. The extreme physical conditions in deep subsea oil reservoirs, with temperatures of 60-100 °C (140-212 °F) and pressures between 200-400 bars (2900-5800 psi), represent an environment unsuitable for most bacterial growth. Resent research have, however, reported on presence of natural bacterial communities in environments with similar physical conditions (Stetter et al.,1987; Huber et al.,1989), and it is therefore likely that a thermophilic barotolerant bacterial community also will develop in the reservoir. The reservoir porewater contains soluble organic compounds, which can support growth of fermentative bacteria when mixed with the carbon limited seawater. Short chained organic acids, as metabolic products from fermentative bacteria, are further utilized by sulfate-reducing bacteria. The production of H_2S from the metabolic activity of SRB generate probably the most severe effects of microbial activity in an oil reservoir. Corrosion of iron-based alloys, souring of crude oil and gas production, plugging of filters and loss of permeability have been reported (Curd-Ruwisch et al. 1987).

Average recovery from the oil reservoirs on the Norwegian Continental Shelf is 30-40 %. Enhanced recovery, e.g. using tertiary production techniques, has therefore great potential, and is under constant consideration and evaluation. Some of the reservoirs in the North Sea exhibit reservoir characteristics suitable for both chemical recovery processes and microbial enhanced methods, with e.g. porosities in the region of 20 %, and permeabilities between 750-2000 mD (Goff,1983). Substantial quantities of organic compounds are injected into the reservoir, both under chemical and microbial enhanced recovery, somtimes also containing elements of nitrogen and phosphate. This addition of nutrients to the reservoir water will increase the activity of fermentative bacteria, and will, in sulfate containing environments, impose high levels of hydrogensulfide.

To investigate the severe impacts of the sulfate-reducing bacteria, we have examined the degradation of carbohydrates and production of hydrogensulfide in a thermophilic anaerob bacterial community both at ambient conditions and at simulated reservoir conditions.

2 MATERIALS AND METHODS

2.1 Isolation and enrichment procedure

Small watersamples, 5 ml, were taken from the water separator tanks on off-shore oil platforms. The pressure in the separators ranged from 1-80 bar (14-1160 psi), the temperature was 70 °C (158 °F) and the pH was 7.8 measured after pressure release. The bacteria were cultivated in an anaerobic marine synthetic medium as

described by Widdel et al.,1983, with 1 ml traceelements SL-10 and 5 ml vitamin solution (Pfennig,1978) per liter medium. The pH in the medium was 7.1 and the incubation temperature was usually 60 °C (140 °F). Carbon and energy source was added aseptically to the medium using 20 mM for fatty acids (sodium salts) and 12 mM glucose.
The sporeforming sulfate-reducing bacteria were isolated by first autoclaving the enriched culture at 121 °C (250 °F) for 20 min at 1 atm overpressure. The surviving spores were germinated in fresh liquid medium. Single colonies were thereafter picked from dilutions of Gelrite Gellan Gum (Kelco Division of Merck and Co, San Diego, Ca).

2.2 Core preparation

The cores used in this study were drilled from a matrix block of the eolic Hopeman sandstone obtained from an outcrop at Elgin in Scotland, the Clashach quarry. Petrographic and mineralogical studies of the sandstone indicate a rather pure quartz composition, 91 % quartz and only trace amounts of clay (muscovite). Grain sizes are in the order of 0.25 mm, with only small variations. The average porosity is 18 % and the permeability in the region of 700-800 mD. Average pore diameter was measured to 26 um by use of mercury injection. Cylindrical cores, 5.2 cm (2.0 in.) in diameter, were drilled with varying lengths, from 10 to 80 cm. To provide even fluid distribution at the end faces of the cores, endcaps were mounted and the core was epoxy coated (Graue et al., 1988).

2.3 The Flow-rig

The main component of the rig is a 1 m (3 feet) long cylindrical steel pressure vessel with an inner diameter of 7 cm (2.8 in.) and 1.1 cm (0.4 in.) wall thickness (Fig. 1). A core prepared as described above was pressurized in transformer oil inside the vessel. The rig can simulate pressure conditions in a reservoir down to a depth of 4 km (15000 ft.), corresponding to a geostatic pressure of 1000 bar (15000 psi). Pressure transducers monitor the simulated overburden pressure and the water injection pressure. By pumping the fluid against a back pressure regulator, pore pressures up to 10500 psi could be obtained. Experimental flow rates may range from 0.01 to 9.99 cm^3/min. The pressure vessel was heated by a heating cable and the temperature was regulated by a thermostate.

Two standard thermocouple K- elements, measured the temperature inside the rig at each end.

2.4 <u>Pressure tube</u>

A 96 cm long steel tube was used as growth vessel. The volume of the tube is 47 cm^3 (18.8 $in.^3$) with an inner diameter of 0.79 cm (0.312 in.) and an outer diameter of 1.43 cm (0.552 in.). The steel tube, grade 316 Seamel SS certificated to sustain a pressure of 1000 bar (15000 psi), was bent to fit into a heating chamber (Fig 1.). The pump and the same type of pressure transducers as described above, was used. The thermostate regulated heating chamber had an accuracy of +-1 ^{o}C (33.8 ^{o}F) at 70 ^{o}C (158 ^{o}F). It was possible to change the liquid without reducing the pressure, by using a back pressure regulator at the outlet end of the tube.

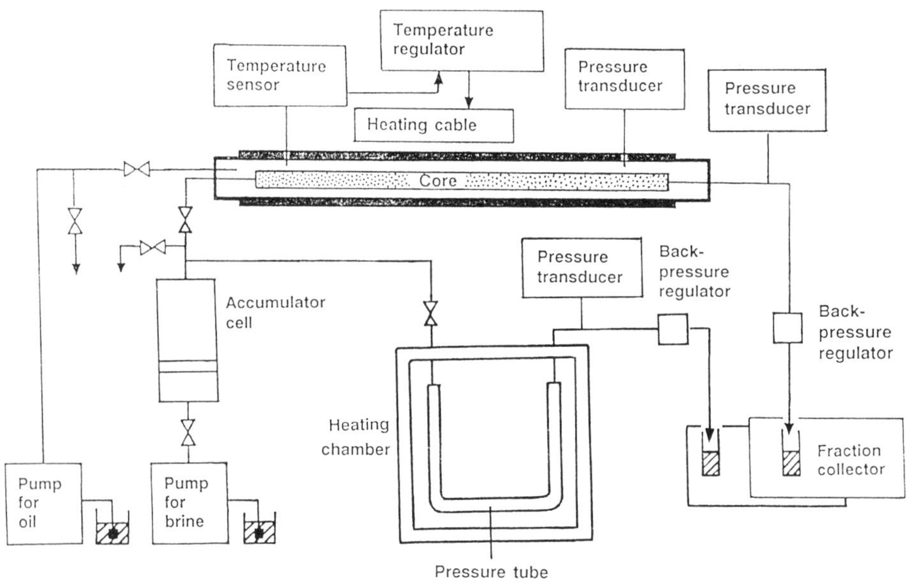

Fig. 1. Schematic drawing of the flow-rig and the pressure tube.

The pressure tube and the flow-rig were used unsterilized. The eperiments were started by inoculating the pressure tube and the core with bacteria, then the following procedure were carried out sequentially: the physical conditions were set for the experiment, at the end of an incubation period, three volumes of growth medium were pumped through the tube and core respectively and the bacterial activity was measured in the effluent. The physical conditions were then changed and the remaining bacteria from last experiment became the inoculum for the next incubation period.

2.5 Microscopic investigations

Morphology of the bacteria was investigated using a phase contrast microscope (Leitz, Labolux K). Cell wall structure was examined by using a JOEL 100 CX transmission electron microscope.

2.6 Methods

Fatty acids were assayed by gas chromatography on a HP-20M (Carbovax 20M) column. A column with length of 15 m and 0.53 um x 1.33 um film thickness, was used in a Hewlett Packard HP 5890 gas chromatograph, with flame ionization detector, and helium carrier gas. Samples were acidified prior to injection with 10 M formic acid to 0.5 M final concentration.

Sulfide was determined by the methylene blue method (19) or with copper sulfate (18). Glucose concentration in the culture was determined by the glucose oxidase method (Sigma Chemicals, Catalog No. 510). After the enzymatic reactions, the colour intensity of the solution was measured at 450 nm (OD_{450}) in a spectrophotometer. The glucose concentration was then determined by comparing the OD_{450} reading with a calibration curve for standard glucose solutions.

3 RESULTS

3.1 The consortium

Thermophilic bacteria from produced water, were enriched at 60 ^{o}C (140 ^{o}F) using different carbon sources in the growth medium. The enrichment cultures were a part of an investigation of thermophilic sulfate-reducing bacteria in produced water on offshore oil platforms. H_2S producing enrichments were achieved on acetate, propionate, lactate, butyrate, a mixture of C_4-C_6 acids and palmitate. At the beginning the bacteria grew slowly, taking months to become dense, but after 2-3 transferences of the bacteria to the same medium, the same density was obtained in about two weeks.

Efforts were made to isolate the different members of the consortium. Several dilution series were carried out in a glucose containing medium without sulfate, but the sulfate-reducing bacteria ultimately appeared in all of the dilution tubes.

The spore forming sulfate-reducing bacteria were successfully isolated by autoclaving as described in Materials and Methods section. A characterization of the strain, designed T93B, was carried out. The results supported the inclusion to the genus *Desulfotomaculum*, Table 1a and b.

a)

Cell characteristics	Morphology	rodshaped 0.8-1 x 1-2 µm
	Motility	non motile
	Cell wall	Gram +
	Endospores	central thermoresistent
	Mol % G + C	51
Growth requirements	NaCl range	25-800 mM
	pH optimum	7.0 -7.2
	Temp. range	43 - 78 °C
	Temp. optimum	68 °C

b)

Growth with sulfate	Carbon source	
Autotrophic growth	$H_2 + CO_2$	
Heterotrophic growth	formate acetate[1] propionate butyrate caproate valeriate pyruvate lactate	methanol[1] ethanol propanol
Not utilized	glucose fructose maltose cellobiose	benzoate

[1]) Poorly utilized.

Table 1 Properties of the thermophilic sulfate-reducing bacterium T93B. a). Cell characteristics and growth requirements. b). Carbon sources tested.

The spores were tested for heat tolerance by autoclaving closed bottles containing spores suspended in medium. The spores germinated in fresh medium after being exposed to 130 °C (266 °F) for 30 min.

The sulfate-reducing bacteria and the accompanied fermentative bacteria had a distinct different cell morphology. The sulfate-reducer was a light dense rod, while the most striking feature of the fermentative bacteria was a long transparent sheat with a small light dense inclusion inside.

3.2 Degradation of glucose in batch cultures

In batch cultures at ambient pressures, the consortium used 35 days to degrade 12 mM glucose, Fig. 2. Both the degradation of glucose and production of H_2S were approximately linear with respect to time and inversely proportional to each other. The production of acetate increased slowly the first 6 days, and thereafter decreased from day 6 to day 14, followed by a considerable increase from day 30 to day 60. Both the total cell density and the ratio between the fermentative and the sulfate reducer were examined during the growth, Fig. 3. This was possible by light microscopy, due to the differences in cell features as described above. At the beginning of the experiment it was found $5.0 * 10^6$ bacteria/ml and the concentration increased to a level of $1.4 * 10^8$ bacteria/ml in 35 days.

The growth of the consortium on C_5, C_6, C_{12}-sugars, was examined in batch cultures. Some of the detected degradation products are listed in Table 2.

Mono- and di-saccharides	substr. mM	H_2S mM	HAc mM	N $*10^8$	final pH
glucose	11.1	7.7	9.6	4.2	6.7
fructose	11.1	6.8	5.0	2.6	6.7
mannose	11.1	7.7	9.3	3.5	6.4
arabinose	13.3	6.6	11.2	5.0	6.5
ribose	13.3	7.2	10.8	4.3	6.5
sucrose	5.8	7.1	6.0	1.6	6.6
maltose	5.8	6.6	5.0	2.0	6.6
lactose	5.8	6.3	0.6	1.9	6.8
cellobiose	5.8	8.5	5.5	2.3	6.6

Table 2. Products obtained in batch cultures with the consortium, incubated for 30 days, 60 °C, ambient pressure.

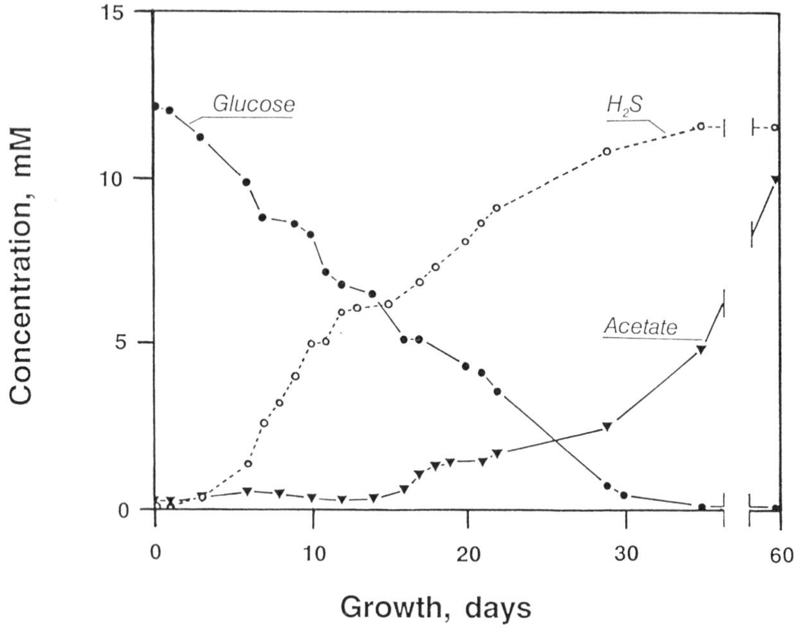

Fig. 2. Time course for glucose degradation by the consortium in batch culture.

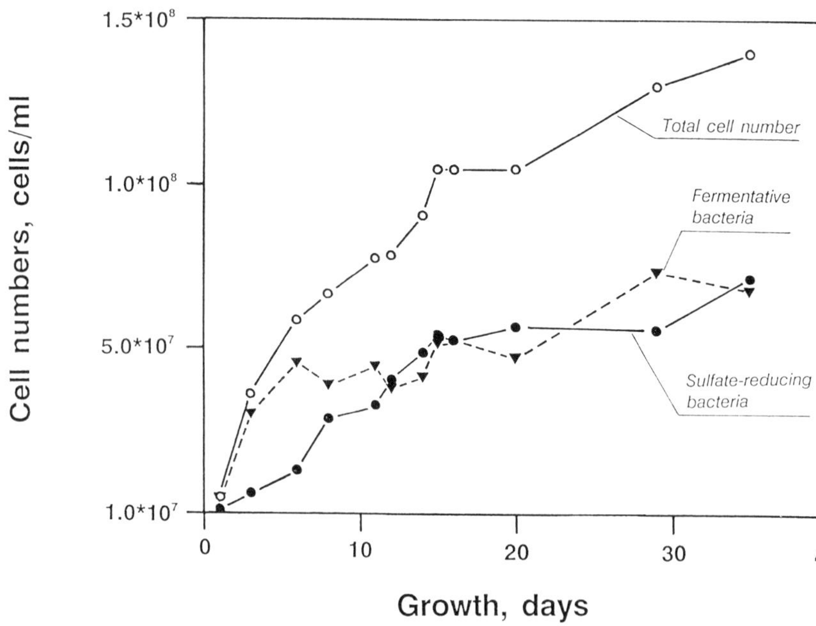

Fig. 3. Cell numbers of the consortium during glucose degradation.

In medium containing glucose, the consortium grew both with and without sulfate. The growth rate was, however, increased when sulfate was present, resulting in higher glucose degradation rates.

3.3 Degradation of glucose at simulated reservoir pressure

The consortium was capable of degrading glucose at 60-70 °C (140-158 °F) at pressures from 1-300 bars (14-4350 psi), Table 3 a and b.

a)

Reservoir conditions		Substrate Glucose		Products	
P bar	T °C	Added mmol	Used mmol	H_2S mmol	HAc mmol
100	60	0.366	0.231	0.252	0.693
300	60	0.366	0.127	0.140	0.194
300	70	0.366	0.099	0.079	0.037

b)

Reservoir conditions		Substrate Glucose		Products	
P bar	T °C	Added mmol	Used mmol	H_2S mmol	HAc mmol
300	60	0.484	0.200	0.545	1.200
300	70	0.484	0.166	0.270	0.536

Table 3. Degradation of glucose at simulated reservoir conditions. a), in the pressure tube, incubation period 12 days, and b), in the core, incubation period 6 days.

This result was obtained both in the pressure tube and in the reservoir core. Although the liquid volumes were relatively similar in the pressure tube and the core, 47 cm^3 and 62 cm^3 respectively, the activity of the bacteria differed significantly when they were exposed to the same physical and chemical conditions. The most important difference is the surface area exposed to the consortium, constituting 240 cm^2 (96 $in.^2$) for the pressure tube and approximately 100 m^2 (300 sq.ft) in the core.

4 DISCUSSION

There are many ways in which microbial systems may be of benefit in the enhancement of oil recovery and some of them may be used in some of the North Sea oil reservoirs. Three different types of injection operations can be identified: a) using selected bacteria with desirable characteristics, b) injecting nutrients to stimulate the natural microflora resident in the oil reservoir formation or c) utilize a purified microbial product. All of these three potential classes of operations will increase the level of organic compounds in the reservoir. Glucose, as an example on an organic compound of carbohydrate that may be degraded, is a component of a viscosity promotor like Xanthan and is also known to be produced by many hydrolytic bacteria under anaerobic conditions.

The enrichments from the produced water system on an oil production platform have shown that thermophilic fermentative and sulfate-reducing bacteria are naturally inhabitants of such environments. The investigation of produced water has resulted in isolation of several straines of sulfate-reducing bacteria, both sporeforming and non-sporeforming. Observations that sulfate-reducing bacteria appear to be indigenous inhabitants of oilfield waters are confirmed by others (Antloga et al., 1985). The physiological characterization of the sulfate-reducing bacteria, utilizing different kinds of carbon sources including authothrophic growth, indicates that the strains are well adapted to reservoir conditions. The ability of forming spores ensures the strain to survive large fluctuations in temperature. Strain T93B has previously been shown to grow on lactate at 80 ^{o}C (176 ^{o}F) and 300 bars (4300 psi) (Rosnes et al., 1990).

The sulfate-reducing bacteria did not grow on glucose or fructose alone. Hydrogen sulfide was only produced when the sulfate-reducing bacteria was supplied with H_2, organic acids or alcohols wich are degradation products from fermentative bacteria. The batch cultures showed a nearly linear consumption of glucose and concomitant production of hydrogen sulfide. For each mole of glucose consumed one mole of hydrogen sulfide was produced, Fig 2. Acetate started to accumulate when glucose was almost consumed.

At the start of a new incubation, the fermentative bacteria accumulated faster than the sulfate-reducing bacteria Fig. 3. After a brief period of exponential growth both bacteria continued to grow linearly until all glucose was consumed.

When the growth of the sulfate-reducer was prevented, by omitting sulfate in the medium, the fermentative bacterium grew slower. It is possible that the sulfate-reducing bacteria function as a hydrogen scavanger for the fermentative bacteria, indicating that the consortium constitute a mutualistic interaction.

An important part of the experiments have been the studies of the growth of the consortium at high temperature and high pressure conditions. The two different experimental set ups, the pressure tube and the sandstone core, differed considerably in microbial activity when exposed to the same physical conditions, Table 3.

In the pressure tube equivalent amount of hydrogen sulfide production and glucose consuption was observed, in accordance with the results obtained from batch cultures, Tab. 3a. The results from the core showed much higher production of hydrogen sulfide and acetate than expected from the consumed amounts of glucose, Tab. 3b, probably due to an incomplete exchange of fluid in the core. The core has a complex pore geometry with a surface area many orders larger than that in the pressure tube. The sulfate-reducing bacteria have previously been shown to grow on surfaces, producing slimy extracellular material which fix the cells to the substatum (Rosnes et al., 1990). Also this extracellular material may influence the exchange of fluid. The experiments with the core mimics more closely microbial growth under reservoir conditions. For quantitative measurements the results from the pressure tube are more reliable.

5 CONCLUSIONS

1. A thermophilic consortium consisting of one fermentative and one sulfate-reducing bacterium was isolated from produced water taken from a North Sea oil production rig.

2. The consortium grows on a broad specter of carbon sources, including carbohydrates, with concomitant production of H_2S.

3. In batch cultures with glucose, one mole of H_2S was produced for each mole of glucose consumed.

4. The consortium degraded carbohydrates and produced H_2S, both at physical conditions which is occuring in waterflooded oil reservoirs.

6 REFERENCES

Antloga, K.M. and Griffin, W.M., 1985. Characterization of Sulfate-Reducing Bacteria Isolated from Oilfield Waters. Developments in Industrial Microbiology, vol.26. 41st general Meeting of the Society for Industrial Microbiology, Aug 11-17,1984. Society for Industrial Microbiology: Arlington, Va., USA., pp. 597-610.

Cord-Ruwish, R., Kleinitz, W. and Widdel, F., 1987. Sulfate-Reducing Bacteria and their activities in oil production. J.Petroleum Technology (Jan. 1978) pp. 97-106

Goff, J.C., 1983. Hydrocarbon Generation and Migration from Jurassic Source Rocks in the East Shetland Basin and Viking Graben of the Northern North Sea. J. of the Geological Society of London. 140:445-474.

Graue, A., Kolltveit, K. and Lien, J.R., 1988. Imaging Fluid Saturating Development in Long Cooreflood Displacements. SPE 17438, pp. 363-372. To appear in SPE Formation Evaluation.

Huber, R., Kurr, M., Jannash, H.W. and Stetter, K.O., 1989. A Novel Group of Abyssal Methanogenic Archaebacteria (*Methanopyrus*) Growing at 110 ^{o}C. Nature, 342:833-834.

Pfennig, N. 1978. *Rhodosystus purpurus* gen.nov. and sp.nov., a ringshaped, Vitamin B_{12}-requiring member of the family *Rhodospirillaceae*. Int.J.Syst.Bacteriol. 28:283-288.

Rosnes, J.T., Graue, A. and Lien, T., 1990. Activity of Sulfate Reducing Bacteria Under Simulated Reservoir Conditions. Ninth Symposium on Formation Damage Control, Lafayette, Louisiana, Febr. 22-23, 1990. Proceedings, SPE 19429, pp. 231-236.

Stetter, K.O., Lauerer, G., Thomm, M. and Neuner, A., 1987. Isolation of Extremely Thermophilic Sulfate Reducers: Evidence for a Noval Branch of Archaebacteria. Science, 236:822-824.

Widdel, F., Kohring, F.G. and Mayer, F. 1983. Studies of Dissimilatory Sulfate-reducing Bacteria that Decompose Fatty Acids III. Characterization of the Filamentous gliding *Desulfnoema limicola* gen.nov. and sp.nov. and *Desulfonema magnum* sp.nov. Arch. Microbiol 134:286-294.

Reprinted from: *Microbial Enhancement of Oil Recovery - Recent Advances,* edited by E.C. Donaldson

CH. R–18

PROSPECTS FOR THERMOPHILIC MICROORGANISMS IN MICROBIAL ENHANCED OIL RECOVERY (MEOR)

E.T. PREMUZIC and M.S. LIN
Biosystems and Process Sciences Division, Department of Applied Science, Brookhaven National Laboratory, Upton, NY 11973

ABSTRACT

Developments in MEOR have shown that successful microbial candidates should be able to metabolize crude oils in the presence of high salt concentrations, as well as be able to produce beneficial metabolites such as acids and emulsifying agents. Furthermore, under reservoir conditions, such microorganisms should be viable over extended periods of time at elevated temperatures and pressures. Depending on the depth and manner of application, the ability to grow under anaerobic to low oxygen conditions and oil as a sole source of carbon would also be advantageous.

Studies in this laboratory have shown that certain species of thermophilic microorganisms may satisfy MEOR requirements. Several species of thermophilic microorganisms have been grown in the presence of varying concentrations of crude oils at different temperatures and pressures. Experimental evidence indicates that some thermophiles are viable over long periods of time (six months) at 70°C and 2000 psi in the presence of crude oils as the sole source of carbon. The interaction between selected thermophilic microorganisms occurs in some cases under aerobic and in others under anaerobic conditions. During the interactions there are pH changes due to production of acids. In addition to acidification, crudes are also emulsified. The extent of emulsification is microbial-species dependent, and considerable variations have been observed when the same crude oil has been biotreated with different types of microorganisms under otherwise identical experimental conditions. The analysis of the emulsified fraction indicated that hydrocarbons in the C_{13}-C_{26} range were present in the emulsion. Under the experimental conditions used, the biotreatment also caused changes in the composition of the organic sulfur fractions, as well as an overall decrease in the content of sulfur-containing compounds of the crude oils tested. Further, biotreatment of crudes caused significant qualitative and quantitative changes in low and high molecular weight fractions of test oils. As in emulsification, the observed changes in different molecular weight fractions are also species dependent. Experimental data will be presented and their significance in terms of MEOR will be discussed.

INTRODUCTION

During the past several years, a considerable amount of work has been carried out showing that microbially enhanced oil recovery (MEOR) is promising and the resulting biotechnology may be deliverable (Donaldson et al., 1989; Grula, 1986; King and Stevens, 1986; King, 1987; and Bryant and Douglas, 1987). Because many mechanistic aspects of interactions between microorganisms and oils are unknown, additional exploratory and basic studies are needed. However, several desirable MEOR properties have been identified. Thus, biochemical, chemical, and physical processes which govern microbial-oil interactions should include production of gases, surfactants, acids, and solvents. Further, successful microorganisms should not require expensive nutrients; they should be anaerobic and/or be capable of growing in the presence of low oxygen concentrations, and they should be able to withstand relatively high pressures and salinities. Also, such microorganisms should be easily grown in facilities aboveground, remain viable over extended periods of time, and be transportable. Once underground, they should continue to be viable and preferably continue their activity upon re-feeding.

Systematic studies in this laboratory which address the biochemical mechanisms that enable microorganisms to grow under harsh conditions, particularly in environments related to energy production (e.g., Premuzic et al., 1985, 1987, and 1989), have indicated that thermophilic, toxic metal resistant microorganisms may be suitable candidates for MEOR, a possibility mentioned at the First International MEOR Workshop in 1986 (King and Stevens, 1986). If the mechanisms by which microorganisms adapt to extreme environments involve processes that convert and/or metabolize toxic and unusable substrates, such heavy crude oils, then such processes may serve as a basis for the development of new biotechnologies. Thermophilic organisms live under harsh conditions, such as extreme pH (acid and alkaline) and high temperatures (up to 110°C) (Brock, 1978 and 1985). These organisms can use inorganic and organic energy sources, and some are also capable of switching from an aerobic to anaerobic metabolism. Further, the natural habitats of these organisms are geothermal brines, which means that they are tolerant of high salt concentrations. Thus, the general properties of thermophilic organisms, although little explored, indicate that they possess a number of desirable properties for MEOR. We have recently reported (Premuzic and Lin, 1990a) early findings of studies dealing with the chemistry and biochemistry of interaction between several thermophilic, acidophilic, chemotrophic organisms from the BNL collection and crude oils. Biotreatment of crude oils with selected thermophilic and temperature and pressure adapted microorganisms (Premuzic and Lin, 1990b) at elevated temperatures and pressures (>70°C, 2000 psi) causes emulsification and lowering of pH.

Analysis of biotreated crude oils has also shown that during the biotreatment, qualitative and quanitative changes in the composition of low and high molecular weight fractions have also occurred. Further, the microorganisms used were viable over extended periods of time (months) and were able to tolerate extreme pH and high salt concentrations. Based on the currently available evidence, the interactions between microorganisms and crude oils are complex and involve multiple processes. Therefore, in order to better understand the chemistry and biochemistry of these processes, extensive studies and analyses of several variable parameters are needed. In this paper results from continuing multiparameter studies of thermophilic or thermo-adapted microorganisms and their utility in MEOR are presented.

2 MATERIALS

2.1. Bioreactors

Small-scale bioreactors have been made out of stainless steel tubes (Premuzic and Lin, 1988). Each bioreactor can handle a total volume of 20 ml of fluid and can be re-used many times. Typically in each experiment, three bioreactors were used. Thus, one bioreactor was charged with nutrients in water, gases (CO_2, N_2) and the experimental organism only. A second bioreactor was charged with oil, gases (CO_2, N_2), and nutrients in water. A third bioreactor was charged with oil, microorganisms, gases (CO_2, N_2), nutrients, and water. All were kept under identical experimental conditions of temperature and pressure.

2.2 Instrumentation

A Perkin-Elmer (PE) model 8700 microprocessor-controlled gas chromatograph (GC) with multiramp temperature programming, has been used in all GC work. The PE 8700 was interphased with a PE-Finnigan Ion Trap Detector (ITD) for mass measurements in the 20-650 mass units range. This gas chromatography-mass spectrometry system (GC-MS) encompasses the NBS/EPA mass spectral library. The PE 8700 GC system is also equipped with a Flame Ionization Detector (FID) and Flame Photometric Detector (FPD).

3. METHODS

3.1. Growth and Adaptation of Microorganisms

Culture media consisted of inorganic salts, e.g., $(NH_4)_2SO_4$, $MgSO_4$, KH_2PO_4, and crude oil or yeast extract as a source of carbon. Incubations were carried out at different temperatures (30-80°C) and pressures (200-25000 psi). Through a sequence of different experimental regimes a methodology has been developed (Premuzic and Lin, 1990b) which makes it possible to adapt microorganisms to different pressures and temperatures. Several different species and strains of microorganisms have been used in the work described in this paper and are listed with the summary of the appropriate treatment in Table 1.

3.2 Metal Complexes

Hewlett-Packard gas chromatograph model HP5921A equipped with atomic emission detector was used for the determination of metal complexes.

TABLE 1

Microbial species and treatments

Microorganism Species	Treatment Conditions			
	Temp. °C range	Medium	Pressure psi	pH
BNL-TH-29 Acidophilic thermophile	60-80	A	up to 2000	1.5-4.5
BNL-TH-31 Acidophilic thermophile	60-80	A	up to 2000	1.5-4.5
BNL-4-21 *Acinetobacter*	25-75	B	Atm.	6-7.5
BNL-4-22 *Arthrobacter*	25-75	B	Atm.	6-7.5
BNL-4-23 *Achromobacter*	25-75	B	Atm.	6-7.5
BNL-4-24 *Pseudomonas*	25-75	B	Atm.	6-7.5
BNL-4-25 *Nocardia*	30	C	Atm.	7
BNL-5-32 *Methanogenium*	55-60	D	Atm.	6-7.5
BNL-TH-1 Acidophilic thermophile	60-80	E	up to 2000	1.5-2.5
BNL-3-25 Acidophilic thermophile	30-60	F	100	1.0-2.5
BNL-4-25 Acidophilic thermophile	30-60	G	100	1.0-2.5

Medium A is medium 1304, supplemented with a non-peptone modified carbon source, B is a nutrient broth containing beef extract supplemented by a non-peptone carbon source. C medium is a yeast extract medium, D and E media are 1442 and 1256, respectively, also supplemented with non-peptone modified carbon source. Medium F is a basal salt solution, and G is a basal salt solution supplemented with iron sulfate. In each case, crude oil becomes the sole carbon source in the final adaptation at the upper temperature limit.

3.3 Sulfur Analysis

Total sulfur was determined by combustion (Huffman Laboratories, Golden, CO). The organic sulfur compounds were monitored by GC from 40°C-300°C equipped with FPD. A J&W, DBI column was used throughout.

3.4 Emulsification

The extent of emuilsification produced by different microorganisms was determined by literature methods (Rosenberg et al., 1979) and expressed in Klett units given by $\frac{1000 \times D}{2}$, where D is the absorbance determined at 545 nm.

4 RESULTS AND DISCUSSION

4.1 Acidification

Earlier experimental observations have shown that there is a lowering of pH from 5-7 to 3 (Premuzic and Lin, 1989 a-d and Premuzic and Lin, 1990a) of the medium during the microbial action on crude oil under our experimental conditions, indicating that the aqueous phase may contain soluble acidic compounds. These observations were further verified by treating a sample of PR3 (Teapot Naval Petroleum Reserve #3) crude with BNL-4-22 strain. Preliminary analysis of the aqueous phase by high pressure liquid chromatography (HPLC) showed that lactic, propionic, and isobutyric acids were present in the aqueous phase. Additional components were also in evidence, of which only one, n-butanol, has been identified so far.

Long chain fatty acids may also be produced during the biotreatment of oil. Because such acids are good emulsifying agents, an experiment was designed to verify this possibility. PR3-crude was treated with BNL-TH-1, under our standard experimental conditions. After the treatment, culture medium was saponified with sodium hydroxide and the free acids methylated with methanol/HCl. The produced fatty acids methyl esters were extracted and analyzed by GC-MS. The first three fatty acids so far identified are heptanoic acid (C-7), decanoic acid (C-10), and octadecanoic acid (C-18).

4.2 Emulsification

Treatment of PR3 with several strains of thermophilic microorganisms resulted in different emulsifying effects (Premuzic and Lin, 1990a), suggesting that certain microbial species may produce more efficient surfactants than others. Further, it may also be possible that interactions between microorganisms and crude oils could exhibit microorganism/crude oil species specificity and/or selectivity. In order to further investigate these possibilities experimentally, several heavy fractions of a suite of crude oils have been treated by different microorganisms under identical experimental conditions. The results are shown in Table 2 and Figure 1.

These results are considered preliminary until detailed chemical and kinetic data about the nature and the production of emulsifying agents have been obtained. However, current experimental evidence strongly indicates that there are significant differences between the effects of the same microbial

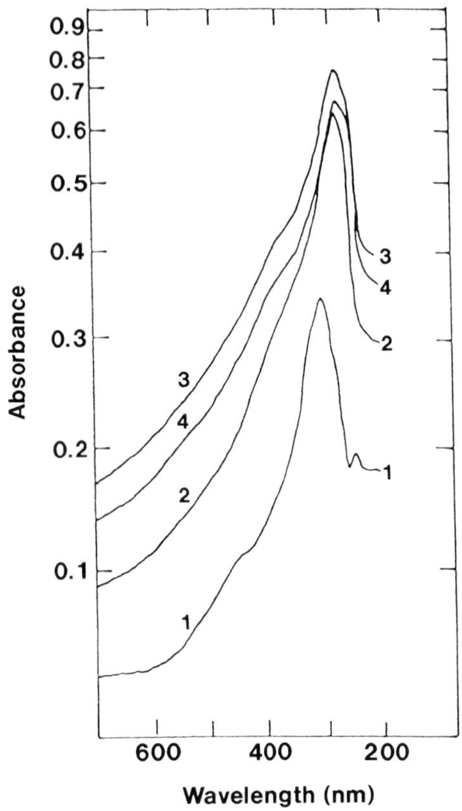

Fig. 1. Extent of emulsification (700-200 nm spectral region) of PR3 by BNL-4-24 (1); BNL-4-23 (2); BNL-4-22 (3); BNL-4-21 (4).

TABLE 2

Extent of emulsification due to the action of various microorganisms on heavy fractions of crude oils. The results are expressed in Klett units

Microorganisms	Heavy Oil Fraction			
	Wilmington[1] (Calif)	Goch Saran[1] (Iran)	Recluse[1] (Wyo)	Prudhoe[1] (Alaska)
BNL-4-24	115	168	250	215
BNL-4-23	290	238	225	195
BNL-4-22	252	320	175	285
BNL-4-21	515	142	600	615

[1]Heavy fractions (200°C) of crude distillate

strains on different crude oils as well as variations in the effects of different strains of microorganisms on the same oil.

4.3 Changes in the Hydrocarbon Composition of Crudes

The possibility that various emulsification effects may also reflect chemical changes brought about by the biotreatment of crudes was explored by treatment of PR3 with BNL-4-21 and BNL-4-22. The changes in the composition of hydrocarbons in the organic phase were followed by gas chromatography/mass spectrometry (GC/MS) (Premuzic and Lin, 1990a, b; Premuzic and Lin, 1989e).

In a typical experiment the organisms were cultured for a period of two to four weeks in the presence of 10-18% by volume of crude oil with no other carbon source in the media. Analysis by GC/MS has shown that there are significant alterations in the distribution of hydrocarbons, particularly in the regions of lower and higher molecular weight compounds. Analogous to the strain dependent variability observed in emulsification, the microbial effects onthe hydrocarbon composition also changed when different strains were used. In order to further explore the biochemical action of microorganisms on crude oils, GC/MS system was used to analyze for diagnostic molecular markers. In the mass spectromatic analysis of mixtures containing organic compounds, such as crude oils, it is customary to use characteristic masses generated during fragmentation of organic molecules. For example, C_4H_9 (m/e 57) for alkanes, C_7H_7 (m/e 91) for substituted aromatics and others, etc., all of which are characteristic molecular markers (Williams et al., 1985). We have added the application of these diagnostic parameters to our systematic studies of oil-microbial interactions. For example, biotreatment of the teapot Naval Petroleum Reserve #3 (PR3) with microbial strain BNL-4-24 at 65°C under 2000 psi of nitrogen and 80 psi of carbon dioxide for two weeks yielded the following results: The single ion chromatogram monitored for mass 57 (Fig. 2), shows that the lighter alkanes (up to C16) were degraded over that period of time, while those hydrocarbons which were larger than C16 and up to C30, based on peak comparison, were about 80% degraded. Similarly, Fig. 3 shows the effect of the biotreatment on alkylaranes (m/e 91), Fig. 4 on naphthalenes (m/e 169), and Fig. 5 on cyclic hydrocarbons, e.g., adamantane type (m/e 135). Considerable biodegradation of other cyclic saturated hydrocarbons, e.g., bicyclic sesquiterpanes (m/e 123) also occurs, as shown in Fig. 6.

Preliminary results shown in Figs. 2-6 show that the treatment of PR3 with a thermophilic pseudomonas, BNL-4-24, affects characteristically both lighter (C16) and heavier (up to C30) alkanes, aromatics, and sesquiterpane types of compounds. These studies have been further expanded to give particular attention to the effects of other microbial species under the same temperature and pressure conditions. Thus, treatment of PR3 with BNL-4-25 and BNL-4-32 resulted in qualitative and quantitative changes of alkanes, alkenes, and aromatics as shown in Figs. 7-9. These results clearly show that there exist considerable differences between the action of BNL-4-24, BNL-4-25, and BNL-4-32

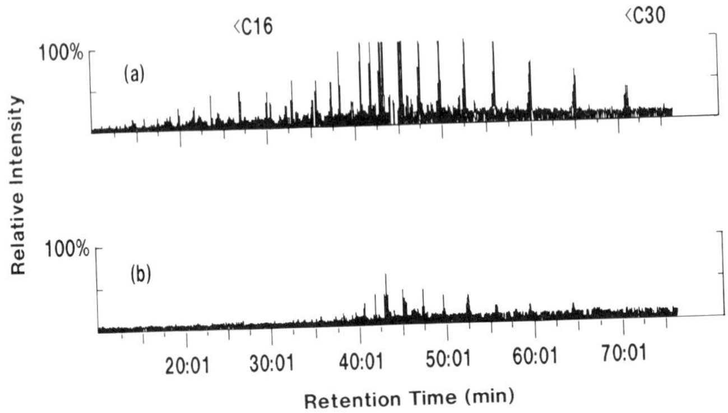

Fig. 2. Degradation of hydrocarbons, (a) before treatment and (b) after treatment, m/e 57 ion trace of PR3 crude treated with BNL-4-24.

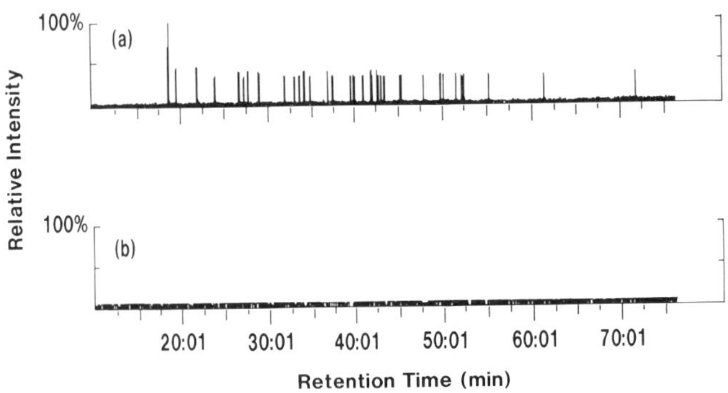

Fig. 3. Degradation of alkylarenes, (a) before treatment and (b) after treatment, m/e 91 ion trace of PR3 crude treated with BNL-4-24.

on PR3. For example, the m/e 57 scan for BNL-4-24 treated crude suggests a larger effect on the oil than that brought about by BNL-4-25 and BNL-4-32 in the same molecular range of hydrocarbons. Because optimum conditions have not been defined, these results must be considered preliminary. It suffices to say that, although the chemical interactions between microorganisms and oils are

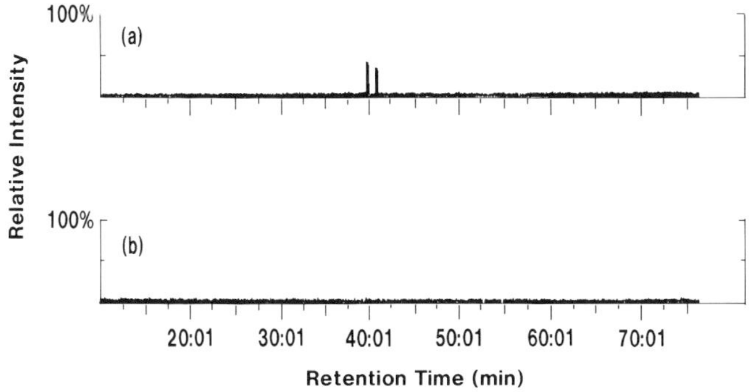

Fig. 4. Degradation of aromatic hydrocarbons, (a) before treatment and (b) after treatment, m/e 169 ion trace of PR3 crude treated with BNL-4-24.

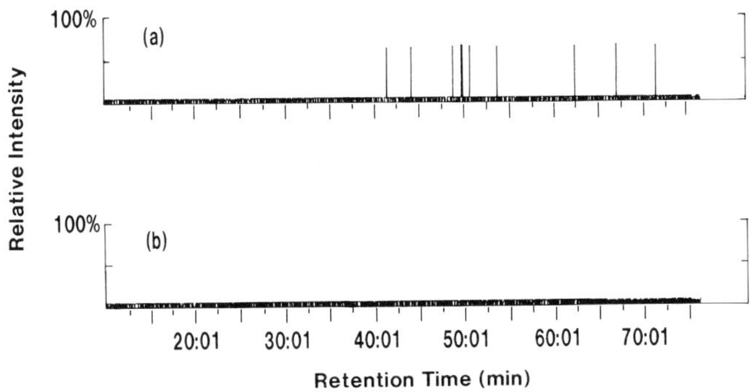

Fig. 5. Degradation of cyclic hydrocarbons, (a) before treatment and (b) after treatment, m/e 135 ion trace of PR3 crude treated with BNL-4-24.

complex, the use of molecular markers makes it possible to observe and follow chemically characteristic patterns associated with these interactions. Further the use of ion-scans and other diagnostic parameters will lead the development of a data base which should ultimately determine trends and variations in the composition of biotreated crude oils.

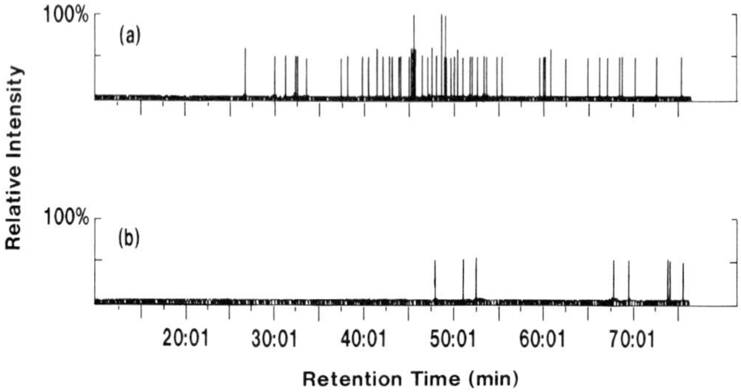

Fig. 6. Degradation of cyclic hydrocarbons, (a) before treatment and (b) after treatment, m/e 123 ion trace of PR3 crude treated with BNL-4-24.

4.4 Changes in the Composition of Sulfur Containing Fractions of Crudes

During routine searches for suitable diagnostic markers as described in the previous section, changes in the m/e 32, GC/MS scans of treated vs. untreated crudes have been noticed. Mass 32 signals may be due to sulfur, a result which would suggest changes in the nature and/or composition of sulfur containing fractions of the crudes. In these experiments hydrogen sulfide was not detected. Because of this observation and the fact that m/e 32 could have been caused by a different mass fragmentation pattern, for example, an abstraction of O_2, an independent verification of the observed data was necessary. To accomplish this, a flame photometric detector (FPD) was installed and calibrated. FPD works as a flame spectrophotometer and for certain elements, e.g., sulfur, phosphorous and chlorine, can be used as a specific, highly sensitive detector. The FPD detector with sulfur filter was then used for the detection of sulfur compounds. An understanding of the changes brought about by biochemical action on the sulfur compounds will make it possible to determine whether indeed sulfur compounds are modified or degraded biochemically in a manner which does not produce volatile products such as H_2S gas. In order to measure such changes on a routine basis the Perkin-Elmer 8700/ITD system has been equipped with a splitter, which allows one portion of the sample to be analyzed by GC/ITD and the other by GC/flame photometric detector (FPD), specifically for sulfur. This capability makes it possible to follow changes in the chemical composition of sulfur-containing compounds caused by microbial treatment. Results of exploratory studies with three different oils and

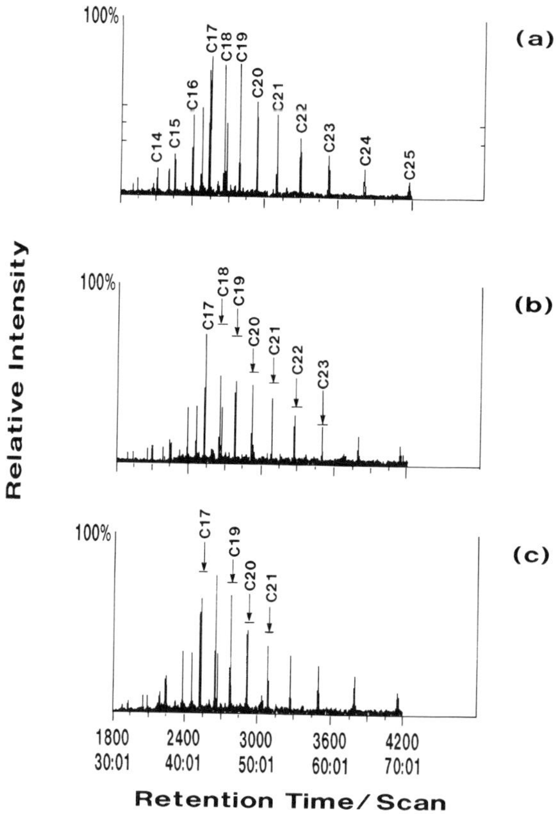

Fig. 7. Changes in the Alkanes composition of PR3 crude. Control (a); PR3 treated BNL-4-25 (b); PR3 treated with BNL-5-32 (c).

several different microorganisms will be briefly discussed (Premuzic and Lin, 1990b). In these studies, three types of oils have been used, one with a relatively low sulfur content, PR3, and two with high sulfur content viz. Venezuelan oils, Cerro Negro and Boscan. Treatment of Teapot Naval Petroleum Reserve #3 crude (PR3) with a temperature and pressure adapted organism, BNL-4-24 at 65-70°C and 1400 psi, resulted in considerable changes in the composition

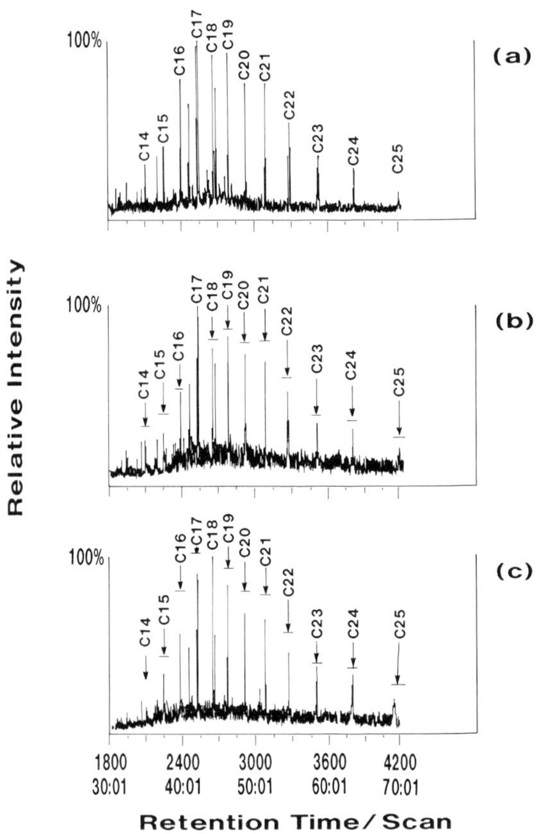

Fig. 8. Changes in the Alkenes composition of PR3 crude. Control (a); PR3 treated BNL-4-25 (b); PR3 treated with BNL-4-32 (c).

of sulfur compounds. A typical FPD sulfur trace is shown in Fig. 10a before treatment and in Fig. 10b after treatment; considerable qualitative and quantitative changes are evident. In view of this unique and promising result, the same biotreatment (patent pending) has been applied to the two high sulfur oils. Changes in the total sulfur contents of the Venezuelan crudes are given

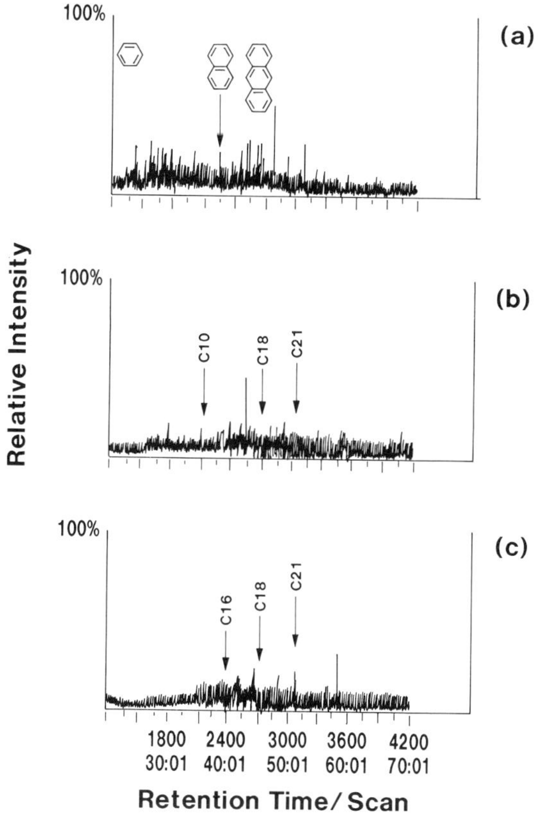

Fig. 9. Changes in the aromatics compositioning of PR3 crude. Control (a); PR3 treated with BNL-4-25 (b); PR3 treated with BNL-5-32 (c).

in Table 3. Significant decrease in sulfur content is evident. In Fig. 11a and b, GC/FPD scans for untreated and treated Cerro Negro oil are shown. In both cases identical experimental conditions have been used. Treatment of Cerro Negro crude with BNL-4-24 results in an almost 50% decrease in sulfur-containing compounds in the region containing less than C20 fractions. Processing under identical experimental conditions (Premuzic and Lin, 1990b) of Cerro Negro crude with several different strains of microorganisms from the BNL

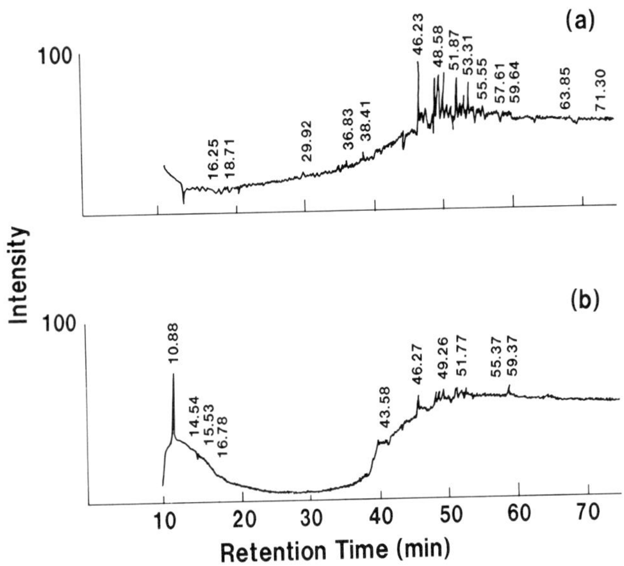

Fig. 10. GC/FPD sulfur trace of BNL-4-24 treated PR3. 10a before treatment and 10b after treatment. Injected (2 μl).

collection indicated that while there is an overall similarity in the decrease in the concentrations of organic sulfur compounds, there may also exist fine structural differentiation depending on which particular microbial species has been used as shown in Fig. 12. These are the first comparative sets of data showing biochemical modification and/or degradation of organic sulfur-containing compounds characterized by molecular markers ranging from thiophene to dibenzothiophene. Extensive studies are being conducted which will enable a better understanding of biochemical processes involved.

4.5 <u>Effects of Biotreatment on the Composition of Metal Complexes</u>

Experimental evidence presented in the preceding paragraphs has shown that considerable biochemical changes occur in crude oils when treated with temperature and pressure adapted organisms. It seemed not an unreasonable assumption that chemical and concentration changes of metal complexes present in crude oils may also occur. In order to test for possible chemical changes in the

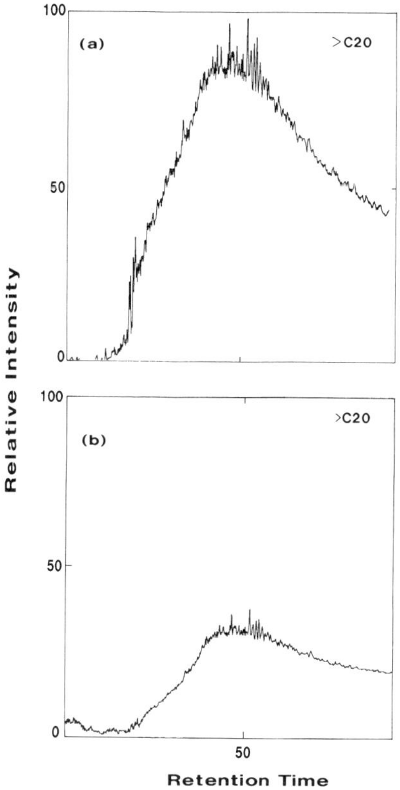

Fig. 11. FPD trace of Cerro Negro (2 μl injection 1/3 split) (a); Cerro Negro treated with BNL-4-24 under idential conditions (b).

composition of metal complexes, several exploratory experiments have been conducted. In all cases, analyses were carried out in which a gas chromatograph with an atomic emission detector has been used (GCAED). In this analytical technique, the metals are detected selectively in the gas chromatogram by their specific emission wavelength. Appropriate calibration allows the determination of metal species. For example, biotreatment of Wilmington, CA crude with BNL-4-22 resulted in a considerable reduction of the nickel porphyrin complex as shown in Fig. 13. In this analysis the GC system was calibrated with nickel octaethyl porphyrin eluting at 18.5 mins and

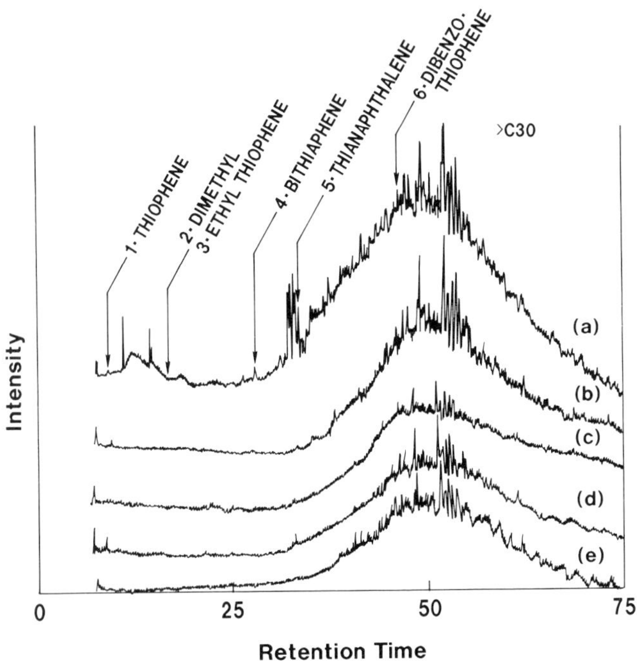

Fig. 12. FPD traces of Cerro Negro crude treated and processed under identical conditions. The data are for <C30 range of compounds.

cobaltoactaethyl porphyrin eluting at 19.1 mins. This is the first clear indication of removal of a trace metal from a crude oil by means of biotreatment under our experimental conditions (patent pending, 1990). We have further explored this technique with another sample of California oil treated with two thermophiles and the results are shown in Fig. 14. These results show an important experimental observation: there is a difference in the biochemistry when the same oil is treated by different microorganisms, an observation consistent with other species-specific changes discussed in this paper.

TABLE 3

Total sulfur content of Boscan and Cerro Negro Venezuelan oils before and after biotreatment

Oil	Total % Sulfur	% Loss
Untreated Boscan	5.49	-
Boscan treated with BNL-4-22	4.14	-25
Boscan treated with BNL-4-23	4.84	-11
Boscan treated with BNL-4-24	4.92	-10
Untreated Cerro Negro	4.37	-
Cerro Negro treated with BNL-4-24	3.10	-29
Cerro Negro treated with BNL-4-23	3.74	-25
Cerro Negro treated with BNL-4-22	3.21	-27

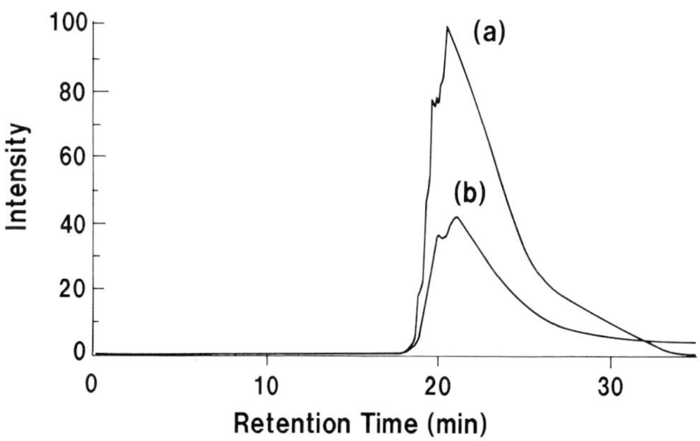

Fig. 13. Reduction of nickel porphyrin content in Wilmington, CA, crude (a) before treatment and (b) after treatment with BNL-4-22.

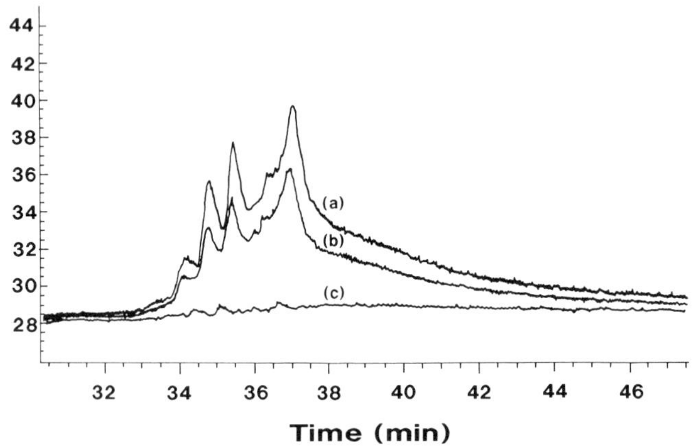

Fig. 14. GCAED trace of California crude untreated (a); and treated with BNL-3-25 (b); and BNL-4-25 (c).

5 CONCLUSIONS

In terms of MEOR, results presented in this paper enumerate a number of highly desirable properties of temperature and pressure-adapted organisms. These can be summarized as follows.

a. Compared to controls there are overall changes in hydrocarbon content of heavy crudes. Preliminary, qualitative analyses indicate that biodegradation of crudes occurs in heavy and lighter fractions of the oils.

b. There are changes in the composition of the organic sulfur components of the crudes, with a net overall effect being a decrease in the content of sulfur compounds.

c. During the biotreatment, a number of organic acids are produced, ranging from small to larger molecular weight acids.

d. Studies of an emulsified phase have shown that hydrocarbons in the range of C13 to C26 are dispersed during the biotreatment.

e. Concurrent with changes in the organic composition of hydrocarbons, there are also changes in the composition of organometallic compounds.

f. Based on the experimental data thus far available, the physiochemical changes brought about by biotreatment are organism specific and vary with different types of microorganisms.

Continuing studies of selected biochemical processes caused by single or mixed cultures of microorganisms which can tolerate high salt concentrations and are also temperature and pressure adapted, will enable definition of fundamental principles applicable in the development of new MEOR biotechnologies.

6 REFERENCES

Brock, T.D., 1985. Life at high temperatures. Science 20: 132-138.

Brock, T.D., 1978. Thermophilic Microorganisms and Life at High Temperature, Springer Verlag, New York, 564 p.

Bryant, R.S. and Douglas, J., 1987. IITRI/NIPER, Society of Petroleum Engineers, SPE 16284, pp. 449-456.

Donaldson, E.C., Chilingarian, G.V., and Yen, T.F., 1989. Microbial Enhanced Oil Recovery. Developments in Petroleum Science 22. Elsevier, New York.

Grula, M.M., 1986. Proc. of the First International MEOR Workshop, April 1-3,1986, pp. 152-213, DOE/BC/10852-1.

King, J.W., 1987. MEOR Technical Status and Assessment of Needs. DOE/BC/-10852, DE7-001227.

King, J.W. and Stevens, D.A. (Editors), 1986. Proc. of the First International MEOR Workshop, April 1-3, 1986, DOE/BC/10852-1.

Premuzic, E.T., Francis, A.J., Lin, M.S. and Schubert, J., 1985. Induced formation of chelating agents by Pseudomonas aeruginosa grown in presence of thorium and uranium. Arch. Environ. Contam. Toxicol.; 14, 759-768.

Premuzic, E.T. and Lin, M.S., 1989. The role of biotechnology in the treatment of geothermal residual sludges. In: J.P. Vernet (Editor), Heavy Metals in the Environment, Vol. 2, CEP Consultants Ltd., Edinburgh, UK, pp. 60-63.

Premuzic, E.T. and Lin, M.S., 1987. Selective accumulation capacities for metals in different microorganisms. In: S.E. Lindberg and T.C. Hutchinson (Editors), Heavy Metals in the Environment, Vol. 2, CEP Consultants, Ltd., Edinburgh, UK, pp. 338-340.

Premuzic, E.T. and Lin, M.S., 1988. Effects of selected microorganisms on crude oil at elevated temperatures and pressures. Annual Report BNL 42048.

Premuzic, E.T. and Lin, M.S., 1989a. Effects of selected thermophilic microorganisms on crude oils at elevated temperatures and pressures, Enhanced Oil Recovery, 55: 79, DOE/BC-88/3 (DE89000730).

Premuzic, E.T. and Lin, M.S., 1989b. Effects of selected thermophilic microorganisms on crude oils at elevated temperatures and pressures, Enhanced Oil Recovery, 56: 106-107, DOE/BC-88/4 (DE89000742).

Premuzic, E.T. and Lin, M.S., 1989c. Effects of selected thermophilic microorganisms on crude oils at elevated temperatures and pressures, Enhanced Oil Recovery, 57: 71, DOE/BC-89/1 (DE89000755).

Premuzic, E.T. and Lin, M.S., 1989d. Effects of selected thermophilic microorganisms on crude oils at elevated temperatures and pressures. Informal Report BNL-43185.

Premuzic, E.T. and Lin, M.S., 1989e. Enhanced Oil Recovery, DOE/BC-89/3, (D-E90000206), May 1990.

Premuzic, E.T. and Lin, M.S., 1990a. Interaction between thermophilic microorganisms and crude oils: Recent Developments, Workshop Proceedings, Bioprocessing of Fossil Fuels Workshop, Vienna, Virginia, 1989, (in press).

Premuzic, E.T. and Lin, M.S., 1990b. Microbial enhanced recovery of oil at elevated temperatures and pressures. Patent pending.

Rosenberg, E., Zuckerberg, A., Rubinowitz, C. and Gutnick, D.L., 1979.Emulsifer of Arthrobacter RAG-1: Isolation and Emulsifying Properties. Appl. and Environ. Microbiol., 37(3): 402-408.

Thompson, C.J., Dooley, J.E., Vogh, J.W. and Hirsch, D.E., 1974. Analyzing heavy ends of crude. Hydrocarbon Processing, 93-98.

Tillman, R.W. and Martinsen, R.S., 1984. The Shannon Shelf-Ridge Sandstone Complex Salt Creek Anticline Area, Powder Basin, Wyoming, Silicastic Shelf Sediments. In: R.W. Tillman and C.T. Siemers (Editors), Soc. of Economic Paleontologists and Mineralogists, Sp. Pub. 34.

Williams, J.A., Bjorøy, M., Dolcater, D.L. and Winters, J.C., 1986. Biodegradation of south Texas Eocene oils - effects on aromatics and biomarkers. In: D. Leythauser and J. Rullkötter (Editors), Part I, Petroleum Geochemistry. Pergamon Journals, Ltd., U.K., pp. 451-461.

7 ACKNOWLEDGMENTS

This work is supported by the U.S. Department of Energy, Division of Fossil Fuels, under contract No. AS-219-ECD and contract No. DE-AC02-76CH00016 with the U.S. Department of Energy. We wish to thank Bettie Sylvester, the curator of the BNL microbial collection, for assistance and valuable discussions. We also wish to thank W.D. Peters and P.W. Woodward of NIPER, Bartlesville, OK, for samples of Cerro Negro and Boscan crude and to TEXACO for the supply of the California crude used in metal complex studies.

Reprinted from: *Microbial Enhancement of Oil Recovery – Recent Advances*, edited by E.C. Donaldson

CH. R–19

FATE OF SURFACTANT VESICLES SURVIVING FROM THERMOPHILIC, HALOTOLERANT, SPORE FORMING, CLOSTRIDIUM THERMOHYDROSULFURICUM

TEH FU YEN, JOON KYU PARK, KWANG IL LEE, AND YUEQI LI
Civil and Environmental Engineering, University of Southern California, Los Angeles, California, 90089-2531

ABSTRACT
MEOR, Clostridia acetylbutylicum, Clostridium hydrosulfuricum, bio-huff-and-puff, surfactant vesicles

Clostridium hydrosulfuricum 39E was found to have surface-active properties during simulated enhanced oil recovery experiments. The resulting surviving surfactant vesicles may have a significant impact on field investigations. Speculation on an idealized system for an ultimate field recovery is discussed.

INTRODUCTION

A myriad of bacterial species have been proven useful in microbial enhanced oil recovery (MEOR). The principle is based on the species' biochemical by-products produced, such as gases, surfactants, solvents, acids, swelling agents, co-surfactants, etc., which greatly facilitate the displacement of oil (Yen, 1990). In field experiments, often in-situ fermentation is desirable for producing a great quantity of gases. This leads to the use of various strains of Clostridia (Johnson, 1979) which are known gas producers. The strain of Clostridium acetylbutylycum JF-2 has been explored extensively by Grula and Russell (1987) and was found to have a high salt tolerance. Simulation for field tests in the laboratory is often termed "bio-huff-and-puff," meaning the injection of gas-producing bacteria with nutrients is followed by given periods of intermittant shut-off for fermentation and then the release of accumulated gas. Usually this cycle can be repeated numerous times (Jang and Yen, 1985).

Initial experiments using Clostridium thermohydrosulfuricum 39E, isolated from Octupus Spring at Yellowstone National Park, yielded encouraging results. From pressure drop experiments in various sandpack columns, permeabilities have increased after bacteria injection (Huang and Yen, 1988). Therefore, this species of bacteria will be useful for enhanced oil recovery.

Actually, the uses of <u>C. thermohydrosulfuricum</u> 39E for oil recovery are many. Not only are they gas- and solvent-producing, but they are stable under high temperature. For example, they can be continually activated near 70°C. Also, they are spore-forming, implying that they are resistant to elevated temperature; thus, under actual reservoir conditions (high temperature and high pressure), <u>C. thermohydrosulfuricum</u> could be ideal.

One of the questions that needs to be answered is whether <u>C. thermohydrosulfuricum</u> is surface active. At this time, little is known about the surfactant properties related to <u>Clostridia</u>. If the property of surface activity is found in these species, the associated emulsification and de-emulsification will certainly add to the value of oil recovery. Consequently, the fate of the resulting interaction of the biosurfactant reaction and the vesicles which affect the recovery process becomes enormously important.

With this objective in mind, experiments were designed to test the possibility and feasibility of the effect of surface activity exhibited by <u>C. thermohydrosulfuricum</u> 39E during oil recovery.

2 MATERIALS AND METHODS

2.1 <u>Bacteria strain and culture medium</u>

<u>Clostridium thermohydrosulfuricum</u> 39E was isolated from Octupus Spring at Yellowstone National Park (Zeikus et al., 1980) and is listed as American Type Culture Collection (Rockville, MD) ATCC 33223 (Figure 1). Enumeration, enrichment and cultivation were carried out with a pre-reduced phosphate buffered tryptone-yeast extract (TYE), slightly modified from ATCC Media Number 1118 as well as 1045. The basal medium as shown in Table 1 was supplemented with TYE and glucose. The TYE-glucose mixture was sterilized separately before addition, as listed in the final proportions. Strictly anaerobic conditions were observed with a nitrogen gas and hydrogen gas mixture (97 to 83 ratio). Deep anaerobic culture tubes sealed with rubber stoppers were used for cultivation studies. Growth was determined by measuring the increase in turbidity at 660 nm using a Beckman model 24 UV-visible spectrophotometer.

TABLE 1

Composition of Special Medium

Components	Amounts
Mineral Solution 1:	
K_2HPO_4	6.00 g/L
Mineral Solution 2:	
KH_2PO_4	6.00 g/L
$(NH_4)_2SO_4$	6.00 g/L
NaCl	12.00 g/L
$MgSO_4$ $7H_2O$	2.40 g/L
$CaCl_2$ $2H_2O$	1.60 g/L
Sodium Carbonate Solution:	
Na_2CO_3	8.00 %
Cysteine-Sulfide Reducing Agent:	
Cysteine HCl H_2O	300.00 mg/10mL
Na_2S $9H_2O$	300.00 mg/10mL
Trace Mineral Solution:	
Nitrilotriacetic Acid:	1.50 g
$MgSO_4$	3.00 g
$MnSO_4$	0.50 g
NaCl	1.00 g
$FeSO_4$	0.10 g
$CaCl_2$	0.10 g
$CoCl_2$	0.10 g
$ZnSO_4$	0.01 g
$CuSO_4$	0.01 g
$AlK(SO_4)_2$	0.01 g
H_3BO_3	0.01 g
Na_2MoO_4	0.01 g
Distilled water	1.00 L
Mixed Components & Proportions:	
Distilled water	84.50 %
Mineral solution 1	2.50 %
Mineral solution 2	5.00 %
8.0% Na_2CO_3 solution	1.00 %
Trace element solution	1.00 %
Cysteine-sulfide reducing agent	2.00 %
Yeast extract solution	1.00 %
Tryptone solution	1.00 %
Glucose	2.00 %

Fig. 1. Scanning electron micrograph of C. thermohydrosulfuricum strains.

2.2 Surfactant properties

Isolation of the surface-active component was as follows. The culture broth was centrifuged to remove cells, and the resulting cell-free spent medium was filtered using a 0.45 μm filter. The filtrate was equilibrated with 50:50 methanol-chloroform where methanol functioned as a wetting agent. The surface-active agent was noted entirely in the chloroform-soluble fraction. This fraction was concentrated using a rotary evaporator. The amount was approximately 0.1% for the crude surfactant.

Thin-layer chromatography (TLC) using a silica gel plate with a developing solvent system of chloroform-methanol-7N ammonia hydroxide (65:35:5; V:V) in contact with iodine vapor revealed two major components (Figure 2a). The top portion, a neutral lipid, after being sprayed with rhodamine B and hydrogen peroxide revealed a bright red color in daylight (Figure 2b). The bottom portions

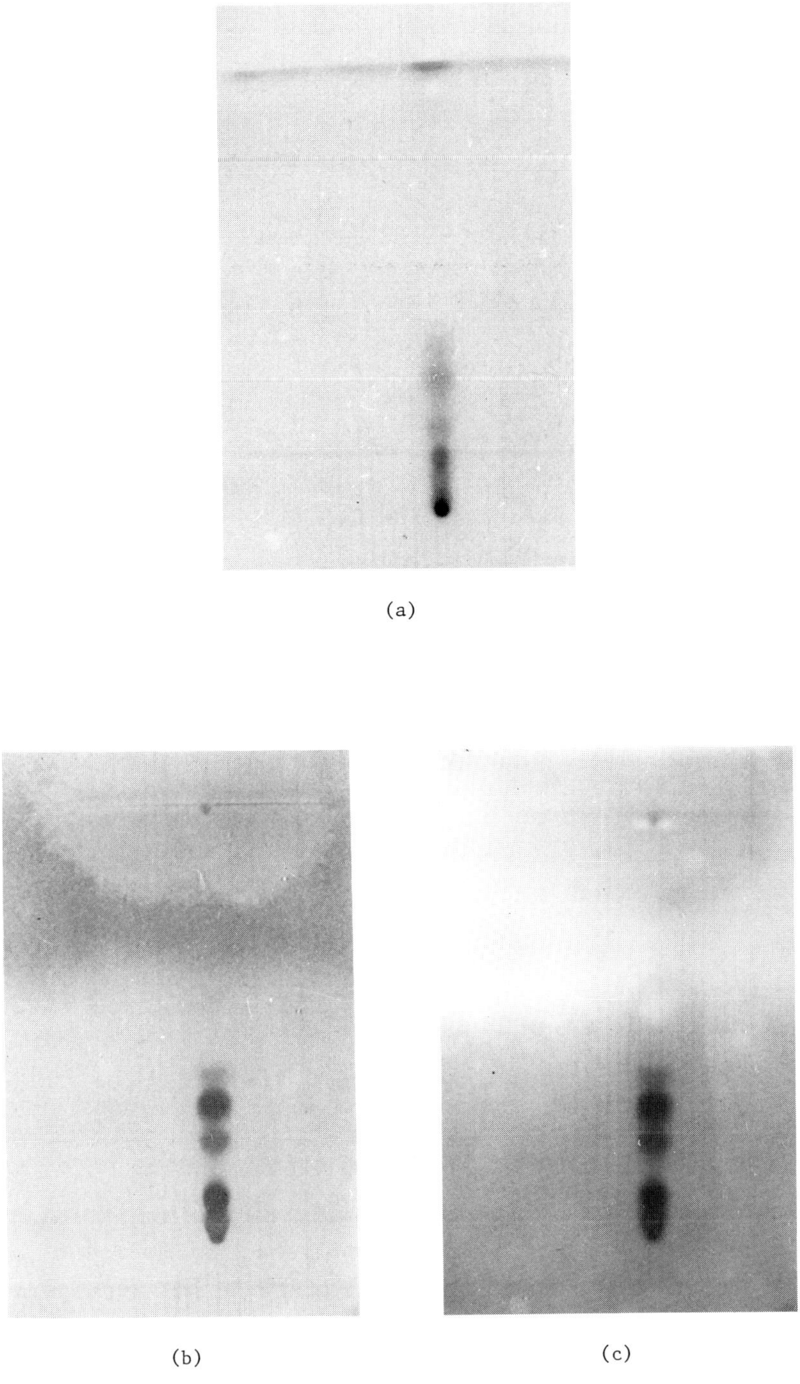

Figure 2. TLC separation of surface-active compounds from culture. (a) from iodine vapor; (b) under daylight; (c) under UV. For (b) and (c), the top portion is stained by rhodamine B and the bottom portion is stained by ninhydrin.

could be a series of glycolipids; when sprayed with ninhydrin and acetic acid, after heat they exhibited a violet color under UV light (Figure 2c). Some of these undoubtedly are glycopeptides.

A spinning drop low interfacial tensiometer constructed by University of Texas at Austin was used for monitoring the surface activities. A Leitz SM-Lux-Pol cross polarizing microscope was used to observe the vesicle formation. Usually a drop of sample to be examined was placed over a clean glass slide and retained with a cover lid. Color film using a WILD MP515 Semiphotomat attachment for both reflection and transmittance light was used.

2.3 Simulation of oil recovery

An experimental apparatus was designed in such a way that actual hydrodynamic conditions for bacteria transport in porous media could be simulated. As shown in Figure 3, assembly mainly

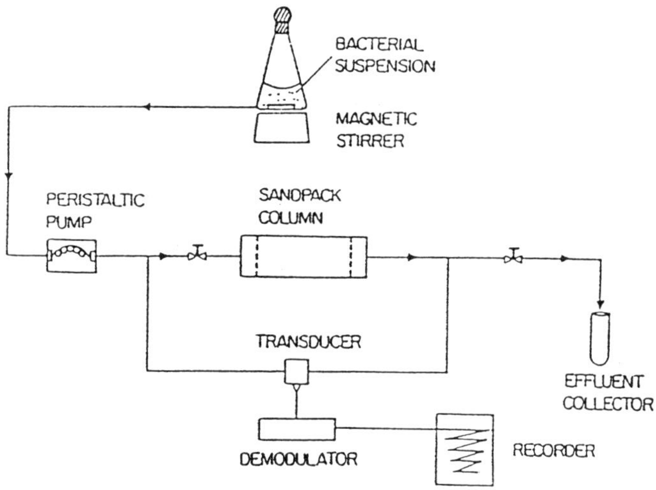

Fig. 3. Schematic diagram of the flow apparatus using a peristaltic pump to supply constant flow and a recorder to monitor pressure change.

consisted of a source flask, a peristaltic pump for constant rate pumping of the bacterial suspension from the source flask, a sandpack column, flow control valves, and an effluent collecter tube. The source flask was kept at a higher elevation to minimize the bacterial settling in the inlet flow lines. Its content could be continuously stirred with a magnetic stirrer. The column was an acrylic cylinder, 28 cm x 3 cm. The packing sand was standard Ottawa sand, 12-20 mesh size, obtained from MCB Manufacturing

Chemical Company. Before packing the column, the sand was thoroughly washed with distilled water to remove fines and then dried in a high temperature oven at 65°C for 24 hrs. The peristaltic pump created a constant flow rate of 90ml/hr. A temperature of 65°C was kept by wrapping a heating tape around the entire column and the inlet tubing. The bacterial suspension was grown in the media expressed by Table 1 for 3 days at 68-70°C. The suspension was injected at the earlier stage of the growing phase (Figure 4).

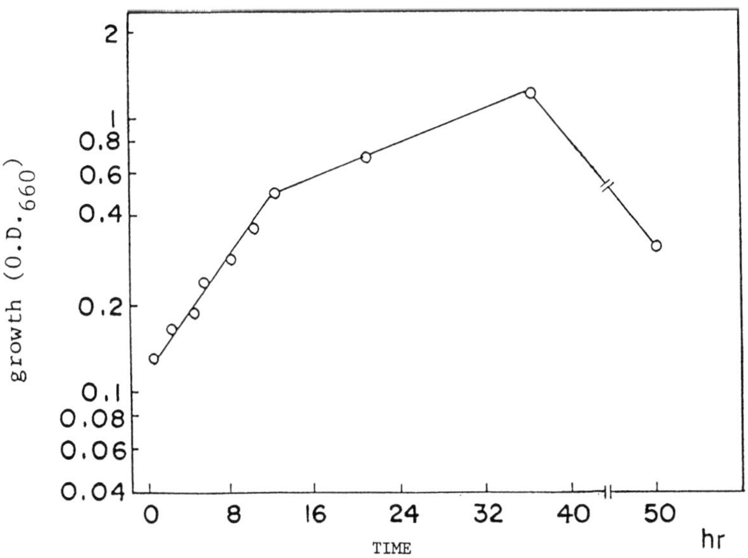

Fig. 4. <u>C. thermohydrosulfuricum</u> grown at 65°C in anaerobic conditions.

The crude oil used for this study was a medium heavy oil (API° 25) from Orcutt Hill, California, obtained from Union Oil Co. (sample taken in 1987). The recovery of oil was done as previously reported (Jang et al., 1989).

3 RESULTS AND DISCUSSION

3.1 <u>Oil recovery mechanism</u>

The use of <u>Clostridium spp.</u> is essentially due to the high yield of gases, mainly CO_2 and H_2, for well-head cleaning and in-situ fermentation for oil recovery. In Table 2, the gas production <u>C. thermohydrosulfuricum</u> is comparable with other <u>Clostridium spp.</u>

TABLE 2

Production of gases and solvents (m moles/100 m moles of glucose)

	CO_2	H_2	ethanol	Ref.
C. acetobutylicum	220	166	4.9	(b)
B. acetoethylicum	215	137	122	(b)
C. thermosaccharolyticum	174	229	---	(b)
C. thermocellum, LQRI	77.9	64.4	35.4	(c)
T. brokii, HTD4	51.8	4.5	50.5	(d)
C. thermohydrosulfuricum 39E	198	23	173	(d)
C. thermohydrosulfuricum 39E	---	11	197	(a)

(a) Zeikus et al., 1980; (b) Moat, 1979; (c) Lamed and Zeikus, 1980; (d) Ng et al., 1981.

It is of importance to note that C. thermohydrosulfuricum has the highest ethanol yield among all of the Clostridium spp. known (Table 2). This fact certainly will aid in oil recovery. It is known that microbial enhanced oil recovery requires a number of coordinated events, such as improvement of relative mobility, partial repressurization, reduction of viscosity, increase in pore size, optimization of aeral sweep efficiency, modification of swelling, wetting and surface tension, etc. These events again depend on the bio-byproducts from a given strain. Enhanced oil recovery data shown in Table 3 represent the residual oil recovery after the completion of secondary water-flooding recovery.

TABLE 3

Percent oil recovery* experiment on C. thermohydrosulfuricum

control	with bacterial culture (90 ml/hr, 12 hrs.)	only surfactant
7.8	70.9	16.7
8.3	59.6	
	68.7	

* Simulation based on residual oil after water-flooding (secondary recovery) in sandpack.

Doubtlessly, the extracellular surfactants are present in C. thermohydrosulfuricum. Otherwise, the surfactant-only efficiency would be similar to the background values given in Table 3.

The fact that simulated in-situ recovery experiments demonstrate that the recovery yield is higher than most alkaline-water flooding (Chan et al., 1982), and even C. acetylbutylicum (Chang and Yen, 1985) strongly supports that C. thermohydrosulfuricum has a combined, concerted and cooperative recovery mechanism.

3.2 Fate of surfactant vesicles

Vesicles are supposedly made from assemblages of amphiphatics originating from the exocellular biosurfactant via membrane-mimetic chemistry. In the case of C. thermohydrosulfuricum, surfactant vesicles were observed both in the presence of hydrocarbon and crude oil. From Figure 5a, evidently the vesicles can be grown proportionally to time (Figures 5b and 5c). Those large vesicles are similar to those derived from tar sand recovery processes where the bitumen is recovered by the in-situ surfactant (Sadeghi et al., 1989). Dill and Flory's model (Yen, 1988) is often used to account for the growth and alignment of the unilamellar vesicle. The growth of multilamellar vesicles is an important feature for possible field applications.

In the present case, a hydrocarbon, n-hexadecane, was used in replacement of oil, upon the addition of the surfactant, the vesicle formation was observed (Figure 5d). For both crude oil and hydrogen studies, the background was devoid of any vesicles (Figure 5e and 5f).

The fate of surfactant vesicles in the field environment may be a mainstream factor for homogeneous stabilization in sweep efficiency. The dynamic Brownian movement, the phase change in the presence of pH and ionic strength, the Kraft point change with temperature and stability at elevated temperature, the transformation of multilamellar vesicles under mechanical shearing and ultrasonic vibration, especially the stability under solvent and cosurfactant interaction, can all exert a certain effect on the vesicles. Considering the last mentioned effect, that T. thermohydrosulfuricum produces a massive amount of ethanol, it is interesting to ponder whether the multilamellar vesicles would survive or break down into individual surfactants in an aqueous system, for, in the past it was proven that ethanol will destroy

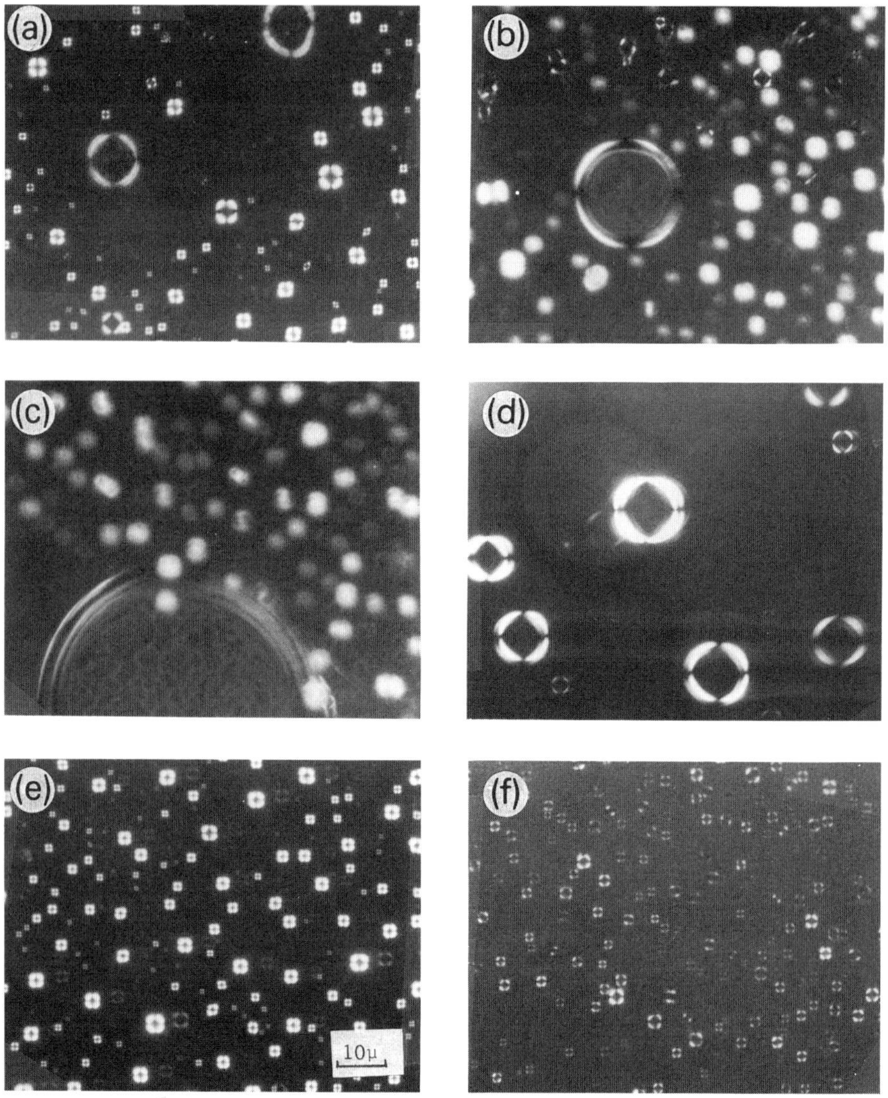

Fig. 5. Micrographs of vesicles derived from biosurfactants of C. thermohydrosulfuricum. (a) crude oil in the presence of biosurfactants; (b) crude oil in the presence of biosurfactants after aging for 2 days; (c) crude oil in the presence of biosurfactants after aging for 10 days; (d) n-hexadecane and biosurfactants in water; (e) reference for crude oil and water; (f) reference for biosurfactants and water.

multilamellar vesicles (Sadeghi et al. 1989).

A positive feature of the surviving vesicles is that they will aggregate and form a gel-like lattice upon interaction with metal ions. For example, gelation can be easily formed from xanthan and Cr (III) (Wilhite et al., 1990). In reservoirs, the trivalent ions can slowly interact with the biosurfactants to form a gelled-polymer network. Such gelled systems can help to modify the permeability of reservoir rock as a thick foam. This will definitely aid in enhanced oil recovery.

3.3 An idealized strategy

C. thermohydrosulfuricum is a spore forming bacteria. The advantages of spores are many, for example, they may be able to tolerate high shears during injection and they also can survive under extreme temperature and pressure; both these qualities are essential for an oil field in the context of geothermal, lithostatic and hydrostatic gradients with regard to the location of the field. Usually, the bottom hole of a producing reservoir experiences both high temperature and high pressure.

Many sporologists claim that pressure is a useful germinating agent because it requires no chemical additives. In reality, the spores of thermophilic bacteria have been isolated from sediments at a depth of over 2000 M in the coastal environment (Marquis, 1982). Furthermore, the Clostridium spp. are extremely heat-resistant, e.g., spores of C. thermohydrosulfuricum have survived heating at 132° and 121°C for short periods, and they were extremely heat resistant below 121°C; the thermal activation energy is ca. 364.4 kJ/mole (Hyun et al, 1983).

Such features in conjunction with the high salt concentration of ~20% saline will make the C. thermohydrosulfuricum ideal for MEOR in field applications. However, there are even more astonishing advantages of C. thermohydrosulfuricum: we will discuss the ultimate microbially enhanced oil recovery.

Most MEOR processes depend on the carbon source and nutrition from added material such as molassas. This is often not economical, especially if the field is extremely large or under a marine environment. The best carbon and nutrition sources are from the crude oil itself. This route, through the years, has had success similar to dewaxing. Yet, if the carbon sources are based on the unusable portion of petroleum, the process is still unknown. In the makeup of a new process it is important to eliminate the bad portion of petroleum as much as possible to retrieve a better quality. Since kerogen is a solid (not pumpable and not flowing) and is widely present in reservoirs, (in conjunction with the rock matrix) it will be of value to expend it.

In a thermal, volcanic environment, the microbial methanogenesis is generally associated with the anaerobic decomposition of algal mats. These algal mats are of high molecular weight organic matter in the absence of certain electron acceptors such as O_2, NO_3^{2-} and SO_4^{2-}. The combination and coordination of diverse bacterial populations comprised of hydrolytic, acetogenic and methanogenic bacteria in which the C. thermohydrosulfuricum, C. thermocellum, and C. thermosaccharolyticum are evolved. These trophic groups can mineralize the biomass from algal mats based on the ecology of thermal environments. C. thermohydrosulfuricum 39E was isolated in conjunction with other trophic bacteria in 12 sites located in hot springs of Yellowstone National Park (Zeikus et al., 1980).

5 ACKNOWLEDGEMENTS

The authors are grateful to National Science Foundation Grant No. CBT-8614019 for support of this work. We would also like to express our appreciation to Dr. R.E. Goodman of the Biological Sciences Laboratory for technical assistance. We thank Ms. N. Yan and L. Raftree for their assistance in preparing this manuscript.

4 REFERENCES

Chan, M., Sharma, M.M. and Yen, T.F., 1982. Generation of surface active acids in crude oil for caustic flooding enhanced oil recovery. Ind. Eng. Chem. Process Design and Dev., 21: 580-583.

Chang, Y.-I. and Yen, T.F., 1985. Effects of sodium pyrophosphate additive on the huff and puff/nutrient flooding MEOR process. Internat. Bioresources J., 1: 247-256.

Grula, M.M. and Russell, H.H., 1987. Genetic, physiological and nutritional studies on Clostridium strains isolated and screened for characterization useful in enhanced oil recovery, with special reference to high salt tolerance. DOE Contract Report AS22-80BC10302, 25 pp. NTIS.

Huang, V. and Yen, T.F., 1988. Enhanced oil recovery through the use of thermophilic bacteria. 196th National American Chemical Society Meeting, Los Angeles, CA.

Hyun, H.H., Zeikus, J.G., Longin, R., 1983. Ultrastructure and extreme heat resistance of spores from thermophilic Clostridium species. J. Bacteriol., 156(3): 1332-1337.

Jang, L.K. and Yen, T.F., 1985. A biotechnological alternative for recovery of heavy oil. Proc. 3rd UNITAR/UNDP International Conference on Heavy Crude and Tar Sands. AOSTRA, pp. 343-350.

Jang, L.K., Yen, T.F., Chilingar, G.V. and Donaldson, E.C., 1989. Bacterial migration through nutrient-enriched sandpack columns for in-situ recovery of oil. In (E.C. Donaldson, G.V. Chilingarian and T.F. Yen, eds.) Microbial Enhanced Oil Recovery. Elsevier, Amsterdam, pp. 151-164.

Johnson, A.C., 1979 Microbial oil release technique for enhanced recovery. Conf. on Microbial Processes Useful in EOR, San Diego, CA, 1979, DOE.

Lamed, R., and Zeikus, J.G., 1980. Ethanol production by thermophilic bacteria: Relationship between fermentation product yields of and catabolic enzyme activities in Clostridium thermoallum and Thermoanaerobium brockii. J. Bacteriol., 144(2): 569-578.

Marquis, R.E., 1982. Microbial Barobiology, Bioscience, 32(4): 267-271.

Moat, A.G., 1979. Microbial Physiology. John Wiley, New York, p. 145.

Ng, T.K., Ben-Bassat, A. and Zeikus, J.G., 1981. Ethanol production by thermophilic bacteria: Fermentation of cellulosic substrate by cultures of Clostridium thermocellum and Clostridium thermohydrosulfuricum. App. Env. Microbiol., 41(6): 1337-1343.

Sadeghi, M.-A., Sadeghi, K.M., Momeni, D. and Yen, T.F., 1989. Microscopic studies of surfactant vesicles formed during tar sand recovery. In (J.K. Borchardt and T.F. Yen, eds.) Oil-Field Chemistry: Enhanced Recovery and Production Stimulation, ACS Sym. Ser. 396 , 391-407.

Wilhite, G.P., Green, D.W., Thiele, J.L., Shan, T., and Mertes, K.B., 1990. Investigation of the application of gelled polymer systems for permeability modification in petroleum reservoirs. DOE Contract Report AC22-85BC10843. 216 pp. NTIS.

Yen, T.F., 1988. Application of membrane-mimetic chemistry to fossil fuel conversion. ACS 20th Central Regional Conference, West Virginia University, Morgantown, WV.

Yen, T.F., 1990. Microbial Enhanced Oil Recovery: Principle and Practice. CRC Press, Boca Raton, FL, 257 pp.

Zeikus, J.G., Ben-Bassat, A. and Hegge, P.W., 1980. Microbiology of methanogenesis in thermol volcanic environments. J. Bacteriol., 143(1): 432-440.

Reprinted from: *Microbial Enhancement of Oil Recovery - Recent Advances*, edited by E.C. Donaldson

CH. R-20

TRANSIENT TRANSPORT OF BACTERIAL SUSPENSIONS IN NATURAL POROUS MEDIA: MODELING OF THE ADSORPTION PHENOMENON

P. CHENEVIERE*†, M.SARDIN*, P. de FARCY** and A. PUTZ***

* Laboratoire des Sciences du Génie Chimique, Centre National de la Recherche Scientifique,1, rue Grandville B.P. 451, 54001, Nancy cédex, France

† currently at the University of Michigan, Dept. of Chem. Eng., H. H. Dow Building, 2300 Hayward, Ann Arbor, MI 48109, USA

** Sanofi-Elf BIO-Recherches, Labege Innopole voie 1 - BP 137, 31328 Labege cédex, France

*** Société Nationale Elf Aquitaine (Production), CSTCS, av. Laribeau, 64018 Pau, France

ABSTRACT

Cheneviere, P., Sardin, M., de Farcy, J.P. and Putz, A.; 1990. Transient transport of bacterial suspensions in natural porous media: modeling of the adsorption phenomenon.

Applications of microparticle systems in the oil industry are numerous: Microbial Enhanced Oil Recovery (MEOR), clay stabilization, fines migration and plugging of water injection wells are some of them.

This paper addresses the importance of microparticle transient transport through natural porous media by proposing a new approach to quantify such transport. Under a slow flow regime, which is often the case in petroleum operations, particle retention due to mechanical effects can be neglected in comparison with colloidal interactions. Nevertheless, colloidal theory is mainly descriptive and quantitative prediction of particle adsorption in natural porous media has often led to a failure, particularly since homogeneous and heterogeneous equilibria were not taken into account.

Batch and column experiments have been performed using a natural sand from Entraygues (France) and a living, non motile and no proliferating bacteria of the species Leuconostoc mesenteroides (L.m.), dispersed in solutions with different ionic compositions. In addition, we modeled microparticle adsorption thanks to three phenomenological mechanisms: surface ionization, ion exchange and interfaces condensation (adsorption) at equilibrium. Most of batch experiments were successfully predicted by the model which provided a comprehensive and predictive approach for microparticle adsorption. The model was used to predict bacterial transport in a column (sandpack), assuming local equilibrium in conjunction with a lumped parameter flow model taking into account hydrodynamical dispersion in the porous medium. Though passable agreement was found between experimental and predicted elution curves, we discuss the validity of the model for predict in bacterial transport in natural porous media.

1 INTRODUCTION

Numerous *in situ* biochemical reactions due to bacterial activity provide chemical products adapted to the petroleum production processes: the production of polysaccharides, surfactants or acids for enhanced oil recovery (Jang et al., 1983; Knabe, 1982) the production of gas for the repressurizing of oil reservoir (Frick, 1972) and the production of polysaccharides for selective plugging of fissured rocks (Updegraff, 1983; Bubela, 1985). To carry out this last application, a limited amount of suspension must be transported efficiently across the reservoir and reach a defined location to induce these reactions *in situ*. Increasing the transport efficiency by controlling cell adsorption is actually a task to be achieved.

It appears that the transport of a non-proliferous bacterial suspension in a natural porous medium is subjected to numerous physicochemical and mechanical interactions and may be considered as the transport of a colloidal suspension (Stumm and Morgan, 1981; Marshall, 1985; Stotzky, 1966). Colloidal transport is a major thematical problem discussed in a great number of recent works: Vaidya and Fogler (1990), Sharma and Yortsos (1987a,b), Rajagolapan and Chu (1982), Mc Dowell-Boyer et al. (1986) and Krishnan (1985). Nevertheless, quantitative prediction of particle retention has not been achieved yet.

Assuming a permeability greater than 100 mD and a linear velocity less than $1 m.day^{-1}$, particle retention due to mechanical interaction is often negligible compared with electrostatic and van der Waals interactions (Hiemenz, 1986; Ruckenstein and Prieve, 1976a,b). Transient transport experiments carried out at laboratory scale have shown a prevailing influence of the ionic strength of the solution on particle retention in porous media (Adamczyk, 1984, 1987a,b).

The DLVO theory (Hiemenz, 1986) is a classical approach to describe transport properties of particles in porous media. Experimental observations are closely related to the energetic diagram based on the characterization of particles and porous media by means of electrokinetic potential or streaming potential, and size distribution measurements. Meanwhile, DLVO theory is mainly qualitative and quantitative prediction of particle retention has often led to adsorbed quantities several orders of magnitude far from experimental observations (Adamczyk, 1984).

One reason for this failure is that in the DLVO theory particles are treated as charged points whose concentration profile in the vicinity of the solid surface is calculated in accordance with the energetic diagram. Indeed, when particle size is higher than the double layer thickness, it is no longer possible to assimilate these particles with charged points. Therefore, the concentration profile has no physical meaning. Possible modification of the ionic composition during transient transport of particles is another reason for this failure. Indeed, most often homogeneous and heterogeneous phase equilibria are not taken into account by the models. This means that for transport experiments, the energetic diagram, defined for a given composition at equilibrium, is a time and space function entirely neglected. As a matter of fact, only electrostatic interaction between a particle and a solid surface varies as a function of the ionic composition; van der Waals attraction force being constant, given a particle-solid surface system. Thus, a variation in ionic composition may lead to a new distribution of ions on the porous medium, on particles and in the aqueous phase. Ion exchange and surface ionization reactions partly control this phenomenon. This, specially when the Ion Exchange Capacity of solids, which is closely related with their surface charge, depends on aqueous phase composition. For instance, it has been demonstrated that cell surface of Leuconostoc mesenteroides is mainly composed of weakly acid and base sites, partially ionized (Cheneviere, 1989).

To account for variations of the ionic composition during bacterial transport, another approach has been proposed (Cheneviere, 1989). It consists in measuring experimentally thermodynamic quantities related to the surface charge of cells and minerals, and cell adsorption on minerals. Those quantities, such as the Cation Exchange Capacity (CEC) of cells and the equilibrium constant for cell adsorption, allow for quantification of elementary reactions that occur during the transient transport of a bacterial suspension. These reactions are mainly, surface ionization, ion exchange and particle adsorption. Then, assuming local equilibrium conditions, the coupling of these elementary reactions with a flow model, may describe the transient transport of the suspension.

Following this second approach, this work examines physicochemical interactions that act between bacteria, mineral surfaces and some major ions encountered in natural environments. The final aim of this experimental and theoretical study was to elaborate a predictive model for cell adsorption and transport. Materials used for this purpose, were the bacterial strain of Leuconostoc mesenteroides (L.m.) in non-proliferous condition, and the natural sand from Entraygues (France), both in contact with NaCl or $CaCl_2$ solutions.

2 DESCRIPTION OF INTERACTION PROCESSES

This description concerns the main mechanisms involved in bacterial adsorption on sand grains, i.e., ionization of surface sites, ion exchange on these sites (for both cells and sand grains) and finally cell adsorption.

2.1 Surface ionization and ion exchange reactions

Our surface charge model stipulates that only potential determining ions (p.d.i) coordinated with solid surface groups, modify the surface charge (Cheneviere, 1989). To meet the constraint for electroneutrality at the solid surface, the surface charge is counter-balanced by ions located in the double layer, but its compact and diffuse parts are not distinguished. Hence, it is assumed that ionization of a surface site S can be represented by the following equilibrium,

$$S = S' + C_{S'} \tag{1}$$

with the associated equilibrium constant, defined as the intrinsic ionization constant:

$$K_{int} = \frac{S'.\ C_{S'}}{S} \tag{2}$$

Where, S is a neutral surface site, S' represents the site S ionized and $C_{S'}$ is a potential determining ion located in the ionic double layer and coordinable with S' ($H^+_{S'}$ or $OH^-_{S'}$). Therefore, ionization of a cation exchanger site A, and an anion exchanger site B, is written as:

$$AH = A^- + H_f^+ \tag{3}$$

$$B + H_2O = BH^+ + OH_f^- \tag{4}$$

In addition, ionization reactions occur generally in presence of salts, therefore the distribution of ions on the solid surface is governed by ion exchange reactions which describe the transfer of electrically charged components between the aqueous and solid phases. Let A_1^{z1} and A_2^{z2} be two exchangeable cations, the ion exchange reaction takes the following form (Helfferich, 1962; Bolt, 1982):

$$z_2\, A_1^{z_1} + z_1\, A_{2f}^{z_2} = z_2\, A_{1f}^{z_1} + z_1\, A_2^{z_2} \tag{5}$$

The equilibrium constant associated with Eq. (5) is expressed as:

$$K_{A1,\,A2} = \left(\frac{A_{1f}}{A_1}\right)^{z_2} \left(\frac{A_2}{A_{2f}}\right)^{z_1} \tag{6}$$

However, this dimensionless coefficient depends on the cation and anion exchange capacity. Hence, the Gaines-Thomas selectivity coefficient (Gaines and Thomas, 1953), expressed in terms of concentration for the aqueous species and ionic fraction for adsorbed species, was preferred:

$$K_{GT\ A1,\,A2} = \left(\frac{x_{1f}}{A_1}\right)^{z_2} \left(\frac{A_2}{x_{2f}}\right)^{z_1} \tag{7}$$

Assuming that the amphoteric behavior of the solid surface originates from sites A and B only, the total ion exchange capacity (IEC), the cation exchange capacity (CEC) and the anion exchange capacity of the solid (AEC), are defined as:

$$\text{CEC} = [A^-] \tag{8}$$

$$\text{AEC} = [BH^+] \tag{9}$$

$$\text{IEC} = \text{CEC} + \text{AEC} \tag{10}$$

In addition, the Ionizable Site Concentration (ISC) of positively or negatively charged sites is expressed as:

$$\text{ISC+} = [\text{B}] + [\text{BH+}] \tag{11}$$

$$\text{ISC-} = [\text{A-}] + [\text{AH}] \tag{12}$$

Furthermore, the differential ion exchange capacity (differential IEC) defines a quantity which represents the theoretical surface charge of the solid:

$$\text{Theor. Surf. Char.} = \text{CEC} - \text{AEC} \tag{13}$$

2.2 Particle adsorption

The following development does not intend to describe the adsorption phenomena at the microscopic level. A phenomenological mechanism is proposed, which accounts for experimental observations related to cell adsorption, allows for its quantification and has a prediction capability.

Let S (porous media) and P (particles) be the two different solids of the system, it is assumed that (i) their surface charge originates from ionization of n_S and n_P surface sites respectively, and (ii) particle adsorption is the result of the condensation of liquid-solid interfaces on the two solids. It is supposed that this phenomenon can be described at the molecular level, as a fast equilibrium between unionized surface sites on S and P, and SP, a new site related to adsorbed particles. Being ionized, this site may exchange anions and cations with the solution. The condensation (adsorption) - relaxation (desorption) equilibrium describing the partition of cells between the solution (suspended cells) and the surface of the porous medium (adsorbed cells) is expressed at the molecular level as:

$$\left\{\begin{matrix} \textit{Linear combination} \\ \textit{of unionized sites of type } S \\ \textit{(Porous Medium)} \end{matrix}\right\} + \left\{\begin{matrix} \textit{Linear combination} \\ \textit{of unionized sites of type } P \\ \textit{(cells)} \end{matrix}\right\} = \left\{\begin{matrix} \textit{New ionized site+SP-} \\ \textit{and counter-balancing ions} \\ \textit{(Adsorbed cells)} \end{matrix}\right\} \tag{14}$$

and is written,

$$\sum_{i=1}^{ns} \alpha_i (S_i) + \sum_{i=1}^{np} \beta_i (P_i) + \gamma H_2O = {}^{+}SP^{-} + H^{+}_{+SP-} + OH^{-}_{+SP-} \quad (14\ a)$$

Where $(S_i) = [S_iH]$ and $(P_i) = [P_iH]$, if i is a cation exchange site, and $(S_i) = [S_i]$, $(P_i) = [P_i]$, if i is an anion exchange site. $\{^{-}SP^{+}\}$ is a ionized site resulting from cell adsorption, its charges are counter-balanced by H^{+}_{+SP-} and OH^{-}_{+SP-}. The coefficients α_i and β_i, are the fraction of site S_i and P_i on the porous media and the cells, respectively. Additionally, correspondence between the two scales of description of solid surfaces, is fully established provided that the following constraints are verified:

$$\sum_{i=1}^{ns} \alpha_i = 1 \quad ; \quad \sum_{i=1}^{ns} (S_i) = ISC_t\,(Sand) \quad (15)$$

$$\sum_{i=1}^{np} \beta_i = 1 \quad ; \quad \sum_{i=1}^{np} (P_i) = ISC_t(Cells) \quad (16)$$

The equilibrium constant associated to the stoichiometric relation (14 a) is expressed as:

$$K_{ads} = \frac{[{}^{+}SP^{-}]\,.\,[H^{+}_{+SP-}]\,.\,[OH^{-}_{+SP-}]}{\prod_{i=1}^{ns} (S_i)^{\alpha_i}\,.\,\prod_{i=1}^{np} (P_i)^{\beta_i}} \quad (17)$$

In addition, classical ion exchange reactions such as Eqs.(3) and (4) take place on the new site $^{-}SP^{+}$.

In section 4.2.3 we demonstrate that thanks to the coupling of this new reaction to surface ionization and ion exchange mechanisms, the main tendencies of the DLVO theory are now accounted for. The determination of equilibrium constant (14 a) from batch experiments by putting into contact a constant mass of sand with different cell concentrations, is also described.

3 EQUILIBRIUM CALCULATIONS

To simulate and predict batch and column experiments, we used the HYBRID code (Cheneviere, 1989), a flexible program of equilibrium calculations which allows the incorporation of homogeneous and heterogeneous phase reactions in polyphasic media. For batch experiments the user can easily simulate a specific experimental procedure or select classical determination methods such as “binary ion-exchange experiment”, “acid-base titration” or “batch adsorption performed with several phases”. Additionally, determination of equilibrium constants and other intensive or extensive constants is facilited by the use of optimization procedures that fit simulated data with experimental measurements.

The following conditions were assumed: reactions are instantaneous and simultaneous, they occur at constant volume, pressure and temperature, only the equilibrium state is considered. The R independent reactions of the reaction scheme involving N species are described by a set of R stoichiometric balance equations expressed as:

$$\sum_{j=1}^{N} \nu_{ij} A_j = 0 \qquad i = 1 \ to \ R \tag{18}$$

Where ν_{ij} is the stoichiometric coefficient for species A_j in reaction i ($\nu_{ij} < 0$ for reactants, $\nu_{ij} > 0$ for products. The mass action law associated to the reaction (18) is (Stumm and Morgan, 1981; Fritz, 1975):

$$K_i = \prod_{j=1}^{N} a_j^{\nu_{ij}} \qquad i = 1 \ to \ R \tag{19}$$

Where K_i is the equilibrium thermodynamic constant and a_j the activity of species A_j. As a first approximation it is assumed that the activity of chemical species can be represented by their molar concentration related to the volume of aqueous phase*. Therefore, mass action laws were written in terms of the concentrations A_j in mol/l of liquid phase, and ideal phase behavior was assumed.

* Except for ion exchange (see Eq.(7)).

3.1 Batch equilibrium calculations

The concentrations A_j may be related to a known and arbitrary reference composition $[A_j]_0$ as (Stumm and Morgan, 1981; van Zeggeren and Storey, 1970):

$$[A_j] = [A_j]_0 + \sum_{i=1}^{R} \nu_{ij} X_i \qquad j = 1 \ to \ N \tag{20}$$

Where X_i is the extent of reaction i with respect to the reference composition. This formulation allows to substitute R extents of reaction for the N concentrations $[A_j]$. The equilibrium concentrations are obtained from Eqs.(19) and (20), by solving the algebraic system:

$$K_i = \prod_{j=1}^{N} \left([A_j]_0 + \sum_{i=1}^{R} \nu_{ij} X_i \right)^{\nu_{ij}} \qquad i = 1 \ to \ R \tag{21}$$

Eq.(21) is a set of R algebraic equations with respect to the R unknowns X_i. For an arbitrary non-equilibrium composition $[A_j]_0$, the solution of Eqs.(20) and (21) yields the equilibrium concentrations. A Newton Raphson procedure was used to solve the system after a logarithm transformation of Eq.(21).

3.2 Equilibrium calculations in a flow system

To model flow patterns in a laboratory sandpack, we used a compartmental approach. This approach is extensively used in chemical reaction engineering to model Residence Time Distributions in real reactors (Levenspiel, 1972, Villermaux, 1982). The mixing tank of uniform composition is used as the basic element in HYBRID code, which allows for elaboration of appropriate networks to simulate observed flow conditions. To model 1-dimensional dispersive flow, as in the laboratory sandpack used in this work, we used the classical mixing tanks in series model. For a non-interactive species, the mass balance equation in a mixing tank is:

$$\frac{d[A]}{dt} = ([A_j]_E - [A])\frac{Q}{V_0} \qquad (22)$$

Where *[A]* is the concentration in the mixing tank and in the outlet stream, $[A]_E$ is the concentration in the feed to the tank. Q is the volumetric flow rate through the tank, and V_0 the volume of aqueous phase in the tank. For reactive species, the mass balance equation involves reaction terms:

$$\frac{d[A_j]}{dt} = ([A_j]_E - [A_j])\frac{Q}{V_0} + \sum_{i=1}^{R} v_{ij}\, r_i \qquad j = 1 \text{ to } N \qquad (23)$$

Where $\{v_{ij} r_i\}$ is the production flux of species j by reaction i at equilibrium. For adsorbed species the first term in Eq.(23) equals zero. At equilibrium, r_i is unknown and must be eliminated from Eq. (23). Let $[A_j]^*$ and X_i be such that:

$$[A_j] = [A_j]^* + \sum_{i=1}^{R} v_{ij}\, X_i \qquad j = 1 \text{ to } N \qquad (24)$$

$$\frac{d[A_j]}{dt} = \frac{d[A_j]^*}{dt} + \sum_{i=1}^{R} v_{ij}\frac{d X_i}{dt} \qquad j = 1 \text{ to } N \qquad (25)$$

Using Eq.(25) into Eq.(23) gives a set of N differential equations with respect to $N+R$ unknowns ($[A_j]^*$ and X_i). Since the choice of $[A_j]^*$ and X_i is arbitrary in Eq.(24), the mass balance equations can be split to give:

$$\frac{d[A]^*}{dt} = ([A_j]_E - [A_j])\frac{Q}{V_0} \qquad (26)$$

$$\frac{d X_i}{dt} = r_i \qquad i = 1 \text{ to } R \qquad (27)$$

Initial condition to Eqs.(26) and (27) is:

$$[A_j]^* = [A_j] \qquad X_i = 0 \qquad (28)$$

Since the flux terms r_i are unknowns at equilibrium, Eq.(27) is replaced by the equilibrium constraints (Eq.(19) and (21)):

$$K_i = \prod_{j=1}^{N}\left([A_j]^* + \sum_{i=1}^{R} v_{ij}\, X_i\right)^{\eta_j} \qquad i = 1 \text{ to } R \qquad (29)$$

Eqs. (24), (26) and (29) are solved simultaneously for $[A_j]^*$ and X_i. The Newton Raphson procedure was used to solve the algebraic equations, and a semi-implicit integration method with variable integration step was used to integrate the differential equations.

4. EXPERIMENTAL RESULTS

4.1 Surface ionization and ion exchange on cells of L.m. and the sand from Entraygues

These experiments have been described extensively by Cheneviere et al. (1989, 1990a) and only the main results will be recalled. Ion exchange mechanism on the sand from Entraygues has been studied by several investigators, specially Poirier (1984). Therefore numerical values of selectivity coefficients were taken from literature. In contrast, data related to ion exchange and surface ionization on living cells of Leuconostoc mesenteroides were determined experimentally. Parameters determined in our previous studies, that were used in this work, are listed in Table I.

TABLE I

Ionization Constants and Ionizable Site Concentration related to bacteria (L.m) and sand (Entraygues)

Material	Sites	Constants	Numerical Values	% of ISC_t
	Site 1			
Bacteria		ISC^-	$1.01\ 10^{-4}$ (mole.g^{-1})	46
		K_1	$5.0\ 10^{-6}$ (mole.l^{-1})	
	Site 2			
		ICS^+	$1.19\ 10^{-4}$ (mole.g^{-1})	54
		K_2	$5.0\ 10^{-4}$ (mole.l^{-1})	
	Site 1			
Sand		ICS^-	$2.12\ 10^{-5}$ (mole.g^{-1})	64
		K_1	10^{-5} (mole.l^{-1})	
	Site 2			
		ICS^+	$1.19\ 10^{-5}$ (mole.g^{-1})	36
		K_2	10^{-4} (mole.l^{-1})	

Two type of surface sites, a cation and an anion exchanger were determined on cells of Leuconostoc mesenteroides, as well as on the sand from Entraygues. Surface ionisation and ion exchange reactionson these sites, numeroted 1 (cation exchanger) and 2 (anion exchanger), were modeled according to Eqs. (3) , (4) and (5) respectively.

4.2 Experimental study and modeling of cell adsorption; Batch Experiments

Two sets of batch adsorptions were performed in polycarbonate flasks with $CaCl_2$ or NaCl solutions. Each provided a set of equilibrium adsorption curves, corresponding to five normalities ranging from 10^{-5} to 10^{-1} N.

4.2.1 *Experimental Methods*

Ten samples of sand (Entraygues, France), each of mass m (m= 5 g for NaCl solutions, m=1.2g for $CaCl_2$ solutions), were saturated by washing with saline (NaCl or $CaCl_2$) solutions of decreasing normality. Final saturation was performed with a 10^{-5} N saline (NaCl or $CaCl_2$) solution, at equilibrium with CO_{2g} (pH = 5.5). A similar procedure was applied to a concentrated bacterial suspension of Leuconostoc mesenteroides, recovered from a batch culture (glucose-based medium). The suspension, was then diluted to obtain ten samples having cell concentrations in the range between 4.0 10^8 and 1.3 10^{10} cells.cm^{-3}. 10 cm^3 of each dilution were put into contact with the samples of sand, in polycarbonate flasks. Mixtures were stirred for 2 minutes and allowed for sedimentation for 5 minutes. Then, the optical density (280nm) of the supernatant was measured in each flask. In addition, pH was measured, and Na^+ or Ca^{++} and Cl^- concentrations were determined in the flask having the highest cell concentration, i.e., about 1.3 10^{10} cells.cm^{-3}.

To raise the normality of the solution to 10^{-4} N, a concentrated saline solution (NaCl or $CaCl_2$, pH 5.5) was added in each flask containing the mixture at equilibrium. This last step along with the measurements, were performed three times sequentially, so as to obtain 10^{-3}, 10^{-2} and 10^{-1} N solutions in contact with the sand. As absorbance measurements were subject to slight variations due to the release of fines already present on sand samples, the overall procedure was applied to two sets of control flasks without cells, to account for the noise.

4.2.2 *Experimental Results*

Having determined the cell concentration in the suspension at equilibrium from optical measurements, the amount of adsorbed cells was calculated according to mass balances. The results related to the experiments performed in NaCl solutions are reported in Figs. 1 and 2. In Fig. 1 are plotted the equilibrium adsorption data and in Fig.2 are reported the pH, ΔCl and ΔNa (or ΔCa) measurements in the flask having the highest cell concentration. The last two quantities are equal to the difference between initial and equilibrium concentrations of Na^+ and Cl^- in this flask .

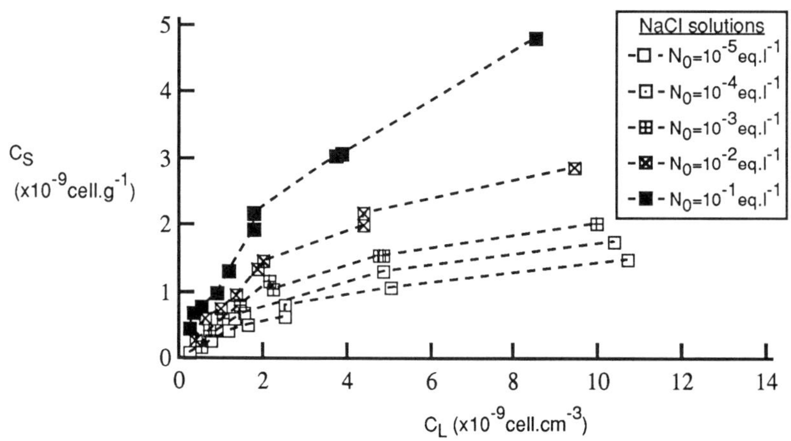

Fig. 1. Equilibrium adsorption curves in NaCl solutions NaCl $\approx 10^{-5}$ to 10^{-1} N.

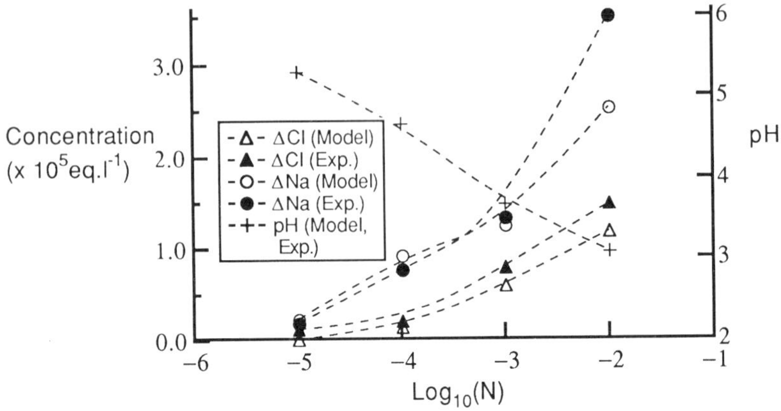

Fig. 2. pH measurements, measured and predicted gaps: ΔCl and ΔNa, in the flask with highest cell concentration. Empty symbols correspond to model prediction and N the normality.

The set of adsorption curves obtained with $CaCl_2$ solutions, shows that for normalities 10^{-2} and 10^{-1} N cell adsorption is increased by a factor 5 to 10, compared with NaCl solutions (see Fig. 3).

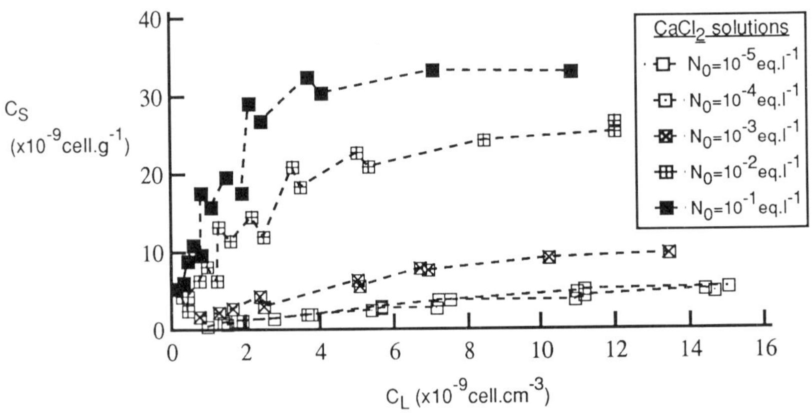

Fig. 3. Equilibrium adsorption curves in $CaCl_2$ solutions, $CaCl_2 \approx 10^{-5}$ to 10^{-1} N.

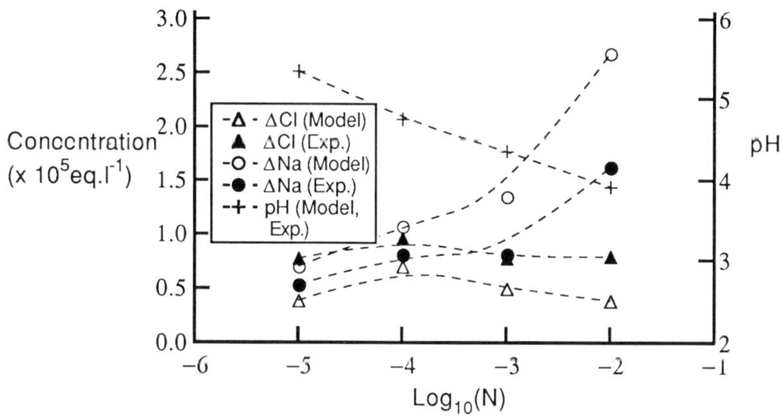

Fig. 4. pH measurements, measured and predicted gaps: ΔCl and ΔCa, in the flask with highest cell concentration. Empty symbols correspond to model prediction.

4.2.3 *Interpretation and modeling*

We only describe experiments performed in NaCl solutions, the procedure being similar with $CaCl_2$ solutions. According to Eqs.(14) and (15) and the procedure described in section 2.2, experimental data of the Fig.1 were converted at the molecular level. For this purpose we used the the following conversion factors:

(i) Specific mass of cells, $m_p = 2.10^{-13}$ g. cell^{-1}
(ii) Ionizable Site concentration for cells, $P_t = 2.210^{-4}$ eq.g^{-1} (Table I)
(iii) Ionisable Site concentration for sand, $S_t = 3.3\ 10^{-5}$ eq.g^{-1} (Table I)

We obtained the converted adsorption curves in the new set of axis (see Fig. 5), where $\{^+SP^-\}$ represents adsorbed cells and $\{P_1H + P_1^- + P_2 + P_2H^+\}$ suspended cells, as described in section 2.2.

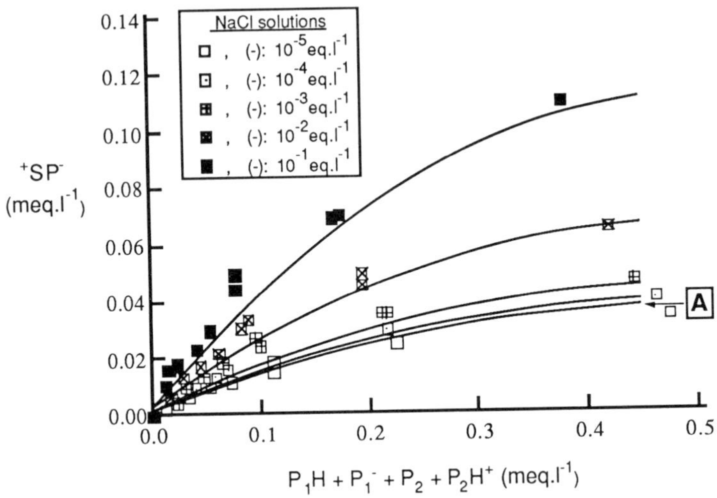

Fig. 5. Converted adsorption curves NaCl $\approx 10^{-5}$ à 10^{-1} N.

Then, the constant of the adsorption reaction (Eq.(14 a)) was determined by simulating the adsorption experiment performed at 10^{-5} N, taking into account the following reactions (Table II):

TABLE II
List of reactions accounting for bacterial adsorption

PHASE CONCERNED	REACTION
Cells (P) and Sand (S)	Surface ionization of Cation exchanger sites (P^-, S^-) Cation Exchange (H^+, Na^+) on (P^-, S^-) sites Surface ionization of Anion exchanger sites (PH^+, SH^+) Anion Exchange (OH^-, Cl^-) on (PH^+, SH^+) sites
Cells (P) + Sand (S)	Particle Adsorption ($^{+}SP^{-}$) Cation Exchange (H^+, Na^+) on ($^{+}SP^{-}$) sites Anion Exchange (OH^-, Cl^-) on ($^{+}SP^{-}$) sites
Aqueous	Water Ionization

We determined the "apparent" equilibrium constant of the adsorption reaction, which is related to the sum of the following reactions: cell adsorption (Eq.(14 a)), cation and anion exchange on adsorbed cells (Eqs.(3) and (4) on $^{+}SP^{-}$ sites). Therefore, K_{app} was expressed as:

$$K_{app} = K_{ads} \cdot K_{GT\,Na,H} \cdot K_{GT\,Cl,OH} \tag{30}$$

with (see Eqs. (14) and (15), and Table I),

$$\alpha_1 = \frac{P_1H^+ + P_1^-}{P_t} = 0.46 \quad ; \quad \alpha_2 = \frac{P_2H^+ + P_2^-}{P_t} = 0.54$$

$$\beta_1 = \frac{S_1H^+ + S_1^-}{S_t} = 0.64 \quad ; \quad \beta_2 = \frac{S_2H^+ + S_2^-}{S_t} = 0.36$$

Where, $P_t = 1.5\ 10^{-5}$ eq.l^{-1} and $S_t = 1.61\ 10^{-2}$ eq.l^{-1} (see Table I). K_{app} was calculated to account for the equilibrium state A (see Fig. 5), assuming $K_{GT\,Na,H} = 1$ and $K_{GT\,Cl,OH} = 1$. It was found:

$K_{app} = 10^{-9}$ mole.l^{-1}

Then, the entire adsorption curve related to the experiment performed in a 10^{-5} NaCl solution was simulated. The simulation procedure included three steps: (i) calculation of initial equilibrium compositions related to sand and particles saturations with 10^{-5} N NaCl solutions (pH = 5.5), (ii) calculation of reference compositions (not at equilibrium), and finally (iii) the calculation of equilibrium compositions, using the global reactive scheme along with reference compositions.

The resulting simulation is reported in Fig. 5 (plain line) and agrees well with experimental data. In addition, it may be seen that predicted Na^+, Cl^- and H^+ (pH) concentrations, agree well with experimental measurements: Fig. 2 (empty symbols). Then, the whole set of equilibrium adsorption curves (10^{-4} to 10^{-1} N NaCl) was simulated, using the adsorption constant determined previously. All predicted curves are in good agreement with experimental data (see Fig. 5). Also, we plotted the predicted gaps ΔNa, ΔCl using the interaction model (see Fig. 2, empty symbols) and obtained a good agreement with experimental data. Differences between predicted and measured pH being negligible they were not reported in this Figure. These last results allowed us to validate quantitatively and qualitatively the adsorption model in sodium solutions.

In Fig. 6 are plotted the predicted adsorption curves (plain line) at five normalities, along with converted data related to experiments performed in 10^{-5}, 10^{-4}, 10^{-3} N $CaCl_2$ solutions

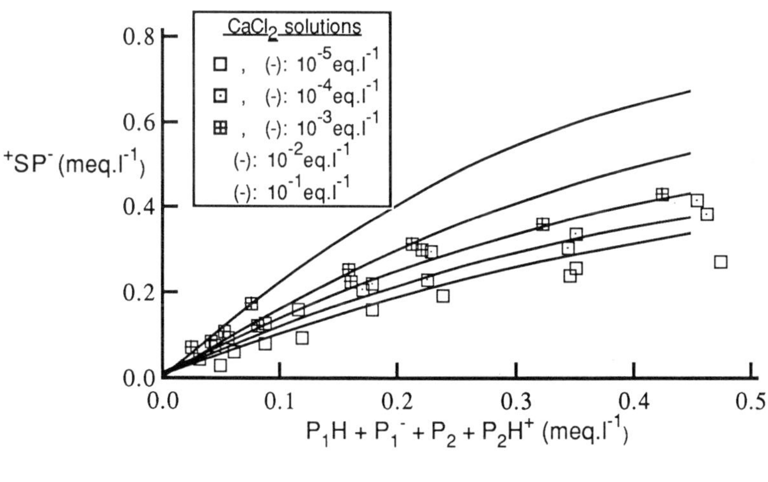

Fig. 6. Converted adsorption curves in $CaCl_2$ solutions.

(Markers). It may be seen that predicted curves agree fairly with experimental data, except for experiments performed at 10^{-2} and 10^{-1} N. Indeed, model predictions were far lower than the measured cell concentration in the suspension. This, may be due in part to flocculation, a phenomena which was not accounted for by the model.

The apparent constant related to the experiments performed with $CaCl_2$ solutions was determined with a similar procedure as described previously for NaCl solutions. The following set of constants was used:

$K_{app}=3\ 10^{-10}$ mole.l^{-1}

$K_{GT\,Ca,\,H} = K_{GT\,Ca,\,Na} \cdot K_{GT\,Na,\,H} = 8$ mole.l^{-1}

$K_{GT\,Cl,OH} = 1$

To examine the reversibility of cell adsorption and validate the equilibrium assumption in Eq.(14 a), two desorption experiments were performed from the equilibrium state "A" of the adsorption experiment conducted with 10^{-2} NaCl solutions (see Fig. 5). The aqueous volume of the flask was doubled by adding 10 cm^3 of a 10^{-2} N NaCl solution (pH = 5.5) in the flask. The mixture was stirred and the cell concentration in the suspension was determined at equilibrium state. From this new equilibrium state, a ten-fold dilution was performed, and again the cell concentration in the suspension was determined at equilibrium state. A similar experiment was performed, using a 10^{-5} N NaCl solution (pH = 5.5) to dilute the sample. Experimental and predicted equilibrium states were also in good agreement, as may be seen in Figs. 7 (a) and (b).

4.3 Experimental study and modeling of cell adsorption: Sandpack Experiments

The first step of this study allowed for elaboration and validation of the adsorption model, in equilibrium conditions. In the second step, model's capability to predict a transport experiment was examined assuming local equilibrium.

Prior to cell injection, the Residence Time Distribution of the fluid in the sandpack was determined from of a tracer experiment performed with a 10^{-3} N $CaCl_2$ solution, using the set-up depicted in Fig. 8. From the conductivity measurements, a graphical interpretation of the RTD curve allowed for the determination of the pore volume of the sandpack and its hydrodynamical dispersion. Then, these two parameters enabled us to build the flow model described in section 2.2 (mixing tanks in series). These results have been reported in Table III.

(a)

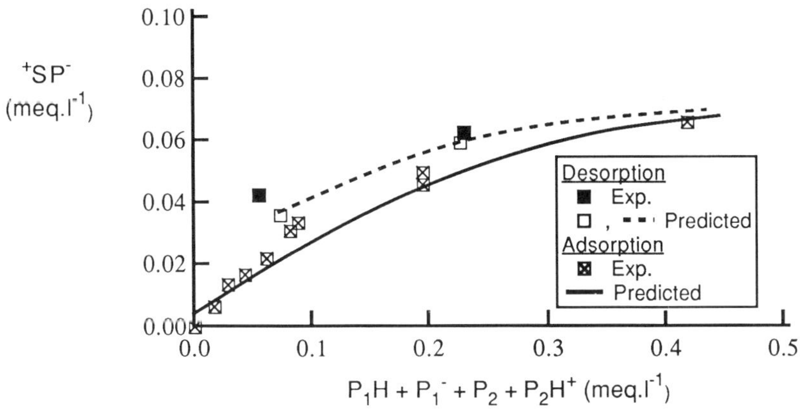

(b)

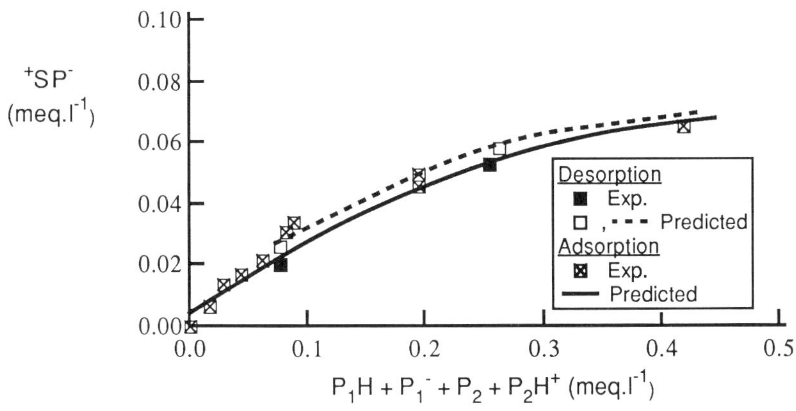

Figs. 7. Dilution experiments: (a) dilution of cell concentration only, (b) simultaneous dilution of cell concentration and salinity.

TABLE III

Characteristics of sandpack columns

Mass of sand (g.)	Pore Volume (cm^3)	J (number of mixing tanks)	Column length (cm)	column diameter (cm)
130	45	250	18	2.54

Two step-injections were performed on the set-up depicted in Fig. 8, using two sandpack columns with similar characteristics (Table III). Injections were conducted with cells of Leuconostoc mesenteroides suspended in 10^{-3} N NaCl solutions, at flow velocities close to those obtained in field conditions far from the well-bore. Experimental conditions for these two injections are reported in Table IV.

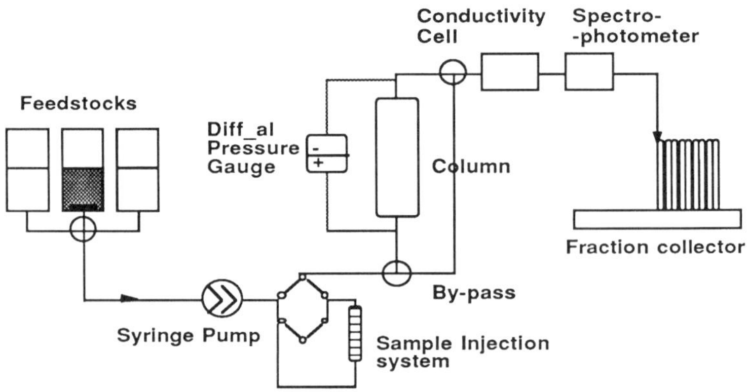

Fig. 8. Schematic of the experimental set-up.

Experimental breakthrough curves for cells are plotted in Fig. 9, along with the model prediction obtained using the interaction mechanism of the Table II along with the flow model. Model parameters determined previously were used, with the exception for J = 50 in the flow model to reduce the calculation time. Nevertheless, this modification induced a negligible increase of dispersion on the predicted breakthrough curve.

TABLE IV

Experimental conditions for sandpack experiments

Step injection Number	Ionic composition (and initial saturation)	Flow conditions $Q(cm^3.min^{-1})$	$V_l(cm.d^{-1})$	Cell Conc. $(cell.cm^{-3})$	Injected Volume (V_p)
1	10^{-3} N NaCl	0.2	130	~ 3.8 10^{10}	6
2	id.	0.1	65	~ 3.0 10^{10}	4

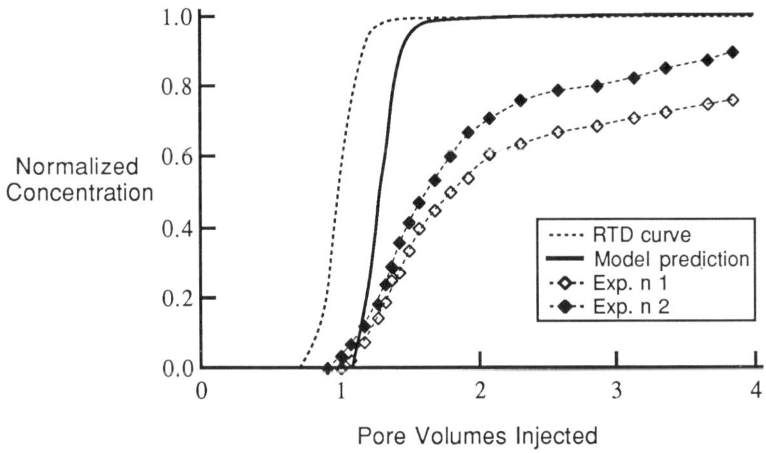

Fig. 9. Experimental and predicted breakthrough curves of cells.

Although gaps between experimental and predicted curves is significant, it is smaller in the case of the lower velocity experiment, i.e., 65 cm.d^{-1}. This means that mechanical interactions and/or non-equilibrium conditions may still occur in these conditions. It is likely that at lower flow velocities (closer to field conditions), i.e., 10 to 30 cm.d^{-1}, this gap may become less significant.

5 CONCLUSIONS

Thanks to a molecular description of bacterial adsorption at equilibrium state, a model of bacterial transport in porous media was developed. Basic interactions accounted for are: (i) Surface ionization, (ii) Ion Exchange and (iii) Particle Adsorption.

The model allows for prediction, to a large extent, of cell adsorption in batch experiments. Most qualitative tenancies established by the DLVO theory are taken into account by a reactive scheme which allows for quantification of adsorption, provided that a conversion between cells and the molecular scale can be established. In flow conditions, local equilibrium assumption appears to be limited to the very low velocity domain (< 0.3 m/d).

6 ACKNOWLEDGMENT

We would like to thank the National Corporation ELF Aquitaine for its financial support. We are also grateful to Dr. J.P. de Farcy, research engineer from Sanofi ELF Bio-research (Toulouse-Labege) and Dr. A. Putz, head of charge of the division for Enhanced Oil Recovery at the corporation ELF Aquitaine (Production), for the scientific consultation.

7 NOTATION

A_j	: chemical species j
a_j	: activity of species j
$[A_j]$	: molar concentration of species j
$[A_j]_0$, $[A_j]^*$	: reference concentration of species j

$[A_j]_E$	: flux average concentration in the inlet stream of a mixing cell
J	: number of mixing cells in the mixing cells in series model
K_i	: thermodynamic constant of the reaction i
Kx	: Gaines-Thomas selectivity coefficient for ion exchange
m	: mass of sand samples in batch experiments
m_P, m_S	: specific cell mass and specific sand mass
N	: number of species
n_S, n_P	: number of sites S and P
Q	: volumetric flow rate
r_i	: production flux by reaction i
P, S	: respectively, the symbols of sites P (particles) and S (sand)
R	: number of reactions
V_0	: volume of aqueous phase
X_i	: extent of the reaction i
x_{jf}	: ionic fraction of the species A_j on the exchanger

Geek letters

α_i, β_i	: fractions of site S_i and P_i on the porous medium and the cells, respectively
v_{ij}	: stoichiometric coefficient of species j in reaction i

8 REFERENCES

Adamczyk, Z., and Petlicki, J., 1987. J. Colloid Interface Sci. 118: 20.
Adamczyk, Z., and van de Ven, T.G.M., 1984. J. Colloid Interface Sci. 97: 68.
Adamczyk, Z., Dabros, T., Czarnecki, J., and van de Ven, T.G.M., 1984. J. Colloid Interface Sci. 97: 91.
Bubela, B., 1985.J. Geomicrobiology, 4: 313.
Bolt, G.H., 1982. "Soil Chemistry : b-Physico-chemical Models". Elsevier Science Publishing Co., Amsterdam, N.Y.
Burger, K., and Noszal, B., 1981. "Advances in Solution Chemistry." Plenum Press, 420 pp.
Cheneviere, P., Sardin, M., de Farcy, J.P. and Putz, A., submitted for publication.
Cheneviere, P., Hlavacek, M., Sardin, M. and Dodds, J.A. submitted for publication.
Cheneviere, P., 1989. Ph.D. Thesis, Lorraine National Polytechnic Institute, Nancy, France.
Frick, J.R., 1972. J. Petrol Technol. 24: 1469.
Fritz, B., 1977. Mem. Inst. Geol., ULP, Strasbourg, 152 pp.
Gaines, L.G., and Thomas, H.C., 1953. J. Chem. Eng. 21: 714.
Helfferich, F.G., 1962. "Ion Exchange" McGraw-Hill, New-York.
Hiemenz, P.C., 1986. “Principles of colloid and surface chemistry”. Dekker.
Jang, L.K., Chang, P.W., Findley, J.E., Yen, T.F., 1983. Appl. Environ. Microbiol., 46: 1066.
Knabe, S., 1982. J. Oil Gas. 52: 59.
Krishnan, C., 1985. "Deposition of colloidal particles from flowing suspensions." M.S. Thesis, Rensselaer Polytechnic institute, Troy, N.Y.
Marshall, K.C., 1985. "Bacterial Adhesion." Plenum Press.
McDowell-Boyer, L.M., Hunt, J.R., and Sitar, N., 1986. Water Resources Research., 22: 1901.
Mende, T.J., 1976. "Biological Foundations of Biomedical Engineering". Little, Brown and Company Boston, 456 pp.
Poirier, J.E., 1984. Ph.D. Thesis, Lorraine National Polytechnic Institute, Nancy, France.
Rajagolapan, R., and Chu, R.Q., 1982. J. Colloid Interface Sci. 86: 299.
Ruckenstein, E., and Prieve, D.C., 1976. AIChE J., 22: 276.

Ruckenstein, E., and Prieve, D.C., 1976. AIChE J., 22: 1145.
Sharma, M.M., and Yortsos, Y.C., 1987. AIChE J., 33: 1655.
Sharma, M.M., and Yortsos, Y.C., 1987. AIChE J., 33: 1673.
Stotzky, G., 1966. Can. J. Microb., 12: 1235.
Stumm, W., and Morgan, J.J., 1981. "Aquatic Chemistry." Wiley-interscience, 430 pp.
Updegraff, D.M., 1983. "Microbial Enhanced oil Recovery." Pennwell publishing Co., Tulsa.
Vaidya, R., and Fogler, H.S., " Formation Damage Due to Colloidally Induced Fines Migration", Colloids and Surfaces, in press.
Van Zeggeren, F., and Storey, S.H., 1970. "The Computation of chemical equilibria." Cambridge University press.
Villermaux, J., 1982. Genie de la reaction chimique. Tech. Doc., Lavoisier, Paris, 401 pp.

Reprinted from: *Microbial Enhancement of Oil Recovery - Recent Advances*,
edited by E.C. Donaldson

CH. R–21

COMPOSITIONAL NUMERICAL SIMULATION OF MEOR PROCESSES

A. K. Sarkar, M. M. Sharma and G. Georgiou*
Department of Petroleum Engineering
*Department of Chemical Engineering
University of Texas at Austin
Austin, Texas-78712

ABSTRACT

From an order-of-magnitude analysis it is shown that biosurfactants have the potential to recover residual oil from a waterflooded reservoir. Biomass and biopolymer may help to recover residual oil in a heterogeneous reservoir by selective plugging, and/or by improving the mobility ratio. A compositional numerical simulator has been developed on this basis. The simulator can be used for predicting growth of bacteria, consumption of nutrient, change in interfacial tension along with production of biosurfactants, oil production rate and cumulative oil recovery as a function of pore volumes of fluid injected.

1 INTRODUCTION

In Microbial Enhanced Oil Recovery (MEOR) processes bacteria and nutrient are injected into a porous medium containing residual oil. The microbes consume nutrient and produce biomass and various metabolic products such as gases, surfactant, acids, alcohols, and polymers. Some of these metabolic products or the biomass, help to mobilize residual oil. It is important to understand the potential mechanisms of MEOR processes before a mathematical model can be developed. The transport of bacteria,and other components such as nutrients, and metabolic products are important issues. Numerical simulation of this highly complex and coupled process is necessary to predict oil recovery and to investigate sensitivity towards different process variations and physico-chemical parameters. Using the oil recovery rates an economic feasibility study can be conducted. To illustrate the results of the model experimental data for *Bacillus licheniformis* JF-2 has been used in this study.

2 MECHANISMS IN MEOR

A better understanding of the mechanisms of oil recovery is extremely useful for selection of appropriate bacterial strains, design of optimal operational procedures and for mathematically modelling the process to predict laboratory and field results. The net oil recovery (E_r)is expressed as the product of the volumetric sweep efficiency (E_v) and the displacement efficiency (E_d),

$$E_r = E_d * E_v$$

In a heterogeneous reservoir, the by-passed residual oil may be recovered by improving the volumetric sweep efficiency by decreasing the mobility ratio, but to recover true residual oil (oil contacted by displacing agent but not displaced) the capillary number must be increased.

Several mechanisms, utilizing these metabolic products, have been proposed for MEOR processes. In this section a order-of-magnitude analysis of some of these mechanisms is presented.

2.1 Effect of gases

Different gases produced during a microbial growth process are CO_2, CH_4, H_2 and N_2. The production of fermentation gases may increase reservoir pressure, improve solution gas or gas cap drive mechanism, miscibly displace oil, or swell the oil phase and reduce the viscosity of the oleic phase.

From our gas build up experiments in Hungate tube it was found that 6.0 SCF/BBL PV (46 mM) of gas was produced (Sarkar, 1990). Based on the available information on the metabolism of *Bacilli* CO_2 and to a smaller extent H_2 are the most likely fermentation gases. Assuming that the gas consists exclusively of CO_2, then:

$$\frac{CO_2 \text{ concentration}}{\text{Glucose consumed}} = 1.53 * 10^{-3} \text{ gm mole } CO_2 / \text{gm glucose}$$

$$\frac{CO_2 \text{ concentration}}{\text{cell produced}} = 3.891 * 10^{-3} \text{ gm mole CO2 /gm cell mass}$$

For a 1 % glucose solution the theoretical limit (if all the carbon atoms in the nutrient are stoichiometrically converted to a desired product) for gas production is 300 mM or 39 SCF/BBL. On the other hand, at 100° F temperature and 1000 psi pressure the gas solubility is 150 SCF CO2 /BBL water (Lake, 1989, Fig.7-23). The huge amount of nutrient that will be required for a significant increase in CO_2 production will make the process uneconomical. From the plot of viscosity ratio of the gas oil mixture to oil as a function of saturation pressure (Lake 1989, Fig.7-21) viscosity reduction can be evaluated to be negligible. From the plot of oil swelling factor as a function of mole fraction of CO_2 in oil (Lake, 1989, Fig.7-22) it can be concluded that this concentration is too low to cause any oil swelling. In laboratory studies conducted with no back-pressure, the produced gas may create a trapped gas saturation that reduces the residual oil saturation (Kyte et. al, 1956, Fig. 4,5). But, this explanation can not be extended to the field.

Under anaerobic conditions, typical high-CO_2-producing microorganisms (for example *Zymomonas mobilis*) can form around 1 mole of CO_2 for every 12.5 grams of cells (wet weight) formed (Gottschalk, 1983). A simple calculation reveals that a large amount of cell mass (approximately 6 % biomass wet weight per unit weight of oil) would have to be generated in order to produce the amount of CO_2 needed for satisfactory oil recovery. Among other things, the generation of such high concentrations of biomass is very likely to cause serious plugging problems. Similarly, Grist (1983) conducted a simple economic analysis and concluded that *in-situ* production of biogas is not economical because of the large amount of nutrients that must be injected.

It has been reported by several authors that the gas produced during microbial growth process helps in the drive mechanism for oil recovery. From a overall material balance (Dake, 1978) two individual drive mechanisms (a) solution Gas Drive and (b) gas cap drive may be analyzed separately. It can be clearly demonstrated that the volume of gas that can be generated by bacteria is insufficient to contribute to either of these mechanisms. It is conceivable, however, that small quantities of gas may buildup localized pressure and substantial gas saturation in the near wellbore region (1 to 2 ft.). But, it is unclear how this will improve oil recovery.

2.2 Effect of acids

Various acids produced during microbial growth include acetic acid, formic acid, lactic acid etc. These acids may cause rock dissolution and improve the permeability of the reservoir. From our experimental study in batch fermentation the maximum concentration of acetic acid is 1.5 g/l (25 mM). The theoretical limit of acid production is 9 g/l (3*50 mM). Very low pH or high acid concentrations cannot be obtained because it will inhibit cell growth.

The acid component can react only with limestone and dolomite and not with clay or silica. Acetic acid has a low dissociation constant and 50 % acid remains unconverted (unreacted). At 0.5% acid concentration the dissolving power of acid (volume of rock dissolved/volume of acid reacted) is 0.0016 for limestone and 0.0014 for dolomite (Williams, 1979, Table 3.4). The reaction products of calcium acetate and magnesium acetate are soluble in spent acid. From the plot (Williams, 1979, Fig. 8.1) of permeability ratio (final to initial permeability) as a function of time, a properly designed mud acid (a mixture of approximately 12 % HCl and 3 % HF) gives a permeability ratio improvement of 4. So, the permeability improvement from acetic acid alone (without HF acid) will be much less and from the definition of capillary number this low improvement cannot cause any order of magnitude increase in capillary number. Grist (1983) also concluded that the production of organic acids is probably not a primary mechanism for enhancing the flow of oil. Since, the permeability (k) improvement is very small microbial acid stimulation is very unlikely to improve productivity in a sandstone reservoir. In carbonate reservoirs, for wells with corrosion problems, acetic acid in the concentration range of 50 to 100 g/l is sometimes used for stimulation purposes. It is not feasible to microbially generate such high acetic acid concentrations. As such microbial acid stimulation is impractical.

2.3 Effect of alcohol

Microbially produced alcohols may act as a solvent to cause oil swelling, reduce oil viscosity, lower IFT in combination with a biosurfactant, act as a cosurfactant to reduce salinity sensitivity of the biosurfactant, or displace oil in a miscible manner. The water solubility of surfactants commonly used in conventional surfactant floods increases with an increase in the concentration of low molecular cosurfactant (n-butanol) and decreases with an increase in high molecular weight cosurfactant. Cosurfactants generally increase optimal IFT, but it has been reported that addition of n-pentanol to a biosurfactant produced by an aerobic species H-13 significantly lowered IFT from 10^{-2} to 10^{-5} mN/m (Brown et al., 1986). This indicates that a particular surfactant and alcohol system must be studied to determine the final IFT of the system against an oil.

The theoretical limit for butanediol (assumed to be the microbially produced alcohol) production is 7 g/l. Butanediol is highly soluble in both water and oil. Since, the amount of alcohol produced is very little, it cannot cause any swelling of oil or reduction of oil viscosity, but it may be a potential agent for near wellbore stimulation.

2.4 Effect of biopolymers

The sweep efficiency of the displacement process may be improved by an increase in the viscosity of the displacing phase or by the biomass selectively plugging high permeability zones.

It is known that good mobility control is achieved when the displacing fluid has a viscosity between 10 to 50 cp at a shear rate of 1 sec^{-1} . It has been reported (Knapp et al., 1987) that a biopolymer, which can be produced anaerobically at 50 °C temperature and at a salinity of 8 %, has the potential of increasing culture viscosity to 4 cp at a shear rate of 150 sec^{-1}. Also some of the strains along with extracellular biopolymers are good plugging agents and can reduce permeability by 70 % or more. The selective reduction of reservoir permeability and increase in the viscosity of the displacing fluid can improve displacement efficiency and mobilize residual oil from a heterogeneous reservoir.

2.5 Effect of biosurfactants

Biosurfactants produced during exponential or stationary phase of bacterial growth may reduce interfacial tension (IFT) between oil and water, and increase capillary number to cause a decrease in the residual oil saturation. The isolation and characterization of the the biosurfactants produced by *Bacillus Licheniformis* JF-2 is currently under investigation. The lowest IFT of the fermentation broth has been found to be 0.03 mN/m. Assuming a surfactant molecule containing 12 carbon atoms and having a molecular weight of 300, the theoretical limit for surfactant production is 7.5 g/l or 7500 ppm. For a good quality surfactant a small fraction of that amount may be sufficient for lowering the IFT of the medium. Since only small quantities of surfactant are required to achieve a high capillary number, this is the most promising agent for MEOR.

It has been estimated (Lake, 1989) that the interfacial tension must be lowered to 10^{-2} to 10^{-3} mN/m (corresponding N_{ca} of 10^{-3} to 10^{-4}) before significant oil recovery is achieved. Biosurfactants, produced anaerobically, which are capable of reducing the IFT to such a low value have been reported (Brown et al., 1986). Furthermore, the production of the surfactant at the location where it is most needed should minimize losses due to adsorption on the rock.

2.6 Effect of Wettability Alteration

Microbes growing on rock surfaces may cause wettability alteration, which may in turn mobilize oil if the change is toward more water wetness. It has been shown (Kianipey et al., 1986), using a flow visualization cell, that the medium became more water wet and oil became emulsified (4 to 80 μm size) as a result of biosurfactants, gas and other metabolites. In the case of JF-2 (mainly a biosurfactant producer) wettability alteration was more prominent with an initially oil wet medium and the oil recovery was slightly higher compared to a water wet medium. With *Clostridia* (mainly CO_2 producers), although the wettability change was of the same magnitude the oil recovery actually decreased (16 % to 12 %). From these results one might conclude that the IFT reduction is more important and the effect of wettability alteration is as yet unclear. The exact mechanism of wettability alteration is an area of further research.

3 MATHEMATICAL FORMULATION OF MEOR PROCESS

The basic equations governing the transport of oil, water, bacteria, nutrient and metabolites in MEOR process are species mass conservation equations. In the formulation (n_c -1) component mass balance equations and one overall mass conservation equation/pressure equation are used. The constitutive relations, initial conditions and boundary conditions have to be specified to provide a complete description of this initial / boundary value problem. The details of this formulation are discussed by Sarkar (1990).

The coupled processes of growth and lysis of bacteria, generation of metabolites, consumption of nutrients and transport of all the components through a permeable medium may be expressed in the form of a general mass balance equation, on a unit bulk volume basis.

3.1 Assumptions

The major assumptions that need to be made are as follows,

(i) Reservoir and fluids are incompressible. No gas component is considered.

(ii) Precipitation/dissolution, cation exchange or redox reactions are not considered.

(iii) Bacteria, nutrient and metabolites are present in the aqueous phase only.

3.2 Material conservation equations

The differential form of material balance equations for isothermal, multicomponent, multiphase transport of chemical/biochemical species in naturally occurring permeable media can be written as,

$$\frac{\partial W_k}{\partial t} = - \nabla . N_k + R_k \tag{1}$$

where W_k , N_k and R_k are the overall concentration, net flux and source terms respectively of a component k.

The overall concentration (W_k)may be expressed as,

$$W_k = \o(1- \sum_{k=1}^{n_c} \hat{C}_k) \sum_{l=1}^{n_p} (\rho_k C_{kl} S_l) + \o \rho_k \hat{C}_k \tag{2}$$

where n_c is the total number of components and n_p is the total number of phases. C_{kl} refers to the volumetric concentration of component k in phase l and $\hat{C}_k$ refers to the volumetric concentration of component k adsorbed on the rock surfaces based on original pore volume respectively. ϕ refers to the porosity of the medium, S_l refers to the saturation of phase l and ρ_k denotes the density of component k.

The flux term(N_k) consists of convective, dispersive and coupled dispersive fluxes. It may be expressed as,

$$N_k = -\sum_{l=1}^{n_p} \{ (u_l \rho_k C_{kl}) + \phi S_l (D_{kl} \cdot \nabla(\rho_k C_{kl}) + D'_{kl} \cdot \nabla(\rho_{k'} C_{k'l}))\} \tag{3}$$

where D_{kl} and D'_{kl} are hydrodynamic and coupled hydrodynamic dispersion coefficients respectively. Subscript k' refers to the indices for the component with which component k is coupled. In MEOR processes there are diffusive fluxes in addition to the molecular diffusive flux. These are due to bacteria which have a chemotactic diffusive flux driven by a nutrient concentration gradient (from a lower nutrient concentration towards a higher nutrient concentration) and a tumbling diffusive flux driven by the bacteria concentration itself (from a higher concentration of bacteria towards a lower concentration of bacteria).

The source terms are a combination of all rate terms for a particular component, e. g. injection, production, growth and death of bacteria component and may be expressed as,

$$R_k = \sum_{l=1}^{n_p} \phi S_l (R_{kl} \rho_k C_{kl} + R'_{kl} \rho_{k'} C_{k'l}) + \frac{Q_k \rho_k}{V_b} \tag{4}$$

where Q_k indicates rate of injection or production and V_b indicates the bulk volume. R_{kl}, R'_{kl} are the specific rate constant and the coupled specific rate constant of component k in phase l. First order rate expressions have been used here for reaction rate terms. For example, in the case of nutrient the coupled rate coefficient indicates that the rate of consumption is dependent upon the concentration of live bacteria and in the case of metabolites the rate of generation depends on the concentration of bacteria (if nutrient is available).

The specific rate constant, considering growth and death processes, can be expressed using the Monod equation (Bailey and Ollis, 1986). The specific growth rate is approximated by the Contois model which was found to provide best fit to experimental data (Goursaud, 1989). This specific growth rate is further modified by introducing inhibition of growth by fermentation acids, which become important towards the end of the growth process when the concentration of inhibitors become significant (Aiba, 1968). The Leudeking-Piret equation is used for modeling generation of metabolites (Bailey and Ollis, 1986).

3.3 Pressure equation/ Overall component balance equation

The continuity equation describing the conservation of total mass can be obtained by summing the species conservation equation over all components.

$$\frac{\partial}{\partial x} \sum_{l=1}^{n_p} u_l = \sum_{l=1}^{n_p} \phi S_l \{ \sum_{k=1}^{n_c} (R_{kl} C_{kl} + R'_{kl} \frac{\rho_{k'}}{\rho_k} C_{k'l} \} + \sum_{k=1}^{n_c} \frac{Q_k}{V_b} \tag{5}$$

Combining the continuity equation, Darcy's law and the capillary pressure relation, and by neglecting gravity the final form of the pressure equation can be obtained. This equation is different from a surfactant flood model in that it contains additional generation/consumption rate terms.

3.4 Constitutive Relations

The constitutive relations used in the model are a saturation constraint, phase concentration constraints, total concentration constraints, adsorption isotherms, phase equilibrium relations, capillary pressure relations, phase densities, phase viscosities and relative permeabilities. Modelling of physico-chemical properties and phase behavior is a very important aspect of a compositional simulator. The approach taken here is similar to that taken in UTCHEM (Bhuyan, 1989). Adsorption of bacteria is treated by using a 3 parameter model.

3.5 Phase behavior

We have taken a simple approach, by assuming that the oleic phase consists of only the oil component and that the bacteria, nutrient and the produced metabolites remain in the aqueous phase. This is a good realistic assumption.

4 SIMULATOR DEVELOPMENT

The mathematical model developed for MEOR processes is solved numerically using a finite difference technique. The finite difference formulation of the species conservation equation and overall material balance equation (pressure equation), treatment of initial and boundary conditions and handling of different rate terms are discussed in detail by Sarkar (1990). The model problem has been solved for a one dimensional medium using a block centered type of grid. An IMPES (Implicit pressure and explicit saturation) technique is used to solve one pressure equation and (n_c-1) species conservation equations. A bandsolver has been used to solve the tri-diagonal pressure matrix. Two phases and ten components are currently considered in the model (Table 1). The simulator can be easily adapted for a different number of components and phases.

The simulator can be used for simulating injection of different sizes and compositions of slugs in the presence of one or more phases. Profiles (concentration as a function of distance) and histories (concentration as a function of time) can be obtained for any component. IFT, which is a function of surfactant concentration and N_c ,which is a function of IFT, flow rate and viscosity of a phase, are evaluated. Rate histories for different phases and oil recovery results can be obtained.

4.1 Numerical experiment

Numerical results are obtained by running the simulator for the following conditions.

(i) Initial condition. Core (1-D system) at waterflood residual oil saturation.

(ii) Operations. The following operations are simulated.

(a) Injection of bacterial slug (2 PV)

(b) Injection of nutrient (2 PV)

(c) Static incubation (63 hrs)

(d) Brine postflush (2 PV)

5 RESULTS & DISCUSSION

In this study the results from a single numerical experiment (simulation run) are presented. In this experiment sufficient PVs of bacteria and nutrient slugs were injected such that concentration of these components were uniform throughout the length of the medium. The results are shown in Figures 1 through 5. The first three figures show the results for the static incubation period, and next 2 figures show the results for the brine postflush period.

Fig. 1 shows growth of bacteria and consumption of nutrient as a function of time during static incubation inside the porous medium, in the presence of oil and water phases. From this result one can predict when the nutrient will be completely consumed. The bacteria concentration increases exponentially in the beginning, but as the inhibitor concentration builds up the rate of growth decreases.

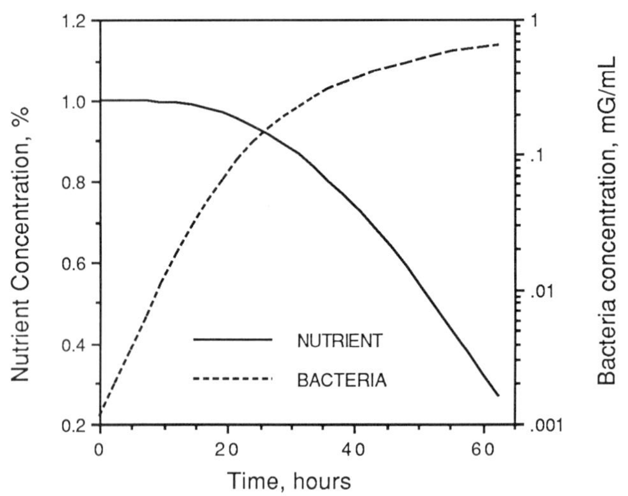

Fig. 1. Growth in porous medium during incubation.

Fig. 2 shows the change in metabolite concentration as the incubation continues. The final metabolite concentration of acids and alcohols match the results from fermentation studies (Goursaud, 1990).

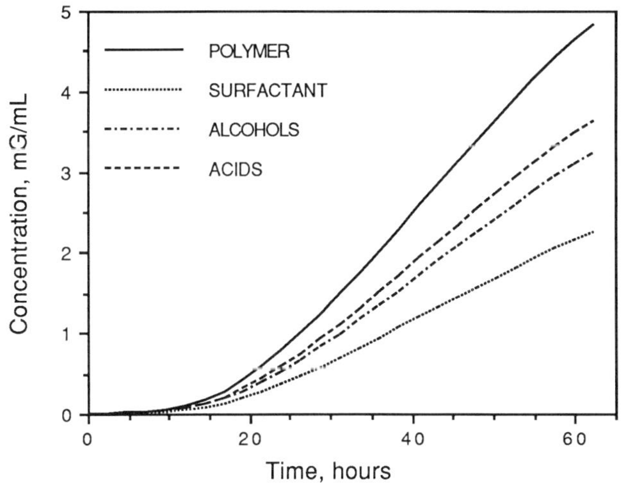

Fig. 2. Metabolite concentration during incubation.

Fig. 3 shows changes in IFT as the surfactant concentration increases during incubation. When the surfactant concentration reaches the CMC the IFT does not change.

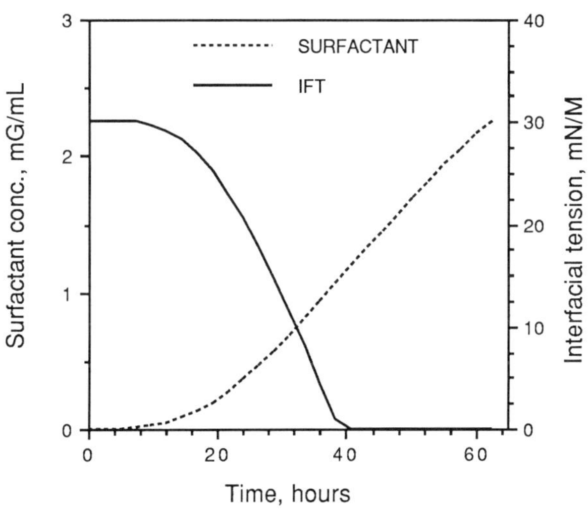

Fig. 3. Reduction of interfacial tension during incubation

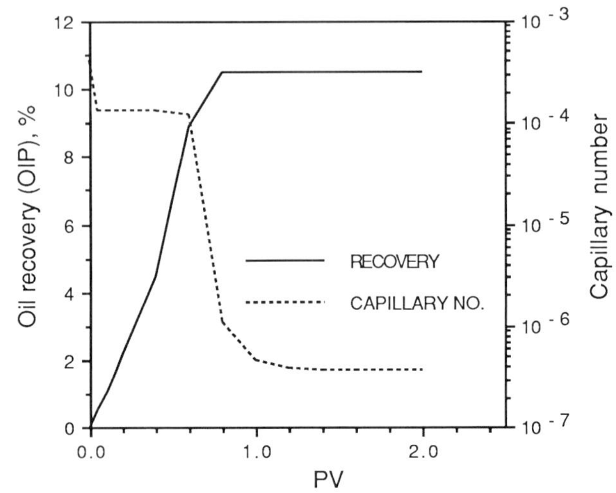

Fig. 4. Oil recovery during brine postflush.

It is possible to predict from this result the time when incubation can be stopped and displacement may be started to recover oil. Fig. 4 shows oil recovery during the brine postflush. It also shows the change in capillary number in the effluent block as the flood continues.

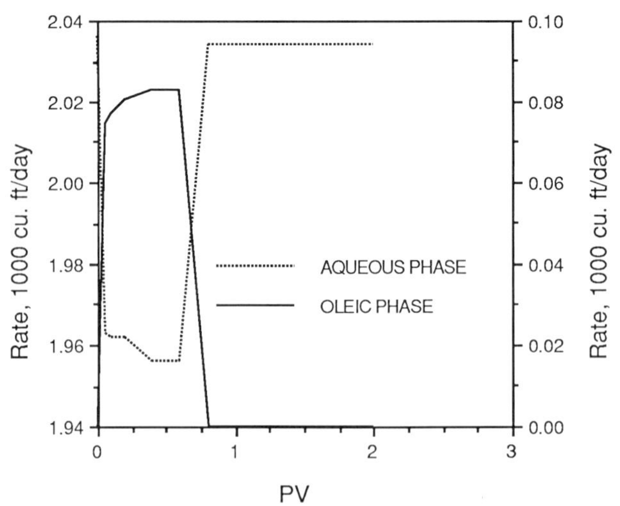

Fig. 5. Oleic and aqueous phase production rates.

As the capillary number decreases oil recovery continues to decline and finally stops below a certain critical capillary number value. The number of PVs needed to obtain the maximum oil recovery can be predicted. Fig. 5 shows oleic and aqueous phase production rates as a function of PV injected.

6 CONCLUSIONS

The following conclusions can be made from this study.

(1) Biosurfactants are potentially the most useful metabolic products for improving displacement efficiency to recover residual oil.

(2) Biomass and polymers may be helpful for improving volumetric sweep efficiency.

(3) A compositional numerical simulator has been developed.

(4) The simulator can be used to predict bacterial growth, nutrient consumption, metabolite production and oil recovery rates.

(5) It will be particularly useful in evaluating various injection strategies and in conducting sensitivity studies.

7 NOMENCLATURE

$\hat{C}_k$	Concentration of a component k adsorbed on the stationary phase.
C_{kl}	Concentration of a component k in phase l.
D'_{kl}	Coupled dispersion coefficient for a component k in phase l.
D_{kl}	Dispersion coefficient for a component k in phase l.
n_c	Number of components.
N_k	Net flux of component k(mass/(area.time)).
n_p	Number of phases.
Q_k	Rate of injection or production of component k.
R'_{kl}	Coupled specific rate constant for component k in phase l.
R_{kl}	Specific rate constant for component k in phase l.
S_l	Saturation of a phase l.
t	Time.
u_l	Darcy velocity(flux) of phase l.
V_b	Bulk volume.
W_k	Overall mass concentration(mass/bulk volume).

Subscript:

k	Index for a component.
k'	Index for the component which is coupled with k.
l	Index for a phase.

Greek Symbols

ø	Porosity of porous medium.
∇	Gradient operator.

TABLE 1

Dispersion coefficients and specific rate constants for MEOR components.

k	k'	Name	Dispersion Coefficient		Specific Rate constant	
			D_{kl}	D'_{kl}	R_{kl}	R'_{kl}
1	*	Water	0.0	0.0	0.0	0.0
2	*	Oil	0.0	0.0	0.0	0.0
3	4	Nutrient	D_{3l}	0.0	0.0	$-(A_N*\mu+B_N)$
4	3	Bacteria	$(D_B+D_T+D_F)_{4l}$	$-D_{C4l}$	$(\mu-k_d)$	0.0
5	4	Polymer	D_{5l}	0.0	0.0	$A_P*\mu+B_P$
6	4	Surfactant	D_{6l}	0.0	0.0	$A_S*\mu+B_S$
7	4	Inhibitor	D_{7l}	0.0	0.0	$A_I*\mu+B_I$
8	4	Spore	D_{8l}	0.0	0.0	0.0
9	4	Alcohols	D_{9l}	0.0	0.0	$A_{CS}*\mu+B_{CS}$
10	4	Acid	D_{10l}	0.0	0.0	$A_A*\mu+B_A$

k Index for a component

k' Index for a component with which k component is coupled

D_{kl} Dispersion Coefficient of component k in phase l

D'_{kl} Coupled dispersion Coefficient of component k in phase l

R_{kl} Specific rate constant of component k in phase l

R'_{kl} Specific rate constant of component k in phase l

μ, k_d Specific growth rate and death constant for bacteria

A, B Growth parameters

Subscripts: N, P, S, I, CS & A Nutrient, Polymer, Surfactant, Inhibitor, Cosurfactant and Acid.

8 REFERENCES

Aiba, S., Shoda, M. and Nagatani, M., 1968. Kinetics of product inhibition. Biotechnology and Bioengineering. 10:845-864.

Bailey, J. E. and Ollis, D. F., 1986. Biochemical Engineering Fundamentals. 2nd Edition, McGraw Hill, New York.

Bhuyan, D., 1989. Development of an Alkaline-Surfactant-Polymer Compositional Reservoir Simulator. Ph. D. Dissertation, Dept. of Petroleum Engineering, U. of Texas, Austin (December).

Brown, M. J., Moses, V., Robinson, J. P., Springham, D. G., 1986. Microbial Enhanced Oil Recovery:Progress and Prospects. CRC Critical Reviews in Biotechnology, 3 : 2, p.159.

Dake, L. P., 1978. Fundamentals of Reservoir Engineering. Elsivier Scientific Publishing Company , New York.

Gottschalk, G., 1983. Bacterial Metabolism. 2nd Edition, Springer-Verlag, New York.

Goursaud, J. C.1989. Production of a Biosurfactant by Bacillus Licheniformis JF-2. M. S. Thesis, Dept. of Chemical Engr., University of Texas, Austin, (December).

Grist, D. M., 1983. Microbial Enhancement of Oil Recovery-An Operator's View. Proceedings of International Conference on Applications of Biotechnologies, p. 463.

Holm, L. W.,.1982. CO_2 Flooding-its time has come. J. of Petroleum Technology, 34:2739.

Kianipey, S. A., Donaldson., E. C., 1986. Mechanisms of Oil Displacement by Microorganisms. paper SPE 15601, presented at the 61 st Annual Technical Conference and Exhibition of the Society of Petroleum Engineers, New Orleans, LA, October 5-8.

Knapp, R. M., McInerney., M. J. and Menzie, D. E., 1987. Microbial Strains and Products for Mobility Control and Oil Displacement. DOE/BC/10300-45.

Kyte, J. R., Stanclift,jr., R. J. , Stephan, jr, S. C. and Rapoport, L. A., 1989. Mechanism of Water Flooding in the Presence of Free Gas. Petroleum Transactions, AIME, 207:215-221.

Lake, L. W., 1989. Enhanced Oil Recovery. Prentice Hall, Englewood Cliffs, New Jersey.

McCain Jr, W. D., 1973. The Properties of Petroleum Fluids. Petroleum Publishing Company , Tulsa.

Sarkar, A. K., 1990. Simulation of Microbial Enhanced Oil Recovery Processes. Ph. D. Dissertation, Dept. of Petroleum Engineering, U. of Texas, Austin (December).

Williams, B. B., Gidley, J. L. and Schechter, R. S., 1979. Acidizing Fundamentals. SPE Monograph, Vol. 6.

FIELD APPLICATIONS

Field applications begin appropriately with work presented by Dr. Lazar who is perhaps the most experienced in field application of MEOR. Dr. Lazar discusses applications in carbonate reservoirs, so his work was followed immediately by Dr. Wagner's presentation of MEOR from carbonate reservoirs with very complex geochemical characteristics.

Papers F-4 to F-6 are applications that involve waterflooding type processes and therefore were grouped together. The next two papers address the control of hydrogen sulfide production and paraffin deposition which are both very severe problems for the industry.

The final paper is a discussion of the application of biopolymers as an MEOR process in the Republic of China.

Reprinted from: *Microbial Enhancement of Oil Recovery - Recent Advances*, edited by E.C. Donaldson

CH. F–1

CHARACTERISTICS OF THE BACTERIAL INOCULUM USED IN SOME RECENT MEOR FIELD TRIALS IN ROMANIA

I. LAZAR, S. DOBROTA, M. STEFANESCU AND V. VELEHORSCHI
Institute of Biological Sciences, Spl. Independentei 296, Bucharest, Romania

ABSTRACT

In the framework of the so called "second generation" of MEOR field trials started in Romania after 1985, several improvements of the method used in the "First generation" of MEOR field trials were made. One such improvement refers to the bacterial inoculum which has been better characterized for its ability to release residual oil from porous media.

The inoculum was represented by adapted mixed enrichment cultures (AMES), selected for the four reservoirs planned to be treated between 1987-1990 by MEOR.

It was found that the most active in gas, acid, solvents and biosurfactants production (as well as in residual oil recovery from porous media) is the AMEC inoculum as compared to pure culture or mixed cultures inoculums.

The bacterial inoculum injected into the formation was the adapted mixed enrichment cultures which are predominated by *Clostridium*, *Bacillus* and some gram negative rods.

INTRODUCTION

For the success of the trial experiments made to stimulate and enhance oil recovery, the bacterial inoculum plays an important part. Within more than 30 important field trial experiments made in several countries in the last 30-35 years, different kinds of bacterial inoculum have been used in hundreds of wells. In many cases, pure cultures were used, in other cases, mixed cultures of bacterial populations. In all of these cases, the bacteria were selected to grow at the reservoir conditions where they are going to be injected.

In some cases only a stimulation of the natural microflora present in the formation water of the reservoir (or in the water injected into the respective reservoir) took place. In a recent paper (Lazar, 1990) the main types of bacterial inoculum used in different countries are presented. Almost 30 MEOR field trial experiments were conducted in these countries. The types of inoculum are presented in Table 1, grouped according to the way the wells were treated for stimulation and enhanced oil recovery.

According to Table 1, the "first generation" field trials in Romania (1975-1980) as well as the experiments of the "second generation" initiated

Table 1 Types of inoculum used in MEOR field trials (after Lazar, 1990)

No.	Country	References	Type of inoculum
		I. Cyclic Microbial Recovery (Single well stimulation method)	
1	USA	Johnson, 1979	Mixed cultures of Bacillus and Clostridium
2	USA	Microbial Systems Corp. 1984 (After Hitzman 1988)	Mixed anaerobic microbial cultures
3	USA	Grula et al., 1985	Clostridium sp.
4	USA	Covington Leases, 1985	Spore suspension of Clostridium sp.
5	USA	Zajic, 1987	Anaerobic and facultative anaerobic bacteria high fermenting sucrose-molasses medium
6	USA	Bryant et al., 1988	Mixed cultures of Clostridium sp., Bacillus sp., B.licheniformis and a Gram-negative rod
7	USA	King, 1989	Culture of Clostridium type
8	Romania	Lazar 1987, Lazar and Constantinescu, 1985, Lazar et al., 1988, 1989, 1990	Adapted mixed enrichment cultures predominated by Clostridium, Bacillus and Gram-negative rods
		II. Microbial Flooding Recovery	
9	Czechoslovakia	Dostalek and Spurny, 1958, 1961	Mixed cultures of sulfat reducing bacteria and Pseudomonas sp. hydrocarbon utilizing
10	Hungary	Jaranyi, 1968	Mixed sewage-sludge cultures
		Dienes and Jaranyi, 1973	Anaerobic thermophilic mixed cultures (predominants: Clostridium, Desulfovibric and Pseudomonas

No.	Country	References	Type of inoculum
11	Poland	Karaskiewicz, 1974	Mixed aerobic and anaerobic bacteria belonging to genera: *Arthrobacter*, *Clostridium*, *Mycobacterium*, *Pseudomonas*, and *Peptococcus*
12	USSR	Senyukov et al., 1970	Mixed aerobic and anaerobic bacteria
13	USA	Petrogen Inc., 1977-1982 (After Hitzman 1983)	unknown
14	USA	Yarbrough and Coty, 1983	*Clostridium acetobutylicum*
15	Romania	Lazar, 1987; Lazar and Constantinescu, 1985; Lazar et al., 1988, 1989 1990	Adapted mixed enrichment cultures predominated by *Clostridium*, *Bacillus* and Gram-negative rods
16	DDR	Wagner and Teumer, 1987	Mixed cultures of thermophilic *Bacillus* and *Clostridium*
17	USA	Bryant et al., 1988	Mixed cultures of *Clostridium* sp., *Bacillus* sp., *B.licheniformis* and a Gram-negative rod
18	USA	Donaldson, 1988	Mixed cultures of *Pseudomonas* and hydrocarbon oxidizing bacteria
19	China	Wang Xiu-Yuan, 1988	Gases, acids and solvents producing bacteria
20	USA	Davidson and Russell, 1988	Special adapted strain of *Clostridium*
21	USA	Oppenheimer and Heibert, 1988	Mixed cultures of hydrocarbon degrading bacteria
	III. Microbial Flooding by Stimulating the Activity of Natural Microflora		
22	Czechoslovakia	Dostalek and Spurny, 1958; Dostalek, 1961	Natural bacterial flora of reservoir brine
23	USSR	Ivanov et al., 1987	Natural bacterial flora from injection water and reservoir brine

No.	Country	References	Type of inoculum
		IV. Selective Plugging Recovery	
24	The Netherlands	Von Heininger et al., 1958 (After Hitzman, 1988)	*Betacocus dextranicus* (Slime-forming bacteria)
25	USSR	Yulbarisov, 1972, 1976	Anaerobic and aerobic bacteria as: sulfat reducing dentrifying, putrefactive and acid butiric fermenting, cellulose digesting
26	USSR	Yulbarisov, 1989	Aerobic and anaerobic activated sludge bacteria
27	Canada	Jack and Stehmeier, 1988	Biopolymer producing bacteria (*Leuconostoc mesenteroides*)
28	China	Wang Xiu-Yuan, 1988	Slime-forming bacteria
29	USA	Berry Energy and Petrol. Bio-Resources, 1983-84 (After Hitzman, 1988)	Surfactant cosurfactant gas producing cultures; Polymer-polysaccharide gas producing cultures
		V. Microbial Fracturing Fluids Recovery	
30	USA	Pure Oil Co., 1957 (After Hitzman, 1983)	*Desulfovibrio hydrocarbonolastus*
31	USA	Micro-Bac-Gulf Coast Inc., 1985-86 (After Hitzman, 1988)	A liquid culture of mixed marine source microorganisms

after 1986, the bacterial inoculum used was the adapted mixed enrichment cultures (AMEC).

The details of the isolation, selection, structure and characteristics of the bacterial inoculum used in Romania in the field trials of the "first generation" (1975-1980) are presented in several review papers (Lazar, 1983, 1987; Lazar and Constantinescu, 1985).

After 1985, in Romania, the activities were intensified, regarding the initiation of the "second generation" of MEOR field trials.

Prior to initiation of these experiments a great deal of attention was devoted to the bacterial inoculum, which had to be used, at least for the first four fields. The fields were chosen for initiation of MEOR tests after 1986-1987.

In some recent papers and patents some of the results of the isolation and selection of the bacterial inoculum by a more rapid and efficient method as well as its performances in the release of the residual oil are presented, i.e. acid, gas, solvents, biosurfactants and biopolymers (Lazar, 1989, 1990; Lazar et al., 1986, 1988; Dobrota et al., 1990; Stefanescu et al., 1990).

At the moment we have already prepared the bacterial inoculum to be used in experiments of cyclic microbial recovery and microbial flooding recovery for four reservoirs. The activity in one of the reservoirs (Sa_3Cl_1) was started in 1986-1987 and is the subject of another paper (Lazar et al., 1990); and at the second reservoir (Sa_3Br_1), the activity was started in 1989.

In this paper we present some of the bacterial inoculum characteristics selected and prepared for the field trials for the four reservoirs we have referred to above.

MATERIAL AND METHOD

Insolation and Selection of the Adapted Mixed Enrichment Cultures

The method of insolation and selection of the adapted mixed enrichment cultures (AMEC) used for the "first generation" (1975-1980) of field trials in Romania is largely presented in papers already published and in some recent papers (Lazar, 1983; Lazar and Constantinescu, 1985; Lazar et al., 1982). After 1980, together with the preparation of the "second generation" of field trial experiments in Romania, there also was initiated, both: (1) research for the simplification of the isolation and a selection method for the bacteria destined to be injected into formations; and (2) better characterization of the selected bacteria, especially for the release of the residual oil from porous media.

TABLE 1a Some geological characteristics of the reservoirs for which adapted mixed enrichment cultures (AMEC) were obtained

Reservoir	Mineralogical structure	T^oC	NaCL g/1	Perme-ability	Porosity %	$CaCO_3$ content %	Depth pay zone m	Co-produced water	Crude oil viscosity	Saturation in crude oil %
Cl_1	Calcareous sandstones with microfissures and microcaves	47	4-6	245-248	25.5	45	750/5	95	26	75
Br_1	Sands and calcareous sandstones dirty porous calcareous with microfissures	36	3-6	100-1000	36	60-80	780/8	90	9	50
Bl	Limestones and calcareous sandstones with microfissures	45	4-6	100-500	25	55	700/14	96	57	65
Gr	Sands and calcareous sandstones dirty calcareous with microfissures	51	6-13	400-1300	30	70	1200/10	86	6	45

In Fig. 1, the main steps of the methodology of selecting the adapted mixed enrichment cultures to the conditions from the chosen reservoir for the microbiological treatment are presented.

Following the steps (presented in Fig. 1) after 1986 the isolation and selection of adapted mixed enrichment cultures for the four reservoirs, were carried out. The main characteristics of the reservoirs are presented in Table 1.a.

Plans for these reservoirs, during 1987-1990, were to use both microbial cyclic and microbial flooding systems (using as inoculum, isolated and selected bacteria, according to the methodology in Fig. 1).

Further details, concerning the methodology in Fig. 1, are presented in recent papers (Lazar, 1987; Lazar et al., 1986, 1988, 1990).

<u>The Characterization of the Selected Bacteria, the Nutrient Support and Sandpacks Column Tests</u>

Using the isolation and selection methodology of the bacteria presented in Fig. 1, we obtained, not only pure cultures, but also mixtures of pure cultures or mixed enrichment cultures adapted to the conditions in the oil reservoirs. These types of bacteria were characterized for the production of gas, acid, solvents, biosurfactants and biopolymers and also for the ability of release residual oil, from porous media or of from aerobic and anaerobic degradation of the crude oil.

The nutrient supports used for testing the importance of the metabolic products in releasing residual oil and for aerobic and anaerobic degradation of the oil, were as follows:

Media M_1 = brine + molasses 2-4% + oil 1-2% (Lazar et al., 1988);

M_2 = Brine + sucrose 3% + molasses 0.5% + K_2HPO_4 0.2% : NH_4NO_3 0.4% + yeast extract 0.05% (Zajic, 1987);

M_3 = Brine + molasses 4%, urea 0.5% + NH_4NO_3 0.5% (King, 1989).

These three media were used to test the capacity of different types of bacterial inoculum to produce substances which play a part in releasing the residual oil. The following media were used for testing the capacity of aerobic and anaerobic degradation of the oil.

M_4 = Molasses 40 g + brine 100 ml + scum from sugar processing plants;

M_5 = KCl 0.5 g + $MgSO_4$ 0.5 g + K_2HPO_4 1.0 g + $FeSO_4$ 0.01 g + $(NH_4)_2SO_4$ 2 g + $CaCO_3$ 3 g + distilled water 1000 ml.

Figure 2 presents the steps of the testing protocol for the collector models represented by sandpack columns and of the ability of the isolated bacteria that were adapted to the conditions of the four oil reservoirs

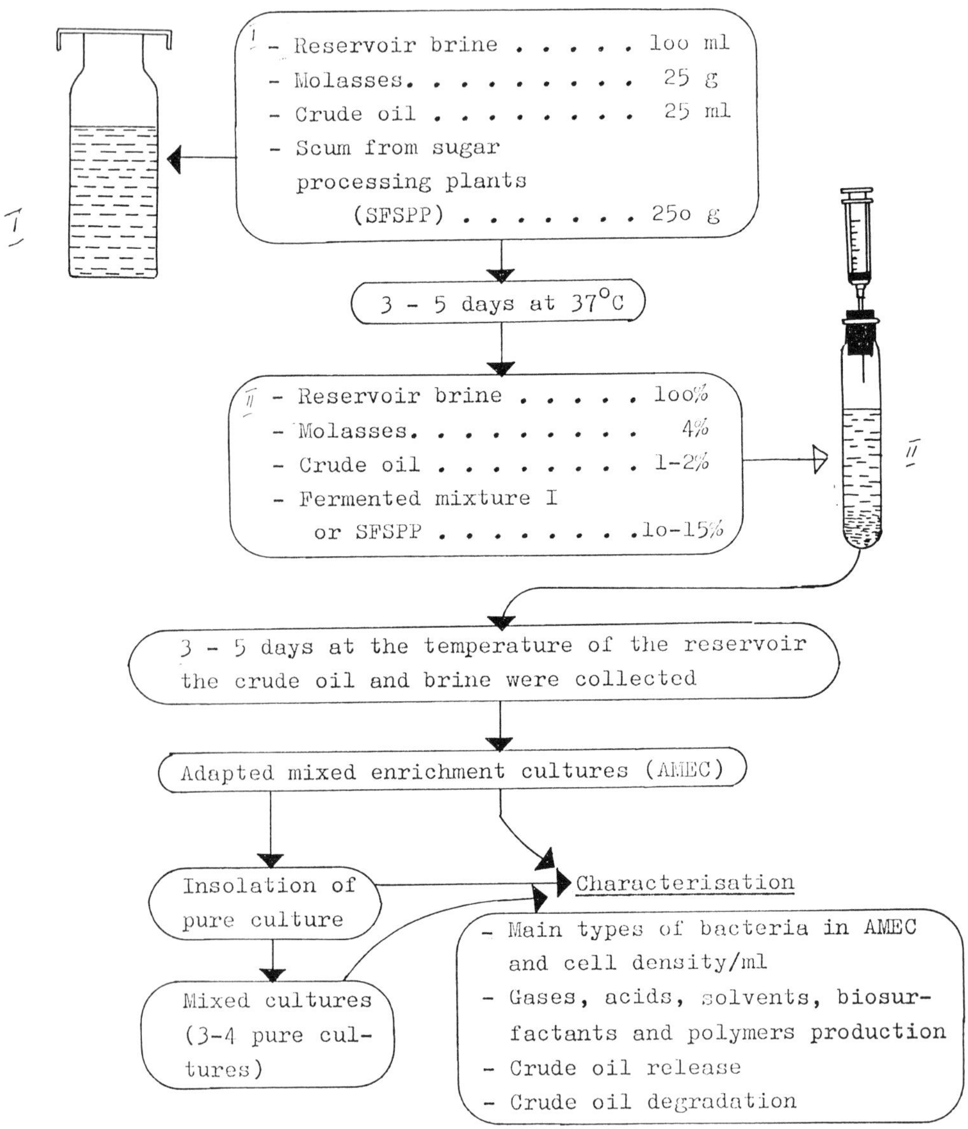

Fig. 1. Selection of the bacteria adapted to oil reservoir conditions.

(Sa_3CL_1); Sa_3Br_1; Sa_3B_1 and Sa_3Gr) to release the residual oil from the porous media.

More details concerning the testing methodology are presented in: Lazar (1980, 1990), Lazar et al., (1986, 1988), Dobrota et al., (1990), Stefanescu et al., (1990).

The Multiplication of the Bacterial Inoculum in the Necessary Quantities for the Field Trial Experiments

After selecting the most active bacterial population (AMEC) the populations that showed the best performance for production of acid, gas, solvents, biosurfactants and biopolymers and with the highest efficiency of recovery of residual oil from a porous medium were selected for use.

These populations were grown in quantities necessary for the MEOR field trials (Lazar et al., 1990).

RESULTS AND DISCUSSIONS

The Performances fo the Bacterial Inoculum in Producing Gases, Acids, Solvents, Biosurfactants and Biopolymers

After going through the stages presented in Fig. 1, bacterial populations were obtained as adapted to the conditions of the four reservoirs chosen for MEOR field trials. The bacterial populations proved to be of great interest for *in situ* processes for stimulating and enhancing the oil recovery.

Within 6-10 days, we could usually obtain "adapted" bacteria. they were characterized afterwards, with respect to concentration (cells/ml), the main types (species or genus level), production of gases, acids, solvents, biosurfactants, biopolymers, and also with respect to the efficiency of residual oil recovery from a porous medium and for the capacity for crude oil utilization.

Within the experiments that have already begun and the future experiments (this year 1990) the adapted mixed enrichment cultures are used (Lazar, et al, 1990). The procedure (Fig. 1) proved to be easy and efficient for selecting adapted mixed enrichment cultures with high efficiency, compared to the pure cultures or the mixtures of pure cultures. Besides, the use of bacteria isolated from the two ecological sources (formation water and the fermented scum from sugar processing plants) proved to be excellent. In this way, we obtained bacteria from the medium in which they will be afterwards injected and from a source where the molasses comes first. Both sources proved to be rich in mesophilic bacteria with a high ability to metabolize the molasses at the oil reservoir conditions.

Table 2 The performances in gases, acids, solvents, biosurfactants and biopolymers production of the bacteria adapted to the conditions of Sa_3Cl_1 and Sa_3Br_1 reservoirs

Type of Bacteria	Reservoir Cl_1					Reservoir Br_1				
	Gases[d]	Acids[e]	Solv.[f]	Bios.[g]	Biop.[h]	Gases[d]	Acids[e]	Solv.[f]	Bios.[g]	Biop.[h]
Pure culture[a] (PC)	0-48	6.7-4.2	0 - +	++-++++	1.018	0-80	7.2-4.0	0 - +	++-+++	0.850
Mixed cultures[b] (MC)	22-56	6.7-4.8	+	++++	0.348	37-92	7.2-4.0	+	++++	0.367
Adapted mixed[c] enrichment cultures (AMEC)	115-160	7.2-4.8	+	++++	0.393	105-147	7.2-4.0	+	++++	0.363

[a]Pseudomonas, Clostridium, Bacillus, Arthrobacter, Mycobacterium, Brevibacterium, Alcaligenes
[b]Clostridium-Bacillus-Pseudomonas; Clostridium-Arthrobacter-Pseudomonas etc.
[c]Bacterial populations as resulted after adaptation process (see Fig. 1)
[d]cc from 15 ml medium (brine + molasses 4% + crude oil 1-2% + bacteria 5-10%) in the first 5 days
[e]pH initial and after 5 days
[f]nigrosin test + - ++++ the intensity of reaction
[g]kerosen test
[h]solvent precipitation test (g%) - growth of bacteria on medium: molasses 50 g, $NaNO_3$ 2g, KH_2PO_4 0.6 g, $MgSO_4$ 0.25 g, $CaCl_2$ 0.12 g, yeast extract 0.06 g, crude oil 100 ml, ethoxilated nonylphenol 6-0 20 ml, tap water 1000 ml

Table 3 The performances in gases, acids, solvents, biosurfactants and biopolymers production of the bacteria adapted to the conditions of Sa_3Bl_1 and Sa_3Gr reservoirs

Type of Bacteria	Reservoir SaBl					Reservoir Sa_3Gr				
	Gases[d]	Acids[e]	Solv.[f]	Bios.[g]	Biop.[h]	Gases[d]	Acids[e]	Solv.[f]	Bios.[g]	Biop.[h]
Pure culture[a] (PC)	0-60	6.7-4.0	0 - +	++-++++	0.725	0-45	7.0-4.5	0 - +	++-++++	0.630
Mixed cultures[b] (MC)	32-87	6.7-5.0	+	++++	0.382	32-65	7.0-5.0	+	++++	0.363
Adapted mixed[c] enrichment cultures (AMEC)	90-145	7.2-4.5	+	++++	0.360	80-112	7.4-4.8	+	++++	0.266

[a]Pseudomonas, Clostridium, Bacillus, Arthrobacter, Mycobacterium, Brevibacterium, Alcaligenes
[b]Clostridium-Bacillus-Pseudomonas; Clostridium-Arthrobacter-Pseudomonas etc.
[c]Bacterial populations as resulted after adaptation process (see Fig. 1)
[d]cc from 15 ml medium (brine + molasses 4% + crude oil 1-2% + bacteria 5-10%) in the first 5 days
[e]pH initial and after 5 days
[f]nigrosin test + - ++++ the intensity of reaction
[g]kerosen test
[h]solvent precipitation test (g%) - growth of bacteria on medium: molasses 50 g, $NaNO_3$ 2g, KH_2PO_4 0.6 g, $MgSO_4$ 0.25 g, $CaCl_2$ 0.12 g, yeast extract 0.06 g, crude oil 100 ml, ethoxilated nonylphenol 6-0 20 ml, tap water 1000 ml

The performances of the isolated and selected bacteria for the four reservoirs are presented in Tables 2 and 3. The protocol shown in Fig. 1 was used.

The values of the production in gases, acids, solvents, biosurfactants and biopolymers for Sa_3Cl_1 and Sa_3Br_1 reservoirs are present in Table 2 and for Sa_3B_1 and Sa_3Gr reservoirs in Table 3. After the analysis of these results it is clear that the adapted mixed enrichment cultures (AMEC) reveal clearly high performances compared with the inoculum, represented by pure cultures or mixtures of pure cultures.

From these results we decided to inject the four reservoirs with the AMEC whose superior performances compared to those of the pure or mixed cultures are noted for the production of substances and also in releasing residual oil.

The Performances of the Bacterial Inoculum in Releasing the Residual Oil

In Fig. 2 are presented the protocols of the testing of sandpack columns and the ability of the selected bacteria to release the residual oil from porous media.

In Tables 4 and 5 are presented the values of the residual oil recovery efficiency, for each reservoir and according to the method used for inoculum preparation (Table 4).

After analyzing the results presented in these two Tables, the following important conclusions are evident: of the four systems used to study the bacterial inoculum (Table 4) selected according to the protocols in Fig. 1, the 4th system (that is the one represented by containers under flowing pressure) exhibited the best efficiency of residual oil recovery. For this reason they were selected for the field trial experiments (Lazar et al., 1990).

Similar results were obtained, as well, in the case of mixed populations (point 5, Table 4) but practically this kind of bacterial inoculum is more difficult to produce in the field. Besides, this mixture has the disadvantage that is more active for the utilization of the oil under both aerobic and anaerobic conditions. We would like to stress that the rates of oil utilization which were sometimes relevant were carried out in excellent lab conditions, while in field trials, in situ, it is hard to achieve the ideal conditions of a laboratory.

The adapted mixed enrichment cultures (AMEC) are at least ten times more active than the pure cultures, for release of residual oil and up to ten times more active than the pure culture mixture, Table 5. Also, Table 5 notes that for oil with viscosities between 6-57 cp and densities between 880-922 Kg/dm^3,

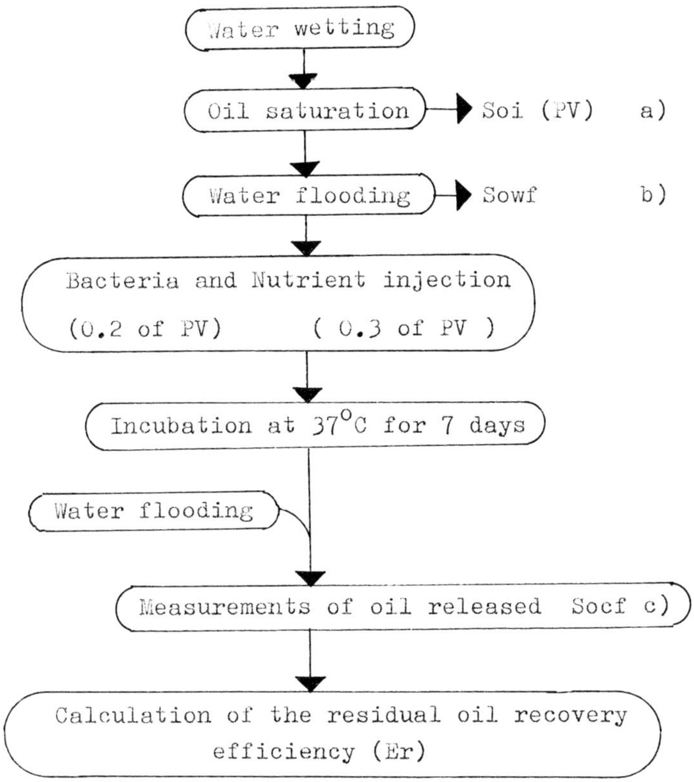

a) residual oil saturation prior to water flood
b) residual oil saturation after water flood
c) residual oil saturation after microbial treatment

Fig.2. The protocol of testing on sandpack columns of the performances for residual oil recovery efficiency of the bacteria adapted to oil reservoir conditions.

Table 4 The performances in residual oil recovery and oil degradation of the adapted mixed enrichment cultures (AMEC) cultivated on several adaption systems

The system for bacteria adaptation[a]	Residual oil recovery efficiency %[b]	Crude oil degradation %[b]	
		aerobic	anaerobic
1. Aerobic (Erlenmeyers flask on rotary shakers)	15.5 - 25.8	37.6	5.6
2. Facultative anaerobic (open vessels covered with a lid)	15.5 - 29.0	26.8	7.9
3. Anaerobic 1 (anaerobic containers under up to 5 atm	13.3 - 27.0	25.0	16.7
4. Anaerobic 2 (anaerobic containers under pressure flow	18.2 - 46.2	30.0	11.5
5. Mixture 1 - 4	23.5 - 40.2	43.3	21.0

[a] depending of the system the cells density/ml from 5×10^9 to 30×10^9
[b] the results are variable due to the reservoir understudy (Sa_3Cl_1, Sa_3Br_1, Sa_3Bl, Sa_3Gr)

Table 5 The performances in residual oil recovery on sandpack columns of the bacteria (AMEC) adapted to Sa_3Cl_1, Sa_3Br_1, Sa_3Bl and Sa_3Gr reservoirs

Crude oil of the reservoir	Crude oil viscosity (cp)	Residual oil recovery efficiency % up to:		
	Density (Kg/dm^3)	Pure cultures	Mixed cultures	AMEC[b]
Sa_3Cl_1	26/0.922	2.6	3.4	45.2
Sa_3Br_1	9/0.880	3.9	9.9	40.2
Sa_3Bl	57/0.918	7.8	18.0	46.2
Sa_3Gr	6/0.890	2.4	3.8	40.4

[a]type A_3 - nonparaffinic

[b]adapted mixed enrichment cultures

no differences can be discerned, concerning the residual oil recovery efficiency from a porous medium, represented sandpack columns.

The results in Tables 4 and 5 led to the decision to use only the adapted mixed enrichment cultures (AMEC) for the field trial experiments in Romania.

CONCLUSIONS

The Romanian experience, during the "first generation" MEOR field trial experiments (1975-1980), in preparing the bacterial inoculum for well injection, was the basis for our research within the "second generation" (after 1985-1986). According to all of these, and simplification of the protocol for the bacterial isolation and selection for the MEOR field trial experiments. We obtained adapted mixed enrichment cultures, which compared to the pure cultures and pure culture mixtures, could yield superior qualities in producing substances with a part in residual oil releasing from the porous media.

The MEOR field trial experiments, initiated in Romania, after 1986-1987 and those that are to be initiated in 1990-1991, will use the adapted mixed enrichment cultures.

REFERENCES

Bryant, S.R., Burchfield, E.T., Dennis, M.D. and Hitzman, O.D., 1988. Microbial Enhanced Waterflooding Mink Unit. Project. SIE 17341, SPE/DOE Enhanced Oil Recovery Symp. held in Tulsa, Oklahoma, April 17-20.

Davidson, S.W. and Russell, H.H., 1988. A MEOR Pilot Test in the Loco Filed. In: Proceedings of the Symp. on Application fo Microorganisms to Petroleum Technology, Bartlesville, Oklahoma, August 12-13, 1987. Eds. Thomas E. Burchfield, Rebecca S., Bartlesville, OK, pp. VII1-VII12.

Dienes, M. and Jaranyi, I., 1973. Increase of Oil Recovery by Introducing Anaerobic Bacteria into the Formation, Demjen Field, Hungary Koolaj as Foldgax, v. 106, No. 7, pp. 205-208.

Dobrota, S., Lazar, I., Stefanescu, M. and Sandulescu, L., 1990 (under press). The influences of several types of bacterial inoculum and nutrient supports on production of gases, acids, solvents and biosurfactants involved in residual oil release. In: Proceedings of the VII-th Symposium on Industrial Microbiology and Biotechnology, Sept. 9-10, 1989, Galati, Romania.

Donaldson, E.C., 1988. Personal communication.

Dostalek, M. and Spurny, M., 1958. Bacterial release of oil. I Preliminary Trials in an Oil Deposit. Fol. Biol. (Praha) 4, 166 p.

Dastalek, M., 1961. Bacterial release of oil. III Areal Distribution of the Effect of Nutrient injection into the Deposit. Folia Microbiologica 6, pp. 10-16.

Grula, A.E., Russell, H.H. and Grula, M.M., 1985. Field trials in central Oklahoma using Clostridium strains of Microbially Enhanced Oil Recovery. In: Microbes and Oil Recovery (vol. 1), Int. Biores. Journal, Ed. by Zajic and Donaldson, El Paso, TX, pp. 144-150.

Hitzman, O.D., 1983. Petroleum Microbiology and the History of Its Role in Enhanced Oil Recovery. In: Proceedings fo Int. Conf. on Microbial Enhancement of Oil Recovery, Shangri-La, Afton, OK, May 16-21, 1982, pp. 162-198.

Hitzman, O.D., 1988. Review of Microbial Enhanced Oil Recovery Field Trials. In: Proceedings of the Symp. on Application of Microorganisms to Petroleum Technology, Bartlesville, OK, August 12-13, 1987, Ed. Thomas, E. Burchfield and Rebecca S. Bryant, Bartlesville Project Office, U.S. Dept. of Energy, Bartlesville, OK, pp. VI1-VI42.

Ivanov, M.V., Beleayev, S.S. and Rozanova, F.E., 1987-1988. Personal communication.

Jack, R.T. and Stehmeier, G.L., 1988. Selective plugging in watered out Oil Reservoirs. In: Proceedings of the Symp. on Application of Microorganisms to Petroleum Technology, Bartlesville, OK, August 12-13, 1987, Eds. Thomas E. Burchfield and Rebecca S. Bryant, Bartlesville Project Office, U.S. DOE, OK, pp. VII1-VII13.

Jaranyi, I., 1968. Beszamolo a nagylengyel terzegeben elvegzett Koolaj mikrobiologiai Kiserletkrol. M. All. Foldtani Intezet Evi Jelentese A., 1968, Evrol, pp. 423-426.

Johnson, A.C., 1979. Microbial Oil Release Technique for Enhanced Recovery Conf. on Microbial Processes useful in Enhanced Oil Recovery, U.S. DOE, San Diego, CA, Aug. 29-Sept. 1, 1979, pp. 30-34.

Karaskiewicz, J., 1974. Zastosovanie metod mikrobiologenneck w intensyfikacji eksploatacji Karpackick Zloz rapy naftowej; Widawnictwo "SLACK", Katowice, pp. 1-67.

King, J., 1989. Personal communication.

Lazar, I., 1980. The use of crude oil by the bacteria from inoculum used in MEOR field trials. In: Proceedings of the 2nd Symposium on Industrial Microbiology May 25-26, 1979, Iasi, Romania, pp. 435-438.

Lazar, I., 1983a. Some Characteristics of the Bacterial Inoculum Used for Oil Release from Reservoirs. In: Microbial Enhanced Oil Recovery, Ed. by Zajic, Cooper, Jack and Kosaric, Tulsa, Oklahoma, 1983, pp. 73-83.

Lazar, I., 1987. Research on the Microbiology of MEOR in Romania. In: Proceedings of the First Intern MEOR Workshop, Ed. by Jeannette King and Debra Stevens, Bartlesville Project Office. U.S. DOE, OK, pp. 124-153.

Lazar, I., 1990. An overview on MEOR field trials. In: Proceedings of the Symposium "Biotechnology for Energy" 16-21 Dec. 1989, Faisalabad, Pakistan, Ed. K. Malik and A. Khalid.

Lazar, I. and Constantinescu, P., 1985. Field Trials Results of Microbial Enhanced Oil Recovery. Microbes and Oil Recovery, vol. 1, International Bioresources Journal, Ed. by Zajic and Donaldson, 1985, pp. 122-144.

Lazar, I., Dobrota, S., Grigoriu, A., Velehorschi, V., 1986. The selection of high gases, acids, biosurfactants and biopolymers producing bacteria for MEOR applications. In: Proceedings of the V-th Symposium on Industrial Microbiology and Biotechnology, Oct. 25-26, 1985, Iasi, Romania, pp. 743-770.

Lazar, I., Dobrota, S., Stefanescu, M., Sandulescu, L., Constantinescu, P., Morosanu, C., Botea, N. and Iliescu, O., 1990. Preliminary results of some recent MEOR field trials in Romania. In the Proceedings of the MEOR International Conference, May 27-June 1, 1990, Norman, OK.

Lazar, I., Grigoriu, A., Mihoc, A., Zamfirescu, I., Dumitru, L. and Trifu, L., 1982. Field trials concerning the use of bacteria for Enhanced Oil Recovery. I. Obtaining of bacterial inoculum used for wells injection. In: Proceedings of the 3rd Symposium on Industrial Microbiology, Bucharest, 11-13 June 1981, pp. 513-518.

Lazar, I., Velehorschi, V., Dobrota, S. and Stefanescu, M., 1988. Characteristics of bacterial inoculum applied on oil reservoir in field experiments. In: Proceedings of the 6th Ind. Micro. and Biotech. Symp., Iasi, 10-12 Sept. 1987, pp. 801-806.

Oppenheimer, H.D. and Heibert, K.F., 1988. Microbial Enhanced Oil Production Field Tests in Texas. In: Proceedings of the Symp. on Application of Microorganisms to Petroleum Technology, Bartlesville, OK, August 12-13, 1987, Eds. Thomas E. Burchfield and REbecca S. Bryant, Bartlesville Project Office, U.S. DOE, Bartlesville, OK, pp. XII1-XII15.

Senyukov, W.M., Yulbarisov, E.M., Taldykina, N.N., Shishenina, E.P., 1970. Mikrobiologiya, 39, pp. 612-616. Translated from Microbiologiay 39, No. 4, pp. 705-710, July-Aug. 1970.

Wagner, M. and Teumerm, P., 1987. Personal communication.

Wang, Xiu Yuan, 1988. Personal communication.

Yarbrough, F.H. and Coty, F.V., 1983. Microbially Enhanced Oil Recovery from the upper cretaceous Nacatoch formation, Union County, Arkansas. In: Proceedings of 1982 Int. Conf. on Microbial Enhancement of Oil Recovery, Ed. by Donaldson and Zajic Published by Bill Linvill, Chief, Technology Transfer Branch, Bartlesville, OK, pp. 149-153.

Yulbarisov, E.M., 1972. Results from Analysis of Petroleum Gas on Introduction fo Biochemical Processes in the Oil Formation, Gazovoe Delo, No. 4, pp. 26-28.

Yulbarisov, E.M., 1976. Evaluation of the effectiveness of the biological method for enhancing oil recovery of a reservoir Neftyanoye Khozyaciystvo, No. 11, pp. 27-30.

Yulbarisov, E.M., 1989. Microbiological method for EOR. In: Proceedings of the Fifth European Symp. on Improved Oil Recovery, Budapest, 25-27 April 1989, pp. 695-704.

Zajic, E.J., 1987. Scale up of Microbes for Single Well Injection. In: Proceedings of the First Int. MEOR Workshop, Ed. by Jeannette King and Debra Stevens, Bartlesville Project Office, U.S. DOE, OK, pp. 241-246.

Reprinted from: *Microbial Enhancement of Oil Recovery - Recent Advances*,
edited by E.C. Donaldson

CH. F–2

PRELIMINARY RESULTS OF SOME RECENT MEOR FIELD TRIALS IN ROMANIA

I. LAZAR[1], S. DOBROTA[1], M. STEFANESCU[1], L. SANDULESCU[1], P. CONSTANTINESCU[2], C. MOROSANU[2], N. BOTEA[3], O. ILIESCU[3]
[1] Institue of Biological Sciences, Spl. Independentei 296, Bucharest, Romania
[2] Research and Design Institue for Petroleum and Gases, Cimpina, Romania
[3] Titu Oil-field Unit, Jud. Dimbovita, Titu, Romania

ABSTRACT

After 1985 a "second generation" of MEOR field trials was started in Romania with several improvements: the reservoirs, the bacterial inoculum, the quantity of nutrients and inoculum used, the protocol of well injection and monitoring of the field trials.

Presently in Romania four reservoirs are under microbial flooding and ten wells are being treated by cyclic microbial recovery (or to clean out of the wellbore area).

Results from a reservoir treated with microbes and from two wells cleaned out by microbes are presented. In the case of microbial flooding the decrease of oil production stopped and in the case of cyclic microbial treatment the wellbore was cleaned out.

INTRODUCTION

At present, the idea is unanimously accepted that microorganisms may play a part in oil release and migration processes in porous media and thus can enhance oil recovery from reservoirs. This possibility is the consequence of the fact that certain microorganisms are able to release, under anaerobic conditions, substances that play a part in oil release and its migration from reservoirs. Such substances are gases, acids, solvents, biosurfactants and biopolymers. The mechanisms by which these substances release residual oil from porous media were first discussed by ZoBell (1947), while today they are largely commented on in specialized literature. The method relies on the ability of the bacterial inoculum to metabolize inexpensive organic substrates such as molasses and yield the above-mentioned substances. Accordingly, 30 years ago, researchers were tempted to apply the method to field trials (Spurny & Dostalek, 1956, Dostalek & Spurny, 1958, Jaranyi, 1968, Karaskiewicz, 1968, 1974; Dienes & Jaranyi, 1973; Hitzman, 1983). During the last few years, several papers have been published which discuss the details of the most representative field trials carried out during the latest 35 years in a few hundred wells (Hitzman, 1983, 1988; Yen, 1986, Donaldson et al., 1989, Lazar, 1989, 1990). The authors of these papers emphasize the in situ method that can be applied for several purposes suggested by the implementing modality, namely cyclic microbial flooding, selective plugging recovery, microbial fracturing recovery, paraffin removal treatment, microbial flooding recovery and microbial flooding by stimulation of the activity of natural microflora. As already confirmed, the first four modalities lead to well stimulation by clean out of the wellbore, or the plugged area neighboring it, while the last two rely on an injection-well/production well systems with both contributing to enhanced oil recovery. In the case of the first five modalities listed above bacteria previously selected and adapted to the field characteristics in question are used while with the last modality indigenous microflora are stimulated in the formation water or in the water used for reservoir injection. Details on these microbial stimulations and on enhanced oil recovery modalities are displayed in the recent papers by Hitzman (1988), Bryant and Burchfield (1988), Donaldson et al., (1989) and Lazar (1990).

The field trials carried out between 1957-1980 in the USA, Holland, Chekoslovakia, Hungary, Poland, USSR and Romania (Dostalek, 1961; Jarahnyi, 1968; Senykov, 1970; Yulbarisov, 1972, 1976; Karaskiewicz, 1974; Johnson, 1979; Lazar et al., 1982; Hitzman, 1983, Lazar and Constantinescu, 1985) belong to the "first generation" of field trials which were often empirical and possibly hazardous.

After 1980, this activity was based on scientific principles developed by several interdisciplinary research groups. Review papers played a significant role in the "second generation" of field trials but the principal guidance came from International Symposia, Conferences and Workshops on Microbial Enhanced Oil Recovery which were mainly sponsored by the Bartlesville Project Office of the United States Department of Energy. The scientific meetings, as well as the concerns of certain researches have resulted in important monographic volumes or scientific meeting proceedings (Moses and Sringham, 1982; Zajic et al., 1983; Donaldson and Clark, 1986, 1989; King and Stevens, 1987; Burchfield and Bryant, 1988). Countries such as the USA, Canada, USSR, Romania, China, East-Germany, Australia, have played an active part in this "second generation" of MEOR field trials, while Great Britain, West Germany, France, and Pakistan, are equally interested in the research being conducted by certain oil companies (Wempe, 1989; Malik and Khalid, 1990).

In Romania (between 1975-1980) a "first generation" of field trials took place yielding encouraging results (Lazar et al., 1982; Lazar and Constantinescu, 1985). After 1980, concerns have been directed towards a "second generation" of field trials, which started in 1986 and are in progress. At present, the two modalities of microbial treatment in cyclic and microbial flooding recovery systems are being tested in four oil fields;

microbial flooding is operated in four zones and cyclic injection in 10 wells. In the future we envisage trials including such modalities as microbial flooding recovery by stimulation of the activity of native microflora, microbial selective plugging and probably fracturing fluids recovery.

The present paper presents the results of one of the fields subjected to microbiological treatment (MEOR) where a zone was subjected to microbial flooding recovery and two wells to single well stimulation treatment to clean out the wellbore area.

MATERIAL AND METHOD

Geological and Physicochemical Features of the Field

Table 1 shows some of the geological and physicochemical features of the Sa_3Cl_1 oil field within which wells in a zone were subjected to MEOR treatment, both for microbial cyclic recovery and for microbial flooding recovery. This reservoir meets the requirements for the microbiological method regarding both temperature, permeability, porosity, salinity, carbonate content, waterflooding, oil characteristic, and rock mineralogic structure.

Table 1 Some geological characteristics of reservoir Sa_3Cl_1

Characteristics	Values
Mineralogical structure	Calcareous sandstones or Calcareous marl (diorite sand) with thin intercalations of sand, with microfissures and microcaves
Temperature (^oC)	47
Depth (m)	750-800
Pay zone (m)	5-6
NaCl (g/l)	4-4.5
Permeability (mD)	245-248
Porosity (%)	25.5
Type of crude oil	A_3, nonparaffinic
Crude oil saturation (%)	75
Crude oil viscosity (CP)	26
Crude oil density (Kg/dm^3)	0.835-0.922
Co-produced water (%)	95-99
pH	6.7-6.8
$CaCO_3$ content (%)	45

Bacterial Inoculum

The bacterial inoculum used for well treatment (both in the microbial flooding recovery system and in the microbial cyclic recovery system) was prepared according to the proceedings presented in another paper (Lazar et al., 1990). This inoculum consisted of bacterial populations adapted to the respective reservoir conditions and was cultured in large quantity using the equipment presented in Fig. 1.

Before being injected into a formation, the bacterial inoculum was characterized both with respect to its performances in gas, acid, solvent, biosurfactant and biopolymer production and to residual oil recovery using sandpack columns. The inoculum also was characterized with respect to the carbonate impact on gas and acid production

Nutrients

The nutrients injected into the formation consisted of the reservoir brine with 2% molasses. Previous papers (Lazar and Constantinescu, 1985; Lazar, 1989, 1990) stated that, with the reservoir brine and molasses used, supplementary sources of nitrogen and phosphorus have proved unnecessary.

Well Injection Protocol

Fig. 2 shows the protocol for injection of bacterial inoculum and nutrients using molasses in the microbial flooding recovery system. While implementing these procedures both our own and American experiences were taken into account, especially that of Yarbrough and Coty (1983).

Fig. 3 shows the protocol for well injection with bacterial inoculum and nutrient support using molasses in a cyclic system. This protocol includes both Romania MEOR experience as well as American, Zajic 1987. Fig. 4 presents the technical facilities for the MEOR field trial experiments in the Sa_3Cl_1 reservoir. Details of the technology of well injection for microbial enhanced oil recovery are presented in a recent patent (Lazar et al., 1989).

Baseline Monitoring

Before initiating the bacteria and molasses injection, we computed both the estimated decline curve of oil production from primary production and the estimated decline curve for oil production after water injection, and then the estimated decline curve of oil production resulting from microbiological treatment based on water injection (Fig. 5). Likewise, before initiating the bacteria and molasses injection, fluid samples were collected for microbiological and physicochemical analyses. The results of these analyses were referred to as basic data for subsequent microbiological and

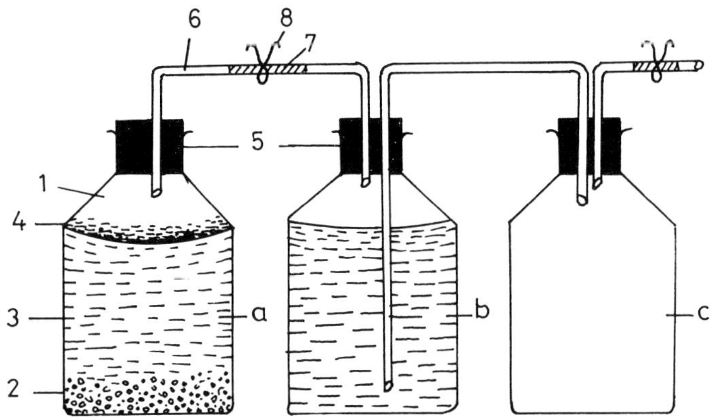

a) fermentor container
b) container for collecting the gases produced in a)
c) container for collecting the water evacuated from b) under gas pressure
1. plastic container (usually 60-100 l)
2. fermented scum from sugar processing plants (10-15%) or bacterial inoculum of 48 h (5-10%)
3. reservoir brine with 4% molasses
4. crude oil (up to 4-5%)
5. rubber plug
6. glass tube
7. rubber tube
8. clamp

Fig.1. The containers for growth of the adapted mixed enrichment cultures (AMEC) used in the field trials at Sa_3Cl_1 reservoir.

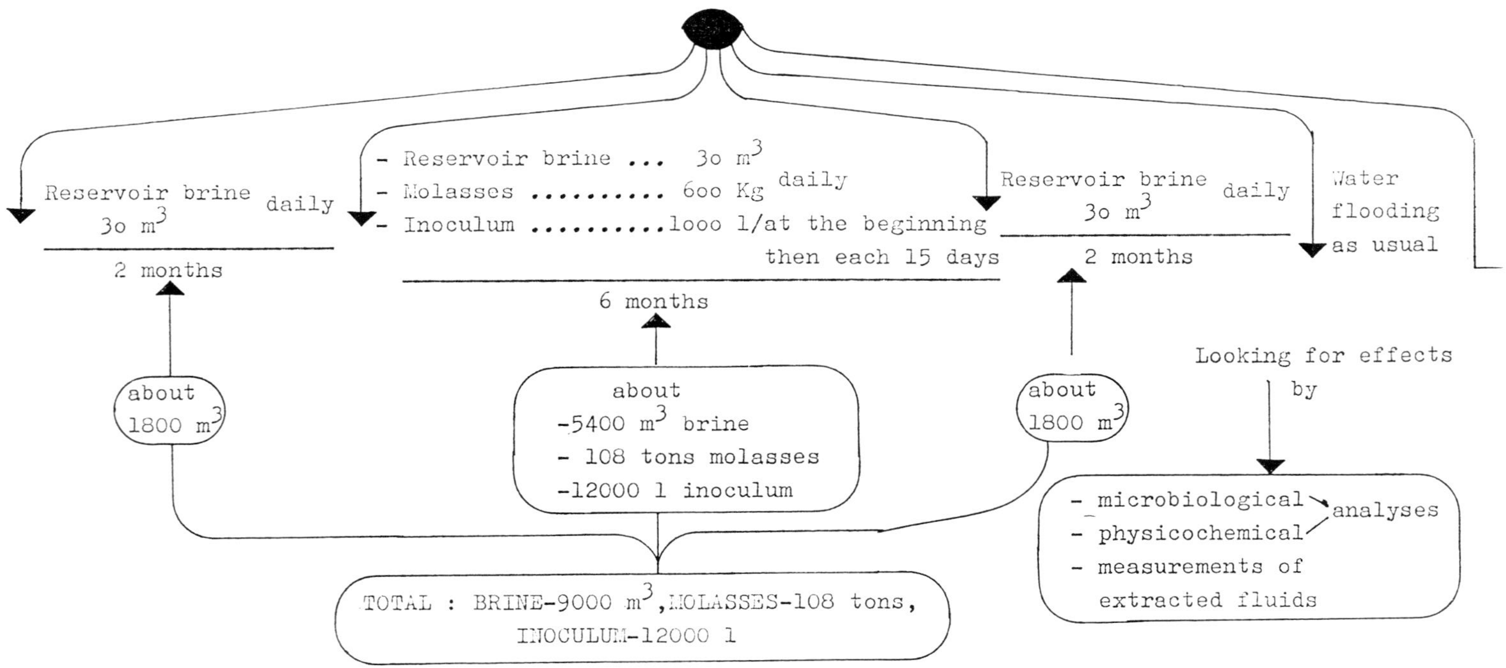

Fig.2. The treatment protocol with bacteria (AMEC) and molasses of the reservoir Sa_3Cl_1, for Microbial Flooding Recovery.

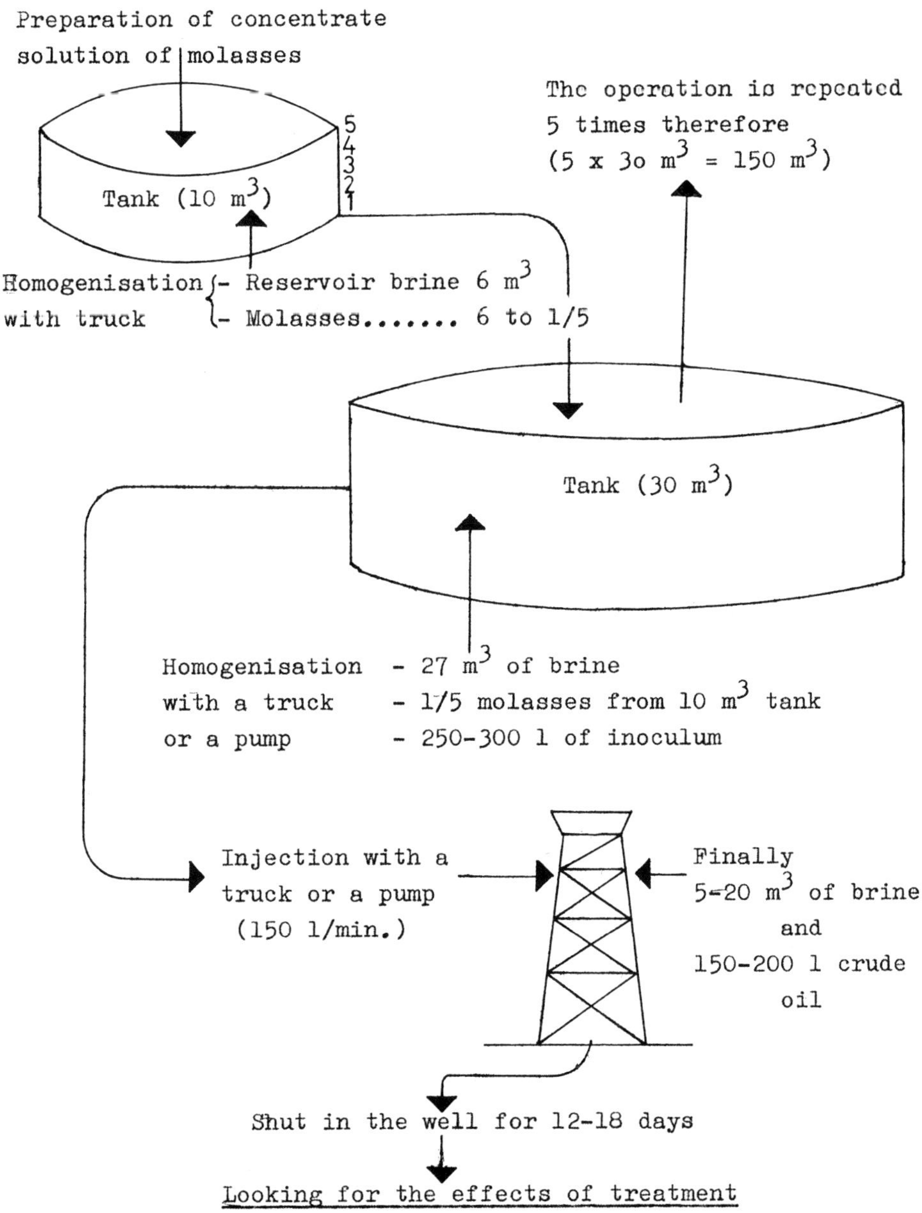

Fig.3. The treatment protocol with bacteria (AMEC) and molasses of injection or reaction wells (reservoir Sa_3Cl_1) for cleaning out the well bore area.

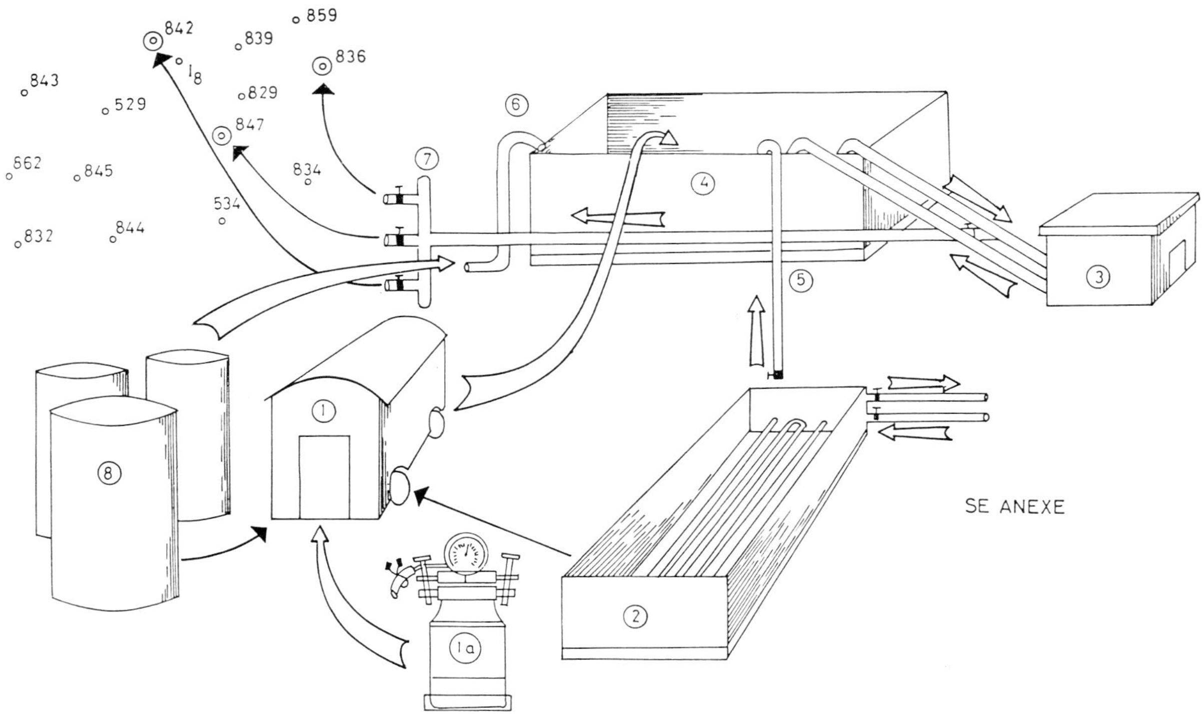

Fig 4. TEHNOLOGICAL FACILITIES AT THE RESERVOIR Cl_1 FOR **MEOR** FIELD TRIALS EITHER BY CYCLIC OR MICROBIAL FLOODING RECOVERY SYSTEM.

Annex to Figure 4.

1. Room for inoculum preparation
1a. Container for inoculum preparation in anaerobic conditions
2. Tank (30 m^3) with steam serpentine for molasses depositing
3. Pump for the omogenisation of the brine-molasses-bacteria, as well as for the injection into the formation.
4. Tank (30 m^3) for the omogenisation of nutrient support (molasses 2%) inoculum 1-2.5% (periodically - each 15 days) and brine
5. Pipe line for the transport of molasses
6. Pipe line for the transport of brine
7. Distributor to the injection wells
8. Tanks for brine depositing

Injection wells

Reaction wells

physicochemical analyses periodically conducted on the fluids extracted by the production wells.

RESULTS AND DISCUSSION

Bacterial Inoculum Performances

The bacterial inoculum injected into the formation was characterized with respect to its ability to produce substances contributing to the releasing of residual oil, namely: gas, acid, solvents, biosurfactants and biopolymers. The results are listed in Table 2. According to these results, the inoculum injected into the formation yields significant amounts of substances that contribute to residual oil recovery. Gases and acids yield substantially increases when carbonates are present in the oil reservoir (Table 3). That was the reason for selecting reservoirs with a minimum $CaCO_3$ content of 20%.

Table 2 The performance of gas, acid, solvents, biosurfactants and biopolymers production by the adapted mixed enrichment cultures (AMEC) used for the field trials at the Sa_3Cl_1 reservoir

Substances Produced	The quantity produced in the first 5 days
Gases (cc from 15 ml medium[a] in the first 5 days)	115-160
Acids (pH initial and after 5 days)	7.2-4.8
Solvents (nicrosin test[b])	+
Biosurfactants (kerosen test[b])	++++
Biopolymers (solvent precipitation test - g%)	0.393

[a] Reservoir brine with molasses 4%, crude oil 1-2% and bacterial inoculum (5-10%) or scum from sugar processing plants (10-15%)

[b] Intensity of reaction

Table 3 The influence of carbonates on gas and acid production of adapted mixed enrichment cultures

Medium	Gas and Acid Production		
	Gas in first 5 days from 15 ml medium (cc)	pH initial	pH after 5 days
M_1: Reservoir brine with molasses 4%, crude oil 1-2% and AMEC 5-10%	60-120	7.8	4.6
M_2: M_1 with carbonates (up to 5%)	105-155	7.8	6.2
M_3: M_1 with SFSPP[a] (10-15%)	125-190	7.8	6.6

[a] Scum from sugar processing plants

As to the capacity of the employed inoculum to release residual oil in porous environments, it has been found that the highest residual oil recovery efficiency occurs with adapted mixed enrichment cultures (AMEC). As Table 4 points out, the residual oil recovery efficiency with AMEC exceed over 10 times those yielded by pure cultures and at least 6 times those yielded by mixed cultures. That is the reason why such adapted mixed enrichment cultures have been used, both with well injection in the microbial flooding system, and with well injection in the cyclic system.

Table 4 The performance of residual oil release from sandpack columns of the adapted mixed enrichment cultures (AMEC) injected into reservoir Sa_3Cl_1

Type of Inoculum Tested	Average of the Residual Oil Release, %				
	Soi	Sowf	Socf	Er	Tor
Pure cultures	82.67	44.35	42.57	3.55	48.62
Mixture of 2-4 pure cultures	82.67	45.82	41.60	7.75	45.67
AMEC	82.67	51.15	28.82	43.00	65.90

Soi : residual oil saturation prior to waterflood
Sowf: residual oil saturation after to waterflood
Socf: residual oil saturation after microbial treatment
Er : residual oil recovery efficiency
Tor : total oil recovery
Er = (sowf - Scof)/Sowf x 100%
Tor = (Soi - Socf)/Soi x 100%

The Effects of the Treatment Applied in Microbial Flooding System

The zone in the Sa_3Cl_1 reservoir where this treatment is conducted with two injection wells and nine production wells is shown in Fig. 5. Starting in the Spring of 1984, a waterflooding process was implemented in three water injection wells injecting daily about 200-250 m^3 brine into the formation. Since the waterflooding failed to increase oil production, probably due to unforeseen injection water drainage through microfissures and microcavernes, in December 1986 (after nearly 3 years of waterflooding) we decided to initiate the microbiological treatment using wells 847 and 836 a injection wells for bacterial inoculum, molasses and brine. The bacteria and molasses injection started December 1, 1986, and ended on July 11, 1988, with several interruptions (Fig. 5) primarily caused by the unfavorable climatic conditions during the winters of 1986-1987 and 1987-1988. During the injection periods a total amount of 6390 m^3 brine, 126.4 tons molasses and 16 m^3 of bacterial inoculum were used (Annex to Fig. 5).

Microbiological analyses show the spread of bacteria throughout the formation by their numerical increase in the co-produced water from values of the 10^2-10^3, measured before the microbiological treatment, to values of the 10^4-10^7, 6-20 months after beginning the treatment, Table 5.

Table 5. Microbiological analyses of the co-produced water before and after microbial treatment at Sa_3Cl_1 reservoir.

Production well	Cells density/ml: Before treatment	After treatment: 6 months	After treatment: 20 months
534	$1.1x10^2$	$1.5x10^3$	$12.0x10^6$
I_8	$1.9x10^2$	$2.4x10^4$	$1.8x10^7$
829	$3.4x10^2$	$2.3x10^4$	$22.8x10^6$
839	$1.7x10^3$	$2.0x10^4$	$4.0x10^6$

The number of bacteria remained at values of 10^5-10^7 one to two years after ceasing the molasses and bacterial inoculum injection.

Table 6 shows some of the results of the physicochemical analyses of the oil extracted. In case of another reservoir (not discussed in this paper) oil viscosity increased. However, in the present case for the Sa_3Cl_1 reservoir,

Annex to Figure 5.

Period	Brine (m^3)	Molasses (tons)	Inoculum (l)
1-25 December 1986	750	15	2000
8-13 January 1987	180	3.6	1000
3 April - 12 Jun3 1987	2910	56.8	5000
10 July - 6 August 1987	840	16.8	2000
22 October - 20 November 1987	930	18.6	2000
23 February - 6 March 1988	420	8.4	2000
4-11 July 1988	360	7.2	2000
TOTAL	6390	126.4	16000

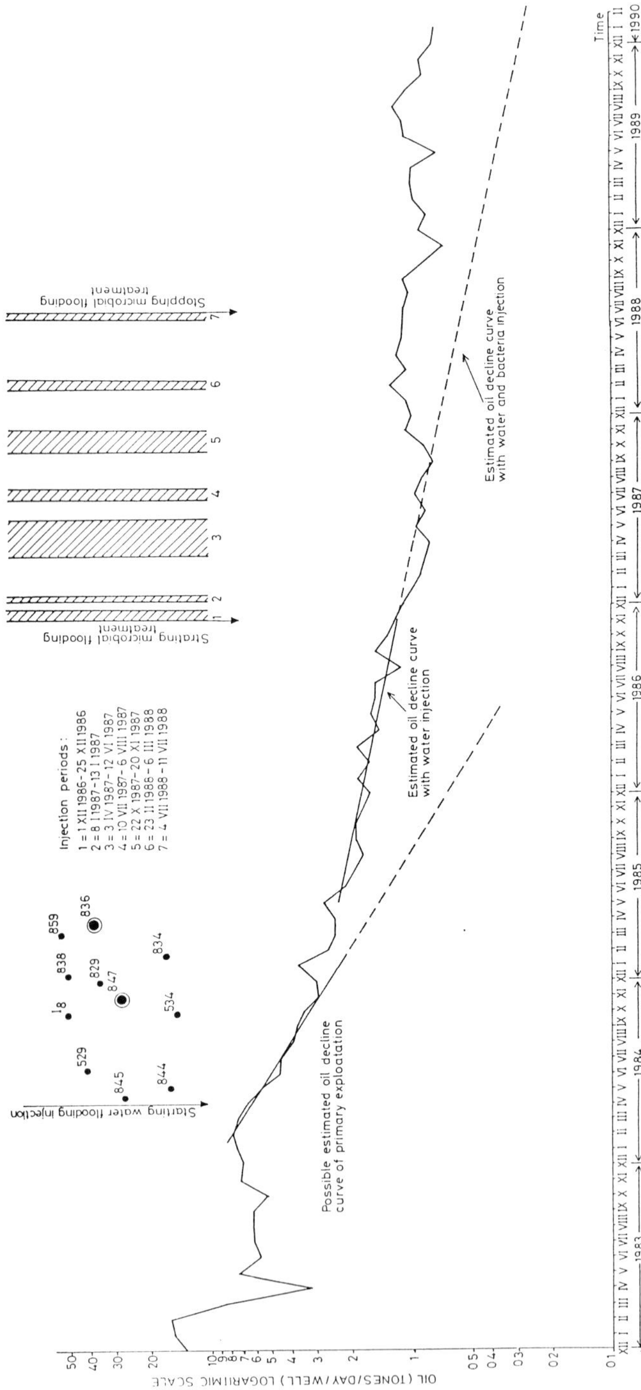

Fig.5 MICROBIAL FLOODING RECOVERY TREATMENT OT THE RESERVOIR $Sa_3\ Cl_1$ (after Lazăr et al.,1990)

TABLE 6 The influence of microbial flooding recovery treatment at Sa_3Cl_1 reservoir on crude oil viscosity

Well	Temp. (°C)	Before microbial treatment viscosity			12 months after microbial treatment viscosity		
		°E	Cst	Cp	°E	Cst	Cp
839	20	24.39	184.83	170.27	25.24	191.82	176.01
	40	7.80	59.68	54.25	8.08	61.25	55.62
	60	3.61	26.10	23.41	5.66	28.60	23.83
859	20	drops	drops	drops	23.31	167.69	162.55
	40	14.00	88.31	80.63	7.31	58.72	53.32
	60	5.55	34.90	31.44	3.51	25.40	22.76
829	20	drops	drops	drops	25.09	190.18	175.16
	40	11.65	88.31	80.63	7.93	60.11	54.64
	60	4.69	34.90	31.44	3.55	25.66	23.02
529	20	drops	drops	drops	23.08	174.95	150.95
	40	13.98	105.32	97.17	7.42	56.16	50.99
	60	5.05	37.63	34.05	3.33	23.80	21.32
834	20	24.95	184.88	170.27	26.36	171.76	158.19
	40	8.13	61.47	55.88	7.27	55.31	50.29
	60	4.11	27.00	24.22	3.22	24.20	21.71
844	20	22.00	166.76	152.50	21.90	166.00	152.06
	40	7.27	55.12	49.77	7.04	53.04	47.95
	60	3.60	26.00	31.92	3.47	25.13	22.41

viscosity underwent an obvious decrease a year after the beginning of the process. According to Dr. Donaldson (1990) this apparent decrease of oil viscosity occurs because dissolved biopolymers decrease the wetting properties of crude oils and thus the friction to fluid flow.

Production standards indicate encouraging results since a decrease of the decline is obvious and accompanied by an increase of the oil production all over the reservoir (Fig. 5). Before the treatment, the well oil production was 0.7-0.8 tons/day, after the treatment it rose constantly to 0.9-1.0 tons/day.

From the estimated decline curves shown in Fig. 5 one may conclude that, since water injection was not affected by addition of molasses and bacterial inoculum, it is possible that the microbial treatment contributed to an improvement of the fluid flow profile, which was not achieved with water injection between 1984-1987.

This standardized flow was only possible due to the impact of bacteria or of their metabolic products over the microfissure and microcavern areas, through which water channeled more easily previous to bacterial and molasses injection. The zone is still under investigation and it is possible to resume the bacteria and molasses injection at some time in the future.

The Effects of the Treatments Applied in the Cyclic System

The two injection wells in the zone subjected to microbial flooding, during 1984-1987, had to be cleaned out periodically by chemical and mechanical treatments. After almost 1-1/2 years from the beginning of the bacteria and molasses injection, the injection pressure increased from 2.0 to 2.5 MPa to 5.0 to 6.0 Mpa. Since a chemical treatment might have an adverse impact on further microbial treatment, it was decided to subject these two wells to a cyclic microbial treatment in order to restore permeability within the wellbore area. Thus, the cyclic treatment in well 836 on March 25, 1988, and on well 847 on March 26, 1988. After injecting about 150 m^3 brine with about 6 tons molasses and 3000 liters of bacterial inoculum at a rate of about 150 liters/min, the wells were shut in for 18 days, then the bacterial injection with nutrients for microbial flooding was resumed. After July 11, 1988, only water injection was applied at a rate of 50-75 m^3/day/well flow rate.

Figures 6 and 7 show the effects of this cyclic treatment. In the case of both wells, it is worth noting that when the wells were re-opened, the injection pressure decreased from about 5.0-6.0 MPa to 2.0-2.5 MPa, then slowly increase again to values up to 4.5-5.0 MPa after almost a year. Two

Yen, T.F., 1986. A State of the ARt REview on Microbial Enhanced Oil Recovery. Unpublished report. Univ. of Southern California, Los Angeles, CAS, 278 p.

Yen, F.T., 1989. Microbial Enhanced Oil Recovery, CRC Press Inc., 2000 Corporate Blvd., NW, Boca Raton, FL 33431.

Zajic, E.J., 1987. Scale up of Microbes for Single Well Injection. In: Proceedings of the First Int. MEOR Workshop, Ed. by Jeannette King and Debra Stevens, Bartlesville Project Office, U.S. DOE, OK, pp. 241-246.

Zajic E.J., Cooper, C.D., Jack, R.T. and Kosaric, N., 1983. Microbial Enhanced Oil Recovery, Penn Well Books, Tulsa, OK, 172 p.

Zajic, E.J., Donaldson, T.E., 1985. Microbes and Oil Recovery Vol. I. International Bioresources Journal, Bioresource Publications, El Paso, TX, 370 p.

ZoBell, C.E., 1947. Bacterial Release of Oil from Oil Bearing Materials, Parts I and II, World Oil, 126(13): 36-47(1):127(1)35-47(11).

Reprinted from: *Microbial Enhancement of Oil Recovery - Recent Advances*, edited by E.C. Donaldson

CH. F–3

MICROBIAL ENHANCEMENT OF OIL RECOVERY FROM CARBONATE RESERVOIRS WITH COMPLEX FORMATION CHARACTERISTICS

WAGNER, M.
Kombinat Erdol-Erdgas, Gommern, GDE

ABSTRACT

Laboratory experiments with microbes adapted to subsurface conditions of temperature, pressure and salinity using oilfield carbonate rocks showed successful enhancement of oil recovery. Based on the laboratory work, field application of the adapted Clostridium species were conducted. A decrease of the percentage of produced water from 80 to 60% and an increase of oil production from an average of 50 tons per day to 150 tons per day occurred after treatment with bacteria.

INTRODUCTION

The German Democratic Republic (GDR) has only limited energy resources; the main source of energy being lignite. The production of natural gas amounts to 10 billion m^3 annually, and oil is found in the north and southeast of the GDR with a production of not more than 40,000 tons per year.

The oil reservoirs in this region developed as a result of an intensive generation of biomasses in a relatively shallow Zechstein sea, which covered todays continent about 250 million years ago. Most important reservoir rocks are the dolomites of the Zechstein formation, encountered today at depths from 1200 to 2500 m. The temperatures in our oil reservoirs are accordingly high (from 50 to 90 °C), Table 1. As relicts of the Zechstein sea, which dried out later, the pay zones have high-salinity formation waters of Nacl, $CaCl_2$ and $MgCl_2$ even the fissures and fractures of the reservoir rocks are partially filled with salt.

Table 1 Reservoir characteristics

Reservoir Beds	
Depth	1240 m
Formation temperature	53°C
Formation pressure	8 MPa
Type of reservoir	fissured-porous carbonate reservoir
Matrix porosity	1-2%
Fissure porosity	0.1-0.5%
Fissure permeability	10-50 mD

In 1967, we began to investigate the possibility of an application of the molasses-in-situ process in these ecologically unfavorable conditions, since other technologies, such as waterflooding and the use of surfactants or biopolymers in fractured-porous carbonate rocks with a low-viscosity oil has not been effective.

This microbial research work, as in most other counties, was stopped in the 1970's and renewed again in 1982. Now we are engaged in the following research topics:

1. Microbial characteristics of the GDR Zechstein formation
2. Cultivation of anaerobic, halophilic, thermoduric, fermentation bacteria
3. Growth and product generation as a function of salinity temperature and pressure
4. Analysis of oil desorption on core models
5. Process engineering for the cultivation of the inoculum at the injection site
6. Field testing

MICROBIOLOGICAL CHARACTERISTICS OF THE ZECHSTEIN FORMATION

Because of the extremely unfavorable initial conditions for ecology, specifically the combination of high formation temperature and high-salinity formation waters, the GDR oil and gas fields are supposed to be sterile.

Bacteriologic analysis in untreated wells showed no bacteria. Bacteria may have been introduced into the formation with the drilling mud or with well treating agents. After essential changes of the formation environment caused by water flooding, a temperature drop and a decrease of salinity growth of microorganisms occurred in the bottomhole area of the reservoir. As a rule, only sulphate-reducing bacteria survived. The formation waters, however, did not contain bacteria suited for the molasses-in-situ processes such as Clostridiae.

To separate appropriate bacteria we carried out an extensive screening program analyzing samples from locations with extreme conditions, such as salt samples from mines, salt gardens, different high-salinity waters from salt lakes and technical units, formation waters from different beds, thermals, hot-water circulation systems, etc.

We selected those microorganisms, which in an anaerobic molasses medium with a salinity of more than 100 g/l, generated gas at a minimum of 100 ml per g molasses. By increasing salinity and temperature stepwise we adapted the bacteria to more unfavorable conditions. Equipment for continuous cultivation of microorganisms was constructed for this purpose. The culture itself controls the nutrient flow in the fermenter by its acid generation through molasses fermentation. This self-regulation by metabolic activity of the wells prevents a wash-out of the culture, which is destroyed with more severe conditions of cultivation.

First we tried to adapt the bacteria to formation temperature, rising it gradually to 55°C, followed by an increase of salinity. The halophilic cultures quickly became accustomed to a salinity of 150 g salt/l, whereas two critical concentration points occurred at 180 g and 250 g of salt per liter. Especially at these concentrations, we observed the well-known transformations such as filamentous growth of cells, partial autolysis and formation of ghosts.

We did not succeed in adapting all isolated halophilic anaerobic fermentation bacteria to a salinity of 320 g/l. *Bacillus polymyxa* and *Clostridium butyricum* developed only up to 180 and 210 g of salt per liter. At higher salinities *Clostridium coccoides spec.* and *Clostridium oroticum spec.* dominated.

TOXICITY OF IONS

High-salinity formation waters often are characterized by different ion-contents. We analyzed only waters of the chloride type.

We found that sodium and potassium may be exchanged without impairing the growth of microorganisms. Magnesium has higher toxicity than Na^+ and K^+, but microorganisms adapt quite readily to it so that we could cultivate them in presence of 160 g/l MaCl and 160 g/l $MgCl_2$. The most toxic formation waters were those with high $CaCl_2$. In the presence of high NaCl (150-200 g/l) adaptation could only be attained with less than 100 g $CaCl_2$/l.

PRESSURE RESISTANCE OF HALOPHILIC BACTERIA

Microorganisms injected into deep reservoir beds are subject to high pressure. In nearly depleted low-pressured formations the stress results from hydrostatic pressure and the injection pressure at the surface. In our test formation at 1200 m, and with a surface injection pressure of 10 MPa the bottom-hole flow pressure may amount to 25 MPa.

The growth experiments were conducted in 2-l autoclaves at 55°C in molasses media with different concentration of salt (Fig. 1). In general, pressure effects decreased with increasing salinity. Sudden pressure increases (15 s) affected the bacteria more than a gradual rise of pressure (2-3 hrs).

Knowledge of the pressure sensitivity of bacteria is essential for design of the injection technology. In high-salinity media bacteria may be destroyed by the injection pump by a sudden pressure increase to 10 MPa. It is essential to use injection media with a salinity of 75 g/l to 150 g/l. A

Figure 1

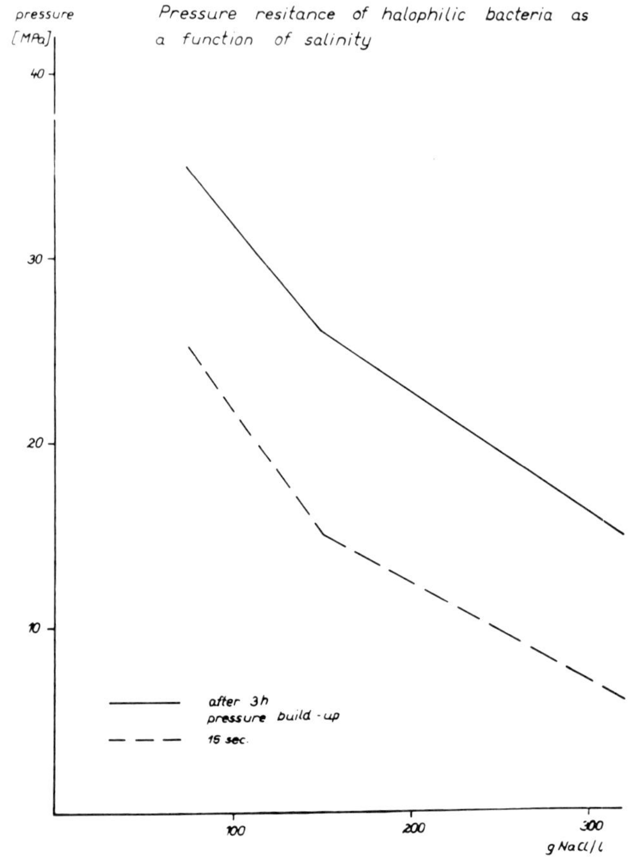

minimum salinity of 75 g/l is dictated for the application of halophilic bacteria, which undergo autolytic decomposition at low salt contents. As to temperature resistivity, thermophilic bacteria of the species Clostridium thermocellum showed better temperature resistance with increasing pressure. The optimum growth temperature increased from 76°C (in low-salinity molasses media with a Nacl content up to 30 g/l) to 82°C with a pressure minimum 2 MPa.

In spite of long-term adaptation experiments we did not succeed in adapting halophilic bacteria to temperatures of more than 55°C, or thermophilic microorganisms to a salinity of more than 50 g/l.

CULTURE MEDIUM

The composition of the culture media was minimized to a degree allowing sufficient growth of the organisms and maximum product generation. The following formulation was developed: Sugar-beet molasses, 40 g/l, polyphosphate 0.3 g/l, NH_4Cl 1.0 g/l, $CaCO_3$ 10 g/l, salinity 75 to 320 g/l, ion content according to the formation water analysis, pH 7.5.

Fig. 2 shows the characteristic parameters of growth and synthesis. With an extremely unfavorable cultivation environment an initial cell content of 1-5*10^7 bacteria/ml is necessary. The maximum cell content of the culture medium was to 4-5*10^8 bacteria/ml. Generation time was 2 hours. The conversion (that is the anaerobic fermentation) of the molasses sugar (nearly 50%), together with the development of organic acids are demonstrated in Fig. 2. A decrease of the pH of the medium to 4.6-4.8 occurred. The acid generation is most intensive at the stage of logarithmic growth.

Gas generation was 300 ml of gas per g of molasses used. The gas phase consists of 60% CO_2 and 40% H_2, with a much higher portion of CO_2 in the production-generation phase than in the growth phase. In the presence of carbonate rock (as in our reservoirs) the CO_2 content of the gas rises to 80% due to acid reaction with the carbonates.

Figure 2

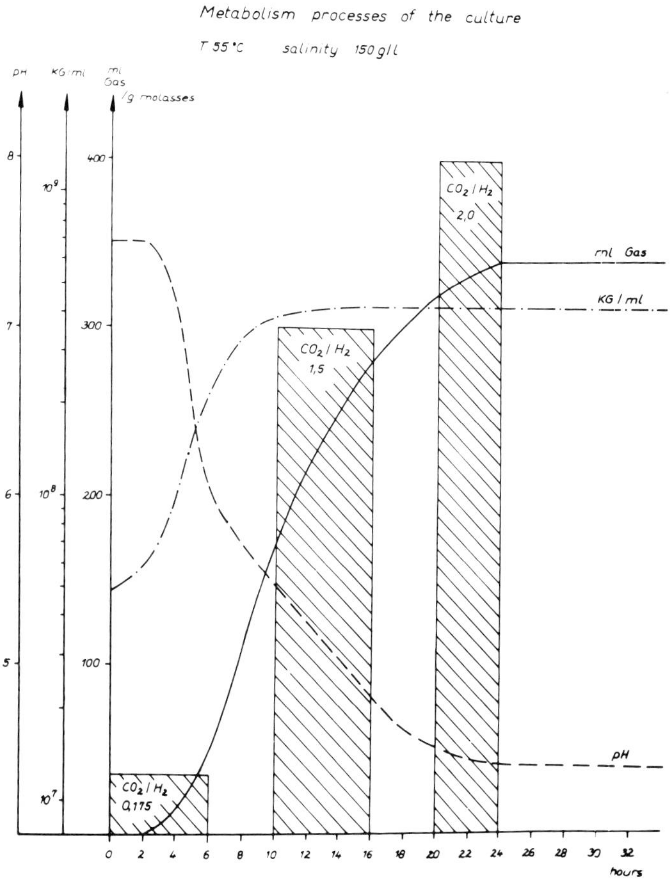

Fig. 3 shows the metabolic products identified in the culture medium during molasses fermentation: CO_2 and H_2, acetic, lactic, propionic, butyric, and valeric acid, acetone, as well as ethanol, propanol, butanol, isopropanol.

SIMULATION OF OIL DESORPTION

Experiments of oil desorption were conducted in static systems (imbibition test) and in flow models under room conditions and under formation pressure.

Six weeks before the beginning of the tests oil-impregnated cores from the Zechstein main dolomite of the Doebern field were stored in formation water to simulate a primary oil recovery, and then subjected to various tests shown in Table 2. After three weeks of reaction time at 55°C we measured the volume of desorbed oil.

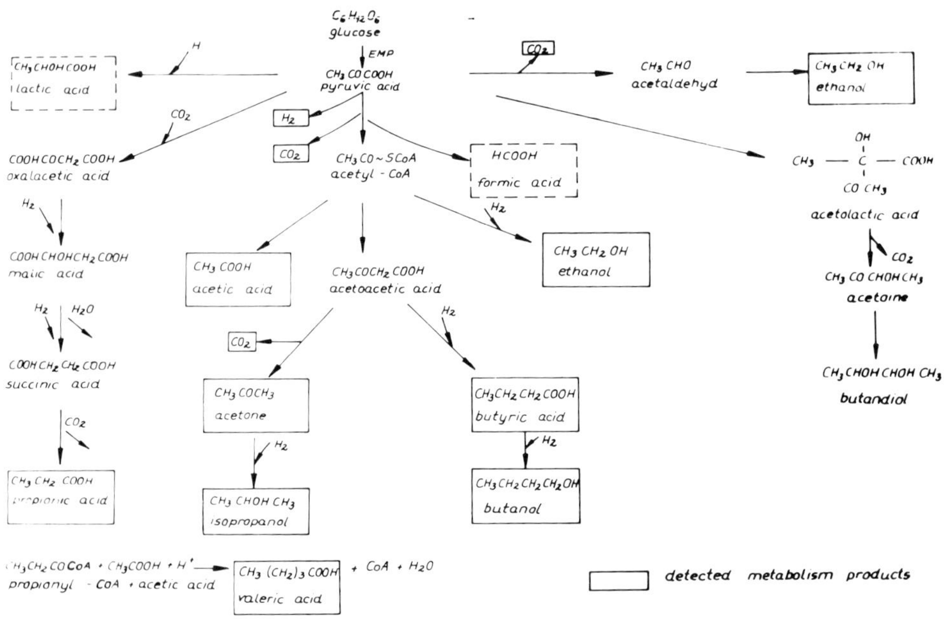

Figure 3

<u>Table 2</u>

Oil:	density:	reservoir rock	
	viscosity:	muscovite lillite	5%
		quartz	3%
		anhydrite	5%
		calcite	67%
		dolomite	20%

The formation water and sterile nutrient solution (Tables 3 and 4) did not affect the core but the sterility filtrated fermented cultural solution had an oil-desorbing effect due to its content of organic acids and alcohols which is caused a change of wettability and rock solution. In the actively growing culture of Clostridium this process was intensified, due to the penetration of bacteria into the rock pores and fractures where they produce pressure change by in-situ generation of biogas, and thus stimulate the oil to flow. In the pressure unit, the solution of biogas in the oil lowers the oil viscosity and improves its fluid flow characteristics.

Table 3

Formation water mg/l			
Na^+	16.670	Cl^-	190.700
K^+	102.690	Br^-	430
Ca^{++}	3.206	J^-	1
MG^{++}	4.4450	SO_4^{--}	1.750
NH_4^+	101		
Fe^{++}	---		

Table 4 Composition of a molasses nutrient medium

	in g/l
sugar-beet molasses	40.0
polyphosphate	0.3
NH_4Cl	1.0
$CaCO_3$	10.0
salinity	75-320
ion content as in formation-water analysis	

FLOW-MODEL EXPERIMENTS

We conducted various experiments with flow models made of sand, limestone granulate, and cores of reservoir beds. Formation conditions are best imitated by tests with original cores and formation pressure carried out in special high-pressure equipment. We used cores which were 6 cm in diameter and 18 cm in length.

After evacuation, the cores were saturated with oil, applying pressure. The tests were carried out at 15 MPa and 55°C. The core was flooded with one pore volume of formation water followed by two volumes of molasses nutrient solution with a salinity of 150 g/l containing $1*10^8$ Clostridiae per ml.

As shown in Fig. 4, water break-through was observed after injection of 1/4 pore volume. The core was flooded until no more oil was displaced. The oil desorption was 25%. After the injection of the nutrient solution and

Table 5 Oil separation of a dolomite core after a reaction time of 3 weeks at 55°C

Type	Preservation medium	ml/desorbed oil
1	formation water	0.0
2	sterile molasses nutrient solution	0.22
3	sterile-filtrated dermented cultural solution	0.6
4	active culture of Clostridiae	1.0
5	active culture of Clostriadiae, 15 MPa	1.6

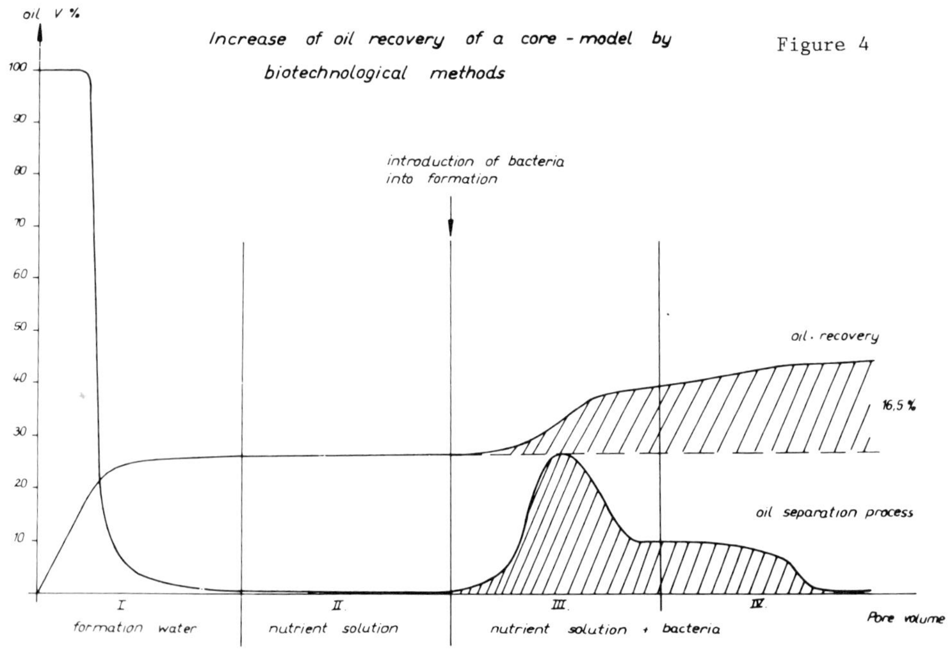

Figure 4

molasses we observed an increase of the oil in the effluent fluid, to a maximum of 30%, and a longer-lasting effect of microbial metabolism. By application of the molasses-in-situ process, the oil recovery from the core increased by 16.5 to 47.5%.

FIELD TESTS FOR ENHANCED OIL RECOVERY

For the direct preparation of the field experiments we conducted small-scale lab experiments in a 2.5 liter fermenter. They served to optimize the molasses-based nutrient media, to calculate the requirement of raw materials and to ascertain technical parameters, such as mixing intensity, turnover time, inoculation interval, development pH, alkali dosage, etc. Thus we obtained the data necessary for the construction and operation of a 2.5 m^3 field fermenter for the cultivation of inoculum at the site. The design of the field testing plant is shown in Fig. 5.

The field plant was initially operated discontinuously but in the last few years continuously with flow rates of 0.4 m^3/day and a cell content of $4*10^8$ bacteria/ml established at a salinity of 150 g/l.

For example, we injected 2000 m^3 of nutrient solution with 50 tons of molasses, with polyphosphate and soda as additives, in a mixture of fresh and formation water, followed by inoculation of an adapted cell mixture in a ratio of 1 to 10, using a batch-wise injection program. After injection the well

Fig 5 Technological diagram of surface equipment for the injection of bacteria and nutrient solution into formations

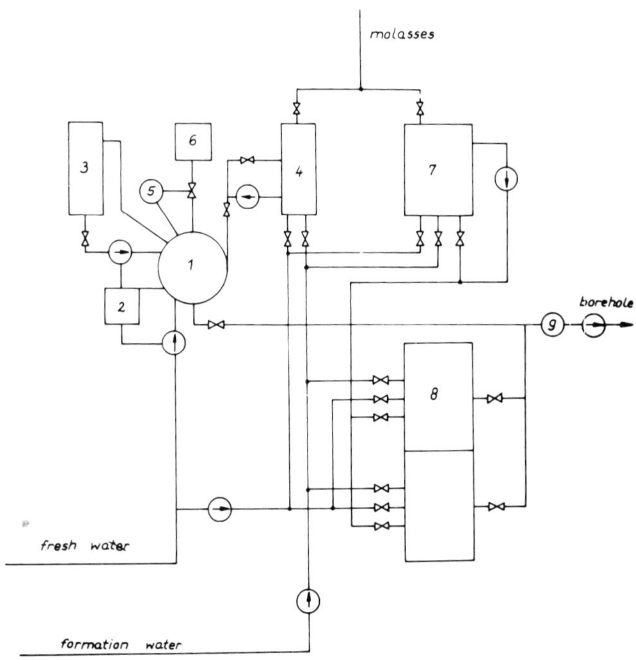

1 - Fermenter for bacterial cultivation
2 - Temperature control
3 - Water heating
4 - Nutrient solution tank for fermenter unit
5 - automatic pH - measuring and control unit
6 - Alkali tank
7 - Mixing unit
8 - Nutient substrate pumping tank
9 - Volume measuring device

was closed in, and the effectiveness of the molasses-in-situ technology observed at a production well located 600 m from the injection well by monitoring the production well shut-in pressure. Fig. 6 shows the percent water of the producing well before and after microbial treatment.

Before the MEOR treatment the producing well flowing with 88% water. After injection the water content decreased substantially, to a minimum of 34%, and to an average of 60%. The water decrease in the output clearly shows increased oil desorption from reservoir by the activity of the microorganisms (Fig. 7).

Before the beginning of the test the average annual oil production was 50 tons/month. Three months after injection of the nutrient molasses solution and the bacteria we observed an increase of oil production to an average of 150 tons/month and also an increased CO_2 content in the gas, which rose from 0 to 0.2% in the beginning to 4% near the end of the project, despite the considerable distance between the injection well and the producing well. About a year after the injection we observed another increase of oil output to an annual average of 300 tons/month. For a period of four months the daily output of 12 to 14 tons of oil reached an eight-fold increase.

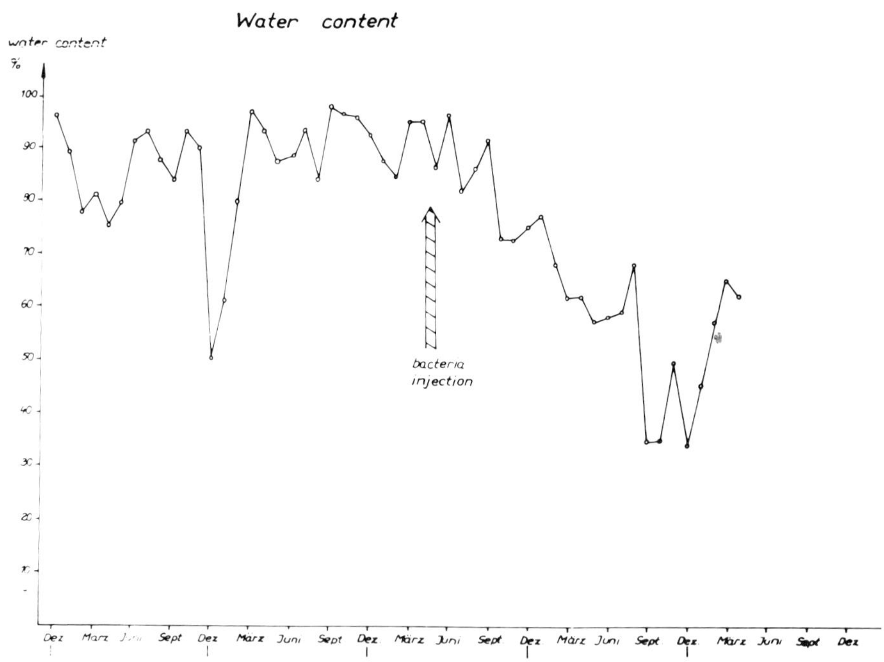

Figure 6

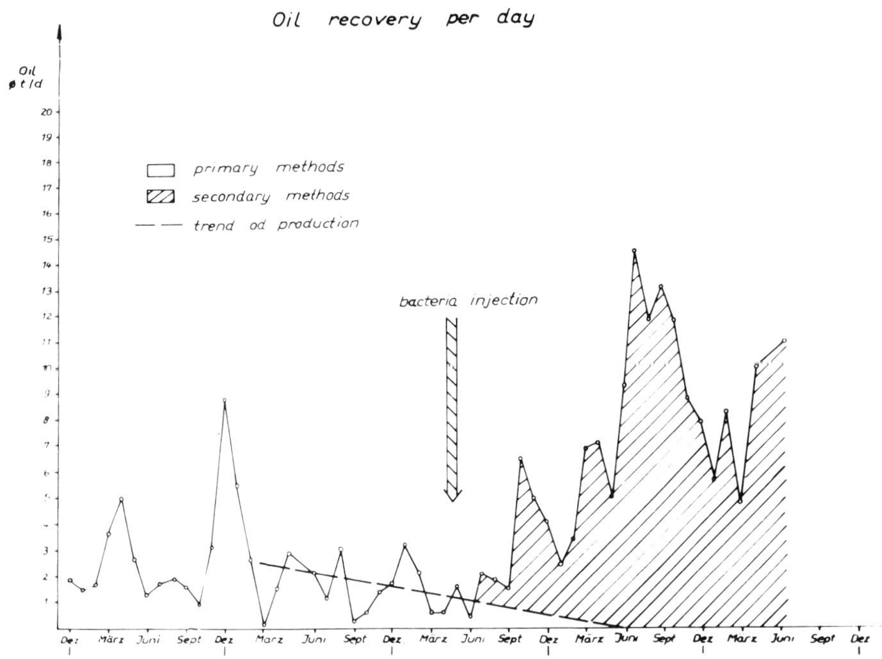

Figure 7

The increase after 12 months is characterized by the first detection of microorganisms at the producing well. The microorganisms obtained from the producing well were identical to the injected bacterial and amounted to $5*10^7$ per md. Therefore we attribute the second increase of oil recovery to the break-through of the injection front. Fig. 8 summarizes the output development of this well.

The oil production increase after the bacteria injection is beyond doubt. We succeeded in increasing not only the prognostic recovery of 2250 tons by 3800 tons of oil, but also the rate of production. The water entrainment did not rise, but dropped during the last year. After microbial treatment the gas production of the reservoir distinctly increased; that is confirmed by observations at the injection well. During the shut-in period the well-head pressure increased from 0.5 to 2.5 MPa. Gas samples taken at the injection well had a biogas component of 60-80% in the natural gas with up to 70% CO_2 and 10% H_2. In the natural gas component the content of nitrogen and methane substantially decreased, whereas the content of oil-typical components such as ethane, propane, and butane had increased. We interpreted this as a greater participation of dead oil in overall production. Production of hydrogen sulfide did not occur.

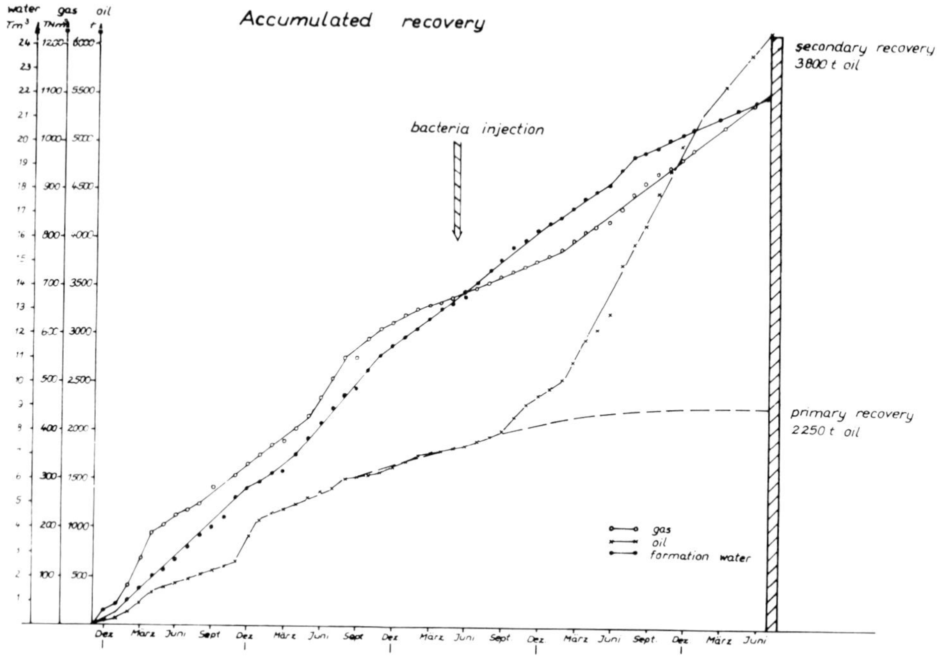

Figure 8

In addition to the current long-time flooding tests, we also tested the huff-and-puff method in the carbonate reservoir rocks. After injection of 20 tons of molasses the well production rate increased (without changing the procedure) from 7 m^3/d to 14-17 m^3/d, and the oil output from 0.5 to 2.5 tons/d with a rise of the wellhead pressure through biogas generation of 2 MPa. The production increase is attributed principally to improvement of flow conditions by microbially enhanced rock solution processes in the formation.

Further development of technologies, extending field application of this method to high-temperature reservoirs, has been successful. The next time we plan to test the microbial reservoir treatment method at a temperature of 82°C with high-salinity waters, using a new technology.

Together with Soviet scientists we have almost completed the development of a technique for flooding the Romashkino field in the USSR with molasses and Clostridiae.

As mentioned earlier, the GDR has only a few oil reservoirs, therefore we are interested in applying our MEOR technology in other countries.

Reprinted from: *Microbial Enhancement of Oil Recovery - Recent Advances*, edited by E.C. Donaldson

CH. F–4

MICROBIAL ENHANCED WATERFLOODING: A PILOT STUDY

R.S. Bryant, T.E. Burchfield, National Institute for Petroleum and Energy Research (NIPER), D.M. Dennis, Microbial Systems Corporation, D.O. Hitzman, INJECTECH, Inc., and R.E. Porter, NIPER.

ABSTRACT

The first microbial-enhanced waterflood field project sponsored by the U.S. Department of Energy (DOE), Microbial Systems Corp. (MSC), and INJECTECH, Inc., and being conducted in cooperation with the National Institute for Petroleum and Energy Research (NIPER) was initiated in October of 1986.

One of the major goals of this project was to develop a technology that could be implemented by independent oil producers; thus, the field site chosen for the pilot test was representative of a mid-continent waterflood operation with stripper wells. The methodology for designing and optimizing MEOR field technology has yet to be established; however, literature information and experience with MEOR processes indicate that there are certain necessary procedures to implement for a microbial waterflooding process.

The site selected for the project is in the Mink Unit of Delaware-Childers field in Nowata County, Oklahoma. This field is typical of mid-continent reservoirs in the United States. The pilot area consisted of four adjacent inverted five-spot patterns drilled on 5-acre spacing. There are 21 injection and 15 production wells on this pilot. Four of the 21 injection wells were treated with NIPER's microbial formulation.

Laboratory screening criteria were developed to evaluate microorganisms for this project. Several different microbial formulations were tested in Berea sandstone cores with reservoir fluids to determine oil recovery efficiency. Baseline monitoring of oil production was conducted to establish pre-pilot conditions, and fluid samples were collected on a weekly basis from producing wells.

Injectivity and microbial field survivability tests were conducted during the baseline period on two off-pattern wells, and a chemical tracer, fluorescein, was injected into the four injection wells during the baseline period. Tracer was observed in production wells about 1.6 years after injection, which corresponded reasonably well with the predicted tracer breakthrough from simulation studies.

Methodologies for field applications of microorganisms in ongoing waterfloods were developed as a result of this project. Results from the field pilot showed that microorganisms could be injected into an ongoing waterflood operation without causing any problems in injectivity. Microorganisms were injected only at the onset of the project, while molasses was injected daily for 2.5 years. Routine injection well backflushing continued to show that the injected microorganisms are thriving in the reservoir. Some of the injected microorganisms were detected at producing wells 32 weeks after injection, which was probably a result of microbial transport through low-volume, high-permeability stringers in the formation.

Microbial treatment did improve oil production rate, and water/oil ratios for producing wells nearest the microbially treated injection wells continue to be more favorable than baseline values. The results from this test are encouraging that microbial-enhanced waterflooding can be applied by independent producers.

INTRODUCTION

A microbial-enhanced waterflood field project sponsored by the U. S. Department of Energy (DOE), Microbial Systems Corp. (MSC), and INJECTECH, Inc., and being conducted in cooperation with the National Institute for Petroleum and Energy Research (NIPER) was initiated in October of 1986. The purpose of the project was to determine the feasibility of injection of a microbial formulation in a mature, ongoing waterflood, and if such an injection could increase oil production.

The concept of the use of microorganisms to recover oil from depleted petroleum reservoirs is not new. Field and laboratory research has been performed, and patents have been granted for this technology since the late 1940s. Early microbial enhanced oil recovery (MEOR) patents by Zobell, Hitzman, and Updegraff and Wren described the use of microorganisms in the reservoir to produce chemicals that could help to mobilize oil. Several literature reviews on MEOR have been published by Bryant (1989), Bryant and Burchfield (1989), Hitzman (1987), and Moses (1983).

In microbial enhanced waterflood applications, it is important that the microbes be capable of moving through the reservoir matrix and producing chemical products that can mobilize oil. The relative rates of transport of the nutrient and microorganisms will affect the injection strategy and design of the microbial system.

A microbial treatment requires careful design and sound reservoir engineering practice, as does any enhanced oil recovery (EOR) method. The methodology for designing and optimizing MEOR field tests has yet to be established; however, the literature and laboratory experience indicates that there are certain necessary procedures to implement for a microbial waterflooding process. This particular microbial enhanced waterflood field experiment was designed to use microoganisms that produced chemicals for improved oil mobilization, i.e., surfactants, gases, alcohols, and fatty acids, and had the ability to move through porous media.

Field Site Selection

After reviewing all laboratory and field data on MEOR processes, the site selection criteria were chosen for this particular field experiment. Over 30 waterflood projects in the Bartlesville, Oklahoma, area were evaluated as potential candidates for the project based upon the reservoir criteria given in table 1, and other desired characteristics, including:

1. Low brine salinity
2. Compatibility of Indigenous microbeswith those selected from NIPER's microbial culture bank
3. Established oil production decline curve
4. Demonstrated volume of target oil
5. Reasonably uniform water injection rates
6. Favorable (<10-20 acre) well spacing and pattern
7. Availability of reservoir and oil production data

TABLE 1 Criteria for reservoir selection for MEOR processes

Total dissolved solids in brine, ppm	<100,000
Usual reservoir temperatures, ° F	<160
Usual reservoir depths, ft	<10,000
Residual oil saturation, %	> 25 to 30
Brine permeability, md	> 100

Many of the waterflood projects fit the above criteria; however, there were only a few waterfloods that had 5-acre spacing instead of 10-acre spacing. We selected those that had the shorter spacing and continued the screening process. When we had narrowed the number of potential candidates to three, brine and oil samples were taken to the laboratory and checked for microbial compatibility. Samples from all three waterfloods showed that the microbes were compatible; however, a B & N Oil Company waterflood in Delaware-Childers field used fresh water from the Verdigris River, and this flood was chosen because the selected microorganisms tend to grow better with lower salinity.

Field Data

The Mink Unit, which includes both the Candy and Sallie Mink leases, selected for the project is in Delaware-Childers field in Nowata County, Oklahoma (figure 1). This particular part of Delaware-Childers field was owned by B & N Oil Company when the project was initiated in 1986. The legal description of the Mink Unit is Section 36, Township 27N, Range 16 E of Nowata County. Delaware-Childers field was discovered in 1906, and by 1911 initial development was essentially complete. The field was produced by primary methods until 1925 when air injection was initiated, and by 1932 air injection was used field-wide. During the 1940s, the field was approaching the economic limit, and waterflooding was begun. By 1945, four small waterfloods were in operation in less prolific areas of the field. During the next 10 years, many waterfloods were implemented throughout the field.

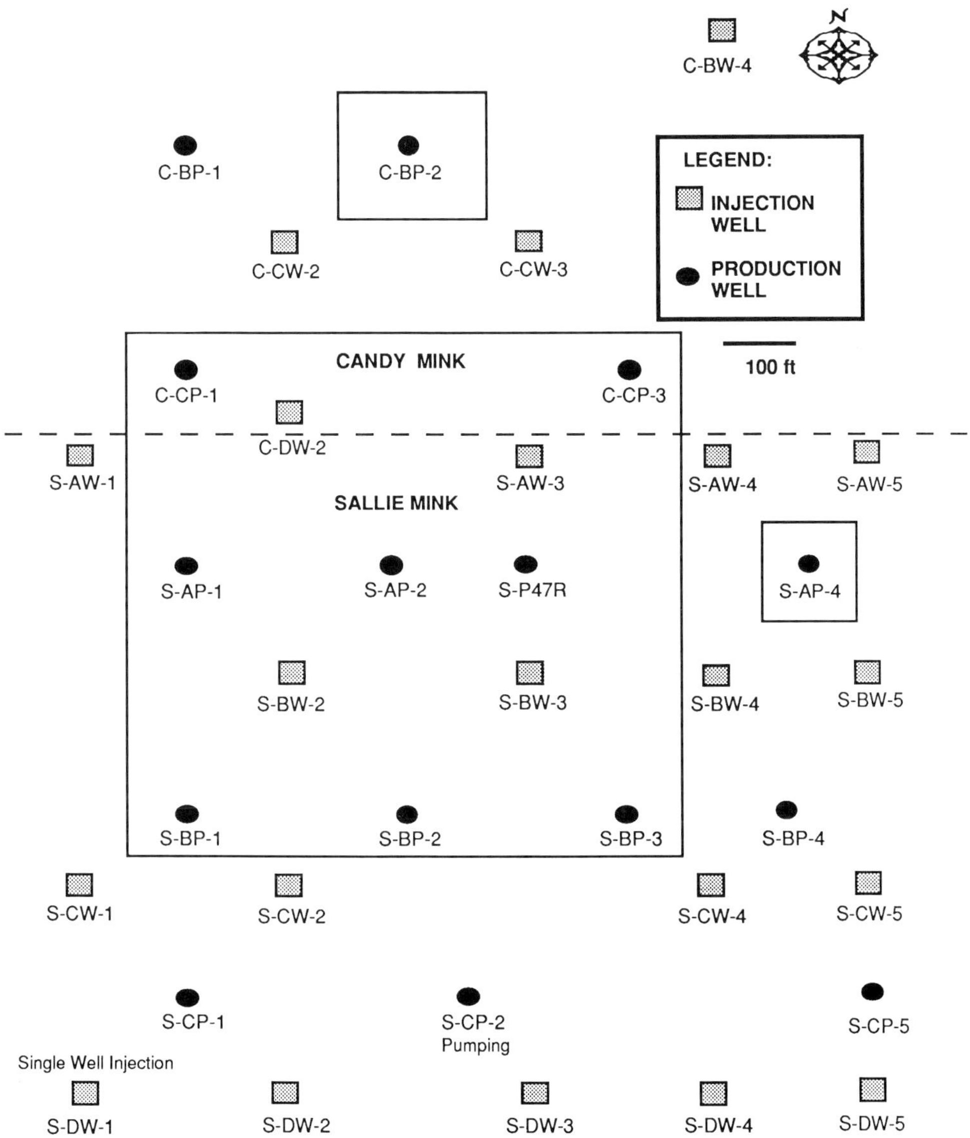

Fig. 1. Map of Mink Unit - Delaware Childers field (36-27N-16E).

Reports from the Bureau of Mines by Eakin, Johnston and Riggs, and Powell provided relevant information about the Mink leases. One report shows that most of the Mink leases were developed late in the primary production stage of the field. Another report by Powell lists the production history of the Sinclair-Tanner Flood (annual totals for the 1,200 acres, but not by lease). Drilling/completion reports for about 75 wells on the Mink leases, many of which have since been plugged, were obtained from NIPER's Oklahoma Well Log Library. B & N Oil Company provided core analyses, from cores drilled in 1935 and 1936, from several wells on the Mink leases. Using all available information about the Mink Unit, a net pay isopach map was constructed, an estimate of initial oil saturation was made, and the production history for the unit was determined from initial development to 1952. Actual lease production records from 1953 to the present have been available for continuing this production history.

The Mink leases were determined to have an average porosity of 20%, an initial average oil saturation of 31%, and a combined net pay volume of 2,900 acre-feet. The estimated cumulative oil production from the two leases has been 341,217 bbl through 1986. The project area has a surface area of 17.8 acres and a net pay volume of 516 acre-feet.

In 1988, as a result of the sale of this oilfield, cores were drilled on the Mink Unit, and the resulting information was provided by the new owners to NIPER. The average oil saturation of Mink Unit cores was 31.9%, which agreed well with our prediction (31%) that had been based on earlier core reports and production history. The average permeability was 90 millidarcies, which was higher than that of some of the earlier core analyses, and the average porosity was 19.1%.

With an estimated irreducible oil saturation of 25%, the recoverable oil within the leases was 76.3 bbl/acre-foot or about 40,000 bbl in the pilot area, at the initiation of the microbial field project. The Mink Unit covers a 160-acre area of which 110 acres are productive, and contains 21 injection wells and 15 production wells drilled on 5-acre spacing. The Mink Unit pilot area well spacing is given in figure 2. Most of the producing wells are flowing, only one of the producing wells (S-CP-2) is being pumped. Well completions are open-hole. Reservoir properties and characteristics are listed in table 2.

TABLE 2 Reservoir properties for Mink Unit

Formation	Bartlesville Sandstone
Depth, ft	600
Average net pay thickness, ft	30
Average permeability, md	90
Porosity, percent	20
Average formation temperature, °F (range 65 to 80)	75
Number of injection wells	21
Number of production wells	15
Average water injection rate, bbl/day	40/well
Average injection pressure, psi	530
Average oil production, bbl/day	6.4
Oil gravity, ° API	34
Oil viscosity, cP @ 77°F	7
Total dissolved solids of injection water, %	0.03
Average total dissolved solids of produced water, %	0.5
Average oil saturation, percent (at start of project)	31

The pilot site for the project was four adjacent inverted 5-spot patterns within the Mink Unit (figure 2). The pilot site covers an area of 17.8 acres and has a net pay volume of 516 acre-feet. The pilot area has four injection and eight production wells. In addition, two off-pattern wells (C-BP-2 and S-AP-4) were monitored as part of this project. Before beginning the microbial waterflood, all possible efforts were made to ensure that no changes in operating conditions or procedures occurred during the pilot test. The normal procedure of backflushing all injection wells each week was continued to maintain good injectivity and water quality.

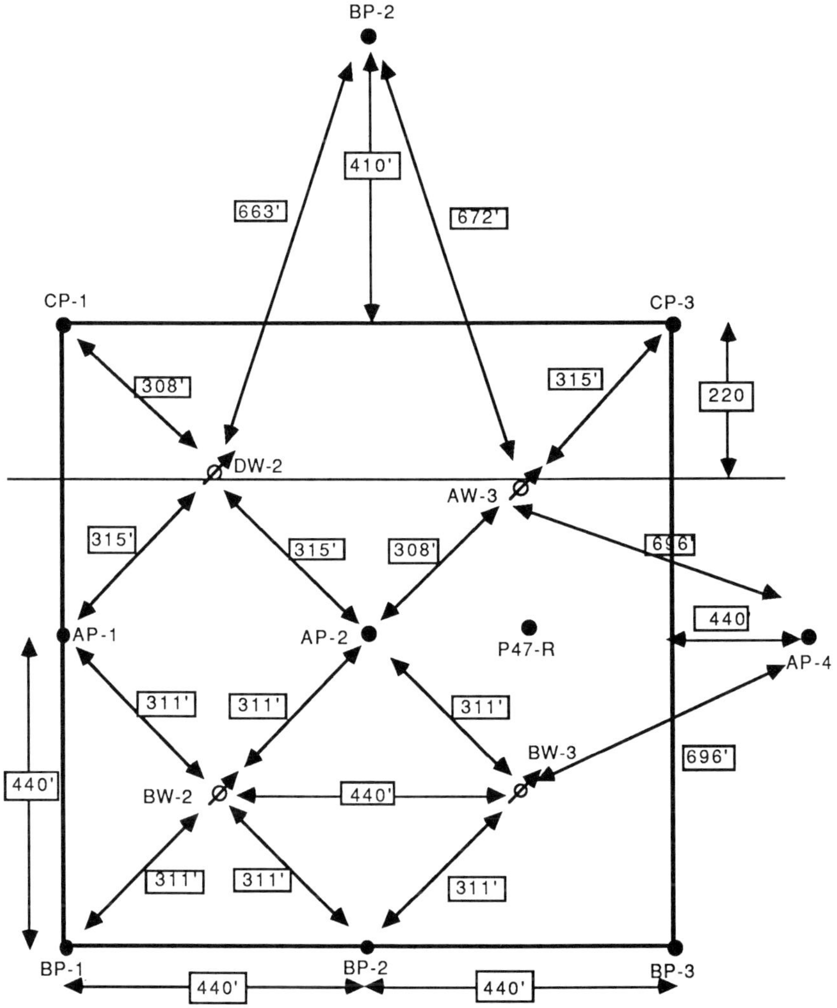

Fig. 2. Pilot area of Mink Unit showing well spacing in ft.

Field sampling began in November 1986, and continued to March 17, 1987. The data from these baseline studies showed that the total dissolved solids (TDS), pH, oil viscosities, and microbial counts were consistent during this period. Field data, including injection pressures and volumes, oil production, and water/oil ratios all remained fairly constant during the baseline monitoring. Since this waterflood had been ongoing since 1953, it is not surprising that the data from the individual wells and the field remained consistent. The same source of injection water, the Verdigris River, has been used since the initiation of the waterflood. The salinity(TDS) of the produced water from each production well and each injection well remained constant to within ± 0.01%. The microbial counts from the producing wells were consistent throughout the monitoring period. The counts were very low, and ranged from 0-100 cells/ml in the producing wells. Sulfate-reducing bacteria were consistently present at low levels in the tank battery water, and intermittently present in the plant injection water. There were sporadic occurrences of sulfate-reducing bacteria in the produced waters. The tabulated data from the baseline studies are compared to the results after microbial treatment and are given later in the paper.

Field data were also monitored during the baseline period. Individual producing well water/oil ratios (WOR), injection plant and injection well pressures, and oil production were tabulated for the baseline monitoring.

During the baseline period, two single-well injectivity/survivability tests were conducted with two microbial formulations. These single-well tests provided valuable information about the ability of the microbial formulation to survive under actual reservoir conditions and showed that no plugging or increase in injection pressures occurred after microorganisms and molasses were injected. The sequence of events that occurred in this project is given below: In May, 1988, a company purchased property in the Delaware-Childers oilfield from B & N Oil Company. The Mink Unit leases were a part of this purchase. The new owners began to conduct an infill drilling program and hydraulic fracturing activity in a lease adjacent to the Mink Unit, along with installation of a new waterflood injection plant.

SEQUENCE OF EVENTS

10/1/86 - SELECTED SITE

10/1/86 - 3/17/87 - BASELINE

10/1/86 - 3/17/87 - MICROBIAL TESTS

1/87 AND 3/5/87 - INJECTED FLUORESCEIN

2/5/87 - SINGLE WELL INJECTION

3/17/87 & 3/24/87 - INJECTED MICROORGANISMS

4/5/87 - INJECTION WELLS ON LINE

4/5/87 - 12/89 - MONITORED MINK UNIT

Chemical Tracer Study

A chemical tracer study was implemented during this baseline period (December, 1986) to determine: (1) the flow patterns of the injected fluids in the Mink Unit; (2) if any gross channeling existed; and (3) if there was communication among all producing wells and the four treated injectors.

Fluorescein was tested in the laboratory with the microbial formulation and reservoir fluids and found to be compatible under reservoir conditions; therefore, it was chosen as the chemical tracer. On Jan. 13, 1987, 27 bbl of a fluorescein solution at a concentration of 714 ppm was injected into wells S-BW-2 and S-BW-3, respectively; and on March 5, 1987, well C-DW-2 was injected with 5.2 bbl of 302 ppm fluorescein, and S-AW-3 was injected with 5.2 bbl of 210 ppm fluorescein solution. Sampling of each producing well was conducted daily for the first 5 days after tracer injection, then biweekly sampling continued until no fluorescein was detected in the production wells. Samples were protected from light and transported to NIPER where the fluorescein concentration was determined using a spectrophotometric method.

The fluorescein concentration curve was plotted against time for each producing well, and these results showed that all production wells communicated with at least one treated injector, since every well showed some fluorescein response. Although some fluorescein was observed in several producers within a few days of injection, there did not appear to be gross channeling because the response persisted for a reasonable period of time of 2 months before leveling to below the detection limit of 100 ppb. Presumably there were some high permeability, low volume stringers in the formation. The area under each curve was integrated, and a value was obtained. This value was divided by the average number of barrels of produced fluid for that well, and the wells were ranked accordingly. The tracer studies seemed to indicate a northeasterly flow pattern (figure 1), because the C-CP-1 and C-CP-3 wells and the S-AP-4 well received fluorescein in greater amounts and more quickly than the other wells. The middle well, S-AP-2, received the highest amount of fluorescein, which was expected since this well is affected by all four injection wells.

Fluorescein was again observed about 1.8 years after injection. The persistence of this response seems to indicate that the tracer had just transported through the matrix of the formation and that the earlier response of tracer was due to low-volume, high-permeability streaks in the formation. Of the wells sampled, the fluorescein appearance again indicates a northeasterly flow pattern, because producers C-CP-1 and C-CP3 showed fluorescein in high amounts, as did S-AP-2, the middle well. Well S-AP-1 was only sampled twice, and thus probably is not representative of its true tracer appearance.

Laboratory Design of the Microbial System

Various microbial formulations isolated by NIPER and INJECTECH, Inc. that satisfied the above criteria were grown with reservoir fluids from the Mink Unit in order to assess the compatibility of the microbial species. Those that appeared to be compatible under reservoir conditions were then tested in Berea sandstone cores to determine their ability to recover crude oil. Data from several microbial corefloods are shown in table 3. Various microbial formulations from INJECTECH, Inc. were primarily *Clostridium* and *Bacillus* species; the composition of NIPER Bac 1 is listed in table 4.

TABLE 3 Microbial coreflooding experiments for selection of microbial formulation

Core Desig.	k^1	Microbes	Amt Inj. (PV)	PV^2	$S_{orwf}{}^3$	$S_{orcf}{}^4$	$\Delta S_{or}{}^5$	$E_r{}^6$
M94*	332	NIPER 1	0.5	54.0	34.8	27.6	7.2	20.7
M68*	180	NIPER 2	0.5	46.0	25.0	11.7	13.3	53.2
M69*	134	NIPER 3	0.5	37.8	22.8	8.7	14.1	61.8
M77*	344	NIPER 3	0.5	56.7	34.0	18.3	15.7	46.2
M44*	304	NIPER 4	0.5	59.6	29.7	19.3	10.4	35.0
M61*	293	NIPER 4	0.5	57.5	27.5	19.1	8.4	30.6
B12*	271	NIPER Bac 1	0.2	57.7	34.7	30.4	4.3	12.4
B16A*	362	NIPER Bac 1	0.2	140.0	35.7	31.9	3.8	10.6
HS2	353	NIPER Bac 1	0.4	568.2	37.8	29.5	8.3	22.0
MSC 10	214	INJECTECH 1	0.5	55.7	30.2	28.8	1.4	4.6
MSC 14	180	INJECTECH 2	0.5	54.6	38.5	36.6	1.9	4.9
MSC 15	181	INJECTECH 3	0.5	54.8	33.9	31.7	2.2	6.5
MSC 21	99	INJECTECH 4	0.5	45.9	36.4	35.1	1.3	3.6
MSC 22	133	NIPER Bac 1	0.5	51.0	36.9	30.0	6.9	18.7
MSC 23	52**	NIPER Bac 1	0.5	43.8	31.5	22.6	8.9	28.3
MSC 24	162	NIPER Bac 1	0.5	53.0	38.6	34.2	4.4	11.4
MSC 25	35**	NIPER Bac 1	0.5	43.8	27.4	23.0	4.4	16.1

[1] Absolute permeability to brine, millidarcies

[2] Pore volume of core in cm^3

[3] Residual oil saturation in core after waterflooding (%PV)

[4] Residual oil saturation in core after microbial treatment (%PV)

[5] $S_{orwf} - S_{orcf}$

[6] $(S_{orwf} - S_{orcf})/S_{orwf} \times 100\%$

* Powdered cattle feed molasses that had been filtered through cheesecloth was used for these corefloods.

** Indicates core from Delaware-Childers field.

Based upon the coreflooding results, NIPER Bac 1 was selected for the field test (table 4). A coreflood test was conducted using core samples from Delaware-Childers field with NIPER Bac 1 to determine its oil recovery efficiency(corefloods MSC 23 and MSC 25). Although these cores had not been preserved, they were somewhat representative of the lithology of the formation. The microbial treatment recovered 28% of the residual oil remaining in Core MSC 23 after waterflooding. Most of the oil was recovered before the first pore volume (PV) of fluid was injected. This was consistent with our earlier observations in Berea sandstone cores(Bryant and Douglas, 1988). The waterflood residual oil saturation (31.5%) obtained for this core corresponded reasonably well with the estimated oil saturation value of 30% for the field. The oil saturation range from the 1988 core analysis of the Mink Unit was 18.6 to 43.9%. The permeability value of 52 md was well within the range for the Mink Unit reported in core analyses (0.5 - 302 md; average was 90 md).

TABLE 4 Composition of NIPER Bac 1

NIPER Designation	Genus Designation	Oxygen Requirement	Microbial Products
NIPER 1	*Bacillus licheniformis*	Facultative	Surfactant, acids
NIPER 2	*Bacillus* species	Facultative	Surfactant, acids
NIPER 3	*Clostridium* species	Anaerobic	Gases, alcohols, acids, surfactant
NIPER 4	Unknown	Facultative	Gases, acids

Facultative - can grow with or without oxygen
Anaerobic - can only grow without oxygen

After the initial microbial coreflood experiment with the field core, the core MSC 23 was flooded with several pore volumes of brine from the Mink Unit, and then re-injected with a consortium of microorganisms isolated from Mink Unit produced water. These microorganisms were characterized as being indigenous to the Mink Unit, and it was important to determine if our microbial formulation, NIPER Bac 1, could survive and out-compete these microbes in porous media. The indigenous microbial consortium was grown in 4% v/v molasses, 0.2 PV was injected into the field core, and the core was shut-in for 3 days. A slug of NIPER Bac 1 (0.2 PV) in molasses was injected, and the core was shut-in for another 3 days. The core was waterflooded, and the effluent was counted for microorganisms and microscopically observed. The results showed that NIPER Bac 1 continued to thrive in the presence of these indigenous microorganisms, even though the indigenous microbes also grew well inside the core.

Environmental Considerations

All reasonable precautions were taken to ensure that there are no adverse effects on the environment from conducting this microbial-enhanced waterflood field experiment. All injected materials were handled in closed containers. A contingency spill plan was formulated by NIPER in the unlikely event of an accidental microbial contamination.

Single-Well Injection Tests

In February 1987, during the baseline period, a single-well injection using an off-pattern well was performed to establish certain parameters before injection of the microbial formulation at the Mink Unit site. The goals of these tests were: (1) to determine if the reservoir experienced any reduced injectivity due to microbial plugging; (2) to ensure that the microbial consortium grew well under reservoir conditions; and (3) to develop adequate sampling procedures for the project. An off-pattern injection well (S-DW-1) was injected with 26 gallons of NIPER Bac 1 (approximately 1.0×10^8 cells/ml) in molasses (4% v/v) and shut-in for 12 days. The well was backflushed and samples were collected every 10 to 15 minutes until NIPER Bac 1 and molasses were detected. Samples were inoculated into sulfate-reducing bacterial growth media to determine if sulfate-reducing bacteria were stimulated by introduction of molasses and other microorganisms. Gas samples were assayed for hydrogen sulfide. Samples and pressures were also taken at another injection well nearby (S-DW-2) to serve as a control well.

The results from these tests showed the following: (1) the injection rates and pressures of the microbially-treated well after the shut-in period were normal and comparable to those of the control well, indicating that no plugging had occurred; (2) all of the injected bacterial species in NIPER Bac 1 were detected in high numbers in the backflush samples, indicating that the microbes were still growing and metabolically active after 12 days of incubation under reservoir conditions. No hydrogen sulfide (limit of detection was < 5 ppm) was found in any of the gas samples, and no sulfate-reducing bacteria were detected from any of the microbially-treated well samples; and (3) our sampling procedures appeared adequate to measure gas and fluid from the injection and production wells.

Injection of NIPER Bac 1 and Molasses

Twenty-six gallons (0.65 bbl) of the microbia! formulation, NIPER Bac 1, was injected into each of the four targeted injection wells, C-DW-2, S-BW-2, S-AW-3, and S-BW-3 (see figure 2). Wells C-DW-2 and S-BW-2 were treated on March 19, 1987, and wells S-AW-3 and S-BW-3, on March 23, 1987. Twenty gallons of undiluted molasses was injected into each well where it was diluted by injection water to a concentration of approximately 4% periodically during and after the microbial injection. The molasses and microorganisms were injected by means of a header bypass system(Bryant, et al. (1990). The header bypass allowed injection water from the injection well to flow into the header, mix with the microorganisms or the molasses, and then flow back into the injection well. This system allowed adequate mixing and dilution of the concentrated molasses, and since the ongoing water was used as the injection driving mechanism, no adverse shearing effects on the microorganisms occurred. The four treated injection wells were shut-in until April 3, 1987, when water injection was resumed. The other 17 injection wells in the Mink Unit were still in operation during the shut-in period. After water injection was resumed, the injection wells were backflushed to determine if microbial activity could be observed. All wells produced foam, indicating surfactant production and that the microbial populations were viable. The four injection wells were injected with 2 gallons of undiluted molasses per well per day until September 21, 1989.

Sampling of the producing wells was conducted on a weekly basis. Samples were collected from a flowing stream in sterile 4-oz flint glass bottles. Each bottle was filled completely and tightly capped. The samples were taken to the laboratories and processed immediately. The parameters monitored after microbial injection are given in table 5. The results given are averages of samples taken during the period of April 3, 1987 through June 1, 1989.

TABLE 5 Parameters monitored for field test

Parameter	Sampling Frequency	Location Sampled
Total dissolved solids	Weekly	Each producer, plant water
pH	Weekly	Each producer, plant water
Surface tension	Weekly	Each producer, plant water
Microorganisms	Weekly	Each producer, plant water
Molasses conc.	Weekly	Each producer, plant water
Oil viscosity	Biweekly	Each producer
Interfacial tensions	Biweekly	Each producer
Microorganisms	Monthly	Each treated injector
Surface tension	Monthly	Each treated injector
pH	Monthly	Each treated injector
Gas analyses	Every few months	Randomly selected producers

Microbial Counts and Molasses Concentration

The microbial counts have always been surprisingly low for this particular field, particularly since the source water is river water. The baseline counts averaged less than 100 cells/ml for most wells. During the test, the microbial populations fluctuated greatly, yet on an average, were about 100 times higher for all production wells after the microbial injection. Some of the injected bacteria from the NIPER Bac 1 consortium were observed at 32 weeks post-injection. Since that time, the same bacterium has been observed periodically from several producing wells. Interestingly, the time from injection to observation of NIPER 1 in the producing wells tends to correlate with the observation of fluorescein. However, a large population of NIPER 1 has never been consistently observed in any producing well or the tank battery.

Sulfate-reducing bacteria (SRB) were monitored using lactate-based API sulfate broth. No significant changes occurred in the SRB population after injection of the microbial formulation and molasses. Additionally, no "black water" has appeared before or after injection from the production wells. The low SRB activity is probably more a reflection of the low sulfate content of the reservoir (< 15 ppm).

Molasses was assayed in the production well waters using an anthrone carbohydrate assay. One week after injection of the microbial formulation and molasses, molasses was detected in seven producing wells and the tank battery. Interestingly, the 4 injection wells had been shut in during this time period. Water was injected into the injection wells 2 weeks after microbial injection and molasses was subsequently injected twice a week to each of the four injection wells. No molasses was detected in any of the produced water samples after 5 weeks from its initial appearance. At week 7 post microbial injection, molasses sporadically appeared in a few production wells for about 4 weeks, then disappeared. The appearance of molasses early in the test indicated the presence of low-volume stringers within the reservoir. The subsequent disappearance of detectable molasses indicated the establishment of a microbial population that consumed all of the molasses. The presence of the molasses occurred primarily in the north wells of the pattern. This directional flow pattern observed with the molasses analyses corresponded well with the chemical tracer studies.

Total Dissolved Solids

The total dissolved solids (TDS) of produced water samples was measured weekly for every producing well, the water from the Mink Unit tank battery, and the plant injection water since the initiation of the baseline monitoring period (October, 1986). Table 6 shows the average values obtained for each well during the baseline and post-microbial injection. The TDS values did not change significantly during the baseline and post-MEOR periods, and are well within their standard deviations.

TABLE 6.Total dissolved solids (TDS, %) from Mink Unit Samples. Sampling average period was Oct. 28 - March 17, 1987 for the baseline, and April 3, 1987 - June 1, 1989

Well	Baseline	Post-MEOR
C-CP-1	1.083 ±0.04	0.967 ±0.07
C-CP-3	1.120 ±0.01	1.074 ±0.06
S-AP-1	0.663 ±0.03	0.660 ±0.05
S-AP-2	1.063 ±0.10	1.014 ±0.08
S-P47R	0.275 ±0.03	0.286 ±0.02
S-BP-1	0.479 ±0.03	0.488 ±0.04
S-BP-2	0.557 ±0.10	0.520 ±0.04
S-BP-3	0.478 ±0.02	0.498 ±0.03
S-AP-4	0.730 ±0.03	0.669 ±0.03
C-BP-2	0.656 ±0.02	0.666 ±0.05
PLANT	0.029 ±0.01	0.034 ±0.01
TANK B.	0.500 ±0.02	0.504 ±0.04

pH

The pH of each sample was recorded for the baseline and post-microbial injection periods, to determine if microbial injection actually altered the pH of the produced water from the individual wells. Table 5 shows the average of the values obtained during these monitoring times. The pH has not changed significantly since the baseline. Although these microorganisms produce short-chained fatty acids such as acetic, propionic, and butyric acids, these are relatively weak acids produced in such low concentrations that they are probably not lowering the pH. During the backflushing of the injection wells, there seemed to be a significant pH effect, which will be detailed later in this paper. The dissolved oxygen content of the injected water for the Mink Unit was determined to be 5 ppm.

TABLE 7 pH averages for samples taken from the Mink Unit. Sampling average period was Oct. 28 - March 17, 1987 for the baseline, and April 3, 1987 - June 1, 1989

Well	Baseline Avg	Post-MEOR Avg
C-CP-1	6.60 ±0.29	6.85 ±0.38
C-CP-3	6.53 ±0.18	6.83 ±0.31
S-AP-1	6.63 ±0.15	6.79 ±0.31
S-AP-2	6.52 ±0.26	6.79 ±0.34
S-P47R	6.53 ±0.11	6.79 ±0.29
S-BP-1	6.65 ±0.18	6.93 ±0.33
S-BP-2	6.56 ±0.21	6.94 ±0.36
S-BP-3	6.64 ±0.30	6.92 ±0.34
S-AP-4	6.55 ±0.13	6.83 ±0.28
C-BP-2	6.62 ±0.26	6.82 ±0.32
PLANT	7.40 ±0.09	7.48 ±0.29
TANK B.	6.67 ±0.28	6.90 ±0.36

Surface and Interfacial Tensions

Surface tension measurements were conducted twice during the baseline period, and weekly on water samples from the Mink Unit post-microbial injection. The interfacial tension (IFT) was not measured during the baseline period, but measurements were initiated soon after microbial injection (May, 1987). Surface and interfacial tension were monitored to provide some indication of microbial surfactant production. There was a decrease in surface tension and IFT for well S-P47R, the well closest to the microbially-treated injection wells. The lower values are probably due to surface active agents produced by the injected microorganisms that traversed quickly through low-volume, high-permeability streaks in the formation. When the values for 1987, 1988, and 1989 were compared, the 1987 values were the lowest, while the injection water surface tension was higher in 1987 than in 1988 or 1989. The lower 1987 values were not due to any injection water change from the plant, but rather the microbial injection. The Mink Unit tank battery was not sampled because an emulsion breaker chemical which affects the surface and IFT measurements was added to the tank battery .

TABLE 13 Predicted and actual oil production rates for the Mink Unit

Year	Predicted Avg bbl/wk	Actual Avg bbl/wk
1981	50.0	50.5
1982	48.8	46.5
1983	47.7	46.8
1984	46.6	46.4
1985	45.6	44.8
1986	44.5	45.1
1987	43.5	48.8
1988A	42.6	48.2
1988B	42.6	46.5
1989	41.7	36.3
1990	40.7	

1988A - Jan. 1 - May 31, 1988
1988B - Jun. 1 - Dec. 31, 1988

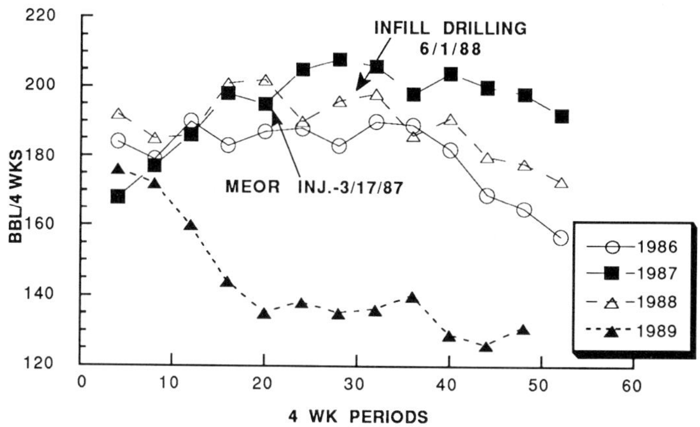

Fig. 9. Oil production from Mink Unit from 1985-1989.

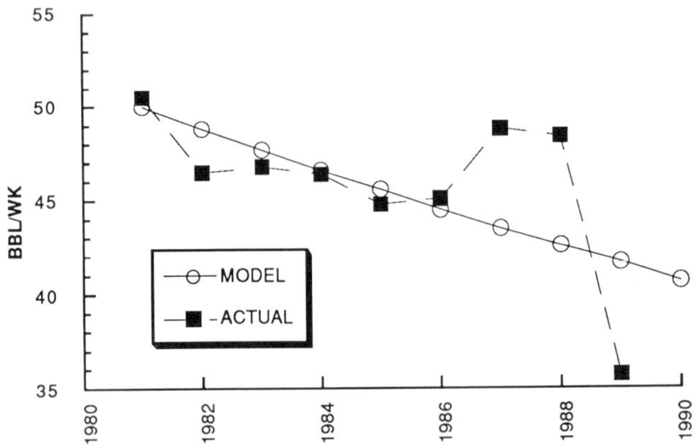

Fig. 10. Predicted and actual average oil production for Mink Unit during 1976-present.

Mink Unit Project Economics

During the 14 months of microbial/nutrient injection prior to infill drilling and hydraulic fracturing activity, 577 incremental bbl of oil were obtained when compared to the predicted oil recovery by waterflooding alone. A total of 18.7 tons of molasses was injected during this period. Using a nutrient cost of $100/ton, this is equivalent to $3.24 per bbl of incremental oil. This does not take into account any other injection costs, although for this particular project, the costs were fairly minimal. However, this cost also does not include any projected recoveries beyond the time of infill drilling. Since fluorescein was detected in the producing wells after the infill drilling, one may assume that the microbial treatment had not yet transported through the formation matrix, and thus the complete effect on incremental oil production that may have occurred would have been masked.

CONCLUSIONS

This microbial-enhanced waterflood field project demonstrated the feasibility of microbial technology in a manner that any independent operator could implement. It is noteworthy that no operating problems were encountered before or during this project, until the initiation of infill drilling and hydraulic fracturing activity. No corrosion problems were experienced; in fact, the sulfate-reducing bacterial populations remained relatively low compared to the baseline counts. There were no problems with injectivity. During the project, laboratory and field data were correlated to develop and document a methodology for conducting microbial-enhanced waterflood field projects. This particular microbial formulation, NIPER Bac 1, and molasses injection improved oil production rate and water/oil ratios for producing wells nearest the injection wells. All in all, this project has shown promise, particularly for mid-continent stripper well oil production.

REFERENCES

Bryant, R.S. and J. Douglas. Evaluation of Microbial Systems in Porous Media for Enhanced Oil Recovery, SPE paper 16284. SPE Reservoir Engineering, May, 1988.

Bryant, R.S. Laboratory and Field Studies on Microbial Enhanced Oil Recovery, J. of Industrial Microbiology, (1989) Volume 30, No. 4, pp. 255-267.

Bryant, R.S. and Burchfield, T.E. Review of Microbial Technology for Improving Oil Recovery, SPE paper 16646. SPE Reservoir Engineering, May, 1989.

Bryant, R.S., T.E. Burchfield, K.L. Chase, K.M. Bertus, and A.K. Stepp. Optimization of Microbial Formulations for Oil Recovery: Mechanisms of Oil Mobilization, Transport of Microbes and Metabolites, and Effects of Additives. SPE paper 19686. Presented at the 64th Annual Technical Conference and Exhibition of the Society for Petroleum Engineers, San Antonio, TX, Oct. 8-11, 1989.

Bryant, R.S., T.E. Burchfield, D.M. Dennis, and D.O. Hitzman. Microbial-enhanced Waterflooding: Mink Unit Project. SPE/DOE paper 17341. SPE Reservoir Engineering, February, 1990.

Eakin, J.L. Recent Developments in Water Flooding in Nowata County, OK, Oil Fields, 1954-1955. Bureau of Mines Report No. RI-5134, 1955.

Hitzman, D.O. Review of Microbial Enhanced Oil Recovery Field Tests, Proc. Symposium on Application of Microorganisms to Petroleum Technology, Bartlesville, OK (Aug. 12-13, 1987) U.S. DOE No. NIPER-351, CONF-870858, VI-1 - VI-42.

Hitzman, D.O. Microbiological Secondary Recovery of Oil, U.S. Patent No. 3,032,472 (May 1,1962).

Johnston, K.H. and C.H. Riggs. Secondary-Recovery Practices and Oil Reserves in the Eastern Part of the Delaware-Childers Field, Nowata County, OK. Bureau of Mines Report No. RI-4019, 1946.

Moses, V. Microbes and Oil Recovery, an Overview, Proc. First World Conference and Exhibition on the Commercial Applications and Implications of Biotechnology, London (May 4-6, 1983) 415-422.

Powell, J.P. Four Waterflooding Projects in Washington and Nowata Counties, Oklahoma, 1959. Bureau of Mines Report No. IC-7896, 1959.

Updegraff, D.M. and Wren, G.B. The Release of Oil From Petroleum-Bearing Materials by Sulfate-Reducing Bacteria, Appl. Microbiol. (1954) vol 2, 309-322.

Zobell, C.E. Bacteriological Process for Treatment of Fluid-Bearing Earth Formations, U.S. Patent No. 2,413,278 (Dec. 24, 1946).

Reprinted from: *Microbial Enhancement of Oil Recovery – Recent Advances,* edited by E.C. Donaldson

CH. F–5

BIOTECHNOLOGY OF ENHANCEMENT OF OIL RECOVERY BASED ON THE GEOCHEMICAL ACTIVITY OF MICROORGANISMS (Field Experiments)

M.V. IVANOV and S.S. BELYAEV
Institute for Microbiology, USSR Academy of Sciences, Prospect 60-letiya Oktyaby, 7, 117811, Moscow (USSR)

ABSTRACT

The indigenous microbial flora of an injection well in a test area of an oil field were activated by injection of surface water containing essential salts. The increase of methanogenesis and decrease of chloride content accompanied an increase of oil production at surrounding production wells. Lack of increased microbial activity or decrease of chloride content at one well indicated that activation of microbes did not occur within the drainage area of the well and explained the fact that oil production did not increase. The increase of production (20-100%) in the test area was the incentive for additional experiments in another oilfield.

INTRODUCTION

Microbiological techniques of enhanced oil recovery (MEOR) can be divided into two groups. The first group includes methods for injecting compounds into reservoirs (biopolymers, surfactants, solvents, emulsifying agents and other products of biotechnology) (Gutnick et al., 1983; Wang Xinyan, 1983). These methods are based on the improvement of the oil displacement properties of injected water, and they are in fact, similar to chemical methods for EOR. The second MEOR group presupposes formation of oil displacing agents as the result of microbial activities in the petroleum reservoir itself. In some cases, microorganisms and nutrients are injected into the strata (Karaskiewicz, 1974; Lazar, 1983), and in other cases a activation of stratal microflora is suggested (Ivanov, Belyaev, 1983; 1985).

There are at least three convincing reasons to consider the possibility of creating new oil recovery technologies based on biogeochemical activities of microorganisms in subsurface petroleum reservoirs. The first one consists of a wide distribution of aerobic and anaerobic microorganisms in oil fields, which are exploited by flooding with surface water. A number of microorganisms do not lose their viability at subsurface reservoir conditions and remain biochemically active. The second reason is based on diverse abilities of microorganisms to transform different oil hydrocarbons into low molecular organic compounds and gases, many of them are oil-release agents such as CO_2, CH_4, fatty acids, alcohols and other solvents. The third reason is the capability of microorganisms to produce these compounds directly from part of residual oil in the reservoir.

OBJECTS AND METHODS OF INVESTIGATION

Our field experiments were conducted in the Bondyuzchskoe oil field, located in the Tatarian Republic. This oil field was described in our previous report at the Shangri-La meeting in 1982 (Ivanov and Belyaev, 1983). Some properties of the oil-bearing rocks of this oil field are presented in Table 1.

Table 1. Some properties of oil-bearing rocks, oil and brine of Bondyuzhskoe Oil Field (After Wolfson and Teleshova, 1966).

Depth of oil-bearing rocks	1500-1700 m
Litology	Sandstones and silty sandstones
Geological age	Upper Devonian
Average porosity of sandstones	21.8%
Average permeability	500 millidarcies
Type of oil	High sulphureous paraffin type
Relative oil density	0.871-0.876
Temperature of brine	30-35°C

Water sampling for the study of microorganisms and their activity in the pre-bottom zone of the injection well were taken from back flow at the wellhead in nitrogen filled sterile glass bottles.

The experiments to investigate the intensity of microbial activity were conducted in the field (Figure 1) using one injection well (No. 446) and six surrounding production wells. Fresh water was pumped into the injection well to sustain stratal pressure. Intensification of microbial activity in the stratum was carried out by supplementary aeration of injected water with the addition of some mineral salts.

Chemical and physicochemical analysis of the injection and formation waters were made during one day of sampling as described earlier by Beyaev et al. (1982). The intensity of bacterial methanogenesis was estimated using the radioisotopic method. Radiocarbon was used as $NaH^{14}CO_3$ and $^{14}CH_3COONa$ compounds. The radioactivity of released methane was registered on a scintillator (Laurinavitchus, Belyaev, 1978).

The isotopic composition of methane and carbon was estimated using a spectrometer by the double-ray compensation method, using CO_2 as the working gas. Methane was purified from its gaseous homologies on molecular sieves; methane purity was controlled gas-chromatographically (Ivanov et al., 1983).

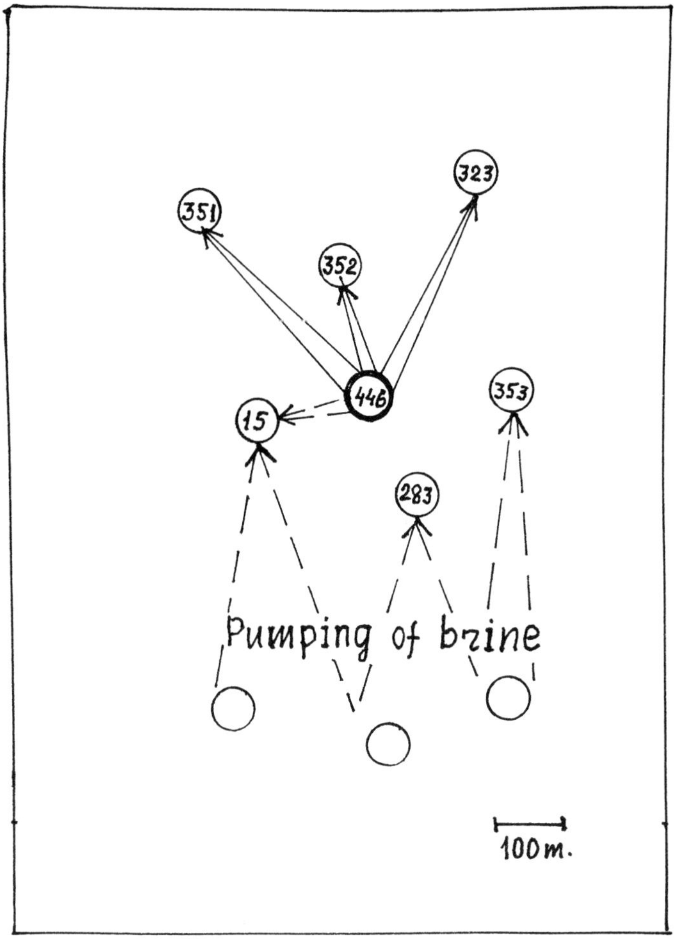

FIGURE 1. The scheme of experimental lot on the exploited oil field.

RESULTS AND DISCUSSION

The studied oil field has been produced for more than 20 years. To sustain stratal pressure, the oil stratum is flooded with surface waters. Undoubtedly, injection of fresh water into the stratum leads to a change in chemical composition and properties of the oil-field brine and to the formation of desalined stratal zones. Another important property of injected surface waters is the fact that they serve as entry for various microorganisms and dissolved oxygen.

Table 2 shows the microbiological characteristics of the injected waters. The total amount of bacteria reached $10x10^9$ cells per ml. Aerobic bacteria, capable of oxidizing oil and individual hydrocarbons, were present in the waters. In addition, sulfate-reducing bacteria were found and even such strict anaerobes such as methanogenous bacteria.

Table 2 Number of microbes in injection water from wells of the oil field

Well No.	Mineralization	Total Count	Oil Oxidizers	$C_{14}H_{30}$ Oxidizers	Sulfate Reducers	Methanogens Cell
428	0.7	8700	600	2500	4000	25000
422	0.8	7600	600	2500	2000	2500
28	0.8	10000	2500	2500	6000	1300
366	0.8	-	25	25	60000	250

The data, obtained while studying the microflora of the formation waters show that the total number of bacteria and the number of different physiological groups, (both aerobic and anaerobic) in the formation waters is considerably lower than in the injected water. The lower count in the produced water is due to the fact that a large number of bacteria cannot survive in the stratal waters - some of them die from the severe conditions existing in stratal waters. Viable cells of methanogenic bacteria are observed in oil-field brine with a total salinity to 280 g/l, but much greater quantities were found in the highly desalinated waters. The intensity of modern bacterial methanogenesis in the formation waters of the pool was $(15-305)*10^{-6}$ ml $CH_4:l^{-1}$ per day. The highest rate was observed in partly desalinated waters. A substantial part of the biogenous methane (30-90%) was formed in the process of CO_2 reduction by hydrogen (Ivanov and Belyaev, 1983).

It was of special interest to investigate microbiological processes occurring in the pre-bottom zone of injection wells because it is the area of the oil pool which is in contact with both surface and formation waters, which of course are different in composition and properties.

During oil stratum flooding dissolved oxygen is carried into the stratum together with the water. It is used rather quickly in the pre-bottom zone and that results in a sharp decrease of the redox potential. Evidently, oxygen is spent on microbial oxidation of the residual oil in the pre-bottom zone. It is indicated by a quantitative increase of oil-oxidizing bacteria and by an increase of water-soluble organic compounds. The decrease of redox potential and the increase of C_{org} content leads to the activation of anaeorbic processes, and of the methanogenesis, in particular.

Hence, activation of aerobic hydrocarbon oxidizing bacteria (which leads to additional production of water-soluble organic compounds due to residual oil oxidation) is the reason for the increase of anaerobic bacteria activity in the oxygen depleted zones of the oil pool (Ivanov and Belyaev, 1983).

The data obtained provided for further experiments on controlled intensification of microbiological activity directly in oil pool.

Figure 1 shows the experimental scheme of the oil field. It consisted of the injection well No. 446, drilled and injected with fresh water two years before the beginning of the experiment, and six production wells. All of the wells were interconnected along layers. Other layers were isolated before the beginning of the experiment.

The field brine of the experimental area was characterized before the beginning of the experiment as follows: (Table 3) weakly-reduced conditions (Eh from +140 to 170 mv), saline (140-175 g/l), and bicarbonate content of about 175 mg per liter.

Table 3 Physico-chemical and microbial characteristics of stratal waters of the experimental area before the field experiment

CHARACTERISTICS	INJECTION WELL NO. 446	PRODUCTION WELLS
Salinity, g/l	0.8	I40-I75
eh, mV	+70	+40-+170
HCO_3^{-I}, mg/l	140	I40-I75
$^{13}C_{HCO_3^{I-}}$, o/oo	-I4/-I6	-II.7 (average)
$^{13}C_{CH_4}$, o/oo	-65/-80	-48.2 (average)
Oil Oxidizers*	20000	6-40
$C_{14}H_{30}$ Oxidizers	2500	2-60
Methanogens**	20000	60-250
Sulphate Reducers**	3700	200-700
Rate of Methanogenesis: $ml \cdot CH_4 i^{-I} \cdot day^{-I} \cdot 10^{-6}$	35.8	6.3-26.8
Rate of Sulphate Reduction: $mg\ H_2S l^{-I} \cdot day^{-I} \cdot 10^{-6}$	96.0	4.0-30.0

*All data - microbial cells per ml.

**All data - microbial cells per liter.

Microorganisms of various physiological groups were present in oil field brine, however, their quantity was about 10^2 lower than those in the pre-bottom zone of the injection well. The rates of current processes of methanogenesis and sulfate-reduction were a hundredth parts of a microliter CH_4 or a microgram of H_2S per liter per day, Table 3.

The carbon isotope composition of methane in the oil-field brine was characterized by ^{13}C to be about 48. The average values of ^{13}C of dissolved carbonate was noticeably lower than the value ^{13}C of carbonates from the pre-bottom zone of injection well.

Thus, these data show that the representatives of various physiological groups of microorganisms were distributed in the formation waters, but their geochemical activity in anaerobic conditions was very low.

For activation of microbial processes in formation waters in the experimental area, were used the injection of aerated fresh water with added mineral salts into well NO. 446. After six injections in the summer of 1983, well No. 464 was under routine flooding. Further observations were made at the surrounding production wells of the experimental area.

Table 4 summarizes the data obtained during a three year observation period following the beginning of the experiment. First of all, note the decrease of salinity in the formation waters; one can conclude that not all the production wells reacted to the activation. Formation waters from wells 351 and 352 desalinated considerably, and those from well 15 desalinated to a lesser degree. Due to stable or even increasing salinity (wells 283, 353), we conclude that fresh water from injection well 446 did not reach the other experimental wells. The analysis of the hydrodynamic condition of the experimental area supports these data (Figure 1). Production wells 282 and 353 are located in the zone of active impact of a strong front from saline water injection, which is responsible for hydrodynamic regime of these wells. Microbiological, chemical and physicochemical characteristics of the formation waters of these wells also point out the absence of noticeable changes compared to the analogous characteristics, obtained before the experimental impact.

According to the data in Table 4, along with the salinity change of formation waters from wells 351, 352 and 323, a number of other changes also were observed. Redox potential in waters from these wells decreased which is indicative of the reduction processes taking place. The quantity of microorganisms in stratal waters (especially that of methanogenic bacteria) increased as well. Along with a quantitative increase of methanogenic bacteria in formation waters of wells 351, 352 and 323, an increase of recent methanogenesis was observed. The quantity of aerobic hydrocarbon-oxidizing and oil-oxidizing microorganisms did not increase markedly, but nevertheless,

Table 4. Changes of physico-chemical and microbial characteristics of waters of production wells during the field experiment.

CHARACTERISTICS	REACTED WELLS			NON-REACTED WELLS		
	351	352	323	15	283	353
Salinity, g/l	140* / 75**	155 / 85	150 / 140	150 / 130	150 / 155	175 / 200
Eh, mV	+150 / +120	+140 / +105	+145 / +165	+155 / +145	+165 / +165	+170 / +175
Methanogens***	60 / 1000	250 / 450	250 / 200	250 / 1700	250 / 250	250 / 230
HCO 1-/3, mg/l	168 / 230	212 / 245	154 / 180	168 / 185	172 / 181	175 / 165
Oil Oxidizers****	25 / 25	25 / 110	20 / 85	25 / 53	40 / 50	6 / 30
$C_{14}H_{30}$ Oxidizers***	15 / 20	10 / 130	60 / 70	25 / 75	2 / 2	6 / 2
Rate of Methanogenesis $ml \cdot CH_4 i^{-I} \cdot day^{-I} \cdot 10^{-6}$	17.5 / 160.6	8.0 / 77.4	6.3 / 34.9	16.9 / 63.7	26.3 / 21.6	19.2 / 11.8
$^{13}C_{HCO_3^{-1}}$, o/oo	-10.8 / -14.4	-11.5 / -19.6	-11.9 / -----	-9.8 / ----	-14.4 / -11.3	---- / -10.8
$^{13}C_{CH_4}$, o/oo	-48.5 / -51.2	-53.7 / -56.3	-42.2 / -48.6	----- / -50.7	-48.3 / -50.6	----- / -43.9

* The numerator - data before experiment

** The denominator - data after experiment

*** This data - number of microbial cells per liter

**** All data - number of microbial cells per ml

the content of aerobic products of metabolism such as bicarbonate-ion increased considerably in the formation waters.

An important proof of oxidizing microbiological process activation at the first stage of impact on stratal microflora is the decrease of isotope content of bicarbonate carbon in formation waters of reacted wells (^{13}C from -

14.4 to -19.6) compared to the isotope composition of carbonates in non-reacted wells No. 283 and No. 353 (^{13}X-11.3 and -10.8). The source of carbonate production with light carbon isotope composition is oil with ^{13}C value -26.

The low isotope composition of carbon of methane in the wells from the activation zone (Table 4) has also indicated more intensive microbial processes in this zone.

In the Figures 2,3 and 4, the data show the changes of oil recovery and some components of the water chemical composition in the course of observation during the two years following completion of the experimental area of the oil field.

In the upper part of Figure 2, the results of the bicarbonate-ion and chloride-ion content are shown. The first index describes a summary of the microbial activity of the oil pool. Chloride-ion concentration defines the general saltiness of the formation water. The lower part of Figure 2 shows

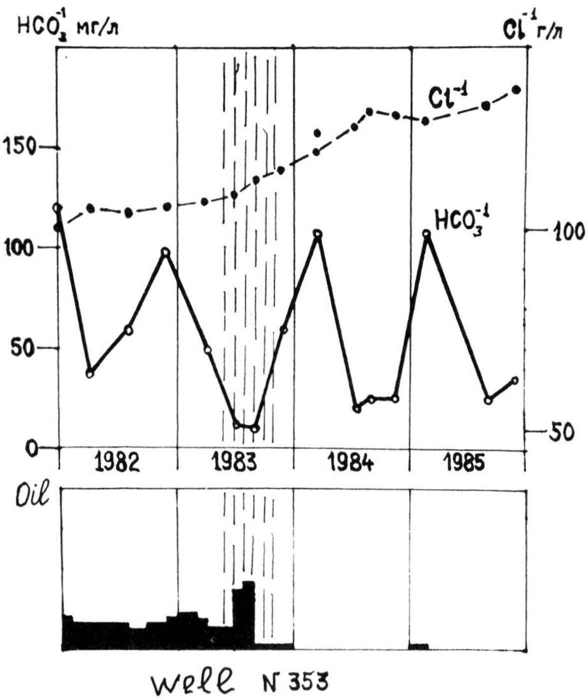

FIGURE 2. Influence of microbiological processes on oil recovery out of the well No. 353.

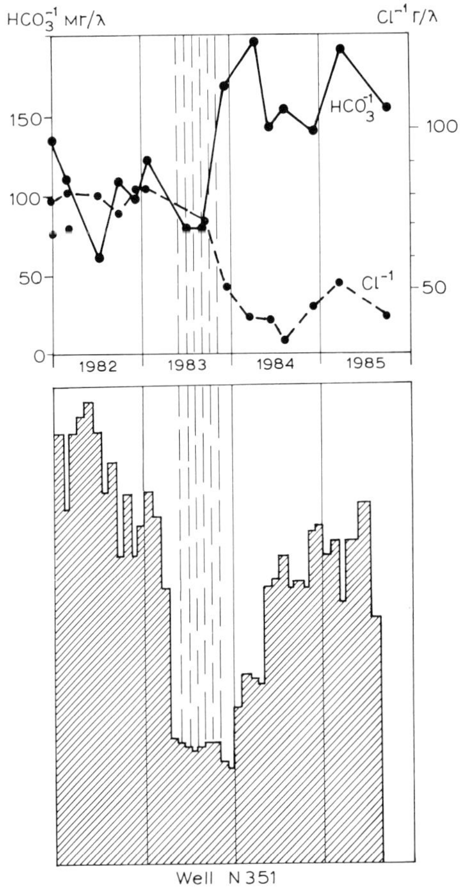

FIGURE 3. Influence of microbiological processes on oil recovery out of the well No. 351.

the monthly oil output in arbitrary units. The time period of the experimental treatment of the oil pool is shown by shading (June-October, 1983).

As the data on Figure 2 shows we failed to achieve an increase of oil recovery from well 353. Progressive increase of salinity of the water indicates that the fresh water containing oil-recovery agents did not reach this well.

Quite the contrary is true of well 351 (Figure 3). In 1982-1983, before the beginning of the field experiment, a constant decline of oil production was observed, and starting in 1984 (about 10 months after the experimental injections) a noticeable increase of oil production occurred. Considerable decrease of chloride-ion content of the water shows that a substantial inflow of fresh water to the experimental well occurred. A growth of bicarbonate-ion during the same period proves that considerable activation of microbial processes took place.

Interesting data were obtained from well 15 (Figure 4). There was no noticeable increasing in oil recovery; however, two small peaks of the oil production coincide in time with the faint water desalinization and noticeable rise of the bicarbonate content in the water from this well. Thus, the data of Figures 2,3 and 4 show the efficiency of the experiments on the activation of the microbiological activity and, to some extent, they interpret the mechanism the of oil recovery increase.

We have concluded that during the injection of aerated water into the flooded oil pool microbial processes were activated in it which were limited by the shortage of oxygen necessary for intensive growth of aerobic microorganisms. The result of injection of aerated water at the first stage is activation of aerobic microorganisms and oxidation of part of residual oil. Products from the oil-oxidizing bacteria, such as CO_2 and low molecular organic compounds, enter the oxygen depleted zone and act as agents of enhancement oil recovery by serving as additional substrates for anaerobic bacteria (for methanogens in particular). The second stage of action consists in the activation of bacterial methanogenesis and formation of additional methane which first is an easily available fuel gas, and second increases the oil fluidity when dissolved in it and thus enhances oil recovery.

The oil flow in the experimental wells of the Bondyzhskoe oil field was increased by 20-100%, and the water content decreased by 20-30%.

According to the calculations carried out by the petroleum engineers, during the three years following the beginning of the experiment, 47,000 tons of additional oil have been recovered. This makes more than one third of the total oil produced in the experimental area during the three years of observations.

During the past two years, the MEOR method of the oil recovery using activation of the oil pool microflora has been tested in several experimental areas of the Romashkinskoye oil field. The increase of the oil recovery in these areas has made up from 12 to 45% from the total oil production during 1.5 years of observation.

Field experiments using the activation of indigenous microorganisms are under way at the present time.

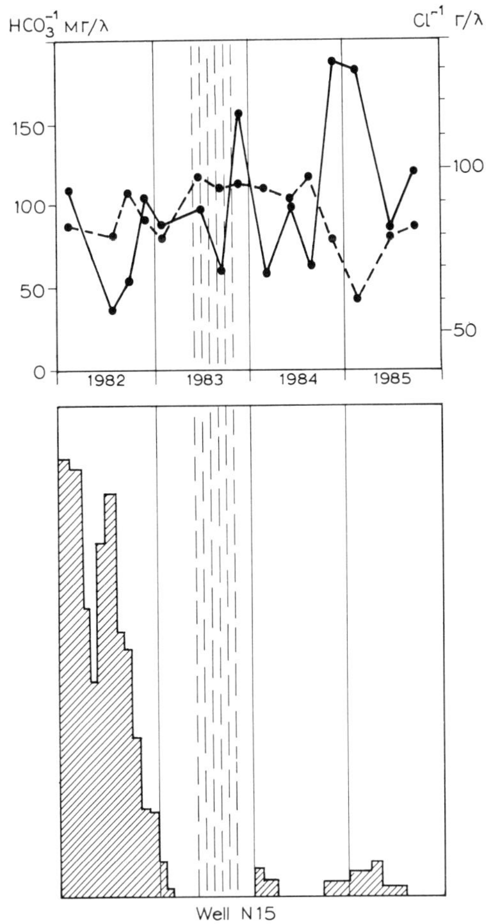

FIGURE 4. Influence of microbiological processes on oil recovery out of the well No. 15.

REFERENCES

Belyaev, S.S., Laurinavichus, K.S., Obrazstova, A.Ya, Gorlatov, S.N., Ivanov, M.V., 1982. Vol. 51, pp. 997-1001.

Gutnick, D.L., Zosim, Z., Rosenberg, E., 1983. Proceedings of 1982 International Conference on MEOR/U.S. Department of Energy Conf. No. 8205140, pp. 6-11.

Ivanov, M.V., Belyaev, S.S., 1983. Proceedings of 1982 International Conference on MEOR/U.S. Department of Energy Conf. No. 8205140, pp. 48-57.

Ivanov, M.V., Belyaev, S.S., 1985. The Solar Energy Transformation, M:/Nauka, pp. 154-160.

Ivanov, M.V., Belyaev, S.S., Zyakun, A.M., Bondar, V.A., Laurinavichus, K.S., 1983. Geochimie, No. 11, pp. 1647-1654.

Karaskiewicz, J., 1974. Prace Institutu Naftowego, Katowice, Wydawnictwo "Slasu", p. 67.

Laurinavichus, K.S., Belyaev, S.S., 1987. Microbiology, vol. 47, pp. 1115-1117.

Lazar, I., 1983. Proceedings of 1982 International Conference on MEOR, U.S. Department of Energy, Conference No. 8205140, pp. 140-148.

Wang, Xiuyuan, 1983. Proceedings of 1982 International Conference on MEOR, U.S. Department of Energy, Conference No. 8205140, pp. 29-37

Wolfson, I.S., Teleshova, M.N., 1966. The Oils of the Tatar ASSR: Chimie, p. 127.

Reprinted from: *Microbial Enhancement of Oil Recovery - Recent Advances*,
edited by E.C. Donaldson

CH. F–6

MICROBIAL SELECTIVE PLUGGING TO CONTROL WATER CHANNELING

T.R. JACK, L.G. STEHMEIER, M.R. ISLAM AND F.G. FERRIS
NOVA HUSKY RESEARCH CORPORATION,
2928 16TH ST. N.E., CALGARY, ALBERTA, CANADA, T2E 7K7

ABSTRACT

Over the past decade, NOVA HUSKY RESEARCH CORPORATION has reported the invention and development of a novel microbial selective plugging system to control undesirable water production in heavy oil fields in the Lloydminster region of Western Canada.

In this paper, the evolution of that system will be summarized along with the results of a major field test carried out in the Standard Hill field of Saskatchewan. The surprising extent of bacterial travel underground in the field will be compared to laboratory observations and to predictions obtained from a particulate transport model. The implications of the findings will be discussed in terms of future work.

INTRODUCTION

NOVA HUSKY RESEARCH CORPORATION first became interested in the purposeful use of microorganisms underground to facilitate or extend oil production a decade ago. At that time, Husky Oil, one of our corporate parents, had operations principally focused on the production of very extensive heavy oil assets located around Lloydminster on the Alberta/Saskatchewan border in Western Canada. An initial review of their holdings suggested that bacterial action was possible in typical Lloydminster reservoirs and that a prime aim of the research should be the selective plugging of high water permeability zones responsible for the "watering out" of most heavy oil wells. In this region, wells do not run out of oil but rather develop excessive water cuts which render further production uneconomical at an oil recovery of less than 10% of the original oil in place. Given the thin pay zones typical of regional sand reservoirs in this area, sophisticated EOR procedures are not and will not be economic in any foreseeable future. The challenge of extending recoveries then falls to finding an economic way to plug off water bearing zones in these highly permeable targets (typically one or more Darcies) so that water cuts could be restored to practical levels at least for a sufficient period to pay for the treatment and return a profit.

While the work entailed in developing and testing a practical microbial system has been reported as the work progressed, an overview the entire project may be useful for those entering this area for the first time. Detailed information on particular aspects of the work can be obtained from the references cited. In the hope of encouraging the development of MEOR systems generally, as much information has been placed in the open literature as was consistent with establishing and maintaining a patent position (Thompson and Jack, 1984, 1985a, 1985b).

2 PRELIMINARY WORK

The project began with a review of existing literature on MEOR (Jack, 1983) as well as existing patents which employed microorganisms in oil production (Jack and Thompson,1983). In the last few years, two review articles have been published describing this and other work (Jack, 1986 and Jack, 1988).

In considering use of organisms underground, preliminary work addressed the question of whether injected bacteria would actually be able to use oil as a food source in the reservoir. This is a key question since such an attack could have led to widespread microbial activity at little cost but would have also entailed degradation of the very asset which was being sought. A study (Jack, Lee and Mueller, 1985) of a range of oils under "anaerobic" conditions led to the conclusion that oil could indeed be attacked with very little oxygen present but that such attack was insignificant in extent, dependent on oxygen and most likely to occur with high paraffin oils. Radioisotopic labelling indicated direct conversion of hydrocarbons into carbon dioxide and methane with coincidental production of hydrogen sulfide. It was also discovered that production of this last gas could be effectively eliminated by addition of small amounts of nitrate, a strategy later proposed by others to control H_2S production for microbial systems injected into sulfate bearing reservoirs.

3 CRITERIA FOR A SUCCESSFUL SYSTEM

Initial lab work proceeded along two lines; identification of potential problems associated with introduction of the microorganisms into the formation face at the bottom of the injection well and selection of microbial isolates able to produce potentially useful agents under the conditions found in the target reservoirs.

Progress along the first line was greatly facilitated by the thesis work of Shaw (1982) done with partial funding support from NOVA HUSKY RESEARCH CORPORATION. In this work, a number of criteria were identified for the successful injection of bacteria into even high permeability formations like the unconsolidated sands found in the target reservoirs (Jack, Shaw, Wardlaw and Costerton, 1989). The conclusion was that organisms had to be small, singly dispersed and not producing polymers, slimes or gas bubbles under the conditions of injection. The ideal organism is spherical with a diameter less than one fifth the pore throat size of the target formation.

Prior to screening for isolates, a practical fermentable food source had to be identified since it was clear that oil would not serve this purpose under the anaerobic conditions of the reservoir. Sugar beet molasses, a byproduct of sugar production from sugar beets in Alberta was chosen for its low stable price and availability. Extensive markets as a food supplement in livestock operations throughout Western Canada has resulted in a widespread distribution system being developed such that large volume truckloads can be ordered to oil well sites in clean trucks diluted and supplemented with sugar as needed.

4 DEVELOPMENT OF THE MICROBIAL SYSTEM

An extensive screening program sought bacterial isolates able to ferment sugar beet molasses into acids, gases, surfactants or polymers under reservoir conditions (Jack, Thompson and DiBlasio, 1982). Consideration of results and opportunities led to the choice of a selective plugging system based on polymer production as the objective of further development. The need for an insoluble polymer for stable plug formation was recognized in lab core studies (Jack, DiBlasio, Thompson and Ward, 1983). Since it was apparent from work described above that organisms cannot be successfully injected if they are producing polymer, it was also apparent that an isolate had to found in which growth on an economical medium which did not support polymer formation was possible to prepare an inoculum which could be injected into the reservoir. Inspection of the literature suggested that use of organisms bearing dextransucrase might meet all criteria. Since they only produce polymer in the presence of sucrose, they can be grown on media containing other sugars for injection and can then be overrun with sucrose rich sugar beet molasses to trip insoluble polymer production *in situ*. This concept is the basis for patent coverage. In a second round of screening wild isolates bearing this enzyme system were separated and identified as members of the genus *Leuconostoc*. This in turn led to the discovery of a ready bank of useful cultures which had been isolated and characterized thirty years before at the U. S. Department of Agriculture, NRRL, Peoria, Illinois for quite different purposes. In fact later field tests were actually based on isolates from the NRRL collection known to produce insoluble dextrans.

5 SELECTION OF THE FIELD TARGET

With the microbial system in hand, consideration was given to the nature of the field target. Catastrophic water breakthrough in heavy oilfields occurs under three different scenarios; bottom water, water channeling and leaks around the casing. Water channeling was selected as the target since bottom water, a scenario in which a contiguous water zone directly underlies the oil zone was considered to be likely to be uneconomical due to the size of plug needed. The channel scenario in which water enters the well via a well defined zone was chosen as the target application (Jack and DiBlasio, 1985).

6 FIELD TESTS

A single well field test was set up to test the practicality of the microbial selective plugging system. In preparing for this test, a series of laboratory core tests were carried out to establish travel rates for cells, essential nutrients and sugar beet molasses medium through the reservoir. This was necessary to ensure that all the components of the microbial system reached the intended site of plugging together. Basically the cells being particles were found to travel relatively slowly, enabling an injected slug of bacteria to be overrun easily by a later injection of sugar beet molasses medium (Jack, 1985). One problem encountered at this late stage was the adsorption of bacteria on clays present in the native reservoir sands. In a potentially disastrous oversight all previous core studies had been done with either artificial cores or cleaned and sorted sands devoid of fines. Fortunately, certain sacrificial agents developed at B. C. Research, Vancouver, B. C. (Howard and Stirling, 1987) under a previous project to prevent adsorption losses of injected surfactants on clays also prevented bacterial adsorption. This technology was based on a treated pulp mill effluent available in Alberta within easy reach of Lloydminster oilfields and was accessible by licence from Canadian Patents Development Ltd., Ottawa for field test purposes.

The single well field test carried out in 1986 failed to give convincing evidence of microbial selective plugging largely due to its simple nature and a serious lack of background data. It did demonstrate the practicality of injecting a microbial system and served as a useful practice for a later large scale field test (Jack, 1986; Jack and Stehmeier, 1988).

In 1988, a serious field test was carried out with the financial assistance of Canmet, EMR, Ottawa through the Extraction and Recovery of Bitumen and Heavy Oil Subprogram, Contract 23440-7-9066/01 with Dr. A. E. George as assigned scientific authority. This field test has recently been described in detail elsewhere (Stehmeier, Jack, Blakely and Campbell, 1990). In brief, declining heavy oil prices required the selection of a target in which a single plugging action could reclaim or save oil production on a field wide basis. Such a field site was found at Standard Hill, Saskatchewan where a series of wells had watered out in a distinct trend across a large and productive field one after the other. The possibility that water had entered the field from an edge aquifer and had channeled across the field from a single well entry point had been recognized by the field operations staff. The intent was to inject the microbial plugging system into the throat of the channel system at the edge of the field with the intent of blocking further water access to the field.

A tracer study was carried out using tritiated water. This served to correctly identify the proper injection point and gather some information on the nature of the channel system. This proved to be very surprising. To account for tracer observations, the channel system volume had to be very small and flow rates very fast. It was estimated that volumes were as low as 50 m^3 for a 240 m length between wells with flow velocities of over 90 m/hour. This exceeded recommended velocities used in lab core studies by two orders of magnitude and caused justified concern for the chance of plugging off such a channel.

Stated briefly injection of the microbial system proceeded smoothly and was preceded by a small volume of sacrificial agent to preserve injectivity in the injection well. The overall effect however was a washout. The injected bacteria and their metabolites were found over 1.8 km from the injection site indicating that the microbial system had been swept deep into the field before polymer formation could occur. A subsequent drawdown test indicated that the throat of the channel system remained unblocked.

Following the salinity of produced waters also indicated passage of the freshwater slug in which the bacterial system had been injected into an otherwise saline environment. Again rapid travel over prodigious distances was observed. The sequential appearance of several slugs of fresh water at a single production well indicated that the channel system must resemble a spider's web in which an injected slug can split up and travel by more than one interconnecting path to a single production well.

Overall, the microbial part of the field test performed well in that the microbial system was injected successfully and did perform as expected underground without serious competition from indigenous organisms. It is apparent however that much more information must be learned about field targets before successful applications can be expected.

7 CURRENT WORK

Current work features two areas of activity and is again financially supported by Canmet, EMR with Dr. F. M. Mourits as assigned scientific authority under Contract 23440 8 9222.

The first component of the work done by Dr. M. R. Islam involved the mathematical modelling of lab core runs using a particulate transport model based on the concept of parallel pathways as derived by Gruesbeck and Collins (1982a, 1982b) for the travel of fines through porous media. Having adjusted the model parameters to fit the observed rates and extent of throughput of bacterial cells in lab sandpacks, an attempt was then made to match the travel seen in the field. Three key parameters were varied; the deposition factor which characterized the retention of cells in the porous media, the reentrainment factor which allowed deposited cells to be returned to flow and the fraction of pluggable channels which assumes that some channels through the porous medium will be plugged by fines. It should be noted that this model was not competent to handle a flow dependent retention mechanism seen in lab core studies since no such term is inherent in the model. Nevertheless, results for continuous injections could be matched in the lab runs.

Attempts to match field observations for bacterial travel over short distances were futile using any reasonable values for the adjustable parameters. Fits for travel over a distance of 1.8 km required a channel of small cross section as well as very low values for the retention factor. Since clay adsorption gave high retentions in lab runs, the most consistent picture is one of well defined small cross section channels swept clear of clays normally found in reservoir sands.

The second area of activity carried out by Dr. F. G. Ferris and L. G. Stehmeier will remain proprietary until patents are in place. To our knowledge Ferris and Stehmeier have invented the first inducible bacterial selective plugging system which does not rely on organic matter to be the plugging agent and which will give thermally stable, durable plugs or sand consolidation capable of withstanding very significant pressure drops.

8 CONCLUSION

Work by NOVA HUSKY RESEARCH CORPORATION has established the practicality of injecting bacterial systems in the field with purposeful intent and that properly designed systems can act as factories underground to convert relatively cheap feedstocks into useful agents to enhance or prolong oil recovery. Obviously much is yet to be learned about effective target selection and placement of such systems.

This overview has been presented with the hope that a detailed account of our efforts will inspire others and enable them to enter this challenging area quickly and effectively.

9 REFERENCES

Gruesbeck, C. and R.E. Collins, Dec. 1982a. Entrainment and Deposition of Fine Particles in Porous Media. Society of Petroleum Engineers Journal, pp.847-856.

Gruesbeck, C. and R.E. Collins, Dec. 1982b. Particle Transport Through Perforations. Society of Petroleum Engineers Journal, pp.857-865.

Howard, J. and M. Stirling, 1987. Sacrificial Agents for Enhanced Oil Recovery, U.S. Patent #4,713,185.

Jack, T.R., 1983. Enhanced Oil Recovery by Microbial Action. Chemical and Geochemical Aspects of Fossil Energy Extraction. T.F. Yen, F.K. Kawahara and R. Hertzberg, editors. Ann Arbor Science, Ann Arbor, Michigan.

Jack, T.R., 1985. Application of Bacteria to the *In Situ* Recovery of Heavy Oil. Proceedings of the Third International Conference on Heavy Crude and Tar Sands. Long Beach, California, UNITAR/UNDP, New York, II: 745-747.

Jack, T.R., 1986. Microbially Enhanced Oil Recovery. Comprehensive Biotechnology, M. Moo-Young, editor. Pergamon Press, Toronto.

Jack, T.R., 1987. Application of Bacteria in the *In Situ* Recovery of Heavy Oil. Proceedings of the First International MEOR Workshop, Abilene, Texas, April 1-3, 1986, J.W. King and D.A. Stevens, editors. U.S. Department of Energy, Bartlesville. pp. 225-232.

Jack, T.R., 1988. Microbially Enhanced Oil Recovery, Biorecovery, 1: 59-73.

Jack, T.R. and DiBlasio, E., 1985. Microbes and Oil Recovery, Proceedings of the International Conference on Microbial Enhancement of Oil Recovery, Fountainhead, Oklahoma, May 20-25, 1984. J.E. Zajic and E.C. Donaldson, editors. Petroleum Bioresources, El Paso, pp. 205-212.

Jack, T.R. DiBlasio, E. Thompson, B.G. and Ward, V., 1983. Bacterial Systems for Selective Plugging in Secondary Oil Production. Preprints, Symposia, Division of Petroleum Chemistry, American Chemical Society, 27: 773-784.

Jack, T.R., Lee, E.G.H. and Mueller, J.C., 1985. Anaerobic Gas Production: Controlling Factors. Microbes and Oil Recovery, Proceedings of the International Conference on Microbial Enhancement of Oil Recovery, Fountainhead, Oklahoma, May 20-25, 1984. J.E. Zajic and E.C. Donaldson, editors. Petroleum Bioresources, El Paso, pp. 167-180.

Jack T.R., Shaw, J.C., Wardlaw, N.C. and Costerton, J.W., 1989. Microbial Plugging in Enhanced Oil Recovery. Chapter 7 in Microbial Enhancement of Oil Recovery, E.C. Donaldson, G.V. Chilingarian and Teh Fu Yen, editors. Elsevier, Amsterdam, pp. 125-149.

Jack, T.R. and Stehmeier, L., 1988. Selective Plugging in Watered Out Wells. In the Proceedings of the Symposium on Application of Microorganisms to Petroleum Technology, Bartlesville, Oklahoma, August 12-13, 1987. (W. Linville, ed.) U.S. Department of Energy, Bartlesville.

Jack, T.R. and B.G. Thompson, 1983. Patents Employing Microorganisms in Oil Production. Microbial Enhanced Oil Recovery, J.E. Zajic, D.G. Cooper, T.R. Jack and N. Kosaric, editors. Pennwell Books, Tulsa, Oklahoma, pp. 14-25.

Jack, T.R., Thompson, B.G., and DiBlasio, E., 1983. The Potential for Use of Microbes in the Production of Heavy Oil. Proceedings of 1982 International Conference on Microbial Enhancement of Oil Recovery, Shangri-la, Oklahoma, May 16-21, 1982. U.S. Department of Energy, BETC, 1983.

Shaw, J.C. 1982. Bacterial Fouling of a Model Core System. M.Sc. thesis, Geology Department, University of Calgary.

Stehmeier, L.G., Jack, T.R., Blakely, B.A. and Campbell, J.M., 1990. Test of a Microbial Plugging System at Standard Hill, Saskatchewan to Inhibit Oilfield Water Encroachment. In Biohydrometallurgy Proceedings of the International Symposium, Jackson Hole, Wyoming, August 13-18, 1989. J. Salley, R.G.L. McCready and P.L. Wichlacz, editors. CANMET, EMR, Ottawa.

Thompson, B.G. and Jack, T.R., 1984. Method of Enhancing Oil Recovery by Use of Exopolymer Producing Microorganisms, U.S. Patent 4,460,043.

Thompson, B.G. and Jack, T.R., 1984. Method of Enhancing Oil Recovery by Use of Exopolymer Producing Microorganisms, Canadian Patent 1,196,303.

Thompson, B.G. and Jack, T.R., 1985. Method of Enhancing Oil Recovery by Use of Exopolymer Producing Microorganisms, Canadian Patent 4,561,500.

Reprinted from: *Microbial Enhancement of Oil Recovery - Recent Advances*, edited by E.C. Donaldson

CH. F–7

MICROBIAL CONTROL OF THE PRODUCTION OF SULFIDE

M. J. McInerney*, K. L. Sublette**, and A. D. Montgomery*
Department of Botany and Microbiology, University of Oklahoma, Norman, OK, 73019-0245* and the Center for Environmental Research and Technology, University of Tulsa, Tulsa, OK 74104**

ABSTRACT

A microbial process to control the production of hydrogen sulfide by sulfate-reducing bacteria using mutant strains of *Thiobacillus denitrificans* was developed. *T. denitrificans* oxidizes sulfide to sulfate using oxygen or nitrate as the electron acceptor, but is inhibited by sulfide concentrations above 100 to 200 μM. A mutant of *T. denitrificans* (strain F) resistant to glutaraldehyde (40 mg/l) and sulfide (1500 μM) was obtained by repeated subculturing at increasing concentrations of the inhibitors. Strain F prevented the accumulation of sulfide by *Desulfovibrio desulfuricans* when both organisms were grown in liquid medium or in Berea sandstone cores. The wild-type strain of *T. denitrificans* did not prevent sulfide accumulation by *D. desulfuricans*. Strain F also prevented the accumulation of sulfide by a mixed population of sulfate-reducing bacteria enriched from an oil field brine. Fermentation balances showed that strain F stoichiometrically oxidized the biogenically produced sulfide to sulfate. Strain F grew at temperatures up to 40°C, salinities up to 2%, and pressures up to 120 atm. These data showed that strain F would be effective in controlling sulfide production in many oil reservoirs and other environments.

INTRODUCTION

In some anaerobic environments such as marine and estuary sediments, sulfate reduction rather than methanogenesis is the dominant electron accepting reaction for the complete mineralization of organic matter (McInerney, 1985; Widdel, 1989). Sulfate-reducing bacteria are a diverse group which oxidize a wide range of organic compounds and H_2 and reduce elemental sulfur and oxyanions of sulfur to sulfide. Sulfide production by sulfate-reducing bacteria is directly or indirectly responsible for major damage in the petroleum industry each year due to the corrosion of of steel and iron tanks, pipelines, values, pumps, etc. (Hamilton, 1985). Also, sulfide production increases the sulfur content of the crude oil which decreases its values and increases refining costs.

Because of the deterimental effects of sulfate-reducing bacteria, the control of their growth and activity is important for the economical recovery of oil. The most common approach is to inhibit the growth of sulfate-reducing bacteria by the addition of biocides (typically glutaraldehyde) to the water distribution system of most waterfloods. These measures are often of limited effectiveness since sulfate-reducing bacteria are associated with other anaerobic bacteria in polysaccharide-containing biofilms which coat the surfaces of pipes and other materials. Within these biofilms, the sulfate-reducing bacteria find themselves in a somewhat protected environment in which water-soluble biocides do not effectively penetrate.

We have developed a biological method to control the production of sulfide by sulfate-reducing bacteria and to treat sulfide-laden waters using a bacterium that can oxidize sulfide using nitrate as the electron acceptor (Montgomery et al., 1990; Sublette et al., 1989). *Thiobacillus denitrificans* is an autotroph and a facultative anaerobe which oxidizes reduced sulfur compounds to sulfate. Under anaerobic condiditions, nitrate is the electron acceptor and is reduced to gaseous nitrogen (N_2). The introduction of viable cells of *T. denitrificans* into environments with sulfate-reducing bacteria has the potential to control sulfide production so long as nitrate concentrations remain high. The application of particular interest is to control sulfide production by sulfate-reducing bacteria at or near the water injection well by inoculating the well-bore area with sufficient numbers of *T. denitrificans.* The ability of a sulfide and gultaraldehyde-tolerant strain of *T. denitrificans* to control biogenic sulfide production in liquid culture and in sandstone cores is described.

2. MATERIALS AND METHODS

2.1 Organisms and Cultivation

The wild-type strain of *Thiobacillus denitrificans* (ATCC 23462) was grown anaerobically in a mineral salts medium with thiosulfate as the electron donor, nitrate as the electron acceptor, bicarbonate as the carbon source and ammonium as the nitrogen source (Sublette and Sylvester, 1987). A spontaneous mutant of *T.*

denitrificans which was resistant to glutaraldehyde and sulfide was isolated by selection for growth in medium with elevated levels of glutaraldehyde and then sulfide (Sublette and Woolsey, 1989). *Desulfovibrio desulfuricans* strain D was obtained from Dr. M. P. Bryant, University of Illinois, Urbana, IL. Sulfate-reducing enrichments were obtained from secondary anaerobic digestor sludge in Norman, OK, two separate aquifer sites underlying the Norman municipal landfill, an aquifer underlying the Savannah River Laboratory, Aiken, South Carolina, and an oil field brine from the Wilmington Field, Long Beach, CA. The thiosulfate mineral medium was modified to grow the sulfate-reducing bacteria and cocultures containing *T. denitrificans* strains by deleting thiosulfate and adding (per 100 ml) 0.1 g Na_2SO_4, 0.05 ml of 60% sodium lactate syrup, and 0.1 ml of Balch vitamin solution (Balch et al., 1979). Thiosulfate was added to this medium when *T. denitrificans* strains were grown in pure culture. The preparation and use of media were according to the anaerobic methods of Hungate as modified (Balch et al., 1979: Bryant, 1972).

2.2 Analytical

Growth was followed spectrophotometrically at 460 nm. Lactate and sulfate were determined using high pressure liquid chromatography (Montgomery et al., 1990). Nitrate, nitrite and sulfide were analyzed colorimetrically (Montgomery et al., 1990). When the amount of sulfide in Berea sandstone cores was determined, the core was crushed in a mortor and pestle, placed in a sparging vessel with concentrated HCl. The vessel was sparged with nitrogen gas for 30 min, and the volatile sulfides present in the gas were trapped in a zinc acetate solution. Portions of this were analyzed colorimetrically for sulfide.

2.3 Preparation of cores

Cylindrical Berea sandstone cores (1.9 in diameter) were prepared as described previously (Jenneman et al., 1985; Raiders et al., 1986). The penetration of *T. denitrificans* and sulfate-reducing strains and cocultures was studied using cores with permeabilities greater than 500 mdarcies. Each core was connected to two flasks of sterile media (Jenneman et al., 1985). One flask was inoculated and the time required for growth to appear in the second flask was determined and defined as the approximate penetration time. The core and flask system was incubated inside of an anaerobic chamber to maintain anaerobic conditions throughout the incubation period.

3. RESULTS

3.1 Liquid cultures

A sulfide and glutaraldehyde resistant strain of *Thiobacillus denitrificans* (strain F) was isolated by spontaneous mutation. The wild-type strain of *T. denitrificans* had a low level of resistance to 1 to 5 ppm levels of glutaraldehyde. By repeated subculturing of this strain in medium with increasing glutaraldehyde concentrations, a strain which was resistant to 40 ppm of glutaraldehyde was isolated. Tests showed that the resistant strain grew in medium with 12 ppm, but the wild-type strain lost 100% of its viability. The glutaraldehyde resistant strain was repeatedly subcultured in thiosulfate medium with increasing concentrations of sulfide (up to 1000 µM). One strain, strain F, was chosen for further study. The growth rate and final absorbance of strain F in thiosulfate medium with up to 1000 µM sulfide was similar to control cultures without sulfide added. In medium with 1500 µM sulfide, strain F grew after a lag of 3 days. Concentrations of 825 µM greatly inhibited the growth of the wild-type strain and no growth was observed in medium with 1000 µM sulfide.

The effect of salt on the growth of strain F was also studied. Strain F grew in the thiosulfate medium with 2% NaCl or 2.6% KCl, but growth was inhibited 82% in medium with these levels of salt. Growth was not detected when the concentration of NaCl or KCl exceeded 3 and 3.7%, respectively.

To determine whether strain F could use sulfide produced by sulfate-reducing bacteria, a medium which would support growth of both types of organisms was developed. The thiosulfate medium used for growth of *T. denitrificans* was modified by adding lactate, sulfate, and a source of organic growth factors required by most sulfate reducers. The growth of the *T. denitrificans* strains was not affected by the addition of 1000 ppm each of yeast extract, trypticase, and sulfate to the thiosulfate medium. However, the addition of sodium lactate greater than 2.65 mM to the medium increased the lag time and decreased the growth rate. The modified medium with 2.65 mM sodium lactate, 1000 ppm sodium sulfate, and a B-vitamin solution (Balch et al., 1979) supported good growth of *D. desulfuricans* alone and in coculture with the *T. denitrificans* strains.

D. desulfuricans was grown alone and in coculture with the wild-type and sulfide resistant strains of *T. denitrificans* to determine the effectiveness of the two *Thiobacillus* strains in using biogenically produced sulfide (Table 1). High concentrations of sulfide were produced

TABLE 1

Sulfide production by *D. desulfuricans* grown with and without *T. denitrificans* strains

Culture[a]	Sulfide concentration (mg/l) after	
	14 days	19 days
DD alone[b]	47.4	42.2
DD + wt	28.0	28.0
DD + F	4.0	<0.1

[a] Cultures were grown anaerobically in the modified medium without thiosulfate and with lactate, sulfate and B-vitamins added. The inoculum size for each strain was 0.1 ml.
[b] Abbreviations: DD, *D. desulfuricans*; wt, with-type strain of *T. denitrificans*; F, sulfide resistant strain of *T. denitrificans*.

by all pure cultures of *D. desulfuricans* and in all cocultures containing the wild-type stain of *T. denitrificans*. The small amount of sulfide present after 14 days in cocultures with strain F disappeared after nineteen days of incubation. When larger inoculum sizes of strain F were used no sulfide was ever detected in the cocultures. This was not the case for cocultures with the wild-type strain of *T. denitrificans*.

To determine whether strain F would be effective in controlling sulfide production by natural populations of sulfate-reducing bacteria, five different environmental samples were collected and grown in modified medium without thiosulfate and nitrate. Each sample was transferred 3 times in this medium in order to select for sulfate-reducing bacteria. In the absence of nitrate, each enrichment produced sulfide as evidenced by the blackening of the medium (Table 2).

TABLE 2
Sulfide production by various environmental samples in the presence and absence of nitrate

Source	Sulfide production in medium	
	Without nitrate	With nitrate
Sewage sludge	+	-
Norman landfill aquifer		
methanogenic site	+	-
sulfate-reducing site	+	-
Savannah River plant aquifer	+	-
Oil field brine	+	+
D. desulfuricans	+	+

When nitrate was present, only the enrichment from the oil field brine produced sulfide. Evidently, the other enrichments had indigenous populations of nitrate-using bacteria which either used the sulfide or inhibited its production by sulfate-reducing bacteria (Jenneman et al., 1986).

The stoichiometry of substrate use and product formation by sulfide-producing cultures in the presence and absence of strain F is shown in Table 3. *D. desulfuricans* grown alone produced stoichiometric

TABLE 3
Fermentation balances sulfate-reducing cultures grown with and without strain F [a]

Culture	Concentration (mM)				
	Lactate	Acetate	Nitrate	Sulfate	Sulfide
DD	<0.1	1.4	39	5.4	0.6
DD + F	<0.1	1.0	33	7.0	<0.1
Brine enrichment	<0.1	<0.2	37	5.8	0.9
Brine enrichment + F	<0.1	<0.2	33	6.6	<0.1
F	1.9	<0.2	8.0	26	<0.1

[a] Medium contained 1.9 mM sodium lactate. Abbreviations are listed in Table 1.

amounts of acetate and sulfide from lactate and sulfate. In coculture with strain F, lactate was completely used and the decrease in nitrate concentration indicated the growth of strain F. No sulfide was detected in these cocultures. The oil field brine enrichment completely degraded lactate to CO_2 without the accumulation of acetate as an intermediate and produced 50% more sulfide per lactate than did the *D. desulfuricans'* pure culture. When strain F was present, sulfide was not detected. Thus, strain F effectively used all of the sulfide produced by the mixed bacterial population.

3.2 Core experiments

Experiments were conducted in Berea sandstone cores to determine the effectiveness of strain F in using biogenically produced sulfide in a porous medium. A small Berea sandstone cores was connected to two flasks of sterile medium containing modified medium and incubated in the anaerobic chamber (Jenneman et al., 1985). The penetration rate of both the wild-type and strain F of *T. denitrificans* was 0.017 cm per h. Penetration of *D. desulfuricans* or of the oil field brine enrichment was not observed unless the cores were autoclaved twice. The penetration rates through twice autoclaved cores of *D. desulfuricans* and the oil field brine enrichment were 0.004 and 0.013 cm per h, respectively. Sulfide concentrations in the liquid medium were 9.6 and 19 ppm; the concentration of sulfide in the cores inoculated with *D. desulfuricans* was 6.4 ppm. The coculture of *D. desulfuricans* with strain F penetrated the core at a rate of 0.012 cm per h. The breakthrough of strain F and *D. desulfuricans* occurred at the same time indicating that they grew together through the core. Sulfide was not detected in either the liquid medium of in the core when strain F was present. It might be possible that the faster penetration time of the sulfate reducer in coculture with strain F was due to the prevention or removal of insoluble iron sulfide precipitates which might plug some of the pores and slow bacterial penetration.

4. DISCUSSION

Strain F of *Thiobacillus denitrificans* effectively prevents the accumulation of biogenically produced sulfide both in liquid culture and in Berea sandstone cores. The effectiveness of strain F is due to its ability to grow and use sulfide at concentrations which are inhibitory to the wild-type strain. Since *T. denitrificans* strain F is an autotrophic bacterium, the addition of organic nutrients is not required. This will limit the growth of indigenous organisms which might require organic nutrients. Strain F grows anaerobically and aerobically so it should be effective in using not only the sulfide produced in anaerobic environments, but also the sulfide that diffuses into the aerobic zone. Strain F readily penetrated sandstones cores at a rate that was 1.5 to 4.5 times faster than a mixed population or a pure culture of sulfate reducers. Thus, it is able to penetrate and be metabolically active in consolidated sandstone formations.

The factors which limit the growth of strain F and its application in the control of biogenic sulfide production in petroleum reservoirs include temperature and salt concentration. Temperatures in excess of 40°C completely inhibit growth of *T. denitrificans*. Thus, the use of strain F would be restricted to shallow reservoirs where the temperatures would not be restrictive. The growth of strain F is inhibited by moderate salinities. However, even though growth is inhibited, strain F can still be effective in the prevention of sulfide accumulation. As an example, the growth of strain F is inhibited by 50% in medium containing 2.6 mM sodium lactate, but it was still effective in using the sulfide produced by the sulfate reducer (Table 1). This indicates that strain F does not have to grow under optimal conditions in order to use sulfide effectively.

The penetration of sulfate reducers through Berea sandstone cores was very slow. Both a pure culture and a mixed population of sulfate-reducing bacteria penetrated the cores at very slow rates. This may be a general property of sulfate reducers since similar rates of penetration were observed for other *Desulfovibrio* species by ZoBell (1958). If this is the case, then the detrimental activity by these organisms may be localized at or near the well-bore injection face. This suggests that a remediation strategy such as the use of strain F to remove sulfides may be feasible since only the area near the well-bore needs to be treated.

5. REFERENCES

Balch, W. E., Fox, G. E., Magnum, L. J., Woese, C. R., and Wolfe, R. S. 1979. Methanogens: reevaluation of a unique biological group. Microbiol. Rev. 43: 260-296.

Bryant, M. P. 1972. Commentary on the Hungate technique for the culture of anaerobic bacteria. Amer. J. Clin. Nutr. 25: 1324-1328.

Hamilton, W. A. 1985. Sulfate-reducing bacteria and anaerobic corrosion. Ann. Rev. Microbiol. 39: 195-217.

Jenneman, G. E., McInerney, M. J., Knapp, R. M. 1985. Microbial penetration through nutrient-saturated Berea sandstone. Appl. Environ. Microbiol. 50:383-396.

Jenneman, G. E., McInerney, M. J., and Knapp, R. M. 1986. Effect of nitrate on biogenic sulfide production. Appl. Environ. Microbiol. 51: 1205-1211.

McInerney, M. J. 1985. Transient and persistant associations among procaryotes. In: E. Leadbetter and J. Piondexter (editors), Bacteria in Nature, Volume 2, Plenum Publications, New York, pp. 293-338.

Montgomery, A. D., M. J. McInerney, and K. L. Sublette. 1990. Microbial control of the production of hydrogen sulfide by sulfate-reducing bacteria. Biotech. Bioeng. 35: 533-539.

Raiders, R. A., McInerney, M. J., Revus, D. E., Torbati, H., Knapp, R. M., and Jenneman, G. E. 1986. Selectivity and depth of microbial plugging in Berea sandstone cores. J. Indust. Microbiol. 1: 195-203.

Sublette, K. L., and Sylvester, N. D. 1987. Oxidation of hydrogen sulfide by *Thiobacillus denitrificans* : Desulfurization of natural gas. Biotech. Bioeng. 29:249-257.

Sublette, K. L., and Woolsey, M. E. 1989. Sulfide and glutaraldehyde resistant strains of *Thiobacillus denitrificans*. Biotech. Bioeng. 34: 565-569.

Sublette, K. L., Woolsey, M. E., Manning, F. S., Montgomery, A. D., McInerney, M. J. 1989. Microbial control of hydrogen sulfide production by sulfate reducing bacteria. U. S. Patent No. 4,879,240.

Widdel, F. 1988. Microbiology and ecology of sulfate- and sulfur-reducing bacteria. In: A. J. B. Zehnder (editor), Biology of Anaerobic Bacteria, John Wiley and Sons, New York, pp.469-585.

ZoBell, C. E. 1958. Ecology of sulfate-reducing bacteria. Prod. Mon. 22:12-29.

Reprinted from: *Microbial Enhancement of Oil Recovery - Recent Advances*, edited by E.C. Donaldson

CH. F-8

MICROBIAL ENHANCED OIL RECOVERY TREATMENTS AND WELLBORE STIMULATION USING MICROORGANISMS TO CONTROL PARAFFIN, EMULSION, CORROSION, AND SCALE FORMATION

JOHN W. PELGER
Bio Tech, Inc., 1235 Sovereign Row, Suite C-9, Oklahoma City, Oklahoma 73108.

INTRODUCTION

The past decade has been filled with many Microbial Enhanced Oil Recovery (MEOR) techniques for stripper wells (wells producing less than 10 BOPD), increasing production through the injection of gas/acid forming microorganisms, i.e. Clostridium, into the formation and adding a nutrient of sucrose to sustain their growth. This technique cleanses the rock matrix of by-products, principally carbon dioxide, acids, ketones, and alcohols. Other techniques using microorganisms have been performed to plug the rock matrix, thereby reducing the production of water and creating a positive capillary pressure. Shut-in times for both techniques are a minimum of 10-12 days, causing production delays or losses.

This paper will explain and demonstrate a method to increase production in individual wells or fields with skin damage, due to paraffin accumulations, corrosion, emulsion and scale formation, utilizing an individual treatment design from a library of facultative anaerobic microorganisms developed by Micro-Bac International, Inc. These neutrally charged marine organisms, in a saline solution of nutrients, are naturally occurring, non-pathogenic, non-toxic, non-carcinogenic, non-combustible and require no permits from the Environmental Protection Agency for storage, transportation or disposal. Environmentally, these microorganisms are as safe as any product in our industry to control and reduce skin damage. Paraffin, corrosion, emulsion and scale formation have resulted in significant economic losses and reduced profitability for producers and operators. The resulting skin damage can be controlled and reduced through proper treatment by locating microorganisms in the immediate formation, perforations, and in the case of most problem wells, in the lifting mechanism or surface equipment, thereby preventing restrictions in your production stream or failures in your lifting equipment. Traditional control methods for treatment involve implementing techniques such as hot oiling, paraffin cutting, chemical treatment and condensate treatment. These techniques have been marginally successful, economically prohibitive and temporary at best. Many of these traditional methods are dangerous from a safety standpoint and often have caused serious long-term damage to the productive life of the well (especially in the case of hot oiling).

The initial step in facultative anaerobic microorganisms treatment, once a problem well or field has been identified and goals for the project have been discussed, is fluid sampling and examination of well histories to determine microbial compatibility. The determination whether biodegradation with microorganisms can be used to control or reduce the existing skin damage and restrictions in the lifting mechanism, wellbore, immediate formation face and the surface equipment must be based upon these specific biochemical parameters:

(a) The survival temperature range is between 34 °F.-208 °F. The microbes are dormant between 34 °F.-40 °F.

(b) The optimum temperature range is between 90 °F.-150 °F.

(c) The chloride content by volume needs to be less than 15%.

(d) The pH needs to be greater than 3.

(e) The H2S content in fluid needs to be less than 250 ppm and 10,000 ppm in a gaseous state.

(f) The formation must produce at least a trace of water, or water must be added to provide the microorganisms with an adequate means of transportation throughout the entire system in the water phase.

When treatment specifications are met, wellbore samples are evaluated in a laboratory to verify that bioremediation can occur and to determine the treatability, presence, and dosage of microorganisms to control and reduce the identified problem. The treatability test measures the compatibility of the microbes with the wellbore fluids and detects any negative toxic influences. Treatability ratings of 1+, 2+, or 3+ indicate fair, good, and excellent prospects for microbial compatibility. The presence test is a microscopic check that determines and measures existing colonies of bacteria, such as sulfate reducing bacteria. The presence test also measures the growth of the treated microorganisms and determines the necessary amount and frequency of treatment. A satisfactory presence test during treatment of a well would fall in the range of 800,000-1,200,000 microorganisms/milliliter. The microbial dosage specifically determines which strains of microorganisms should be used to control and reduce the identified problems. Table 1 lists the microorganisms by their trade names and the problems controlled or reduced. Table 2 lists the hydrocarbon chains biodegraded and the melting points of the associated paraffins.

TABLE 1

Microorganisms and Problems Controlled

Microbial Trade Names	Problems Controlled
Para-Bac	Paraffin, Emulsion, Scale, Corrosion
Para-Bac +	Paraffin, Emulsion, Carbonate/Bicarb., Corrosion
Para-Bac x	Paraffin, Emulsion, Scale, Corrosion
Para-Bac xx	Paraffin, Emulsion, Scale, Corrosion
Para-Bac xxx	Paraffin (Naphthenic), Emulsion, Scale, Corrosion
Litho-Bac	Calcium Carbonate & Bicarbonate, Gypsum, Iron Rust
Corroso-Bac	Calcium Carbonate, Calcium Sulfate, Barium Sulfate
Sulfo-Bac	Hydrogen Sulfide, Iron Sulfide
P.B. M-1000*	Toluene, Benzene Ring Structures

All of these microorganisms protect corrodible surfaces by sequestration. Holding the metal ions tightly, chelate calcium, magnesium, and many other ions.

TABLE 2

Microorganisms and Hydrocarbon Chains Degraded with Melting Points of Associated Paraffins

Microbial Trade Names	Hydrocarbon Chain Degraded	Melting Point of Paraffins
Para-Bac	C16-C22	68 °F.-113 °F.
Para-Bac +	C16-C22	68 °F.-113 °F.
Para-Bac x	C16-C40	68 °F.-176 °F.
Para-Bac xx	C40-C58	176 °F.-208 °F.
Para-Bac xxx	C45-C60+	185 °F.-208 °F.

The Para-Bac products are mixed cultures of naturally occurring microorganisms and have the ability to selectively degrade saturated paraffins into non-saturated oleffins. Hydrocarbons with saturated fractions include N-alkanes, branched alkanes, and cycloalkanes. They degrade, metabolize and absorb certain substances, such as paraffin, in a manner similar to our own digestive process. The differences lie in the unique structure of these single cell microorganisms and their biodegrading process. This process occurs on the surface of their cell membrane where a continuous secretion of protein molecules, such as enzymes, catalyze biochemical reactions. Fig. 1 shows a partial microbial degradation process on the hydrocarbon chain, and Fig. 2 graphically shows the same process.

```
CH3(CH2)nCH3 -> CH3(CH2)nCH2OH --> CH3(CH2)nCHO --> CH3(CH2)nCOOH

                -2C                -2C                   -2C
                --> CH3(CH2)n-2COOH --> CH3(CH2)n-4COOH --> etc.

         ----------> HOOC(CH2)nCOOH

----> CH3(CH)n-1COCH3
```

Fig. 1. Partial microbial degradation process on the hydrocarbon chain.

"The microbial degradation of paraffins is normally proceeded by a monoterminal attack. Usually, a primary alcohol is formed followed by an aldehyde and a mono-carboxylic acid. Further degradation of the carboxylic acids are proceeded by beta-oxidation with the subsequent formation of two-carbon-unit shorter fatty acids and acetyl coenzyme A, with the eventual liberation of carbon dioxide" (Schneider, 1989, p. 1). Other factors produced during this degradation process creating positive flow include the non-enzymatic surface components and excreted components, which bind the fluids improving the flowing mechanics for production, transportation and storage.

INDIVIDUAL WELL TREATMENTS

Bio Tech, Inc., has treated over 400 wells in six states (Colorado, Kansas, Michigan, Oklahoma, Texas, and Utah) with the library of microorganisms discussed for paraffin accumulations, emulsion, corrosion, and scale formation. Treatment depths have ranged from a 500-ft. Bartlesville sand to a 13,500-ft. Bromide sand. The types of reservoirs treated are shown in Table 3. Over half (51%) of the wells treated by Bio Tech, Inc., to date have

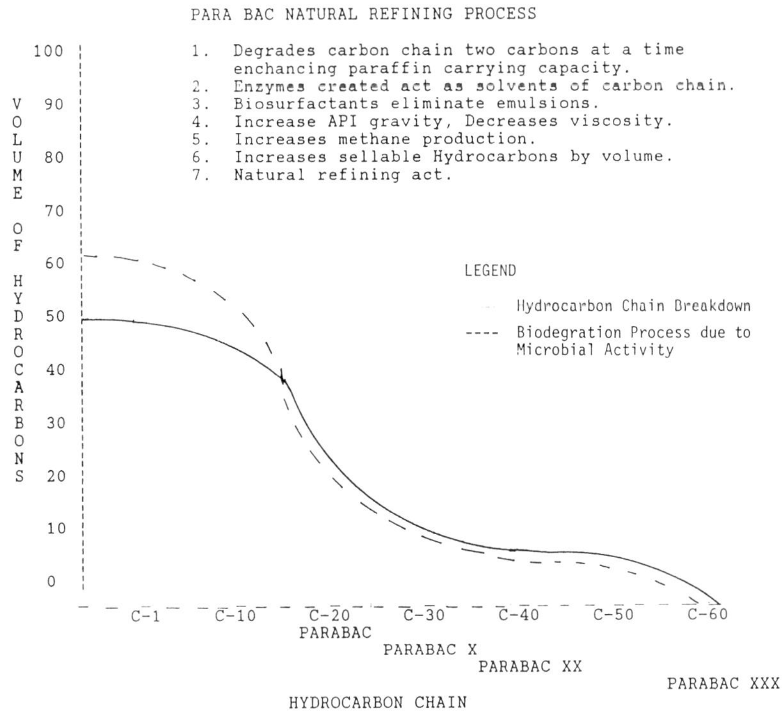

Fig. 2. Graphic display of microbial degradation process on the hydrocarbon chain.

had increases in production. Many of these production increases have occurred due to the fact that the decline curve has either stabilized or slightly increased by just avoiding the downtime associated with problem wells. Many traditional hot oiling/hot water treatments and many chemical treatments cause production to be suspended due to circulation times and production of the treatment water. However, many of the wells treated have responded tremendously to microbial treatments by eliminating wellbore resrictions, plugged perforations, formation damage, and increasing pump life and well producing capabilities. The case studies chosen will illustrate a variety of production problems and the microbial treatments for their control, including the associated decline curves and economic summaries.

TABLE 3

Types of Reservoirs Treated

Reservoir Rock Types	Percentage of Total Wells Treated
Dolomites/Limestones	38%
Granite Wash	27%
Reef Structure	3%
Sandstone	32%

CASE STUDY NUMBER 1

Field Data

The Creek Indian Unit (Section Nos. 20, 21, 28, and 32, T13N, R13E) is located in Okmulgee County, Oklahoma, and producing from the Dutcher sandstone formation at a depth of approximately 2,800 ft.. The 13 leases chosen for treatment (see Fig. 3) had an extensive history of hot oiling since 1980 to avoid paraffin plugging in order to maintain well production. Seven of these wells were hot oiled monthly or on a regular basis.

Objective

The objective of this treatment was to increase and maintain the field production by individually treating each of the 13 wells to try to break through the perforation and formation skin damage. Downhole skin damage and longer carbon chain paraffins were anticipated due to the extent of natural paraffin build-up and the frequency of previous no oil/water treatments.

Sampling of Wellbore Fluids

The treatability testing of all the fluid samples determined that the wells were compatible with the microorganisms (see Fig. 4). The dosage for the first month's treatment would involve Para-Bac (60%) and Para-Bac x (40%).

Treatment Procedure

Each well was batch treated down the backside with 6 gallons of Para-Bac and 4 gallons of Para-Bac x, and 10 barrels of 2% KCl were flushed behind the microbial treatment. The wells were then

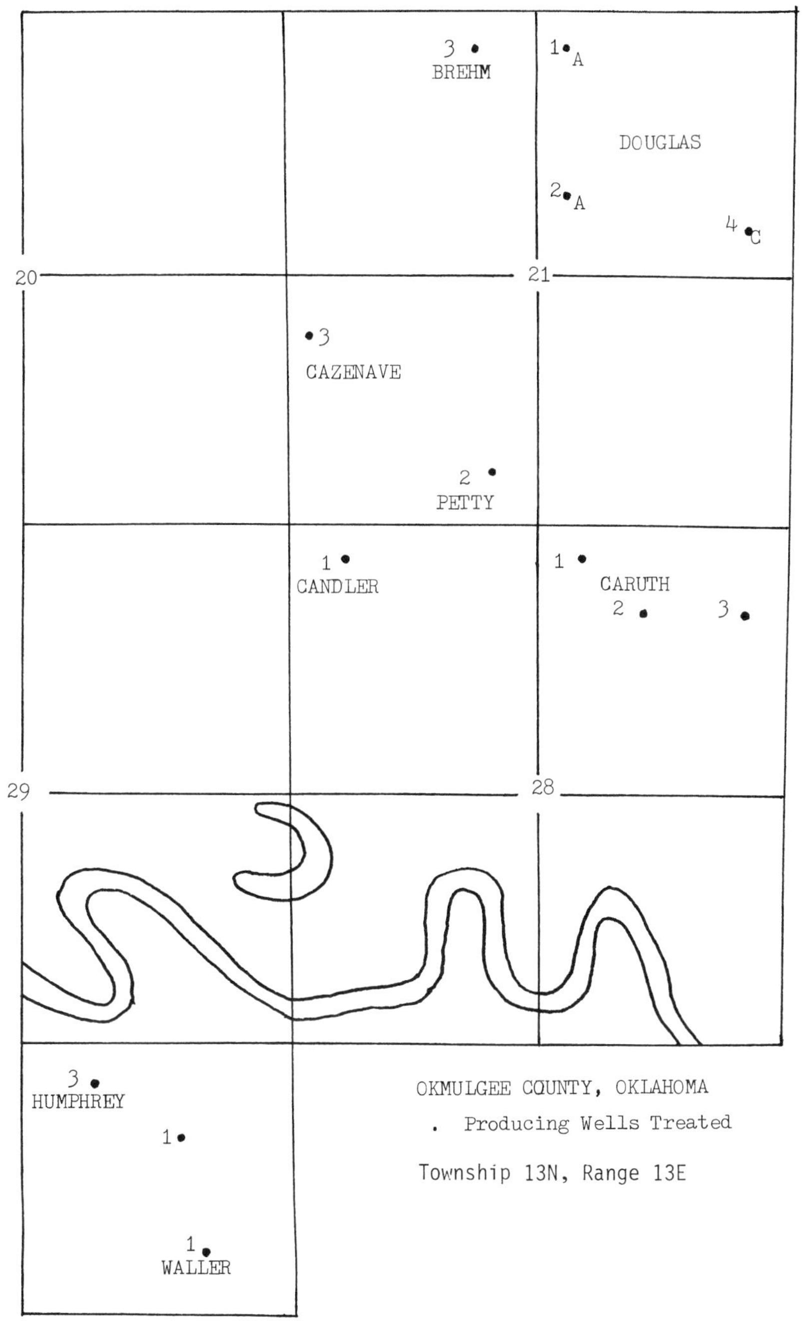

Fig. 3. Location of 13 wells selected for microbial treatment.

shut-in for 14 days to allow microorganism colonization. An extended shut-in time was proposed for adequate microbial migration into the formation and possible degradation of existing paraffin skin damage. The field of wells were produced from September 9, 1989, until October 6, 1989, when a presence test was analyzed on each of the wells. A second batch treatment was performed with 2 gallons of Para-Bac x and flushed with 10 gallons of 2% KCl. Four more microbial treatments occurred as shown in Fig. 4.

DUTCHER SAND UNIT
PRESENCE TESTS AND TREATMENT DATES

		PRESENCE TEST (MICROORGANISMS/MILLILITER)					
	INITIAL	*	**	***	****	*****	******
WELL NAME	TREATABILITY	08/11/89	10/06/89	11/13/89	12/29/89	01/26/90	02/28/90
BREHM # 3	1+		700	1,000	1,100	1,200	1,000
CANDLER # 1	1+		500	1,300	1,200	1,200	1,200
CARRUTH # 1	3+		500	No wtr	800	800	800
CARRUTH # 2	1+		600	No wtr	700	700	800
CARRUTH # 3	1+		300	1,100	1,100	1,100	1,000
F. CAZENAVE # 3	1+		900	1,100	800	800	800
DOUGLAS # 1-A	2+		800	1,200	1,100	800	800
DOUGLAS # 2-A	2+		No wtr	900	No wtr	600	700
DOUGLAS # 4-C	2+		No wtr	800	800	1,300	900
HUMPHREY # 1	2+		700	No wtr	No wtr	900	900
HUMPHREY # 3	1+		900	900	700	900	900
PETTY # 2	1+		500	800	800	800	800
WALLER # 1	1+		700	1,000	1,000	1,000	1,000

	DOSAGE	TREATMENT DATES	SHUT-IN TIME
*	6 Gallons PB & 4 Gallons PB x	08/06/89	14 DAYS
**	2 Gallons PB x	10/06/89	Overnight
***	4 Gallons PB x & 1 Gallon M-1000	11/13/89	Overnight
****	4 Gallons PB x & 1 gallon M-1000	12/29/89	Overnight
*****	4 Gallons PB x	01/26/90	Overnight
******	4 GALLONS PB x	02/28/90	Overnight

Fig. 4. Presence tests, dosage, and the treatment dates in Dutcher sand unit.

Economics

Prior to well treatment, a four-month average production figure was 129 barrels of oil per month (BOPM) and 1,290 million cubic feet per month (MCFM). The average field production increase for the following seven months after initial microbial treatment was 256 BOPM and 1,350 MCFM. An average of four hot treatments per month were eliminated to reduce field operation expenses. Table 4 summarizes the economics of this case study while Fig. 5 illustrates the field's decline curve.

TABLE 4

Economic Summary

Operating Expenses Saved:	
7 months (4 hot oil/month @ $250/hot oil)	= $7,000.00
Increase in Production:	
7 months [(256 BOPM - 129 BOPM) $18.50/Bbl.]	= $16,446.50
7 months [(1350 MCFM - 1290 MCFM) $1.50/MCF]	= $ 630.00
Expenses Incurred:	
Microbial treatment and service for 6 months	- <$8,223.25>
Total Economic Savings	= $15,853.25

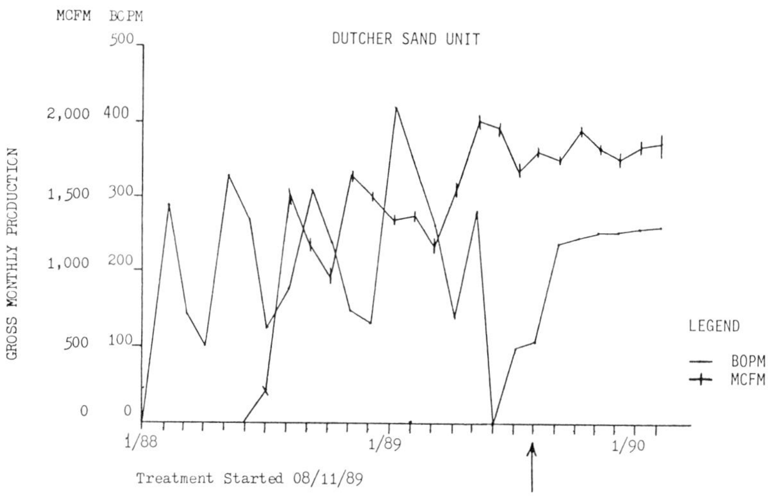

Fig. 5. Production curve showing response due to microbial treatments.

CASE STUDY NUMBER 2

Well Data

The Burgess lease is located approximately 12 miles southeast of Dumas in Moore County, Texas. The well is producing from a white and brown dolomitic limestone formation at approximately a depth of 3,500 ft. This well was chosen for treatment since production had fallen drastically in the past 18 years despite treat ment with traditional, historically favorable paraffin solvents.

Objective

The objective of this treatment was to increase production to previous levels while maintaining or improving cost efficiency. Downhole paraffin damage was anticipated since prior paraffin treatments had shown temporary production increases.

Sampling of Wellbore Fluids

The treatability testing of the fluid samples determined that the microorganisms were fairly compatible (1+) with the identified paraffin problems. The low treatability was due to trace toxic substances, i.e., solvents and bacteriacides found in the tested samples. The dosage for the well treatment involved 75% Para-Bac and 25% Para-Bac x.

Treatment Procedure

This well was batch treated down the backside three times in the initial month at 15-day intervals and twice monthly thereafter. The well was treated bimonthly to maintain a satisfactory presence of microbes due to a high level of water production (70-80 BWPD). Each treatment consisted of 3 gallons of Para-Bac, 1 gallon of Para-Bac x, and was flushed with 20 gallons of 2% KCl. Heavier paraffins were recovered three months into our microbial treatment indicating that formation or immediate wellbore paraffins had been released. This caused oil production to surge to its highest point in April, 1989 (see Fig. 6). At this time, to control the longer chain hydrocarbons, the microorganism treatment was changed to 3 gallons Para-Bac, 1 gallon Para-Bac x, and 1 gallon Para-Bac xx.

Economics

Prior to microbial treatments, the well oil production average was 57 BOPM in nine months. The production average increased in the following nine months after initial microbial treatment to 310

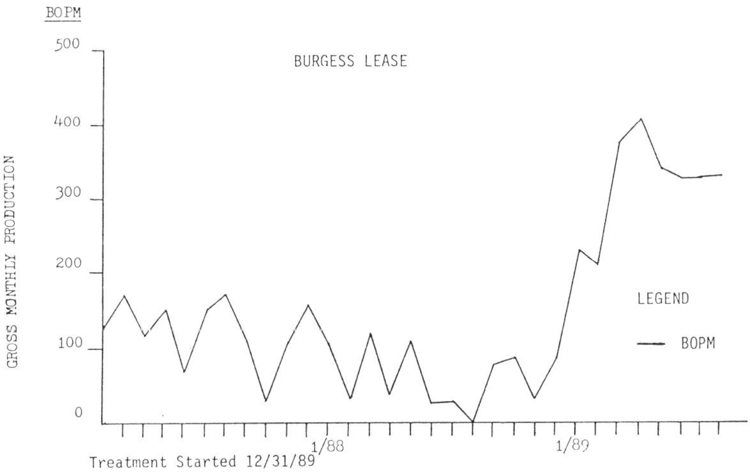

Fig. 6. Production curve showing response to microbial treatment.

BOPM. Table 5 summarizes the economics of this case study.

TABLE 5

Economic Summary

Operating Expenses Saved:	
9 months ($400/month for paraffin solvent)	= $ 3,600.00
Increase in Production:	
9 months [(310 BOPM-57 BOPM) $18.50/Bbl.]	= $42,124.50
Expenses Incurred:	
Microbial treatment and service for 9 months -	<$5,796.00>
Total Economic Savings	$39,928.50

CASE STUDY NUMBER 3

Well Data

The Norwich Unit is located in Missaukee County, Michigan, and is producing from the Richfield sandstone formation at a depth of approximately 5,000 ft. This well previously had received a minimum of four hot oil treatments per year and had a history of annual workovers due to paraffin problems. It also produced a fluid with a pour point of 104 °F.

Objective

The objective of this treatment was to increase well profitability through an elimination of hot oil treatments and a reduction in rig work due to paraffins and the problems associated with 40 °F. fresh water injection. Reducing the paraffin restriction, the pour point, and decreasing the viscosity would improve the flow of the wellbore fluids. A production increase due to paraffin control was anticipated with the accomplishment of the listed objectives.

Sampling of the Wellbore Fluids

The treatability testing of the fluid sampled was determined to be a 1+. The dosage for the well treatment to control the paraffin problems was 50% Para-Bac x and 50% of Para-Bac xxx. Fig. 7 shows the typical information that is kept on all wells treated with microorganisms by Bio Tech, Inc.

Treatment Procedure

This well was batch treated once a month, three times in the initial month to establish microbial colonization, and shut-in for 24 hours after the first two treatments. The well was then shut-in for four hours after treatment during the winter months since the snow blocked access to the well to jump start the units. Fig. 7 illustrates the treatment amounts, microbial presence, and pour points.

Economics

Prior to well treatment, the well averaged 6 BOPD and 8 MCFD over a three-month period. The well production increased to an average of 11 BOPD and 13 MCFD. Two hot oil treatments were eliminated during the microbial treatment and the rods remained clean as shown in Fig. 8. The pour point of the fluid had dropped to less than 45 °F. (see Fig. 7). Injection screens which normally had to be cleaned of paraffin and scale twice a day were now remaining clean for much longer periods of time. Table 6 summarizes the economics of this case study.

WELL NAME	TREATABILITY	BOPD	BWPD	MCFD	TREATMENT AMOUNT P	Px	Pxx	Pxxx	PRESENCE (000)	DOWN Hrs.	POUR Pt.	TREAT DATES	REMARKS
NORWICH # 6-14		9	3	9.7								7/25/89	
		7	5	8.3							104 F	8/21/89	
		4	6	7								9/30/89	
												10/25/89	Paraffined off, Hot oiled, Photo (Fig. 8)
	1+	20	5	9	0%	50%	0%	50%	000			10/28/89	
					0.0	1.2	0.0	1.2		24		11/03/89	
		10	1	10	0.0	1.2	0.0	1.2		24		11/20/89	
									900		50 F	11/29/89	SAMPLE
					0.0	1.2	0.0	1.2		4		12/04/89	
					0.0	1.6	0.0	1.5	1,300	2		12/19/89	SAMPLE, Photo (Fig. 8)
					0.0	1.6	0.0	1.5	900	4		01/19/90	SAMPLE
					0.0	1.6	0.0	1.5	1,100	4		02/20/90	SAMPLE
					0.0	1.6	0.0	1.5	900	4	45 F	03/20/90	SAMPLE, Photo, (Fig. 8) Surf. water temp. (54 F)
					0.0	1.6	0.0	1.5		4		04/25/90	SAMPLE

Fig. 7. Illustrates typical information recorded in microbial treatment monitoring.

TABLE 6

Economic Summary

Operating Expenses Saved:

2 hot oil jobs x $350/hot oil = $ 700.00

Increase in Production:

7 months [(11 BOPD-6 BOPD) 30 days/mo.]$18.50/Bbl. = $19,425.00
7 months [(13 MCFD-8 MCFD) 30 days/mo.]$2.25/MCF = $ 2,315.25

Expenses Incurred:

Microbial treatment and service for 7 months - <$1,667.80>

Total Economic Savings = $21,472.20

CONCLUSION

Wellbore MEOR treatments using facultative anaerobic microorganisms to control and reduce paraffin, emulsion, corrosion, and scale formation are extremely cost-effective. Even with our current industry's unfavorable economic stature, Bio Tech, Inc., has treated over 400 producing wells to control and reduce skin damage with 51% of the wells noticing production increases. Individual wellbore treatments require less investment capital than previous

Figure 8A. Oct. 18, 1989 Paraffin on rods prior to microbial treatment. Pour point -105°, BOPD-4, BWPD-6, MCFD-7.

Figure 8B. Dec. 19, 1989 62 Days since microbial treatment started. BOPD-10, BWPD-1, MCFD-10.

Figure 8C. March 20, 1989 153 Days since microbial treatment started. Pour point -45°, BOPD-12, BWPD-1, MCFD-18.

techniques, and the return on investment is relatively short-term. Over 95% of the wellbores treated to date have realized control of the identified problems and increased the ultimate economic production life of the well. The procedural recommendations for optimal wellbore MEOR techniques utilizing facultative anaerobic microorganisms must include:

(a) Sampling of wellbore fluids to determine the treatability with microorganisms, including compatibility with well characteristics and the optimal inoculation percentages from the microbial library.

(b) Treatment dosages designed specifically for individual wells. Initial and continued sampling of each well before and during treatment is highly recommended, whether treatment is to be rendered to the entire field or isolated wells. Although, traditionally, fields of wells have been thought to have similar rock matrices, fluids, etc., individual dosage, presence and treatability tests have evidenced tremendous differences within the same field.

(c) Well monitoring by the operator and service company is essential to effectively communicate the changes in well operations. The operator must regularly monitor production for changes in the production stream. The service company must monitor microbial presence on a monthly basis to accurately adjust dosage amounts and note hydrocarbon changes, yielding maximum productivity with minimum downtime and optimal cost efficiency. Complete and detailed well charting by the service company is integral to effective system monitoring and treatment for MEOR.

REFERENCES

Bryant, R. S. and Burchfield, T. E.: Review of microbial technology for improving oil recovery, National Institute for Petroleum & Energy Research, SPE Reservoir Engineering, May (1989).

Donaldson, E. C.: Use of capillary pressure curve for analysis of production well formation damage, paper SPE 13809.

Schneider, D.: Para-Bac, Micro-Bac International, Inc., literature, September (1989).

Reprinted from: *Microbial Enhancement of Oil Recovery - Recent Advances*, edited by E.C. Donaldson

CH. F–9

ADVANCES IN RESEARCH, PRODUCTION AND APPLICATION OF BIOPOLYMERS USED FOR EOR IN CHINA

XIU-YUAN WANG
Institute of Microbiology, Academia Sinica, P.O. Box 2714, Beijing100080, China

ABSTRACT
Wang, X. Y., 1990. Advances in research, production and application of biopolymers used for EOR in China

Studies on the biopolymers used for EOR in China began in the sixties with glucan production by *Leuconostoc mesenteroides* from molasses at Yumen oil field. Since that time, the following works have been done: biopolymers produced from crude oil, hydrocarbon or carbohydrates by different bacterial species; model tests on the displacement efficiency of thermally generated gel-agent produced by *Pseudomonas aeruginosa* YA-144; compatibilities of xanthan L4 with some chemical flooding systems; flow mechanisms of the oil displaced by xanthan fluid L4 in micromodels; and field pilot trials of the crosslinked xanthan gel. All these works will be briefly reviewed in this paper.

INTRODUCTION

It was well known that the rheological properties, sensitivity to salts etc. of biopolymersarebetter than those of chemical polymers, and that they are one of the very useful bioproducts for industries, especially for petroleum development. Therefore, their production and application have attracted our attention, and we are making considerable progress.

Three kinds of microbial exopolysaccharide have been put into production in great scale or pilot fermenters. A thermally generated gel agent produced by *Pseudomonas aeruginosa* YA 144 from crude oil or liquid paraffin was found. The fermentation fluid as a plugging agent for adjusting the injection profile was used in heterogeneous model tests. The oil recovery was increased by about 20-40 percent of the original reserve. Xanthan solution as an adjusting agent of the injection profile has been tested at Nanyang oil field and showed a promising result. Some essential results, especially recent progresses, will be discussed.

2 BIOPOLYMER PRODUCTION AND ITS APPLICATION IN PETROLEUM DEVELOPMENT

2.1 *Biopolymers produced from hydrocarbons*

Up to now, all of the microbial exopolysaccharides applied in industries are produced from carbohydrates. From the viewpoint of MEOR, Wang et al. devoted to study the production of biopolymers directly from crude oil and its fractions, thus what is taken from the oil is used in the interests of oil development. They reported that some of the following bacterial strains were able to produce exopolysaccharide from crude oil or liquid paraffin: *Brevibacterium viscogenes* n. sp. 74-230 (1980a,b; 1982b),*Corynebacterium flavoaurantiacum* D9004 (1982 a), and an oil-positive bacterium 437 (1985).*Corynebacterium gummiform* n. sp. S114 was identified by the Research Group of Systematic Bacteriology and others (1979).

(A) *Polysaccharide BV 74-230.* The yields of polysaccharide BV 74-230 were 8.0 g/L from 12% (w/v) crude oil (Wang, 1983) and about 18 g/L from 40 g/L heavy liquid paraffin (mainly composed of C_{12}-C_{19} n-alkanes) for 72h. fermentation period (Fig. 1). The conversion rate of added paraffin was about 45% (Wang et al., 1990 c).

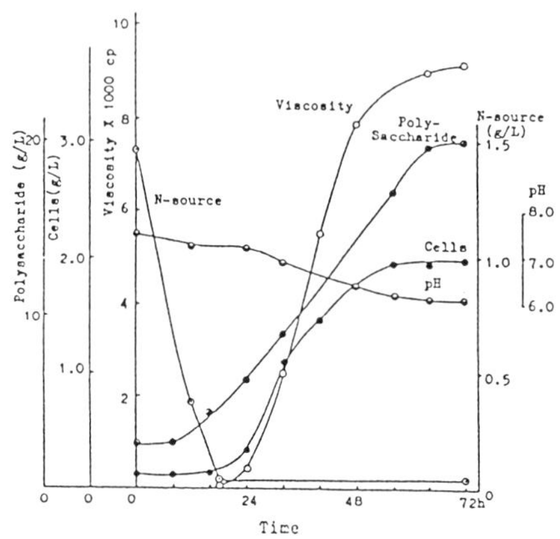

Fig.1 Time course of polysaccharide synthesis from liquid paraffin by *Brevibacterium viscogenes* n.sp.74-230 in 2t fermenter. Medium(g/L) : $NaNO_3$ 2.0, $Na_2HPO_4 \cdot 12H_2O$ 2.0, KH_2PO_4 0.6, $MgSO_4 \cdot 7H_2O$ 0.25, yeast extract 2.0,$CaCl_2 \cdot 2H_2O$ 0.06, heavy liquid paraffin 4%(v/v), pH 7.5, Inoculum: 5%. Aeration: 1:1. Agitation: 200 rpm.

Based on the chemical and physical analyses, both the polysaccharide BV74-230 (Wang et al., 1984) and D-9004 (Wang et al., 1982a) may be two kinds of new bioproduct (Table 1).

TABLE 1.
Some Physical and Chemical Properties of Polysaccharides

Polysaccharide	Monosaccharide constituent (molar ratio)	Content of pyruvic acid %	Glycuronate	$S^0_{20,w}$	η dl/g	$\overline{V}$ ml/g	Molecular weight
D-9004	D-Mannose:D-galactose(2:1)	+		6.28S	0.46	0.67	275,000
BV74-230	D-Glucose: D-mannose:D-galactose:L-rhamnose:D-arabinose (4:3:2:1:0.25)	0.5 - 0.7		10S	6.3	0.775	3,488,000
Xanthan-L4	D-Glucose: D-mannose (1:1)	4.23	4.66 %	11.5-11.9S	11.8	0.55	2,225,600-3,783,000

The rheological properties, shear thining effect, shearing resistance, sensitivity to salts etc. of polysaccharade BV74-230 were determined and compared with xanthan, partially hydrolized polyacrylamide, and carboxylmethylcellulose applied in petroleum industry. The results indicated that the properties of this polysaccharide were comparable to those of xanthan, and much better than those of the two other compounds. It is noteworthy that this polysaccharide is produced from crude oil or its fraction, while xanthan is produced from carbohydrate (Wang, 1983).

The diluted fermentation fluid was used as driving fluid in laboratory scale model experiments. When the injection volume corresponded to 20% of the pore volume, the ultimate recovery for Daqing crude oil was increased by 9%(Wang et al., 1980b). When displaced with the thickened water prepared from the polysaccharide Cg-S114, the ultimate recovery for the Dakang crude oil and Shengli crude oil was increased by 9.5 - 29.4 % and 8.3 - 27.6 %, respectively (Zhang and Qin, 1983).

The polysaccharide BV74-230 is a suitable drilling mud additive (Wang et al., 1982c). Preliminary drilling experiments showed a promising result. Right now, the exopolysacchride BV74-230 is the sole applicable and producible biopolymer produced from crude oil and its fractions.

(B) *Exopolysaccharide produced from mixed substrates by mixed bacterial strains*. At present, the fermentative process has been performed in 2 t fermenters. Two bacterial strains: *Brevib. viscogenes 74-230* and *X. campestris* L5 were mixed as inocula to produce biopolymer from mixed substrates. The yields of polysaccharide were about 20 g/L from 20 g/L starch and 20 g/L liquid paraffin mixed as C-source for 72 h. fermentation period (Fig. 2). The conversion rate of added substrates was about 50 %.

The relationship between plastic viscosity and shear rate of the polymer (Fig.3) indicated that its property was well comparable to those of two xanthan products as well as BV74-230 (Wang et al., 1990b).

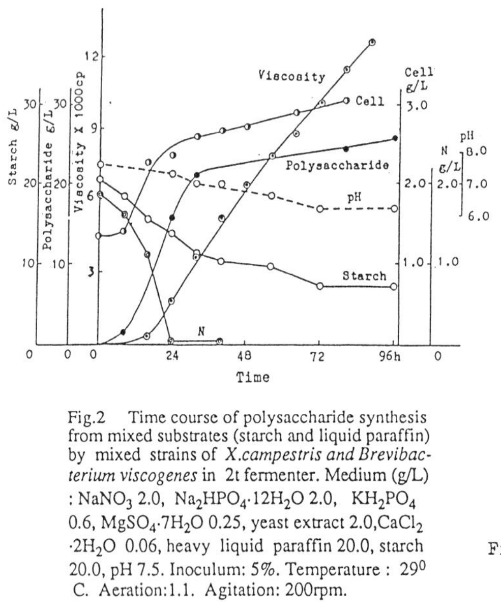

Fig.2 Time course of polysaccharide synthesis from mixed substrates (starch and liquid paraffin) by mixed strains of *X.campestris and Brevibacterium viscogenes* in 2t fermenter. Medium (g/L) : $NaNO_3$ 2.0, $Na_2HPO_4 \cdot 12H_2O$ 2.0, KH_2PO_4 0.6, $MgSO_4 \cdot 7H_2O$ 0.25, yeast extract 2.0,$CaCl_2 \cdot 2H_2O$ 0.06, heavy liquid paraffin 20.0, starch 20.0, pH 7.5. Inoculum: 5%. Temperature : 29^0 C. Aeration:1.1. Agitation: 200rpm.

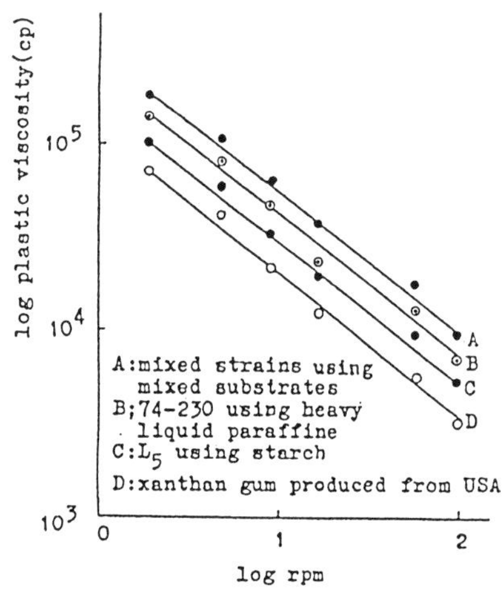

Fig.3 The relationship between plastic viscosity and shear rate

2.2 *Xanthan production*

The famous Xanthan was produced by different strains of *Xanthomonas campestris* in our country, e.g. NK-01 (Lian et al., 1986; Diao et al., 1988; Liu et al., 1989'), ST (Jiang et al., 1988 a,b), L4 and L5 (Liu et al., 1990; Cui et al., 1988, 1990) etc.

Fig. 4 is the electron micrograph of *X.-campestris* L4 and L5. The strain L5 used starch as substrate for xanthan production. The time course of polysaccharide synthesized by strain L4 in 20 t fermenters (Fig. 5) indicated that its yield from 40 g/L saccharose was about 24 g/L . The conversion rate of added substrate was about 60% (Wang et al.,1990a). Their physical and chemical properties are coincided with xanthan product from U.S.A.(Table 1, Cui et al., 1988, 1990).

Fig. 4 Electron micrograph of srain L4 & L5 grown for 24h on nutrient agar slant x 17,000

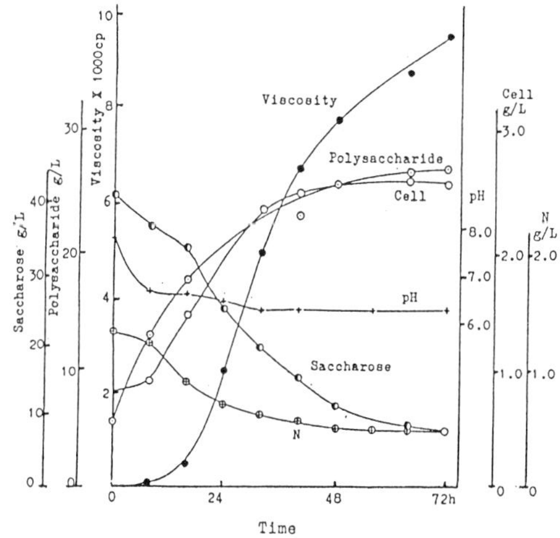

Fig.5 Time course of polysaccharide production by Xanthomonas campestris L_4 in 20t fermenter

Medium(g/L): Saccharose 40.0, yeast extract 2.0, mixed N-sources 1.3, $Na_2HPO_4 \cdot 12H_2O$ 0.1, $MgSO_4 \cdot 7H_2O$ 0.1, $CaCO_3$ 2.0, pH 7.2. Inoculum: 5%. Temperature: 28-30°C. Aeration:1:0.5-1(v/v).Agitation:200 rpm.

Wang et al. (1988) have studied on the behaviour of lyotropic liquid crystal of xanthan NK-01 in the aqueous solution etc. (Wang et al., 1987).

3. THE THERMALLY GENERATED GEL-AGENT PRODUCED BY *PSEUDOMONAS AERUGINOSA*

The bacterial strain YA144 selected from more than 700 strains was able to produce a thermally generated gel-agent from crude oil and its fractions, and was identified as *Pseudomonas aeruginosa* YA-144 (Fig. 6) . The fermented fluid was changed from waterlike into visco-plastic form under heated condition (Fig. 7). It is undoubtedly a perfect property for selective plugging the thermal formation with high permeability.

Using heterogeneous models, we evaluated its efficiency of EOR under 60^0 - 70^0

C by collaboration with Shengli oil field. After water flooding, the fermented fluid was injected as a selective plugging agent. Experiments (pore volume: 162 ml ; saturated oil : 122 ml; 70^0 C) showed that the water-intake capacity and pressure

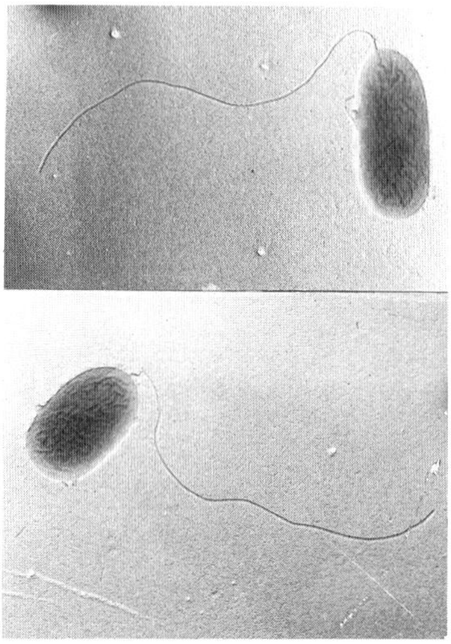

Fig. 6 Electro micrograph of strain YA-144 grown for 24h 0n nutrient agar slant x 10,000

Fig. 7 The thermally generated gel-agent

were obviously decreased and oil recovery increased by more than 20 % of the original reserve (Fig. 8A) .After the water-oil ratio reached to 98 % during water-

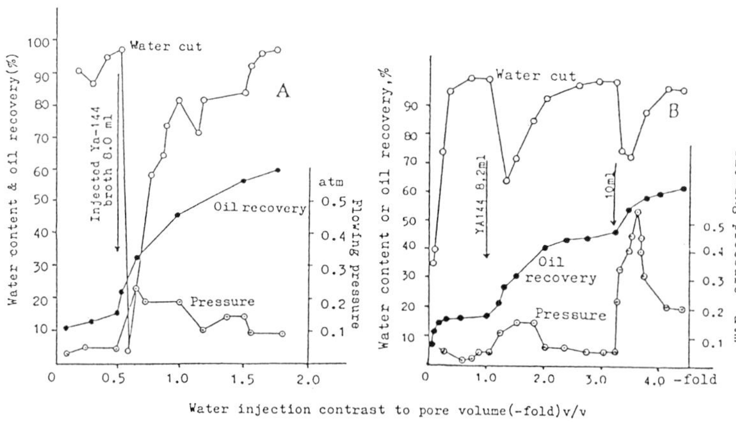

Fig.8 Performance curves of the fermentation broth produced from crude oil by <u>Pseudomonas aeruginosa</u> Ya-144

flooding, the model was injected by fermentation fluid once again. The oil recovery (pore volume : 198 ml, saturated oil: 168 ml, 60^0 C) was increased about 20 % of the original reserve (Fig. 8B) in addition.

After removal of cells by centrifugation and extraction of the product from cell-free fluid by ethylene glycol, glucoprotein was detected by polyacrylamide gel electropholysis. The product was hydrolyzed in 2 N HCl for 5 h and analysed. Results indicated that it is composed mainly of amino acids, glucose, mannose, protein and DNA. But the cell-free product could not transform into visco-plastic form. It is possible that the cell-free product is coupled with cells into a glycocalyx-like product under heated condition. Because the gel-steady duration could not be over 6 months, the field trial has not been performed. But the phenomenon would be clearly warranting careful consideration (Wang et al., 1990d).

4. COMPATIBILITIES OF XANTHAN L4 WITH CHEMICAL FLOODING SYSTEMS

Huang et al. (1988) reported research results on the dynamic interficial tension (DIFT) of acidic crude oil / alkali-polymer systems. The effects of polymer type on the DIFT behaviors for the NaOH-polymer system (Fig. 9A) and for the Na_2CO_3-polymer system (Fig. 9B) showed that the steady-state values of DIFT (γ_{st}) can be lower with the combined alkali-polymer systems than those with the simple alkali systems, and that xanthan L4 is suprior to PHPAM (partial hydrolysed polyacrylamide).

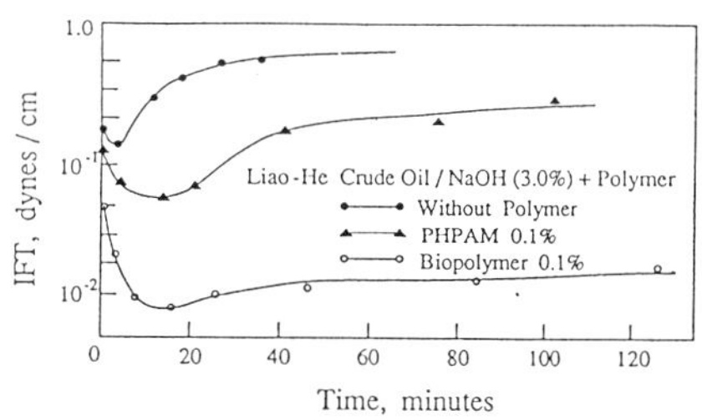

Fig. 9A. Effect of polymer type on the DIFT behavior for the NaOH - polymer system.

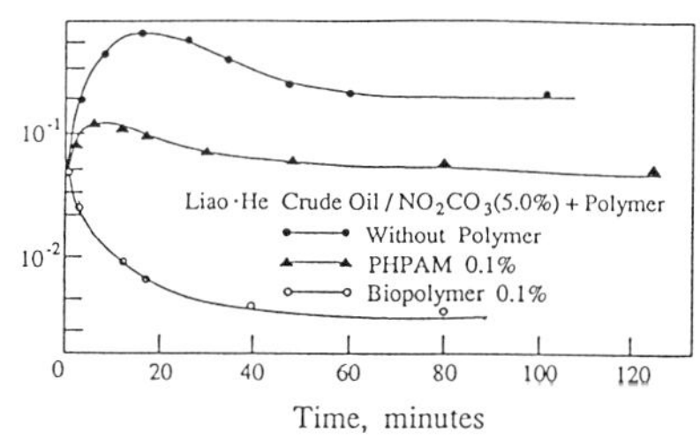

Fig. 9B. Effect of polymer type on the DIFT behavior for the Na_2CO_3 -polymer system.

The comparision of IFT for the simple alkali system with IFT for the combined alkali-polymer system (Fig. 10) indicated that the steady-state values of DIFT can be reduced about two orders of magnitude with xanthan L4 and more than one order of magnitude with PHPAM in optimum combined alkali-polymer systems compared with corresponding simple alkali systems. They suggested that two possible reasons for the fact that biopolymer was better than PHPAM for reducing IFT are: (1) the surface-active by-products from fermentation reaction may have adsorbed onto the oil / water interface, and (2) the salt resistivity of xanthan L4 is higher than that of PHPAM.

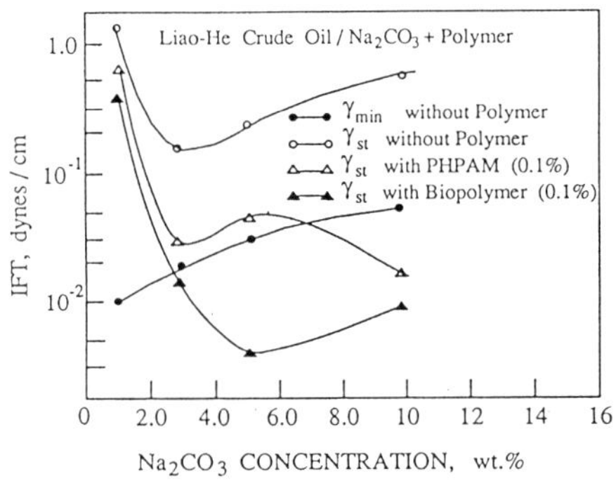

Fig.10 Comparison of IFT for the simple alkali system with IFT for the combined alkali-polymer system.

The effect of xanthan L4 on the extra-low interfacial tension (Liu et al., 1988) showed that the low interfacial tension area was enlarged by xanthan L4 and kept at 10^{-3} level (Fig. 11) in an anion surfactant-polymer system. The viscosity of xanthan solution L4 could be affected by the surfactant and NaOH, and changed from lower to higher, then reduced to original viscosity, gradually. Comparing with a petroleum sulphonate solution without xanthan, the minimum value of IFT was increased from 13.8×10^{-4} to 6.0×10^{-4} mN / m. After the surfactant concentration was more than 0.4 %, its IFT with xanthan 1000 ppm was kept at 10^{-3} level. The effect of xanthan L4 on the mutual solubility in a surfactant-polymer system (Fig. 12) was better than that of HPAM (Jin et al., 1989).

As mentioned above, the compatibilities of xanthan are better than HPAM. Some phenomena could not be clearly explained and need to go further into this topic .

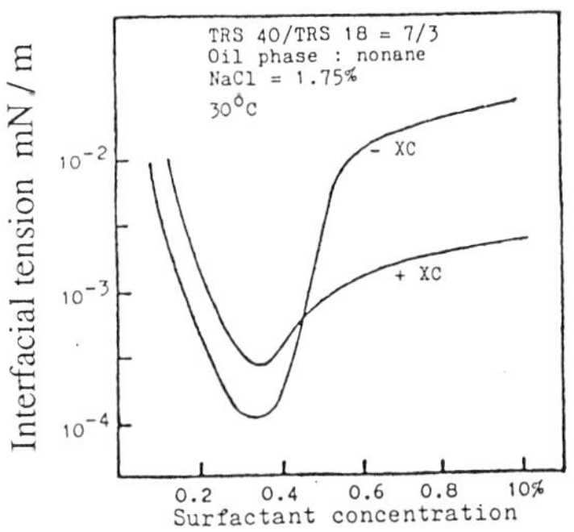

Fig. 11 Effect of xanthan on the extra-low interfacial tension

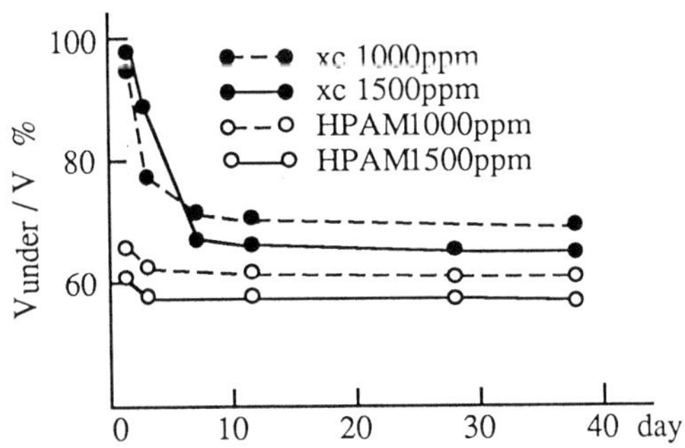

Fig.12 Effect of polymers on the mutual solubility

5. FLOW MECHANISMS OF OIL DISPLACED BY XANTHAN L4 FLUID

The research works are just now carried out by Yu and Zhang (1990) at the Institute of Porous Flow and Fluid Mechanics under China National Petroleum Co. and the Chinese Academy of Sciences.

The technique of making the micromodels lies in packing one mono-layered tiny bead into the fissure between two sheets of glass: 4.0 x 2.5 cm x tens μ; pore volume 50-70 μl; pathway size tens-hundreds μ x tens μ. The viscosity of test oil is about 20 mp / s, flowing pressure ranged from 0.3 to 0.4 Kg / cm^2.

Comparing the existence of residual oil after water flooding (Fig. 13A) with that after xanthan solution drive (Fig. 13B) in same views, the latter's displacement efficiency resulted from the enlarged sweep area was obviously higher than that of the water drive. The xanthan solution flowed along the surface of grains in water-wet media (Fig. 13C), that is similar to water in flow behavior.

In the case of oil-wet models (Fig. 14 A-D), on the contrary to the cases of water-wet models, the injected xanthan solution after water flooding was moved along the axial part of the pathway, and the oil flowed along the grain surface. It was discovered that oil distributed mainly in groups of larger pores surrounded by smaller pores, and, secondly, it was entrapped in small pores and pores with a dead end. Some times it adhered to the pore walls, especially to the rough pore surface in

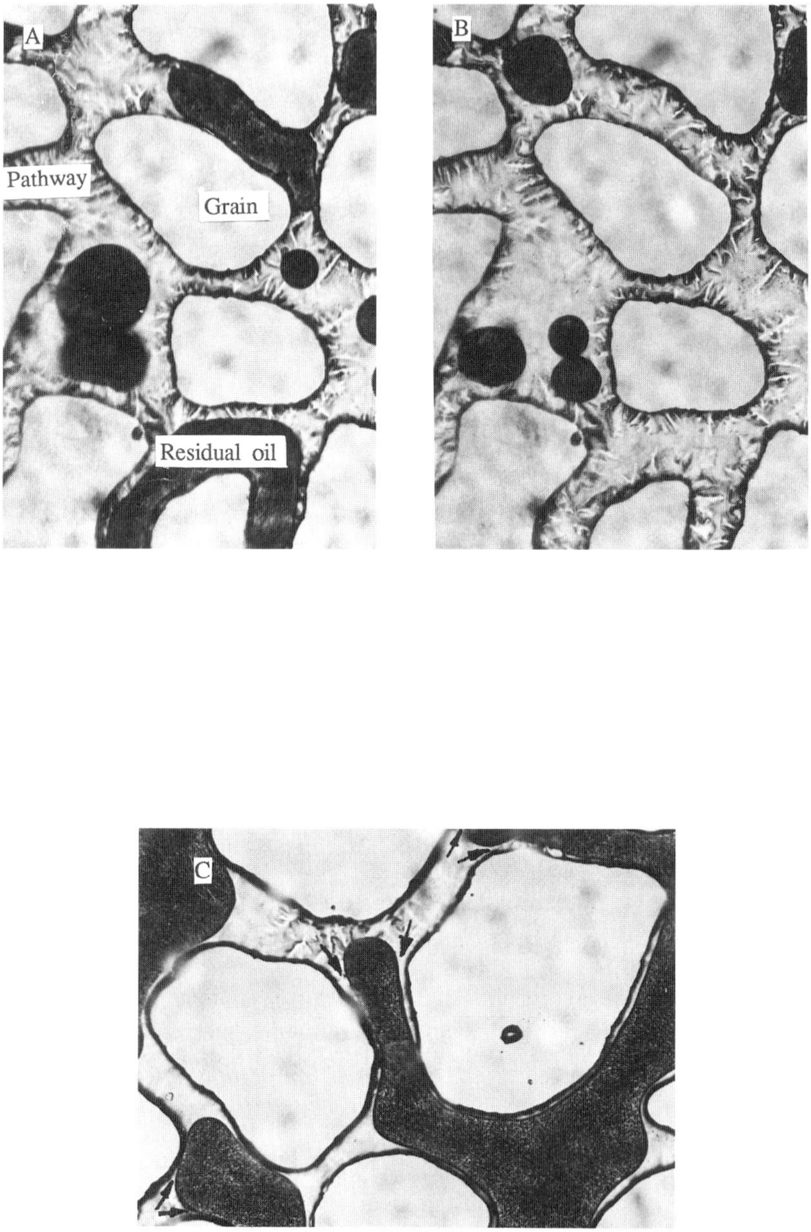

Fig. 13 Micrographs of the flow behavior and distribution of oil displaced by water flooding and xanthan L4-flooding in water-wet media

13A: Pattern of existance of residual oil after water drive

13B: Residual oil after xanthan fluid drive

13C: Oil removement displaced by xanthan fluid

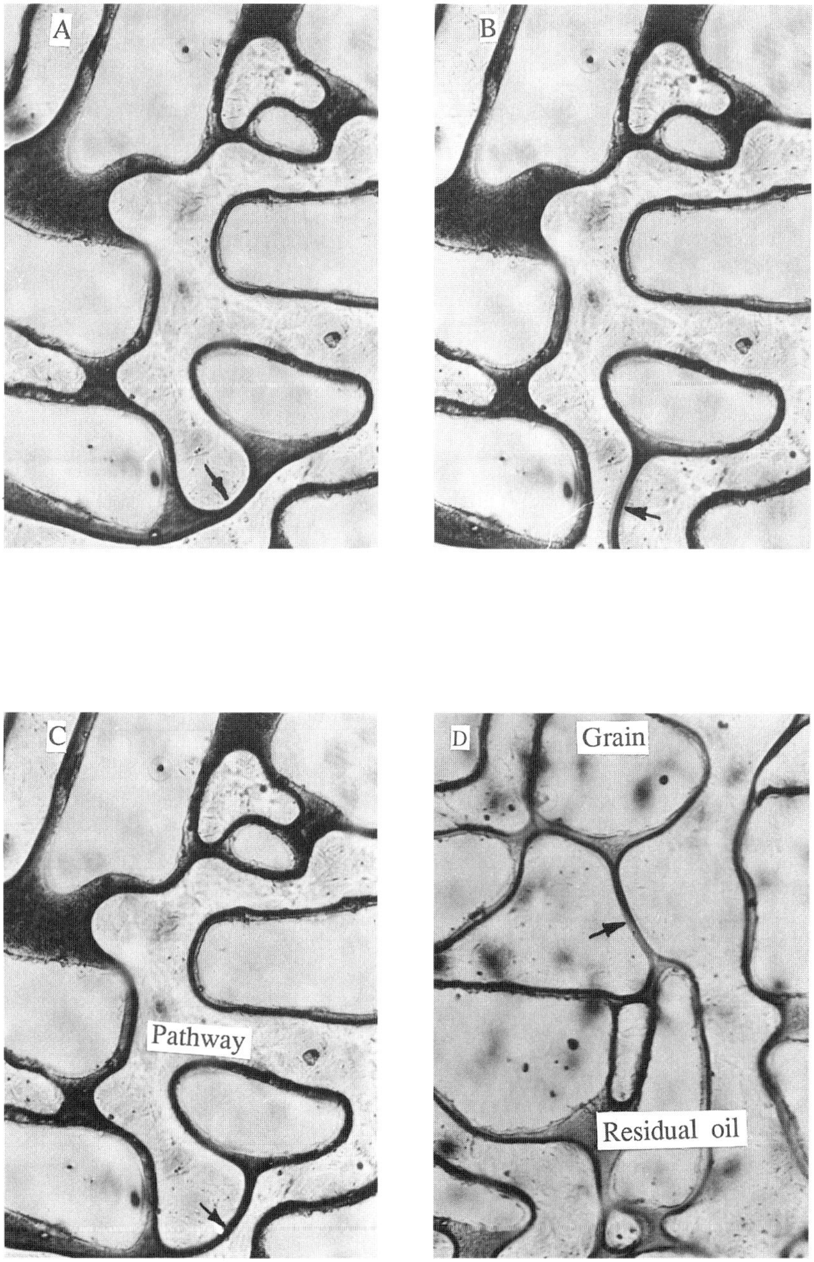

Fig. 14 Micrographs of the flow behavior and distribution of oil displaced by xanthan L4-flooding in oil-wet media

14 A-C: Oil removement displaced by xanthan fluid

14D: Residual oil after xanthan fruid drive

the form of oil film or oil drops. Thirdly, it was found that some oil- filament-bridges existed in pathway among the grains. The oil-filament-bridges did not exist in water-wet models. They did not present in steady state but in dynamic state, and vacillated now to the left and to the right (see arrows in Fig. 14 A-C) , thereby collected the fine oil drops and enriched themself, gradually. In the end, they were driven out by xanthan solution. The residual oil content was less than that of waterflooding (Fig. 14D).

The heterogeneous sand-packed pipe tests were carried out. Using the diluted polysaccharide L4 as driving fluid after water flooding, the averaged oil recovery increased about 10 % of the original reserve.

6. FIELD PILOT TRIALS FOR THE IMPROVEMENT OF INJECTION FLUID PROFILE WITH CROSS-LINKED XANTHAN GEL.

In the middle of the sixties, we had been collaborated with Yumen oil field to study on the glucan fluid flooding produced from molasses by *Leuconostoc mesenteroides*. After laboratory experiments, field tests were performed in two pair wells. Unfortunately, the results were not obtained during the culture revolution (Wang, 1988).

The suitability of cross-linked xanthan L4 gel as an adjusting agent for improving the injection profile (e.g. gel strength, gel stability, sensitivity to temperature etc.) was evaluated by Li Yu-Xiang et al.(personal commucation) at the Scientific Research Institute of Petroleum Exploration and Development.

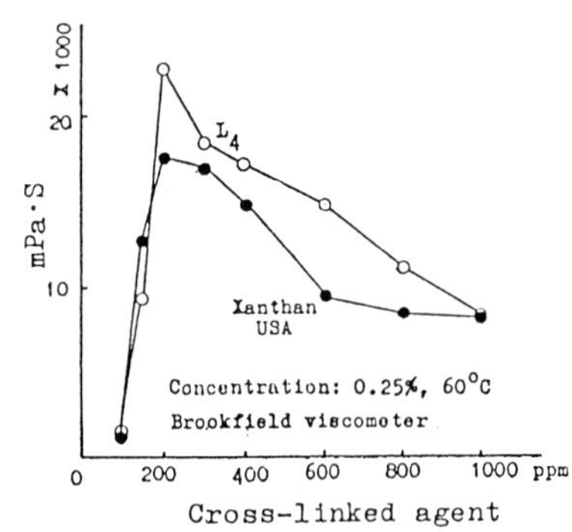

Fig. 15 Effect of concentrations of the cross-linking agent on the gel strength

Fig. 15 is the effect of concentrations of a cross-linking agent on the gel strength. Its suitable concentration is about 200 ppm for 2.5g xanthan / L. The field pilot test has been done at a well pattern (1 injection well, 6 production wells) of Nanyang oil field (Henan province). The data and results obtained at present are as follows:

Test conditions :		**Results :**	
Formation temperature	65^0 C	Water cut decreased	1200 m^3
Well deep	1360 m	Oil recovery increased	845 t
Total injection volume	400 m^3		(for 4 months)
Xanthan content	0.3 %		

7. CONCLUTIONS

The production and application of exopolysaccharides in petroleum industry have been making considerable progress in China. Results mentioned above are promising, showing the possibility of its field application. Studies on EOR using biopolymers will continue in our country. The preparation of biopolymers with better heat resistance is now contemplating. It is reasonably expected that biopolymers will be able to provide a beneficial aid for EOR.

Acknowledgement

The author would like to express his deep thank to all colleagues for their permission to quote their new results in a
biopolymer research.

8 REFERENCES

Cui, W. H., Wang X.Y., Liu X.F., 1988. Gas chromatographic analysis of the xanthan XA5. in "Analytical Microbiology" (special issue), Science Press, Beijing, pp. 128-131.

Cui, W. H., Wang, X. Y., 1990. Physical and chemical analysis of the exopolysaccharide produced by *Xanthomonas campestdris* L_4. Acta Microbiologica Sinica, 30(6) : .

Diao, H. X., Liu R. L., et al., 1988. Studies on the xanthan. II. Fermentative experiment in shaking flask and 1200 L, 4000 L fermenters. Acta scientiarum Naturatium Universitypatis Nan Kaiensis, No. 2 : 34-44.

Huang, Y. D., Yang, P. H., and Qin T. L., 1988. A study on the dynamic interfacial tension of acidic crude oil / alkali (alkali-polymer) systems. Proceedings of the 4th European Symposium on Enhanced Oil Recovery, 27-29 Oct. 1987, Hamburg, GDMK, pp. 171-182.

Jiang, B. Y., et al., 1988a. Chinese Journal of Bioitechnology, (4)3 : 241-243.

Jiang, B.Y., et al., 1988b. Acta Scientiarium Naturalium Universitatis Shang Dongensis, 23(3) : 113-119.

Jin, C. W., Li, S., Zhang, J., Liu, X. Q., and Cheng, Y. X. et al., 1989. Effects of polymers on the mutual solubility in a surfactent-polymer system (in preparation).

Liang, F. L., Liu, R. L., et al., 1986. Studies on the xanthan. I. Identification of the producing strain NK-01, Acta Scient. Naturalium Universitatis Nan Kaiensis No.2 : 105-109.

Liu , H., Lu, Y. X., Dong, M. Z., Tu, Y., Yuan, P. and Liu, P., 1988. Studies on the waterflooding systems with low interficial tension.(in preparation).

Liu, R. L., Zhang, Y. Z. et al., 1989. Studies on the xanthan gum. III. Recovery of product and Its properties analysis. Acta Scientiarum Naturalium Universitatis Nan Kaiensis, No.1: 57-64.

Liu, X. F., Wang, X. Y., Cui, W. H., and He, W. H., 1990. Identificaton of two xanthan-producing strains : L_4 and L_5. (in preparation).

Research Group of Systematic Bacteriology, Institute of Microbiology et al., 1979. Studies on the production of polysaccharide by petroleum fermentation of *Corynebacterium gummiferum* nov.sp. I. Identification of organism. Acta Microbiologica Sinica , 19(4), 341-346.

Wang, D. R., Zuo, J., et al., 1988. Studies on the behaviour of lyotropic liquid crystal of xanthan in the aqueous solution, Acta Polymerica Sinica, No.1 : 23-28.

Wang, D. R., Zhang, W. D. et al., 1987. Chemical J. of Chinese Universities, 8(1) : 93-95.

Wang, X. Y., Liu X. F., Wang, X. J., Tian, X. Y., 1980a. Studies on microbial exopolysacchariеds produced from crude oil and oilproducts. I. Identification of strain 74-230. Acta Microbiologica Sinica, 20(4) : 345-350.

Wang, X. Y., Wang, C. Z., 1980b. Microbial polysaccharide produced from crude oil and its applicability in secondary oil recovery. Acta petrolei Sinica, 1(4) : 77-85.

Wang, X. Y., Yu, N. X., Liu, X. F., Tian, X . Y. and Cui, W. H.,1982a. Studies on microbial exopolysaccharides produced from crude oil and oilproducts. II. Exopolysaccharide produced from crude oil by *Corynebacterium flavo-aurantiacum* D9004. Acta Microbiologica Sinica, 22(1) : 71-78.

Wang, X. Y., Liu, X. F., Shi, Z. J. and Wu, C. H., 1982b. Studies on microbial exopolysaccharides produced from crude oil and oilproducts. III. A suitable medium for exopolysaccharide synthesis by *Brevibacterium viscogenes n. sp.* 74-230. Acta Microbiologica Sinica, 22(4) : 367-372.

Wang, X. Y., Zhang, K. Q. et al., 1982c. Studies on a microbial exopolysaccharides produced by *Brevibacterium viscogenes* 74-230 as a drilling mud additive. Acta Petrolei Sinica, 3(4) : 367-372.

Wang, X. Y., 1983. A microbial polysaccharide produced fromcrude oil or liquid paraffin and its application in petroleum industry. Proceedings of 1982 "International Conference on Microbial Enhancement of Oil Recovery", May

16-21, 1982, Shangri-La, Afton, Oklahoma, U.S. Department of Energy, pp. 29-37.

Wang, X. Y., Cui, W. H. and Yu, N. X., 1984. A new microbial exopolysaccharide. Scientia Sinica (Series B, Chinese), 27(9) : 799-804; (English), 28(2) : 139-145, 1985.

Wang, X. Y., and Schwartz, W., 1985. Conditions for the exopolysaccharide synthesis from n-hexadecane by an oil-positive bactyerium 437. J. Basic Microbiology, 25 (3) : 213-219.

Wang, X. Y., 1988. Development of petroleum microbiology in China. In " Achievements in Industrial Microbiology " ed. by Zhang, S. Z. and Wang, X. Y., Science Press, Beijing, pp. 213-219

Wang, X. Y., Liu, Q. P., Cui, W. H., Miao, X. L., He, J. M., Liu, X. F. and Shi, N. Q., 1990a. Fermentative, Production of xanthan in 2 t and 20 t fermenters (in preparation).

Wang, X. Y., Cui, W. H. and Ma, Y. H., 1990b. Polysaccharide fluid produced by microorganisms. Chinese Patent (in application).

Wang, X. Y., Cui, W. H., 1990c. Exopolysaccharide BV74-230 produced from heavy liquid paraffin in 2 t fermenters (in preparation).

Wang, X. Y., Cui, W. H., Tian, X. Y., Yu, N. X. and Liu, X. F.,1990d. Studies on the thermally generated gel-agent produced by Pseudomonas aeruginosa YA-144 for EOR (in preparation).

Yu, D. S. and Zhang, G. F., 1990. Studies on the mechanism of oil displacement by xanthan fluid in the micromodels (in preparetion).

Zhang, Z. C. and Qin, T. L., 1983. A survey of research on the application of microbial techniques to the petroleum production in China. " Proceedings of 1982 Internatinal Conference on Microbial Enhancement of Oil Recovery", May 16-21, 1982, Shangri-La, Afton, Oklahoma, U.S. Department of Energy, pp. 135-139.

APPENDIX

Dr. Lazar prepared this review of all of the well documented field applications as a special inclusion to the Proceedings. A similar inclusion, by Dr. Hitzman, was entered in the Proceeds of the first (1982 International Conference on MEOR). This makes a very useful quick reference to the MEOR processes that have been used in the field and the results obtained.

Reprinted from: *Microbial Enhancement of Oil Recovery - Recent Advances*, edited by E.C. Donaldson

CH. A–1

MEOR FIELD TRIALS CARRIED OUT OVER THE WORLD DURING THE LAST 35 YEARS

I. LAZAR
Institute of Biological Sciences, Spl. Independentei 296, Bucharest, Romania

ABSTRACT

During the last 35 years, in countries such as the USA, The Netherlands, Hungary, Czechoslovakia, Poland, USSR and Romania almost 30 MEOR field trials were carried out. Presented herein is information concerning the types of inoculum, nutrient supports, protocol of wells injection and the response to the microbiological treatment of these field trials.

By comparison of the information presented herein anyone interested in MEOR can have a large choice of possibilities from which to select the most suitable MEOR in situ technique that has already been tested.

INTRODUCTION

This paper includes about 20 appendices presented in previous papers (Lazar, 1989,1990) as well as another 10 appendices of recent MEOR field trials.

As recorded both in Tables 1, 2 and 3 and in the 29 flow diagrams, the MEOR field trials carried out during the last 30-35 years can be grouped in the following six categories in accord with the method of treatment of wells for stimulation and enhanced oil recovery, namely:

1. Cyclic microbial recovery (a single well stimulation method);
2. Microbial flooding recovery;
3. MEOR by stimulation of reservoir microflora;
4. Microbial selective plugging recovery;
5. Microbial fracturing fluids recovery;
6. Paraffin removal treatment.

THE BACTERIAL INOCULUM USED

In Table 1 and the 29 diagrams, the frameworks of the MEOR field trials are grouped based on the six categories mentioned above. Several types of bacterial inoculum were used; the most important ones being the pure culture, mixed cultures and adapted mixed enrichment cultures, all of which were previously selected as adapted to the conditions of the reservoir proposed for microbiological treatment. As mentioned in a recent paper (Lazar et al., 1990), from the three main types of bacterial inoculum the most active in residual oil release from a porous medium is the last one, namely the adapted mixed enrichment cultures. These results are according to those obtained by other authors as recorded in Table 3. Generally, in the case of categories 4, 5 and 6 pure cultures only are used, while in the case of categories 1 and 2 mixed cultures or adapted mixed enrichment cultures are used. For category 3 a bacterial inoculum previous selected is not necessary since this technique is based on the stimulation of the activity of the native microflora from both formation and injection waters.

NUTRIENT SUPPORT

In Table 2, as well as the 29 diagrams, information concerning the nutrient supports injected into the formation together with the bacterial inoculum are included. Many field trials are based on using 2-4% molasses without addition of other ingredients as supplementary sources of nitrogen and phosphorus. In some field trials several organic and inorganic ingredients were added along with molasses. We believe that, at least for the techniques 4, 5 and 6, no complex nutrient supports seem to be necessary as they were used by Jaranyi (1968, 1973), Zajic (1987), and King (1989).

THE RESPONSE TO MICROBIOLOGICAL TREATMENT

In Table 3, as well as in the 29 diagrams, detailed information about the response of the reservoir subjected to treatment with bacterial inoculum and nutrient support, are presented.

For many field trial experiments the response, including increase of oil production, is very promising. These results stimulated an increase of interest in MEOR during the last ten years, when the work in this field became very active. In the context of a world petroleum crisis, this is now the time for the MEOR to be developed and to become a general technology of enhanced oil recovery. There are some advantages of this method with respect to simplicity and cost.

THE PROTOCOL OF WELL INJECTIONS

On the basis of the information included in those 29 diagrams, the protocol of well injection developed constantly. At the beginning, for the pioneering MEOR field trials, 25-30 years ago, the protocol of well injection was a very simple one often consisting of a single injection with modest quantities of inoculum and nutrients; during the last ten years, however, the protocol has become more complex. This change has occurred, in our opinion, due to a better understanding of the process as a result of participating interdisciplinary research teams. For example the protocol presented in flow diagram 21 seems to give much better results compared to the protocol used in Romania during 1975-1980. At the same time the protocol presented in diagram 22 seems to be of great interest because it is based on a different idea then all of the other MEOR field trials.

The technique based on the idea of selective plugging recovery seems to have a great future based on results from diagrams 24, 26 and 27.

CONCLUSIONS

1. During the last 30-35 years in several countries such as the USA, Czechoslovakia, Hungary, The Netherlands, Poland, USSR, Romania and Canada, almost 30 MEOR field trials were carried out. These field experiments are based on the idea of injecting the reservoirs with bacterial inoculums and nutrient supports with cause fermentation and lead to generation in situ of gases, acids, solvents, biosurfactants and biopolymers, all of which are well known to be involved in releasing oil from porous media.

2. During the years, as a result of the MEOR field trials, a very fruitful library experience has been accumulated with respect to the bacterial inoculum, nutrient supports and the protocol of well injection with these ingredients. In the last few years valuable contributions were added to the quantity of ingredients (brine, inoculum and molasses) to be injected into formations, the length of time for injection, as well as monitoring of field trials.

3. The results obtained to date have proved to be of interest for the field trials of the microbial cyclic system, and for the field trials of the microbial flooding systems.

REFERENCES

Bryant, S.R., Burchfield, E.T., Dennis, M.D. and Hitzman, O.D., 1988. Microbial Enhanced Waterflooding Mink Unit. Project. SIE 17341, SPE/DOE Enhanced Oil Recovery Symp. held in Tulsa, Oklahoma, April 17-20.

Davidson, S.W. and Russell, H.H., 1988. A MEOR Pilot Test in the Loco Filed. In: Proceedings of the Symp. on Application fo Microorganisms to Petroleum Technology, Bartlesville, Oklahoma, August 12-13, 1987. Eds. Thomas E. Burchfield, Rebecca S., Bartlesville, OK, pp. VII1-VII12.

Dienes, M. and Jaranyi, I., 1973. Increase of Oil Recovery by Introducing Anaerobic Bacteria into the Formation, Demjen Field, Hungary Koolaj as Foldgax, v. 106, No. 7, pp. 205-208.

Donaldson, E.C., 1988. Personal communication.

Dostalek, M. and Spurny, M., 1958. Bacterial release of oil. I Preliminary Trials in an Oil Deposit. FOl. Biol. (Praha) 4, 166 p.

Dastalek, M., 1961. Bacterial release of oil. III Areal Distribution of the Effect of Nutrient injection into the Deposit. Folia Microbiologica 6, pp. 10-16.

Grula, A.E., Russell, H.H. and Grula, M.M., 1985. Field trials in central Oklahoma using Clostridium strains of Microbially Enhanced Oil Recovery. In: Microbes and Oil Recovery (vol. 1), Int. Biores. Journal, Ed. by Zajic and Donaldson, El Paso, TX, pp. 144-150.

Hitzman, O.D., 1983. Petroleum Microbiology and the History of Its Role in Enhanced Oil Recovery. In: Proceedings fo Int. Conf. on Microbial Enhancement of Oil Recovery, Shangri-La, Afton, OK, May 16-21, 1982, pp. 162-198.

Hitzman, O.D., 1988. Review of Microbial Enhanced Oil Recovery Field Trials. In: Proceedings of the Symp. on Application of Microorganisms to Petroleum Technology, Bartlesville, OK, August 12-13, 1987, Ed. Thomas, E. Burchfield and Rebecca S. Bryant, Bartlesville Project Office, U.S. Dept. of Energy, Bartlesville, OK, pp. VI1-VI42.

Ivanov, M.V., Beleayev, S.S. and Rozanova, F.E., 1987-1988. Personal communication.

Jack, R.T. and Stehmeier, G.L., 1988. Selective plugging in watered out Oil Reservoirs. In: Proceedings of the Symp. on Application of Microorganisms to Petroleum Technology, Bartlesville, OK, August 12-13, 1987, Eds. Thomas E. Burchfield and Rebecca S. Bryant, Bartlesville Project Office, U.S. DOE, OK, pp. VII1-VII13.

Jaranyi, I., 1968. Beszamolo a nagylengyel terzegeben elvegzett Koolaj mikrobiologiai Kiserletkrol. M. All. Foldtani Intezet Evi Jelentese A., 1968, Evrol, pp. 423-426.

Johnson, A.C., 1979. Microbial Oil Release Technique for Enhanced Recovery Conf. on Microbial Processes useful in Enhanced Oil Recovery, U.S. DOE, San Diego, CA, Aug. 29-Sept. 1, 1979, pp. 30-34.

Karaskiewicz, J., 1974. Zastosovanie metod mikrobiologenneck w intensyfikacji eksploatacji Karpackick Zloz rapy naftowej; Widawnictwo "SLACK", Katowice, pp. 1-67.

King, J., 1989. Personal communication.

Lazar, I., 1987. Research on the Microbiology of MEOR in Romania. In: Proceedings of the First Intern MEOR Workshop, Ed. by Jeannette King and Debra Stevens, Bartlesville Project Office. U.S. DOE, OK, pp. 124-153.

Lazar, I. and Constantinescu, P., 1985. Field Trials Results of Microbial Enhanced Oil Recovery. Microbes and Oil Recovery, vol. 1, International Bioresources Journal, Ed. by Zajic and Donaldson, 1985, pp. 122-144.

Lazar, I., Velehorschi, V., Dobrota, S. and Stefanescu, M., 1988. Characteristics of bacterial inoculum applied on oil reservoir in field experiments. In: Proceedings of the 6th Ind. Micro. and Biotech. Symp., Iasi, 10-12 Sept. 1987, pp. 801-806.

Lazar, I., Dobrota, S., Stefanescu, M., Constantinescu, P. and Iliescu, O., 1989. Procedeu de injectare a sondelor in vederea cresterii recuperarii titeiului din zacaminte. Brevet de inventie nr. 98578, acordat prin Hotarirea Nr. 528-30.04.1989.

Lazar, I., Dobrota, S., Stefanescu, M. and Velehorschi, V., 1990. Characteristics fo the Bacterial Inoculum used in some recent MEOR field trials in Romania. In: Proceedings of the MEOR International Conference, May 27-June 1, 1990, Norman, OK.

Lazar, I., Dobrota, S., Stefanescu, M., Sandulescu, L., Constantinescu, P., Morosanu, C., Botea, N. and Iliescu, O., 1990. Preliminary results of some recent MEOR field trials in Romania. In the Proceedings of the MEOR International Conference, May 27-June 1, 1990, Norman, OK.

Oppenheimer, H.D. and Heibert, K.F., 1988. Microbial Enhanced Oil Production Field Tests in Texas. In: Proceedings of the Symp. on Application of Microorganisms to Petroleum Technology, Bartlesville, OK, August 12-13, 1987, Eds. Thomas E. Burchfield and REbecca S. Bryant, Bartlesville Project Office, U.S. DOE, Bartlesville, OK, pp. XII1-XII15.

Senyukov, W.M., Yulbarisov, E.M., Taldykina, N.N., Shishenina, E.P., 1970. Mikrobiologiya, 39, pp. 612-616. Translated from Microbiologiay 39, No. 4, pp. 705-710, July-Aug. 1970.

Wagner, M. and Teumerm, P., 1987. Personal communication.
Wang, Xiu Yuan, 1988. Personal communication.
Yarbrough, F.H. and Coty, F.V., 1983. Microbially Enhanced Oil Recovery from the upper cretaceous Nacatoch formation, Union County, Arkansas. In: Proceedings of 1982 Int. Conf. on Microbial Enhancement of Oil Recovery, Ed. by Donaldson and Zajic Published by Bill Linvill, Chief, Technology Transfer Branch, Bartlesville, OK, pp. 149-153.
Yulbarisov, E.M., 1972. Results from Analysis of Petroleum Gas on Introduction fo Biochemical Processes in the Oil Formation, Gazovoe Delo, No. 4, pp. 26-28.
Yulbarisov, E.M., 1989. Microbiological method for EOR. In: Proceedings of the Fifth European Symp. on Improved Oil Recovery, Budapest, 25-27 April 1989, pp. 695-704.
Zajic, E.J., 1987. Scale up of Microbes for Single Well Injection. In: Proceedings of the First Int. MEOR Workshop, Ed. by Jeannette King and Debra Stevens, Bartlesville Project Office, U.S. DOE, OK, pp. 241-246.

Table 1. Types of inoculum used in MEOR field trials.

No.	Country	References	Type of inoculum
		I. Cyclic Microbial Recovery (Single well stimulation method)	
1	USA	Johnson, 1979	Mixed cultures of *Bacillus* and *Clostridium*
2	USA	Microbial Systems Corp. 1984 (After Hitzman 1988)	Mixed anaerobic microbial cultures
3	USA	Grula et al., 1985	*Clostridium sp.*
4	USA	Covington Leases, 1985	Spore suspension of *Clostridium sp.*
5	USA	Zajic, 1987	Anaerobic and facultative anaerobic bacteria high fermenting sucrose-molasses medium
6	USA	Bryant et al., 1988	Mixed cultures of *Clostridium sp.*, *Bacillus sp.*, *B.licheniformis* and a *Gram-negative* rod
7	USA	King, 1989	Culture of *Clostridium* type
8	Romania	Lazar 1987, Lazar and Constantinescu, 1985, Lazar et al., 1988, 1989, 1990	Adapted mixed enrichment cultures predominated by *Clostridium*, *Bacillus* and *Gram-negative* rods
		II. Microbial Flooding Recovery	
9	Czechoslovakia	Dostalek and Spurny, 1958, 1961	Mixed cultures of sulfat reducing bacteria and *Pseudomonas sp.* hydrocarbon utilizing
10	Hungary	Jaranyi, 1968	Mixed sewage-sludge cultures
		Dienes and Jaranyi, 1973	Anaerobic thermophilic mixed cultures (predominants: *Clostridium*, *Desulfovibric* and *Pseudomonas*
11	Poland	Karaskiewicz, 1974	Mixed aerobic and anaerobic bacteria belonging to genera: *Arthrobacter*, *Clostridium*, *Mycobacterium*, *Pseudomonas*, and *Peptococcus*

No.	Country	References	Type of inoculum
12	USSR	Senyukov et al., 1970	Mixed aerobic and anaerobic bacteria
13	USA	Petrogen Inc., 1977-1982 (After Hitzman 1983)	unknown
14	USA	Yarbrough and Coty, 1983	*Clostridium acetobutylicum*
15	Romania	Lazar, 1987; Lazar and Constantinescu, 1985; Lazar et al., 1988, 1989 1990	Adapted mixed enrichment cultures predominated by *Clostridium*, *Bacillus* and *Gram-negative* rods
16	DDR	Wagner and Teumer, 1987	Mixed cultures of thermophilic *Bacillus* and *Clostridium*
17	USA	Bryant et al., 1988	Mixed cultures of *Clostridium sp.*, *Bacillus sp.*, *B.licheniformis* and a *Gram-negative* rod
18	USA	Donaldson, 1988	Mixed cultures of *Pseudomonas* and hydrocarbon oxidizing bacteria
19	China	Wang Xiu-Yuan, 1988	Gases, acids and solvents producing bacteria
20	USA	Davidson and Russell, 1988	Special adapted strain of *Clostridium*
21	USA	Oppenheimer and Heibert, 1988	Mixed cultures of hydrocarbon degrading bacteria
	III. Microbial Flooding by Stimulating the Activity of Natural Microflora		
22	Czechoslovakia	Dostalek and Spurny, 1958; Dostalek, 1961	Natural bacterial flora of reservoir brine
23	USSR	Ivanov et al., 1987	Natural bacterial flora from injection water and reservoir brine
	IV. Selective Plugging Recovery		
24	The Netherlands	Von Heininger et al., 1958 (After Hitzman, 1988)	*Betacocus dextranicus* (Slime-forming bacteria)

No.	Country	References	Type of inoculum
25	USSR	Yulbarisov, 1972, 1976	Anaerobic and aerobic bacteria as: sulfat reducing dentrifying, putrefactive and acid butiric fermenting, cellulose digesting
26	USSR	Yulbarisov, 1989	Aerobic and anaerobic activated sludge bacteria
27	Canada	Jack and Stehmeier, 1988	Biopolymer producing bacteria (*Leuconostoc mesenteroides*)
28	China	Wang Xiu-Yuan, 1988	Slime-forming bacteria
29	USA	Berry Energy and Petrol. Bio-Resources, 1983-84 (After Hitzman, 1988)	Surfactant cosurfactant gas producing cultures; Polymer-polysaccharide gas producing cultures
		V. Microbial Fracturing Fluids Recovery	
30	USA	Pure Oil Co., 1957 (After Hitzman, 1983)	*Desulfovibrio hydrocarbonolastus*
31	USA	Micro-Bac-Gulf Coast Inc., 1985-86 (After Hitzman, 1988)	A liquid culture of mixed marine source microorganisms

Table 2. Types of nutrient supports used in MEOR field trials.

No.	Country	References	Type of nutrient support		
		I. Cyclic Microbial Recovery			
1	USA	Johnson, 1979	Molasses 4% with suitable mineral nutrients as $(NH_4)_3PE_4$		
2	USA	Microbial Systems Corp. 1984 (After Hitzman 1988)	Molasses		
3	USA	Grula et al., 1985	Molasses (4-8%) Dry milk solids (0.09%)		
4	USA	Covington Leases, 1985 (After Hitzman, 1988)	Injection water...100 bbl Sucrose..........1000 lbs Molasses...........25 gal Dry milk solids....25 lbs 30% NH_4NO_3 fer-tilizer...........33 lbs		
5	USA	Zajic, 1987	Sucrose 3.0 g, molasses 0.5 g, phosphate salts 0.2 g, nitrate salts 0.4 g, yeast extract 0.05 g, reservoirs brine 100 ml pH 8.5-9.3		
6	USA	Bryant et al., 1988	Molasses 4%		
7	USA	King, 1989	Molasses 4-10% Salts: urea 0.5%; ammonium nitrate 0.5%; glicolate acetate 0.5%		
8	Romania	Lazar et al., 1990	Molasses 4%		
		II. Microbial Flooding Recovery			
9	Czechoslo-vakia	Dostalek, 1961	Molasses		
10	Hungary	Jaranyi, 1968, 1973		Before 1967	After 1967
			Molasses	4 tons	2 tons
			KNO_3	120 kg	70 kg
			Na_3PO_4	40 kg	10-20 kg
			Sucrose	100 kg	50 kg
			Brine	100 m^3	100 m^3
11	Poland	Karaskiewicz, 1974	Molasses 4%		
12	USSR	Senyukov, 1970	Molasses 4%		

No.	Country	References	Type of inoculum
13	USA	Petrogen Inc., 1977-1982 (After Hitzman 1983)	Molasses and mineral salts
14	USA	Yarbrough and Coty, 1983	Molasses 2%
15	Romania	Lazar et al., 1985, 1988 1989	<u>Before 1977</u> Same as Jaranyi, 1968 <u>After 1977</u> Molasses 2-4%
16	DDR	Wagner and Teumer, 1987	Molasses 4%, polyphosphate, soda and a mixture of sweet water and reservoir brine (70 g NaCl/l)
17	USA	Bryant et al., 1988	Molasses 2-4%
18	USA	Davidson and Russell, 1988	water..............10,000 bbl free corn syrup....200,000 lbs some mineral salts
19	USA	Oppenheimer and Heibert, 1988	Inorganic nitrogen and phosphate nutrients Biocatalyst (natural liquid product)

<u>III. Microbial Flooding by Stimulating the Activity of Natural Microflora</u>

No.	Country	References	Type of inoculum
20	USSR	Ivanov et al., 1987, 1988	Waterflood along with air injection and mineral salts addition at the beginning

<u>IV. Microbial Selective Plugging Recovery</u>

No.	Country	References	Type of inoculum
21	The Netherlands	Von Heininger et al., 1958 (After Hitzman, 1988)	Sucrose - molasses 10%
22	USSR	Yulbarisov, 1972, 1976	Nutrient solution containing peat biomass and silt reach in hydrolysable substrates
23	USSR	Yulbarisov, 1989	Waste waters with addition of some biostimulators and chemical additives
24	USA	Berry Energy and Petroleum Bio-Resources, 1983-84 (After Hitzman, 1988)	injection medium (ingredients not mentioned)

No.	Country	References	Type of inoculum	
25	Canada	Jack and Stehmeier, 1988	Dry sucrose...........3600 kg Sugar beet molasses...3860 kg Fresh water...........7370 kg	
		V. Microbial Fracturing Fluids Recovery		
26	USA	Pure Oil Co., 1957		g/l
			Ca or Na lactate	3.5
			Ascorbic acid	0.5
			Yeast extract	1.0
			$MgSO_4$	0.2
			K_2HPO_4	0.5
			NaCl	10.0
			Agar gel agent (CMC)	2%
		VI. Paraffin Removal Treatment		
27	USA	Micro-Bac-Gulf Coast Inc., 1985-86 (After Hitzman, 1988)	Saline solution of nutrients for use in crude oil to control the deposition of paraffin	

Table 3. Some results reported from MEOR field trials.

No.	Country	References	Type of nutrient support
		I. Cyclic Microbial Recovery (Single well stimulation method)	
1	USA	Johnson, 1978	Increase the oil production up to 350 percent
2	USA	Microbial Systems Corp. 1984 (After Hitzman 1988)	Oil production increase 230% for 7 months
3	USA	Grula et al., 1985	Oil production increased from 1 to 2 bbl/day for 30 days
4	USA	Covington Leases, 1985 (After Hitzman, 1988)	Slight increase in oil production
5	USA	Zajic, 1987	unknown
6	USA	Bryant et al., 1988	Increase the oil production up to 79 percent
7	USA	King, 1989	Significant increase of oil production for 5 months
8	Romania	Lazar et al., 1988, 1990	Increase the oil production with 100-200% up to 5 months; Decreased the injection pressure of waterflooding from 50-70 atm to 20-25 atm
		II. Microbial Flooding Recovery	
9	Czechoslovakia	Dostalek and Spurny, 1958; Dostalek, 1961	Microbial growth detected at well up to 500 m in high permeable zones
10	Hungary	Jaranyi, 1968; Dienes and Jaranyi, 1973	Increase of oil production between 12-60 percent for a few weeks up to 18 months as well as gas production (CO_2)
11	Poland	Karaskiewicz, 1974	Significant increase of oil production (up to 300-360 percent for 2-8 years) and important changes of fluids characteristics

No.	Country	References	Type of inoculum
12	USSR	Senyukov et al., 1970	Oil production increased from 37 tons to 40 tons for 4 months
13	USA	Petrogen Inc., 1977-1982 (After Hitzman 1983)	From 24 wells: 4 doublet production for 6 months; 12 increased production 50 percent for 3 months; for all the wells average of increase oil production 42 percent
14	USA	Yarbrough and Coty, 1983	Increased oil production with 250%
15	Romania	Lazar and Constantinescu, 1985; Lazar et al., 1990	In 2 of 6 reservoir treated between 1975-79, increased of oil production up to 200 percent for 1-4 years In 2 reservoirs under testing encouraging increase of oil production
16	DDR	Wagner and Teumer, 1987	After 5 months - from 50-150 tons/month After 12 months to 300 tons/month and water/oil ratio decreased from 88% to 34%
17	USA	Bryant et al., 1988	After 10 months oil production increased with 13% and water/oil ratio decreased with 30%
18	USA	Davidson and Russell, 1988	Reduction of oil viscosity caused by CO_2 production
19	USA	Oppenheimer and Heibert, 1988	For several months an improvement of oil recovery through production of surfactants, CO_2 and a decrease in pH Reduction of paraffin and an increase in API gravity
		III. Microbial Flooding by Stimulating the Activity of Natural Microflora	
20	USSR	Ivanov et al., 1987	Significant increase of oil production

No.	Country	References	Type of inoculum
		IV. Selective Plugging Recovery	
21	The Nether-lands	Von Heininger et al., 1958 (After Hitzman, 1988)	Significant increase of oil production comparatively with the well only water had been injected The oil to water ratio changed to 1:20 compared to 1:50 untreated
22	USSR	Yulbarisov, 1972, 1976	Increase in oil production from 180 to 200-300 tons oil/day and decrease water factor for a few years, (but, intensive fresh water injection coincide with microbial growth!)
23	USSR	Yulbarisov, 1989	About 600 producing wells have been stimulated; additional oil recovery amounts 1000-2000 tons/year per well treatment
24	USA	Berry Energy and Petroleum Bio-Resources, 1983-84 (After Hitzman, 1988)	Oil production increased from 15 to 40 BOPD (for short time)
25	Canada	Jack and Stehmeier, 1988	Restarting good fluid flow pH from 6.4 to 6.0 Surface tension from 66.5 to 59.6 dynes/cm Fermentation products as: acetic, lactic acids and ethanol, propanol
		V. Microbial Fracturing Fluids Recovery	
26	USA	Pure Oil Co., 1957	Production rate increased from 15 bbl/day to 25 bbl/day
		VI. Paraffin Removal Treatment	
27	USA	Micro-Bac-Gulf Coast Inc., 1985-86 (After Hitzman, 1988)	Increase of oil production from 1.5 to 4.0 BOPD for 3 months

DIAGRAM 1

CYCLIC MICROBIAL RECOVERY

(a well-stimulation method)

(USA)

(Johnson, 1979)

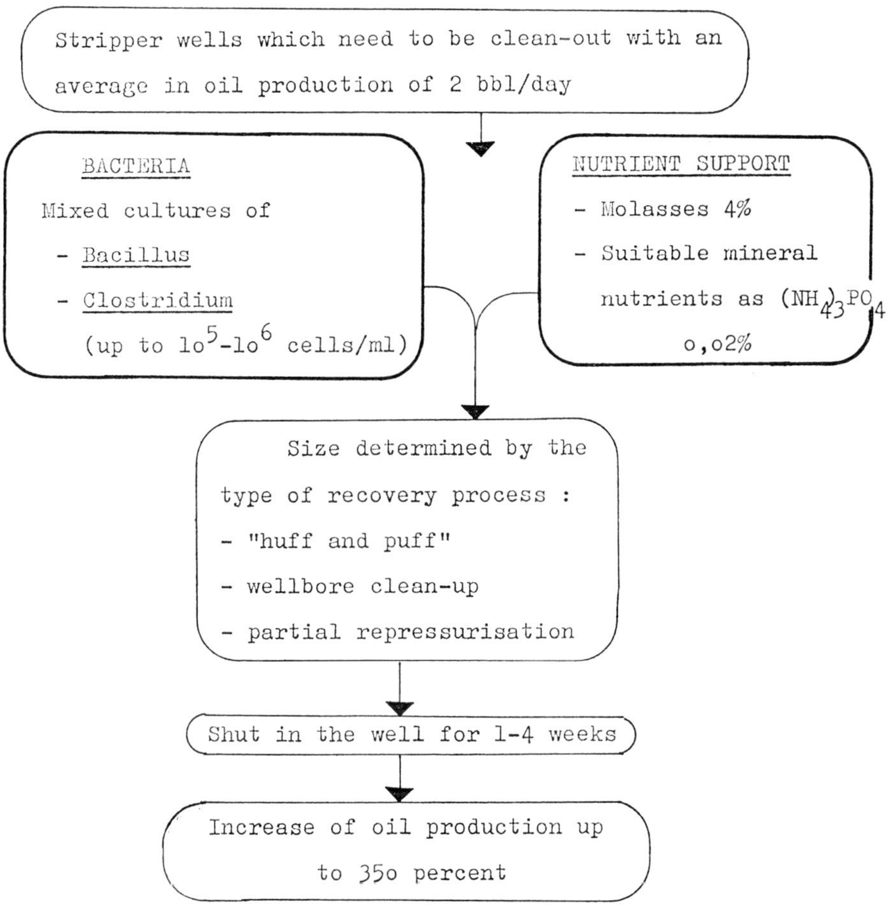

DIAGRAM 2

CYCLIC MICROBIAL RECOVERY

(A single well stimulation method)

USA, Microbial Systems Corp., Okla., 1984

(After Hitzman, 1988)

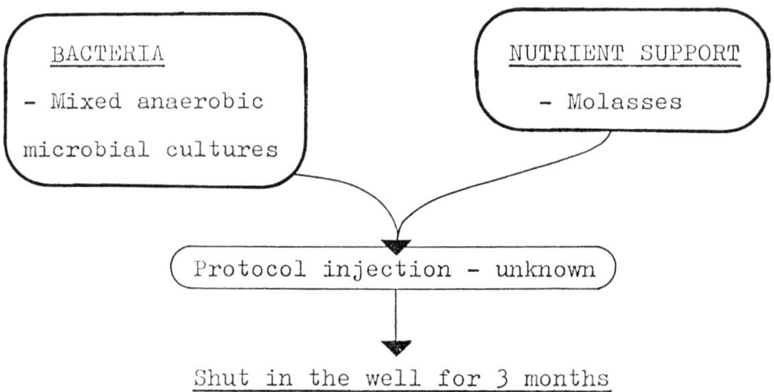

	Before test	After test
Response: Avg.Prod. (BOPD) per well	0.5	2.0
Avg.Prod. (BWPD) per well	13	40

- Oil Production increased 230% for 7 months
- Water production increased 300% then gradually decreased

DIAGRAM 3

CYCLIC MICROBIAL RECOVERY

(A well stimulation method)

USA - Oklahoma State Univ.Field Test,1983

(Grula et al. 1985)

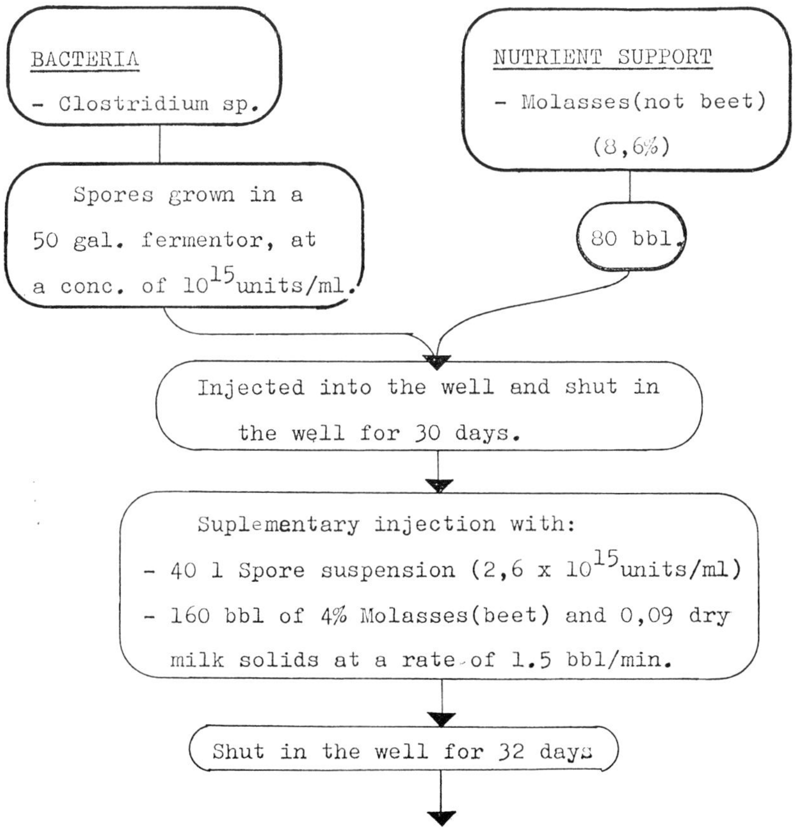

Response: - Increase of gases, acids and solvents production

- Decrease the pH

- Oil production from 1 to 2 bbl/day for 30 days.

DIAGRAM 4

CYCLIC MICROBIAL RECOVERY

(A well stimulation method)

USA - Covington Leases, 1985

(After Hitzman, 1988)

BACTERIA

- Clostridium sp.
  (Spore suspension)

NUTRIENT SUPPORT

- Injection water. . . . 1oo bbl
- Sucrose. . . . . 1ooo lbs
- Beet molasses. . . . . 25 gal
- Powdered milk solids .. 25 lbs
- 3o% NH_4NO_3 fertilizer.. 33 lbs

Protocol injection

Amount injected (gal)

	Cov.18	Cov.22	Cov.24
Inj. water Spacer	889	6o2	28o
Bacterial spores	55,5	55,5	41,5
Nutrient mixture	385o	4195	3654
Water displacement	1196	2oo5	o

Shut in the wells for 3 months

Response : The cov.24 showed a slight increase in oil production

DIAGRAM 5

CYCLIC MICROBIAL RECOVERY

(a well stimulation method)

(Zajic, 1987 - USA)

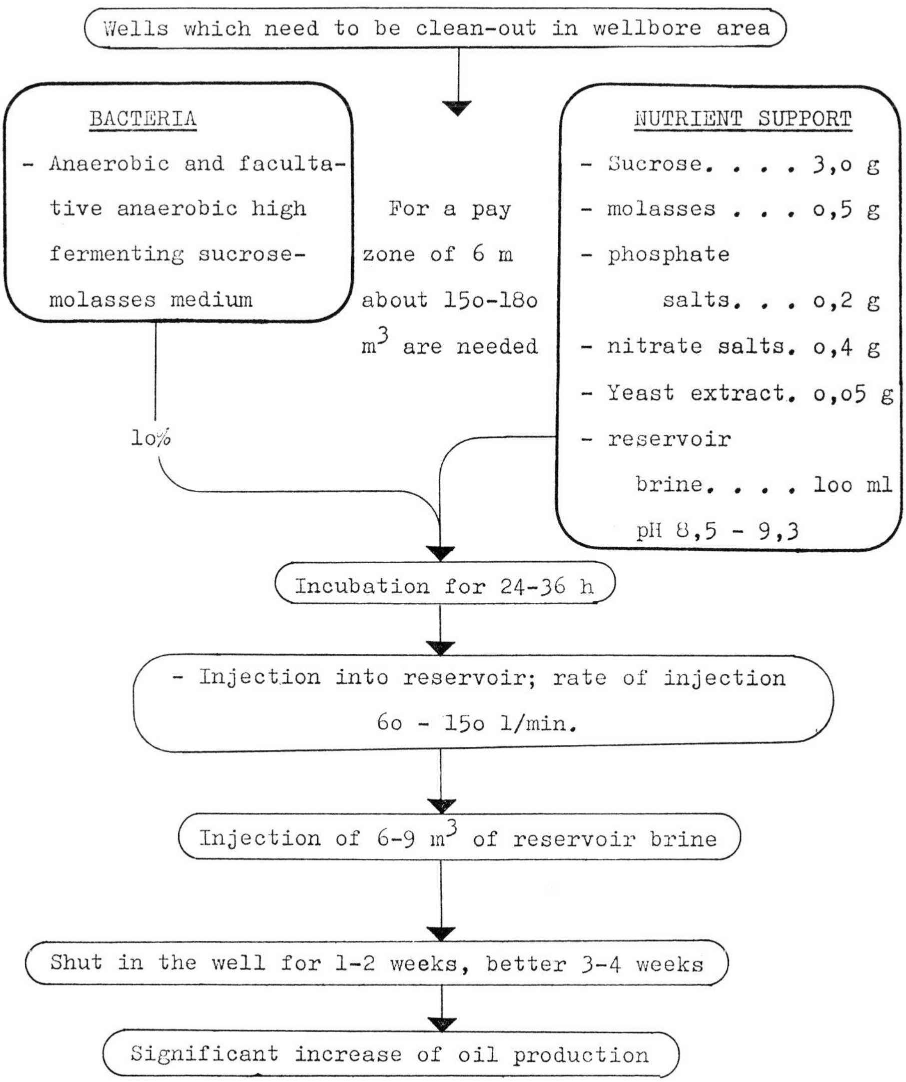

DIAGRAM 6

CYCLIC MICROBIAL RECOVERY

(a well stimulation method)

(Bryant et al., 1988)

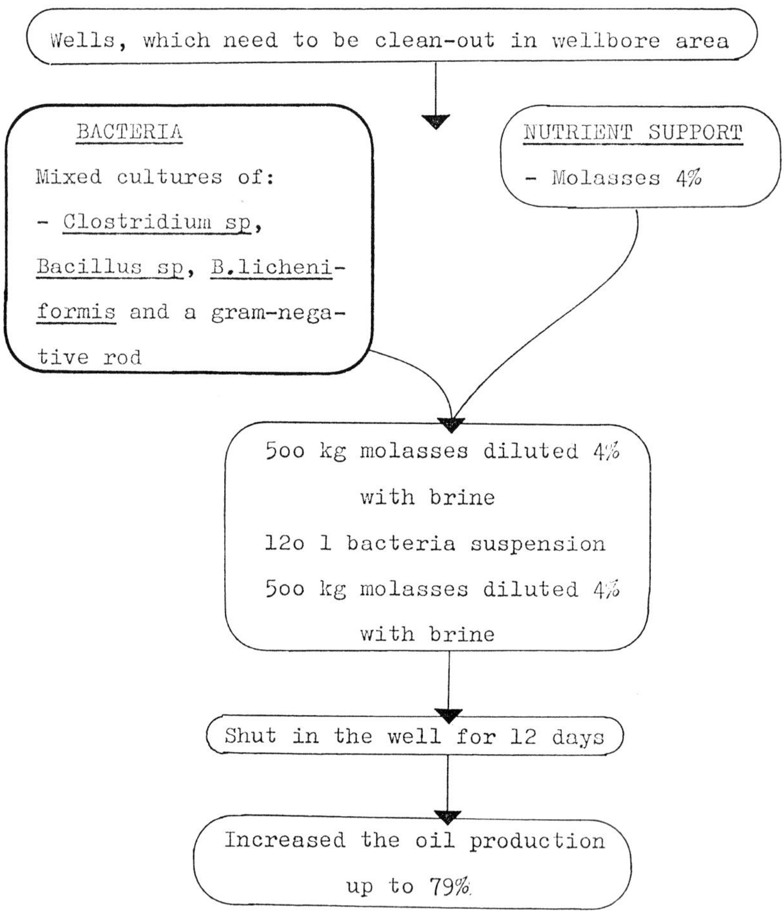

DIAGRAM 7

CYCLIC MICROBIAL RECOVERY

(Single Well Stimulation Method)

(King, 1989)

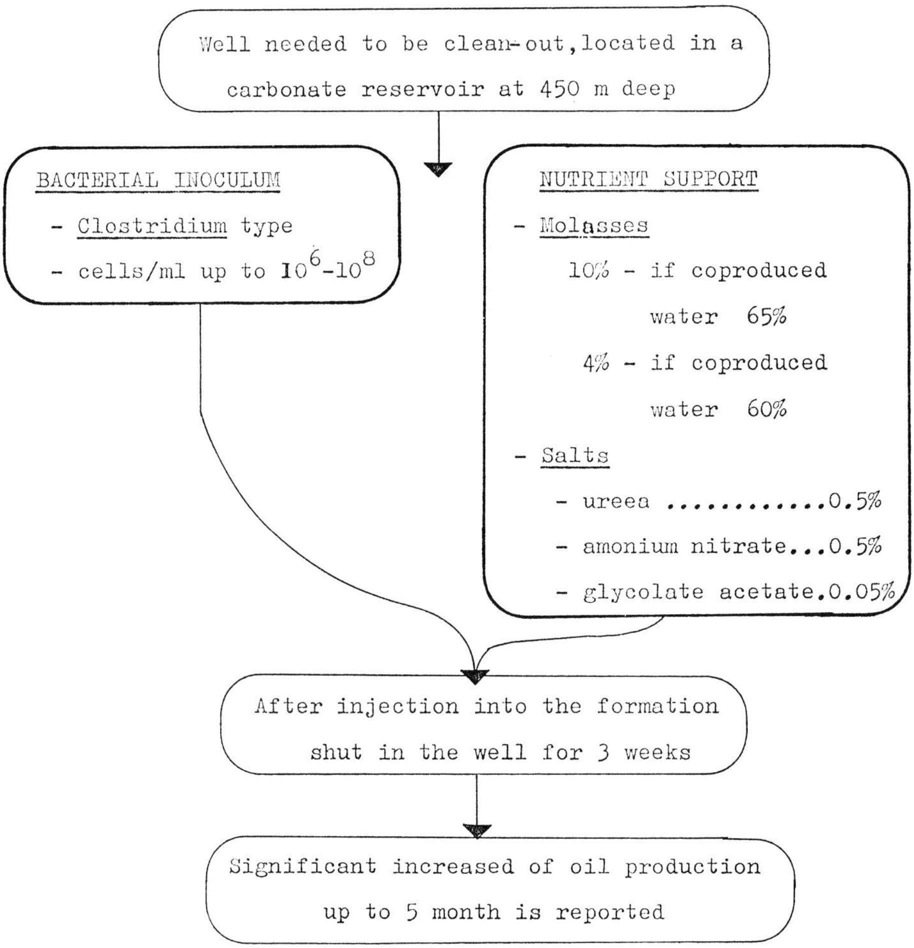

DIAGRAM 8

CYCLIC MICROBIAL RECOVERY

(Single Well Stimulation Method)

(Lazăr et al. 1989)

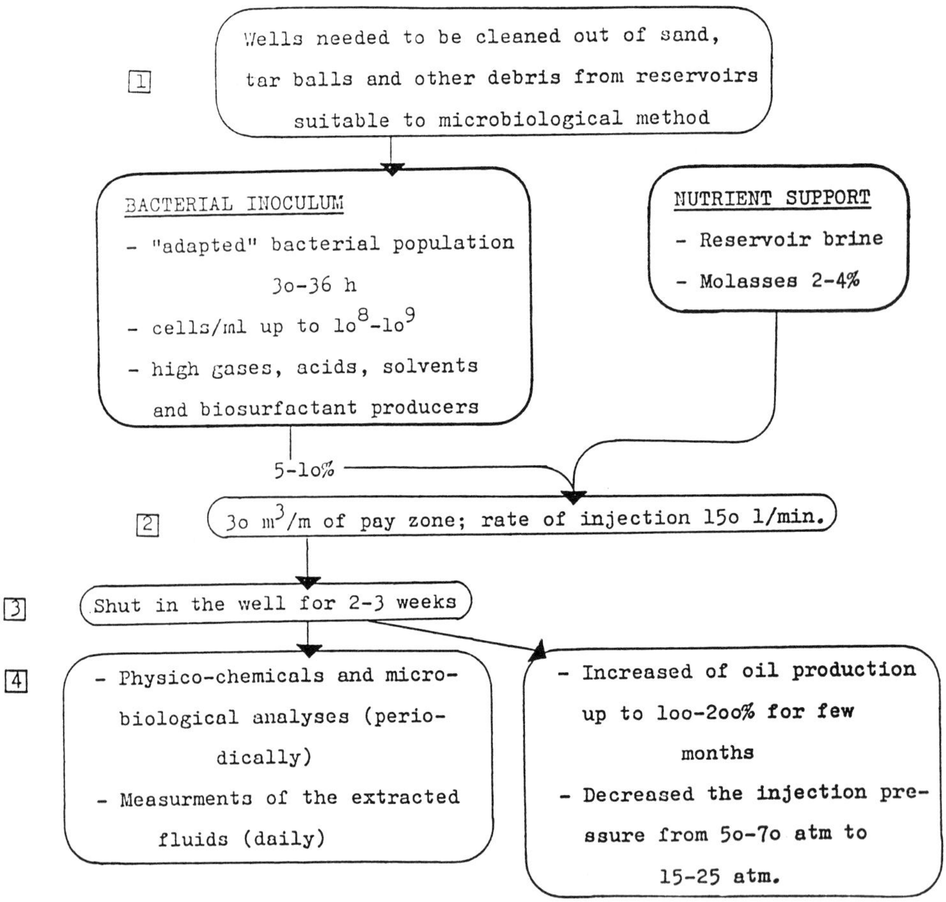

DIAGRAM 9

MICROBIAL FLOODING RECOVERY

(Czechoslovakia, Tests I-VII)

(Dostalek and Spurny, 1958; Dostalek, 1961)

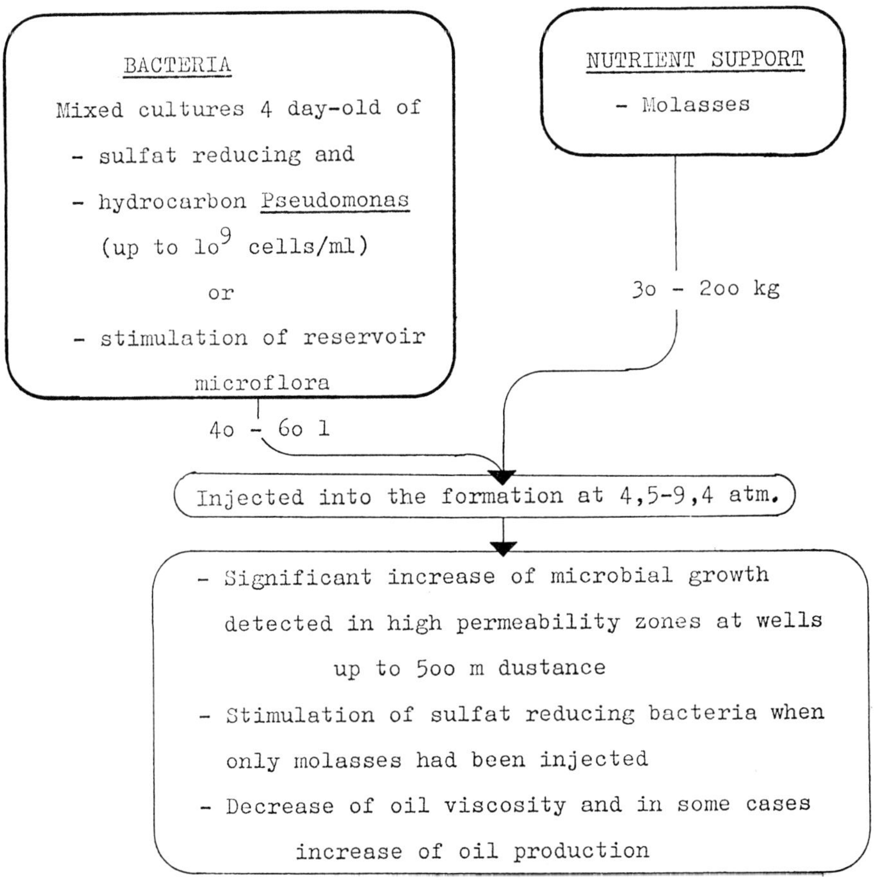

DIAGRAM 10

MICROBIAL FLOODING RECOVERY

(Jaranyi, 1968, Dienis and Jaranyi, 1973)

(Hungary, Tests I-X)

Injection into the well of 2o m^3 of reservoir brine

↓

BACTERIA

- Mixed sewage-sludge bacteria or
- Anaerobic thermophilic mixed cultures of *Clostridium*, *Desulfovibrio* and *Pseudomonas*

NUTRIENT SUPPORT

	Before 1967	After 1967
	kg	
- Molasses	4ooo	4ooo
- Sucrose	1oo	5o
- KNO_3	12o	7o
- Na_3PO_4	4o	1o-2o
- Brine	1oom^3	1oo m^3

↓

NUTRIENT 5o m^3 → BACTERIA 1oo 1 → NUTRIENTS 5o m^3 (up to 10^6 cell/ml)

↓

Shut in the well for 4-6 months

↓

2-3 suplimentary injections at 2-4 months intervals with molasses (2 t), KNO_3(5o kg), Na_3PO_4 (5o kg) and reservoir brine 1oo m^3

↓

Keep the well shut in as long as positive results are noticed in reaction wells

↓

- Increased of oil production between 12-6o percent for a few weeks up to 18 months, as well as gases (mainly CO_2)
- Decreased of pH, oil viscosity and oil/water ratio

DIAGRAM 11

MICROBIAL FLOODING RECOVERY

(Karaskiewicz, 1973, Poland)

Test I 1961 to test XVI, 1969

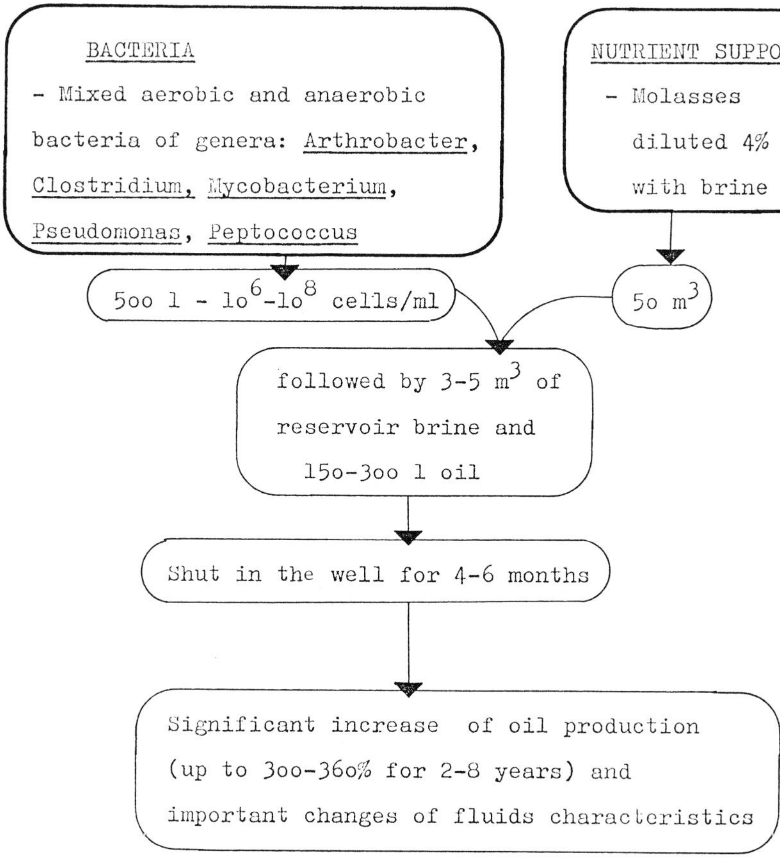

DIAGRAM 12

MICROBIAL FLOODING RECOVERY

USSR, 197o

(Senyukov et al., 197o)

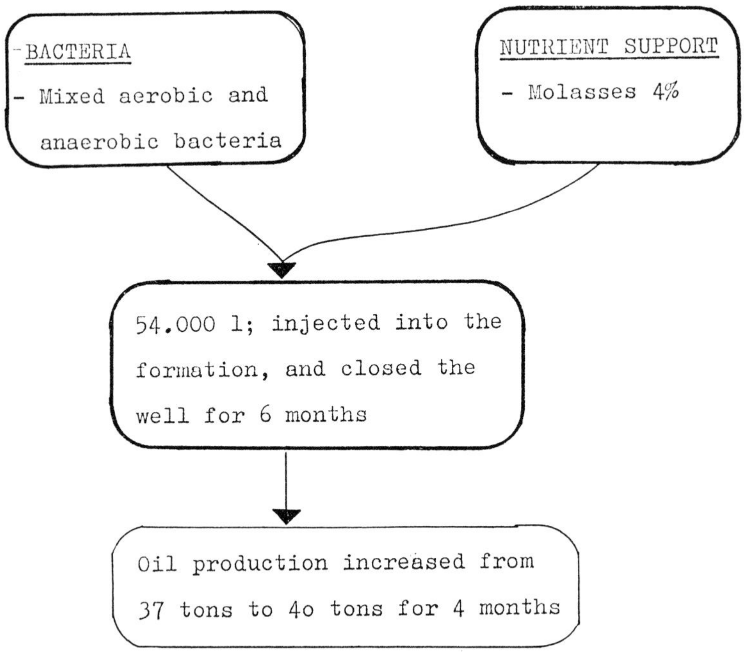

DIAGRAM 13

MICROBIAL FLOODING RECOVERY

(USA)

(Petrogen Inc., 1977-1982, After Hitzman, 1983)

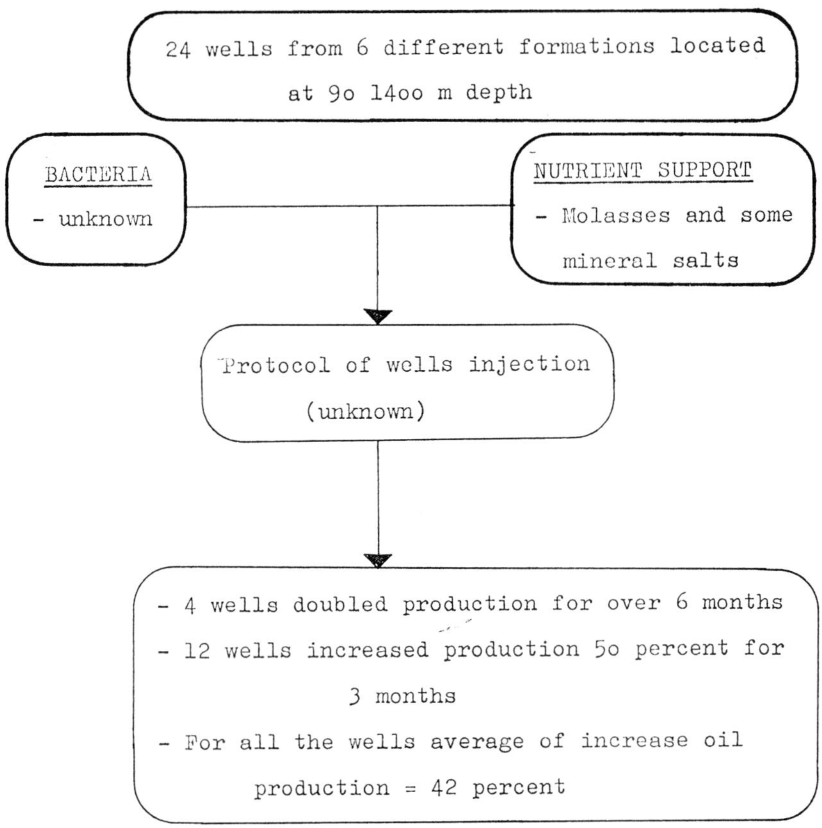

DIAGRAM 14

MICROBIAL FLOODING RECOVERY

(USA, Yarbrough and Cotty, 1983)

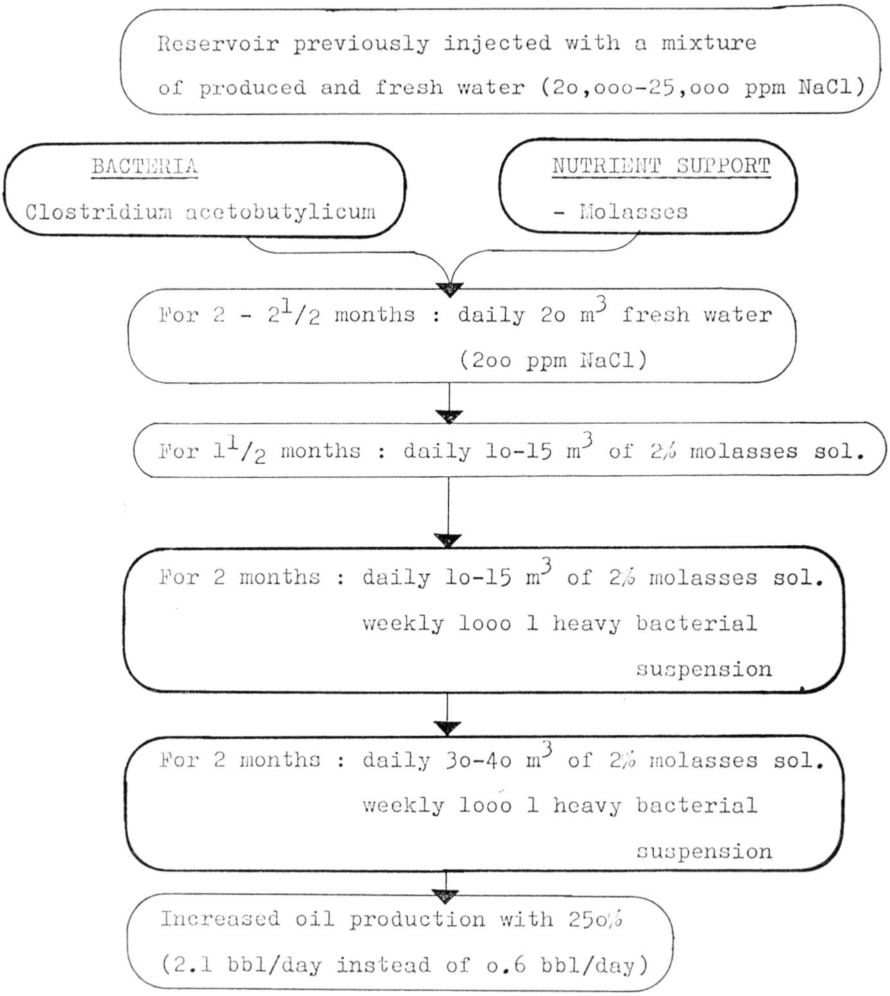

Notice : Total materials injected : - Water 4,8oo m^3

- Molasses 7o.5
- Bacteria 16 m^3

DIAGRAM 15

MICROBIAL FLOODING RECOVERY

România, Lazăr and Constantinescu, 1985

(Test I, 1975 to test VI, 1979)

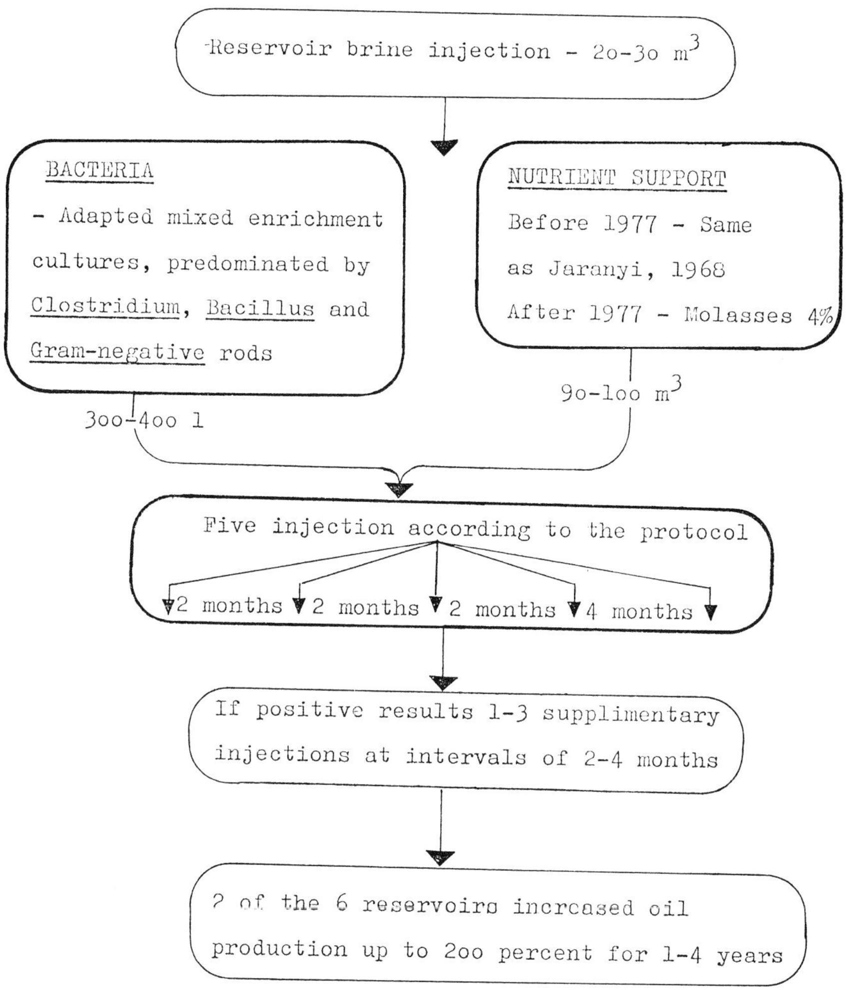

DIAGRAM 16

MICROBIAL FLOODING RECOVERY

România, Lazăr and Constantinescu, 1985

Test VII, 1977

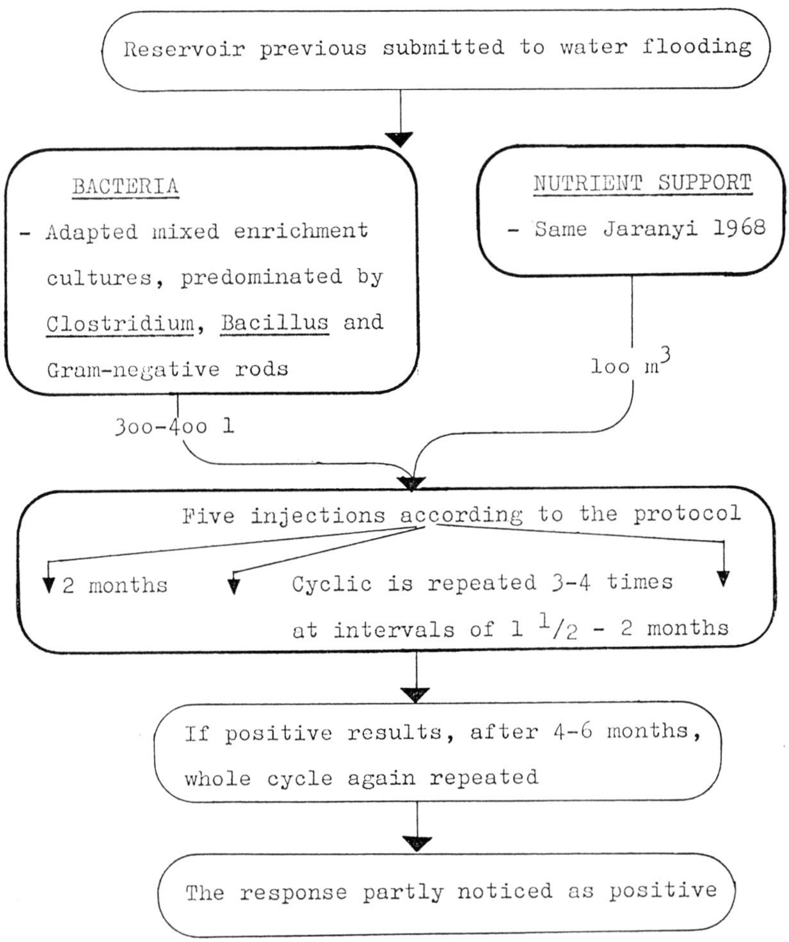

DIAGRAM 17

MICROBIAL FLOODING RECOVERY

DDR, Wagner and Teumer, 1987

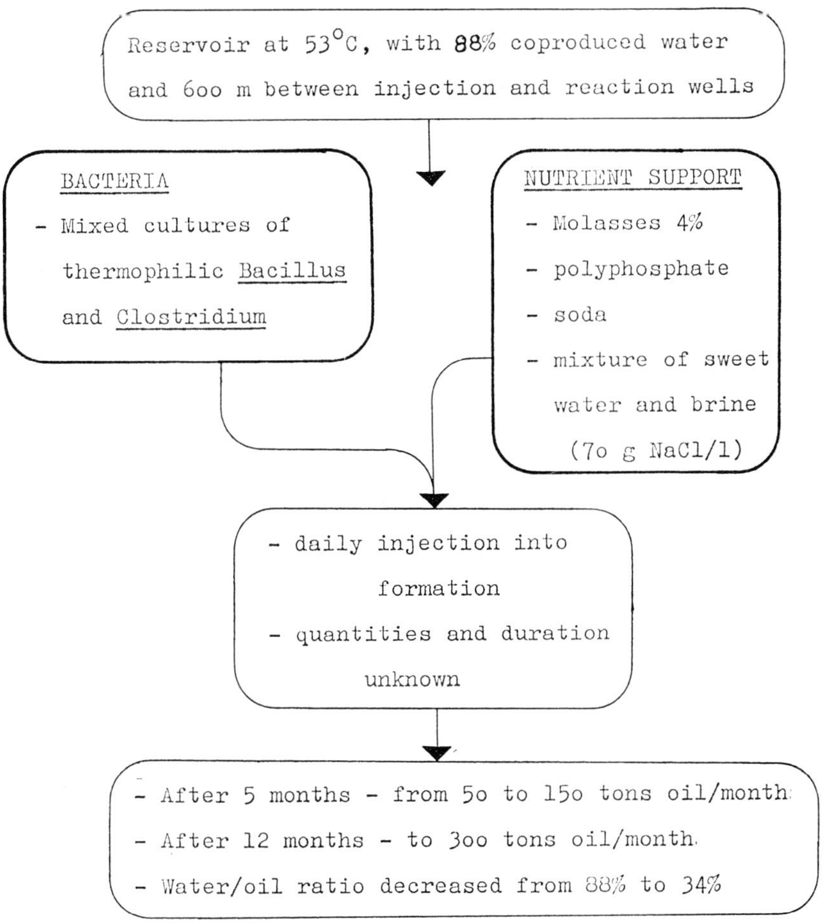

DIAGRAM 18

MICROBIAL FLOODING RECOVERY

USA - NIPER, Microbial Systems Inc, Injectech Inc, 1986-1987(After Hitzman,1988)

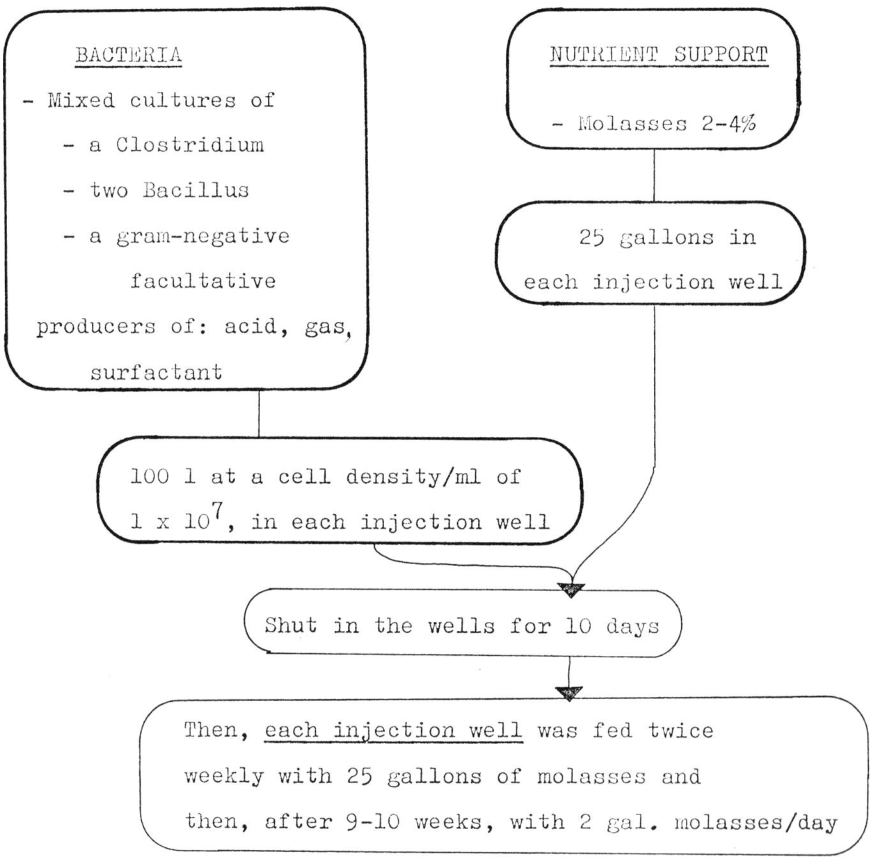

DIAGRAM 18A

MICROBIAL FLOODING RECOVERY

USA, Bryant et al., 1988

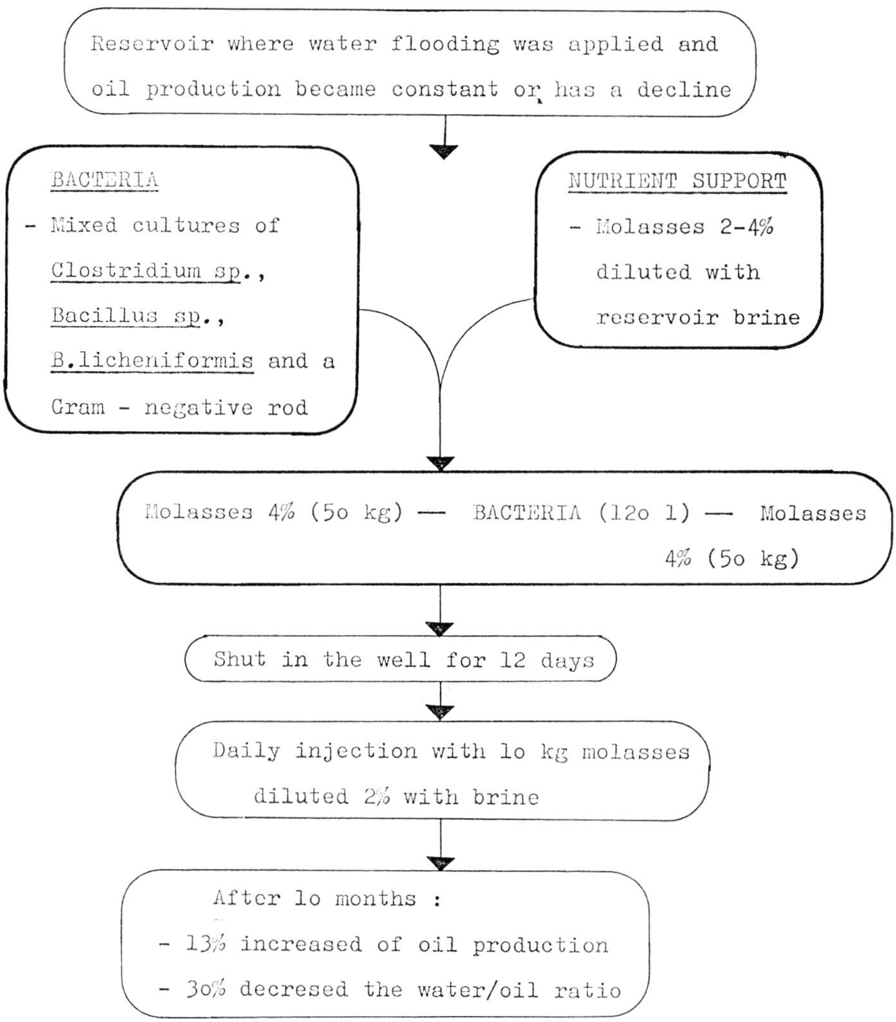

DIAGRAM 19

MICROBIAL FLOODING RECOVERY

USA, Davidson and Russell, 1988

(Imperial Energy Corporation Test)

A heavy oil reservoir (21°API) where the water-flooding operations were inefficient because of the adverse mobility ratio

BACTERIA

- a special adapted strain of *Clostridium*, high producer of CO_2 and alcohols.

NUTRIENT SUPPORT

- water . . . . . 10,000 bbls
- free corn syrup. . . . .200,000 lbs
- some mineral salts.

40 gal. of spores in 20 bbls of potable water and injected 1/2 bbl/min.

308 bbls - 1/2 bbls/min.

The lease was shut in seven days after the injection, then all wells were returned to production

- Reduction of oil viscosity caused by CO_2 production
- - butanol and Surfactants production improved mobility control and sweep efficiency

DIAGRAM 20

MICROBIAL FLOODING RECOVERY

USA, Oppenheimer and Heibert, 1988

(Alpha Environmental Field Test, Texas)

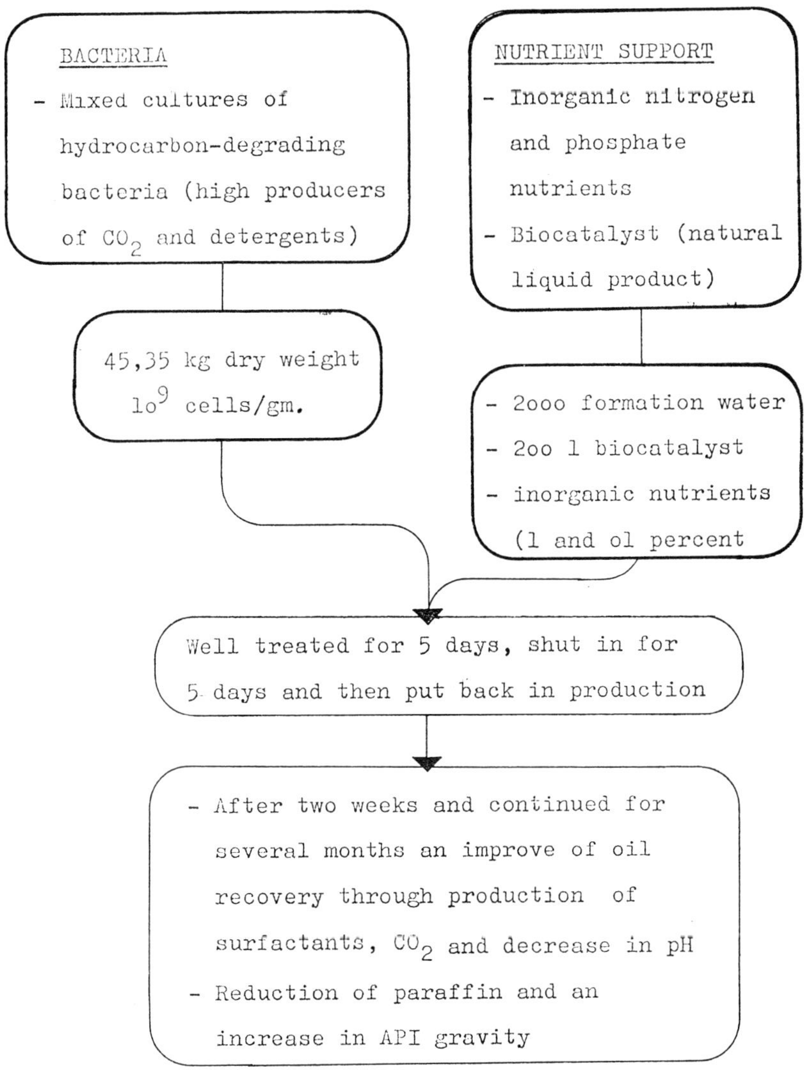

DIAGRAM 21

MICROBIAL FLOODING RECOVERY

Romania, Lazar et al., 1990, Test Sa_3Cl reservoir

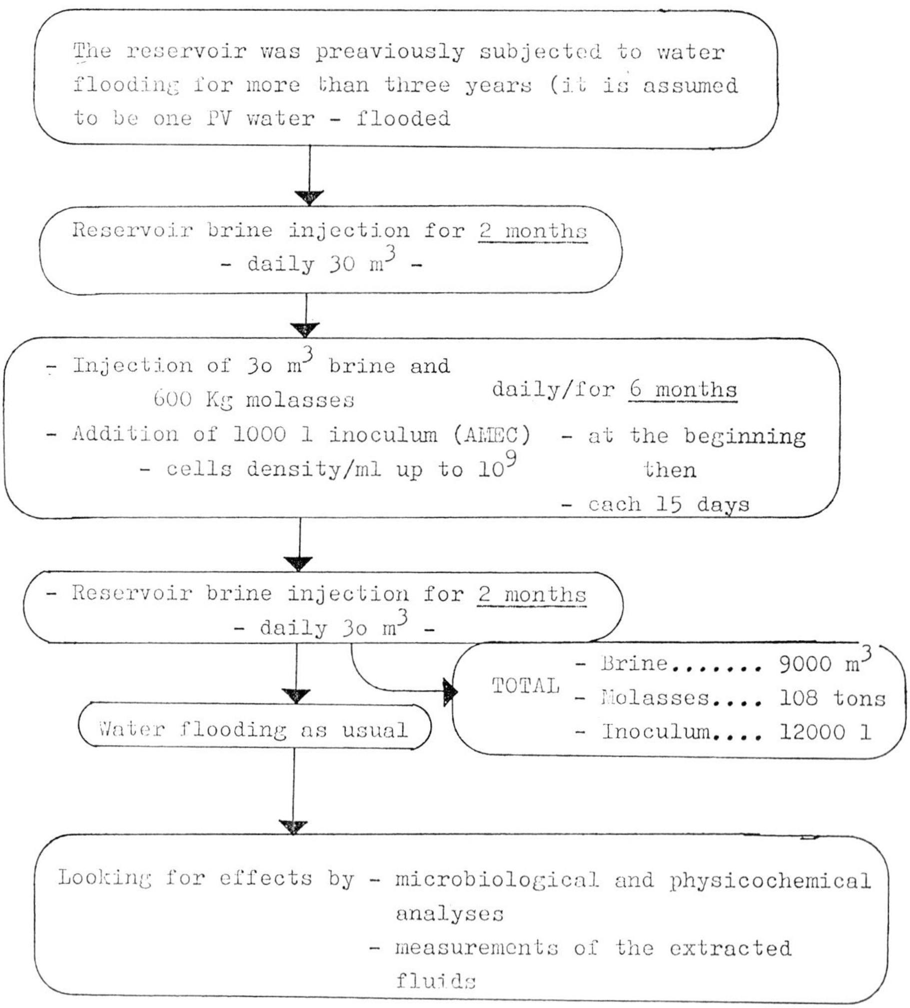

DIAGRAM 21A

MICROBIAL FLOODING RECOVERY

Romania, Lazar et al., 1989, Test Sa_3Br_1 reservoir

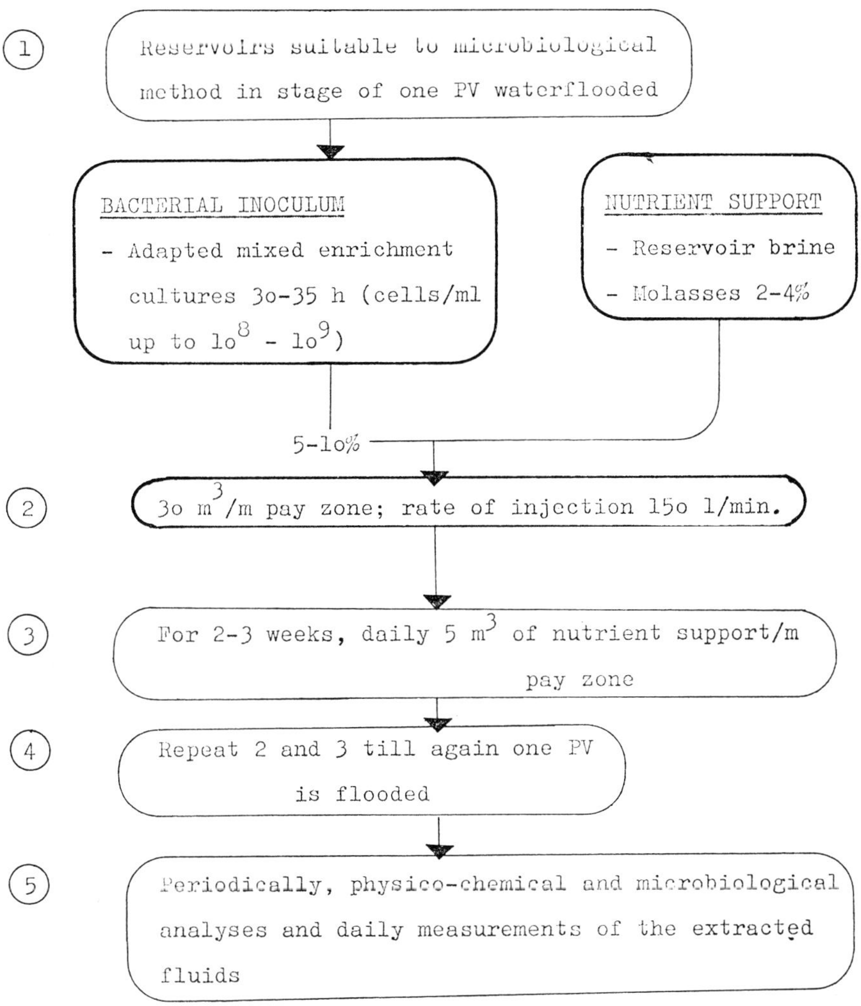

DIAGRAM 22

MEOR BY STIMULATION OF RESERVOIR MICROFLORA

URSS, Ivanov, 1987, References 1987, Seria "Nefte-promîslovoe Delo"

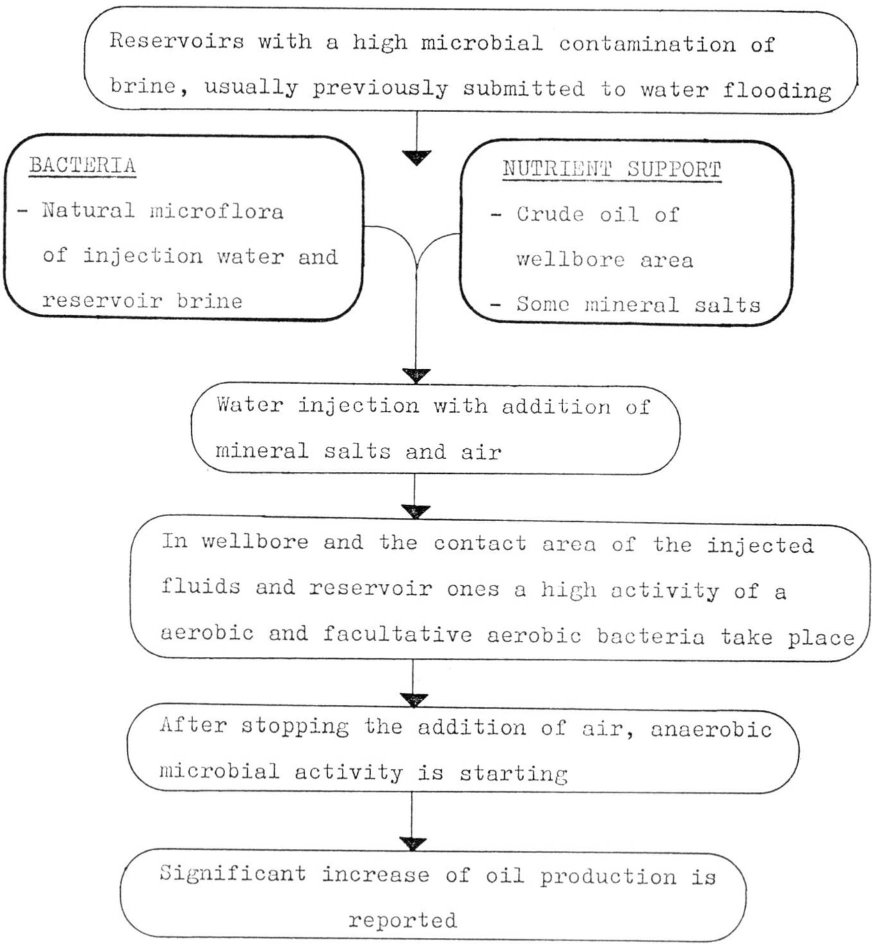

DIAGRAM 23

MICROBIAL SELECTIVE PLUGGING RECOVERY

(THE NETHERLANDS)

(Von Heiningen et al., 1958)

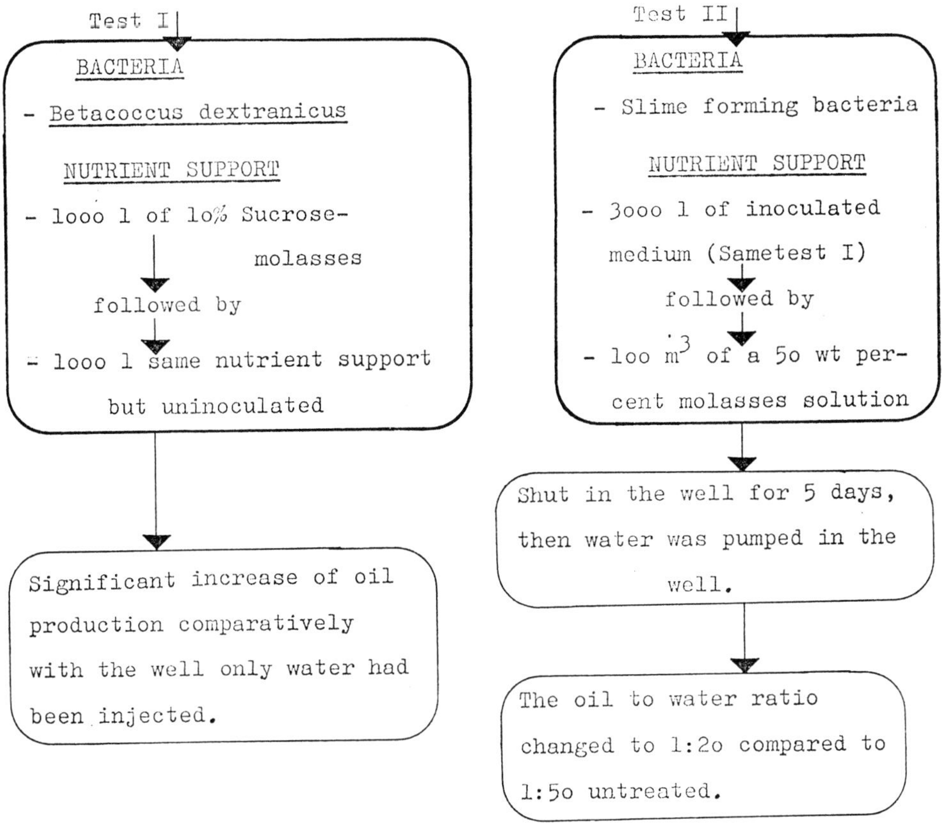

DIAGRAM 24

MICROBIAL SELECTIVE PLUGGING RECOVERY

USSR - 1966

(Yulbarisov, 1972, 1976)

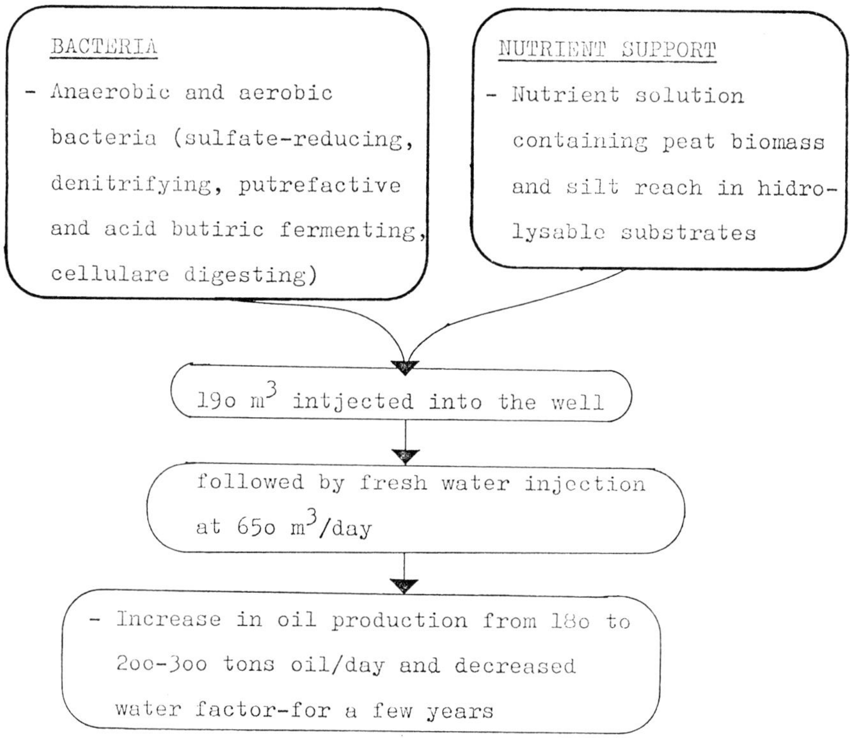

Notice : Intensive fresh water injection coincid with microbial growth

DIAGRAM 25

MICROBIAL SELECTIVE PLUGGING RECOVERY

USA-Berry Energy and Petroleum Bio-Resources,1983-1984

(After Hitzman, 1988)

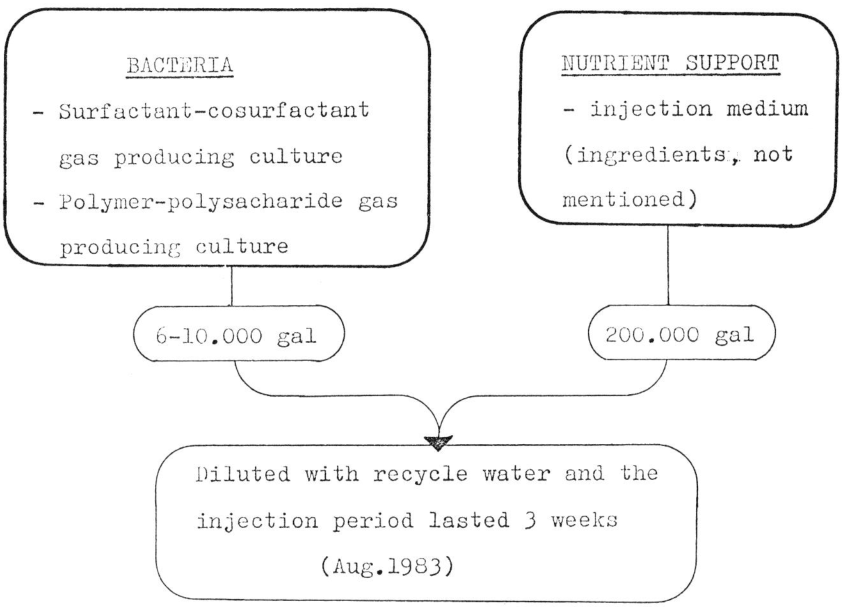

Response: In December 1983 the production increased from 15 BOPD to 40 BOPD - but for short time

DIAGRAM 26

MICROBIAL SELECTIVE PLUGGING RECOVERY

URSS, Yulbarisov, 1989

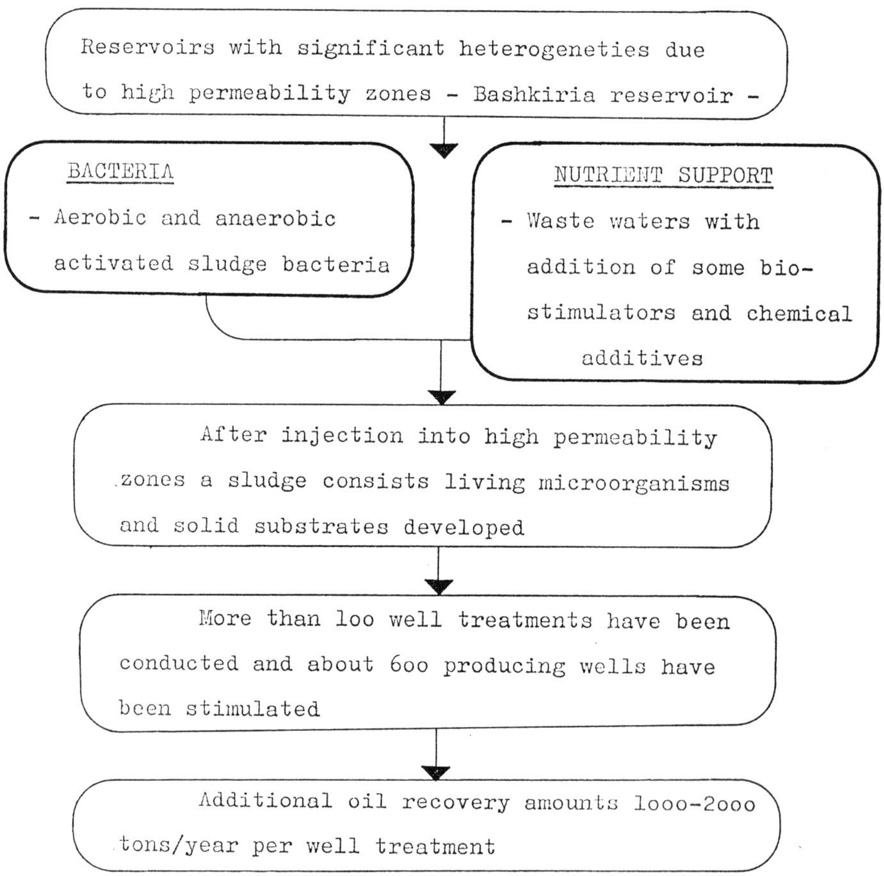

DIAGRAM 26A

MICROBIAL SELECTIVE PLUGGING RECOVERY

Nova/Husky Operations Ltd. - 1985 - 1986

(After Hitzman, 1988)

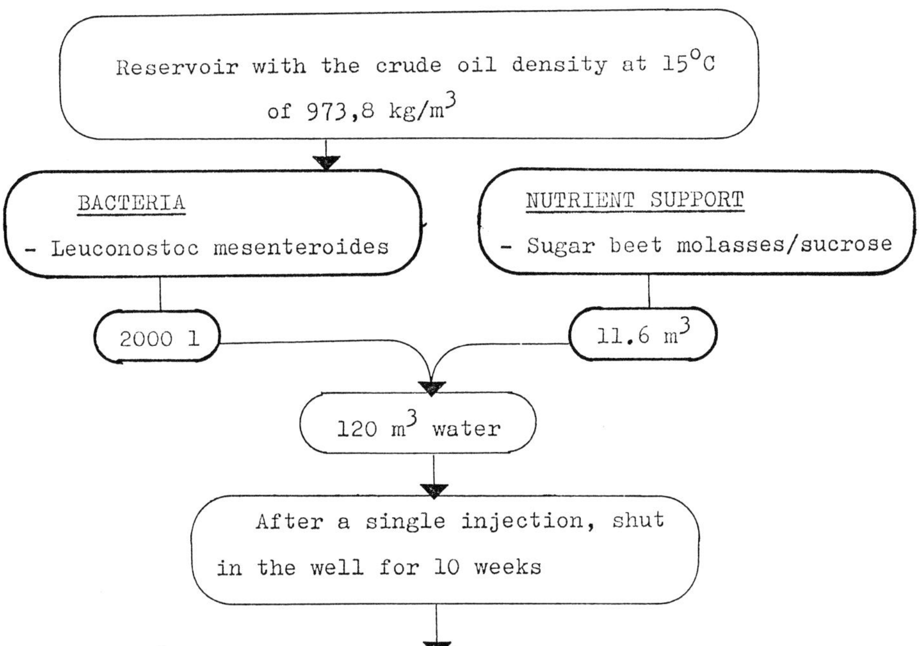

Response:
- pH of the fluid decreased from 6.3 to 6.0
- Surface tension decreased from 66 to 59 dynes/cm
- Fermentation products detected:
  - fatty acids(lactic,acetic,pyruvic)
  - alcohols(ethanol,propanol)

DIAGRAM 27

<u>MICROBIAL SELECTIVE PLUGGING RECOVERY</u>

Canada, Jack and Stehmeier, 1988

(Nova/Huscky Operations Ltd. Tests 1985-86)

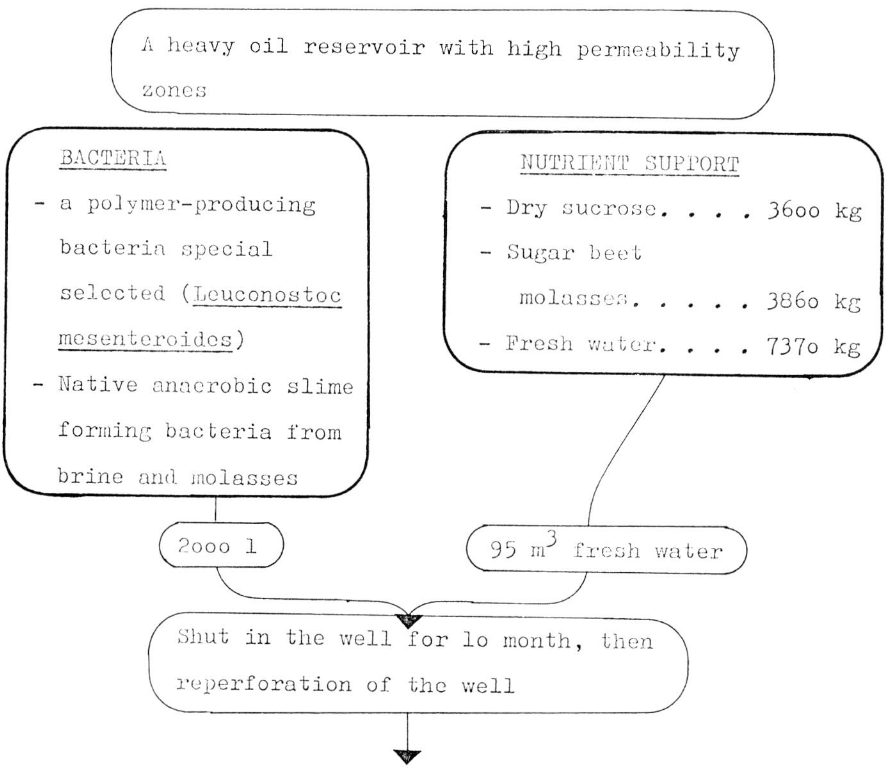

- Restoring good fluid flow
- pH - from 6.4 to 6.0
- Surface tension from 66.5 to 59.6 dynes/cm
- Fermentation products as :lactic,acetic acids and ethanol, propanol

DIAGRAM 28

MICROBIAL FRACTURING FLUIDS RECOVERY

(USA)

(Pure oil Co., 1957, after Hitzman, 1983)

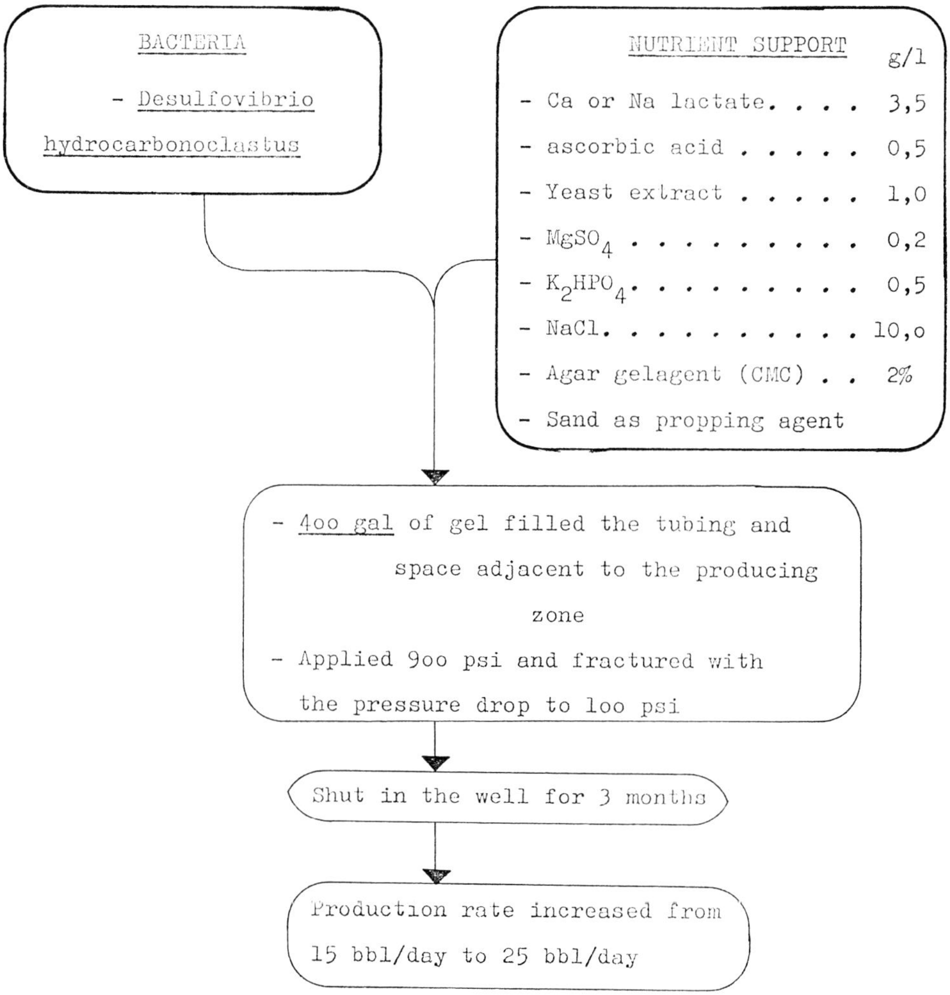

DIAGRAM 29

PARAFFIN REMOVAL TREATMENT

USA-Micro-Bac-Gulf Coast Inc.,1985-1986

(After Hitzman,1988)

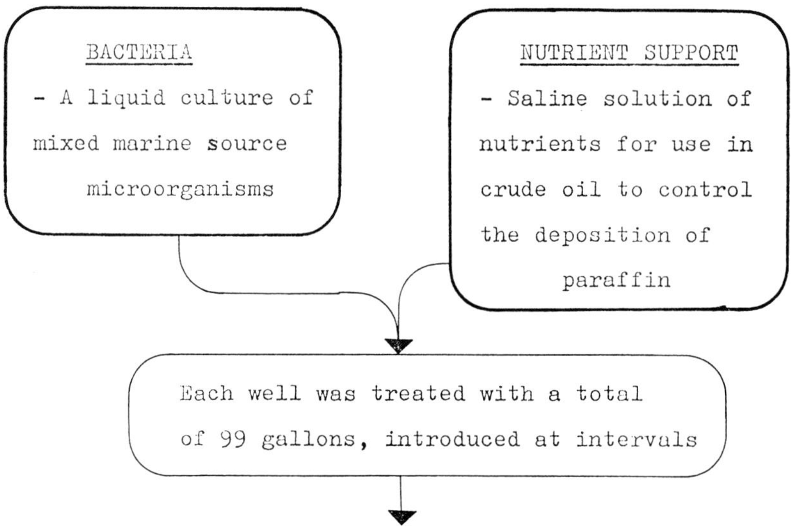

Response: Increase of oil production from 1.5 BOPD to 4.o BOPD for 3 months

Notice: Prior to the treatment, the wells required servicing every month as production declined from 1.5 to 0.5 BOPD